Foundations of Engineering Mechanics

Polina S. Landa, Regular and Chaotic Oscillations

Springer
Berlin
Heidelberg
New York
Barcelona
Hong Kong
London
Milano
Paris
Singapore
Tokyo

Polina S. Landa

Regular and Chaotic Oscillations

With 162 Figures

Springer

Series Editors:
V. I. Babitsky, DSc
Loughborough University
Department of Mechanical Engineering
LE11 3TU Loughborough, Leicestershire
United Kingdom

J. Wittenburg
Universität Karlsruhe (TH)
Institut für Technische Mechanik
Kaiserstraße 12
D-76128 Karlsruhe / Germany

Author
Polina S. Landa
Moscow State University
Dept. of Physics
Vorobyovy Gory
119899 Moscow
Russia

Library of Congress Cataloging-in-Publication Data
Landa, P.S. (Polina Solomonova)
Regular and chaotic oscillations / Polina S. Landa.
(Foundations of engineering mechanics)
Includes bibliographicla references and index.
ISBN 3540410015
1. Vibration. 2. Nonlinear oscillations. I. Title. II. Series
TA 355. L265 2001
620.3--dc21 00-046993

ISBN 3-540-41001-5 Springer-Verlag Berlin Heidelberg New York

Springer-Verlag Berlin Heidelberg New York
a member of BertelsmannSpringer Science + Business Media GmbH

http://www.springer.de

Printed in Germany

Typesetting: Camera-ready copy from author
Cover-Design: de'blik, Berlin
Printed on acid-free paper SPIN 10782688 62/3020 5 4 3 2 1 0

Preface

In this book the modern theory of both regular and chaotic nonlinear oscillations is set out, primarily, as applied to mechanical engineering problems. The main purpose of the book is to convince the reader of the necessity of a thorough study of this theory, to enable him understanding of the general oscillation laws and to show that this theory can be very useful in engineering research.

The primary audience for this book is researchers working on different oscillatory processes, and students interested in studying the general laws and applications of the theory of nonlinear oscillations.

Moscow, June 2000 *Polina Landa*

Contents

1. Introduction 1
1.1 The importance of oscillation theory for engineering mechanics 1
1.2 Classification of dynamical systems. Systems with conservation of phase volume and dissipative systems 2
1.3 Different types of mathematical models and their functions in studies of concrete systems 5
1.4 Phase space of autonomous dynamical systems and the number of degrees of freedom 6
1.5 The subject matter of the book 7

2. The main analytical methods of studies of nonlinear oscillations in near-conservative systems 9
2.1 The van der Pol method 10
2.2 The asymptotic Krylov–Bogolyubov method 12
2.3 The averaging method 16
2.4 The averaging method in systems incorporating fast and slow variables 19
2.5 The Whitham method 21

Part I. OSCILLATIONS IN AUTONOMOUS DYNAMICAL SYSTEMS

3. General properties of autonomous dynamical systems 27
3.1 Phase space of autonomous dynamical systems and its structure. Singular points and limit sets 27
3.1.1 Singular points and their classification 27
3.1.2 Stability criterion of singular points 30
3.2 Attractors and repellers 32
3.3 The stability of limit cycles and their classification 33
3.4 Strange attractors: stochastic and chaotic attractors 35
3.4.1 Quantitative characteristics of attractors 36
3.4.2 Reconstruction of attractors from experimental data 41

3.5 Poincaré cutting surface and point maps ... 42
3.6 Some routes for the loss of stability of simple attractors and the appearance of strange attractors ... 44
3.7 Integrable and nonintegrable systems. Action–angle variables 45

4. Examples of natural oscillations in systems with one degree of freedom ... 49
4.1 Oscillator with nonlinear restoring force ... 49
4.1.1 Pendulum oscillations ... 49
4.1.2 Oscillations of a pendulum placed between the opposite poles of a magnet ... 51
4.1.3 Oscillations described by Duffing equations ... 52
4.1.4 Oscillations of a material point in a force field with the Toda potential ... 54
4.2 Oscillations of a bubble in fluid ... 56
4.3 Oscillations of species populations described by the Lotka–Volterra equations ... 58
4.4 Natural oscillations in a system with slowly time-varying natural frequency ... 61

5. Natural oscillations in systems with many degrees of freedom. Normal oscillations ... 63
5.1 Normal oscillations in linear conservative systems ... 63
5.2 Normal oscillations in nonlinear conservative systems ... 64
5.3 Examples of normal oscillations in linear and nonlinear conservative systems ... 65
5.3.1 Two coupled linear oscillators with gyroscopic forces .. 65
5.3.2 Examples of normal oscillations in two coupled nonlinear oscillators ... 66
5.3.3 An example of normal oscillations in three coupled nonlinear oscillators ... 67
5.3.4 Normal oscillations in linear homogeneous and periodically inhomogeneous chains ... 71
5.3.5 Examples of natural oscillations in nonlinear homogeneous chains ... 78
5.4 Stochasticity in Hamiltonian systems close to integrable ones. 84
5.4.1 The ring Toda chain and the Henon–Heiles system ... 84
5.4.2 Stochastization of oscillations in the Yang–Mills equations ... 87

6. Self-oscillatory systems with one degree of freedom ... 89
6.1 The van der Pol, Rayleigh and Bautin equations ... 89
6.1.1 The Kaidanovsky–Khaikin frictional generator and the Froude pendulum ... 92
6.2 Soft and hard excitation of self-oscillations ... 93

6.3 Truncated equations for the oscillation amplitude and phase . 95
6.3.1 Quasi-linear systems . . . 95
6.3.2 Transient processes in the van der Pol generator . . . 98
6.3.3 Essentially nonlinear quasi-conservative systems . . . 98
6.4 The Rayleigh relaxation generator . . . 100
6.5 Clock movement mechanisms and the Neimark pendulum. The energetic criterion of chaotization of self-oscillations . . . 102

7. Self-oscillatory systems with one and a half degrees of freedom . . . 107
7.1 Self-oscillatory systems with inertial excitation . . . 107
7.1.1 The model equations of self-oscillatory systems with inertial excitation . . . 107
7.1.2 Examples of self-oscillatory systems with inertial excitation . . . 110
7.2 Self-oscillatory systems with inertial nonlinearity . . . 127
7.3 Some other systems with one and a half degrees of freedom . . 132
7.3.1 The Rössler equations . . . 132
7.3.2 A three-dimensional model of an immune reaction illustrating the oscillatory course of some chronic diseases . . . 134

8. Examples of self-oscillatory systems with two or more degrees of freedom . . . 137
8.1 Generator with an additional circuit . . . 137
8.2 A lumped model of bending-torsion flutter of an aircraft wing 141
8.3 A model of the vocal source . . . 145
8.4 The lumped model of a 'singing' flame . . . 152
8.5 A self-oscillatory system based on a ring Toda chain . . . 155

9. Synchronization and chaotization of self-oscillatory systems by an external harmonic force . . . 161
9.1 Synchronization of self-oscillations by an external periodic force in a system with one degree of freedom with soft excitation. Two mechanisms of synchronization . . . 161
9.1.1 The main resonance . . . 162
9.1.2 Resonances of the nth kind . . . 167
9.2 Synchronization of a generator with hard excitation. Asynchronous excitation of self-oscillations . . . 176
9.2.1 Asynchronous excitation of self-oscillations . . . 179
9.3 Synchronization of the van der Pol generator with modulated natural frequency . . . 180
9.4 Synchronization of periodic oscillations in systems with inertial nonlinearity . . . 188
9.5 Chaotization of periodic self-oscillations by an external force . 193

9.6 Synchronization of chaotic self-oscillations. The synchronization threshold and its relation to the quantitative characteristics of the attractor 195
9.7 Synchronization of vortex formation in the case of transverse flow around a vibrated cylinder 196
9.8 Synchronization of relaxation self-oscillations 199

10. Interaction of two self-oscillatory systems. Synchronization and chaotization of self-oscillations 205
10.1 Mutual synchronization of periodic self-oscillations with close frequencies 205
10.1.1 The case of weak linear coupling 206
10.1.2 The case of strong linear coupling 213
10.2 Mutual synchronization of self-oscillations with multiple frequencies 216
10.3 Parametric synchronization of two generators with different frequencies 219
10.4 Chaotization of self-oscillations in two coupled generators 220
10.5 Interaction of generators of periodic and chaotic oscillations .. 223
10.6 Interaction of generators of chaotic oscillations 225
10.7 Mutual synchronization of two relaxation generators 231
10.7.1 Mutual synchronization of two coupled relaxation generators of triangular oscillations 231
10.7.2 Mutual synchronization of two Rayleigh relaxation generators 233

11. Interaction of three or more self-oscillatory systems 237
11.1 Mutual synchronization of three generators 237
11.1.1 The case of close frequencies 237
11.1.2 The case of close differences of the frequencies of neighboring generators 243
11.2 Synchronization of N coupled generators with close frequencies 244
11.2.1 Synchronization of N coupled van der Pol generators . 244
11.2.2 Synchronization of pendulum clocks suspended from a common beam 246
11.3 Synchronization and chaotization of self-oscillations in chains of coupled generators 248
11.3.1 Synchronization of N van der Pol generators coupled in a chain 248
11.3.2 Synchronization and chaotization of self-oscillations in a chain of N coupled van der Pol–Duffing generators .. 249
11.3.3 Synchronization of chaotic oscillations in a chain of generators with inertial nonlinearity 250

Part II. OSCILLATIONS IN NONAUTONOMOUS SYSTEMS

12. Oscillations of nonlinear systems excited by external periodic forces 255
12.1 A periodically driven nonlinear oscillator 255
12.1.1 The main resonance 257
12.1.2 Subharmonic resonances 260
12.1.3 Superharmonic resonances 262
12.2 Oscillations excited by an external force with a slowly time-varying frequency 263
12.3 Chaotic regimes in periodically driven nonlinear oscillators ... 266
12.3.1 Chaotic regimes in the Duffing oscillator 267
12.3.2 Chaotic oscillations of a gas bubble in liquid under the action of a sound field 267
12.3.3 Chaotic oscillations in the Vallis model 268
12.4 Two coupled harmonically driven nonlinear oscillators 269
12.4.1 The main resonance 270
12.4.2 The combination resonance 275
12.5 Electro-mechanical vibrators and capacitative sensors of small displacements 283

13. Parametric excitation of oscillations 289
13.1 Parametrically excited nonlinear oscillators 289
13.1.1 Slightly nonlinear oscillator with small damping and small harmonic parametric action 289
13.2 Chaotization of a parametrically excited nonlinear oscillator . 292
13.3 Parametric excitation of pendulum oscillations by noise 294
13.3.1 The results of a numerical simulation of the oscillations of a pendulum with a randomly vibrated suspension axis 298
13.3.2 On-off intermittency 299
13.3.3 Correlation dimension 302
13.3.4 Power spectra 302
13.3.5 The Rytov–Dimentberg criterion 303
13.4 Parametric resonance in a system of two coupled oscillators .. 305
13.5 Simultaneous forced and parametric excitation of an oscillator 312
13.5.1 Parametric amplifier 312
13.5.2 Regular and chaotic oscillations in a model of childhood infections 314

14. Changes in the dynamical behavior of nonlinear systems induced by high-frequency vibration or by noise 323
14.1 The appearance and disappearance of attractors and repellers induced by high-frequency vibration or noise 323

14.2 Vibrational transport and electrical rectification 331
14.2.1 Vibrational transport 332
14.2.2 Rectification of fluctuations 335
14.3 Noise-induced transport of Brownian particles (stochastic ratchets) 336
14.3.1 Noise-induced transport of light Brownian particles in a viscous medium with a saw-tooth potential 337
14.3.2 The effect of the particle mass 344
14.4 Stochastic and vibrational resonances: similarities and distinctions 359
14.4.1 Stochastic resonance in an overdamped oscillator 362
14.4.2 Vibrational resonance in an overdamped oscillator 366
14.4.3 Stochastic and vibrational resonances in a weakly damped bistable oscillator. Control of resonance 367

A. Derivation of the approximate equation for the one-dimensional probability density 373

References 377

Index 393

1. Introduction

1.1 The importance of oscillation theory for engineering mechanics

Engineers routinely deal with oscillatory processes. In mechanics such processes are often called *vibration*. When applied to mechanical processes we will adhere to this term. However, we will use the term *oscillations* for general problems of the theory. Vibration can be both deleterious and useful. For example, in complex constructions undesirable vibration can arise because of inevitable clearances. Alternatively, there is a class of machines, called *vibrational machines*, where vibration is the main working process. To make sense of these complicated processes, it is desirable to know the basics of modern oscillation theory.

It should be noted that the foundations of oscillation theory were laid by the greatest scientist in the field of analytical mechanics J. Lagrange, who introduced generalized coordinates and momenta and, by doing so, in fact digressed from traditional mechanics. The equations derived by him can be applied to systems of any nature. It is no accident that many of the fundamental ideas of the present-day oscillation theory are expounded through the use of the Lagrange equations (or of their counterpart, the Hamilton equations).

The theory of oscillations is the science that sudies oscillatory motions irrespective of their physical nature. In common with any other science the theory of oscillations has laws of its own. It is essential that these laws are general and do not depend on the actual type of system studied. In this respect the outstanding Russian scientist L.I. Mandelshtam, who, as early as 1930, delivered the first lectures on oscillation theory in Moscow State University [229], told students in one of these lectures, "All of you know such systems as a pendulum and an oscillatory circuit, and also know that from the oscillatory point of view they are similar. Now all this is trivial, but it is excellent that this is trivial." These ideas have not yet become fashionable. In the paper "L.I. Mandelshtam and the theory of nonlinear oscillations", Andronov [7] wrote that "The lectures and seminars of Mandelshtam sometimes contained new scientific results which had not been published. But, perhaps, the greatest significance of these lectures lies in the methodical inculcating of habits of *oscillatory thinking*, in the general rise of *the oscillatory culture*". Unfortunately, many prominent scientists studying concrete problems still lack this 'oscillatory culture'. This is the reason why that up to the present the scientific works, being erroneous from the point of view of the general laws of oscillation theory, still appear from time to time.

Unfortunately, up till now the general laws of oscillation theory have not been well formulated. However, many of the specialists in this theory apply these laws, sometimes intuitively, to their studies. With knowledge of the general laws one can not only profitably predict different phenomena from diversified areas of science but also study oscillatory phenomena using simple models being sure that in more complicated systems these phenomena will be of the same character. This fact is used tacitly by all researchers.

The availability of analogies between oscillatory processes in systems of diversified physical nature is why the theory of oscillations got its subject of investigation, and thereby took on the status of an original science. According to Neimark [255], dynamical systems are such a subject. A dynamical system is a system whose behavior is predetermined by a set of rules (by an algorithm). In particular, and most frequently, the behavior of a dynamical system is described by differential, integral or finite-difference equations. In addition to study of dynamical systems in themselves, an important part of oscillation theory is the study of the response of a dynamical system to external actions, both periodic and random. Obviously, a dynamical system is a model of a real system. Thus, we can say that the theory of oscillations studies abstract models, but not concrete systems.

1.2 Classification of dynamical systems. Systems with conservation of phase volume and dissipative systems

All dynamical systems having a physical meaning can be separated into two main categories: systems with conservation of phase volume and systems with a decrease of phase volume; the latter are called *dissipative systems* (see Fig. 1.1). If a system is described by differential equations of the type

$$\dot{x}_j = f_j(x_1, x_2, \ldots, x_n) \quad (j = 1, 2, \ldots, n)\,, \tag{1.1}$$

then it can be shown that, based on the divergence theorem, the variation of the system phase volume $\mathrm{d}V$ in a time $\mathrm{d}t$ is

$$\mathrm{d}V = \mathrm{d}t \int \left(\frac{\mathrm{d}\dot{x}_1}{\mathrm{d}x_1} + \frac{\mathrm{d}\dot{x}_2}{\mathrm{d}x_2} + \ldots + \frac{\mathrm{d}\dot{x}_n}{\mathrm{d}x_n} \right) \mathrm{d}x_1\, \mathrm{d}x_2 \ldots \mathrm{d}x_n = \mathrm{d}t \int \operatorname{div} \dot{\boldsymbol{x}}\, \mathrm{d}\boldsymbol{x}\,, \tag{1.2}$$

where $\dot{\boldsymbol{x}}$ is a vector with components $\dot{x}_1, \dot{x}_2, \ldots, \dot{x}_n$. It follows from this that the sufficient condition for the conservation of the phase volume is

$$\operatorname{div} \dot{\boldsymbol{x}} = 0\,. \tag{1.3}$$

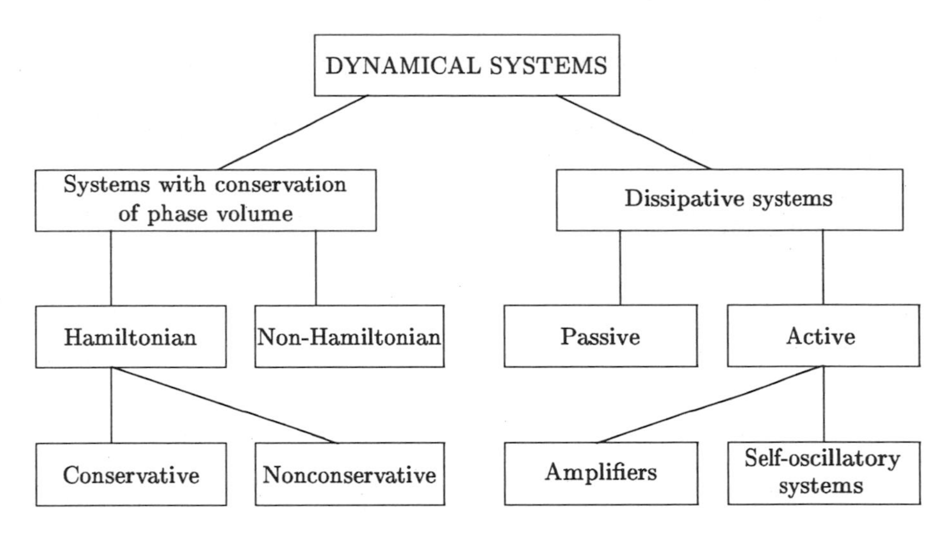

Fig. 1.1. The classification of dynamical systems

Similarly, the sufficient condition for the decrease of the phase volume is

$$\operatorname{div} \dot{\boldsymbol{x}} < 0 . \tag{1.4}$$

Systems with conservation of the phase volume in turn can be broken down into Hamiltonian and non-Hamiltonian systems. Systems are called Hamiltonian if their equations can be written in canonical form by means of a Hamiltonian $H(\boldsymbol{q}, \boldsymbol{p}, t)$, where $\boldsymbol{q}$ and $\boldsymbol{p}$ are generalized coordinates and momenta. These canonical equations are

$$\dot{\boldsymbol{q}} = \frac{\partial H}{\partial \boldsymbol{p}}, \qquad \dot{\boldsymbol{p}} = -\frac{\partial H}{\partial \boldsymbol{q}} . \tag{1.5}$$

For Hamiltonian systems there is an important conservation law described by the so-called Hamilton–Jacobi equation [203]

$$\frac{\partial S}{\partial t} + H(t, \boldsymbol{q}, \boldsymbol{p}) = 0 , \tag{1.6}$$

where $S(t, \boldsymbol{q})$ is the action. In the case when the Hamiltonian H does not depend on time explicitly, i.e., the corresponding system is autonomous, then

$$S(t, \boldsymbol{q}) = S_0(\boldsymbol{q}) - Et , \tag{1.7}$$

where E is the *energy* of the system,[1] and Eq. (1.6) takes the form of the energy conservation equation:

$$H(\boldsymbol{q}, \boldsymbol{p}) = E . \tag{1.8}$$

[1] The definition of energy for mechanical systems is introduced independently of the Hamilton–Jacobi equation as a sum of the kinetic and potential energies.

Systems for which the energy conservation equation (1.8) is valid are called *conservative* systems.

If the original equations of the system are written as the Lagrange equations

$$\frac{\mathrm{d}}{\mathrm{d}t}\frac{\partial L}{\partial \dot{\boldsymbol{q}}} - \frac{\partial L}{\partial \boldsymbol{q}} = 0\,, \tag{1.9}$$

where $L = L(\boldsymbol{q}, \dot{\boldsymbol{q}}, t)$ is the Lagrangian, then the Hamiltonian can be easily found from the relation

$$H(\boldsymbol{q}, \boldsymbol{p}, t) = (\dot{\boldsymbol{q}} \cdot \boldsymbol{p}) - L(\boldsymbol{q}, \dot{\boldsymbol{q}}, t)\,, \tag{1.10}$$

where $\dot{\boldsymbol{q}}$ on the right-hand side must be expressed in terms of $\boldsymbol{p}$ by using the equation

$$\boldsymbol{p} = \frac{\partial L}{\partial \dot{\boldsymbol{q}}}\,. \tag{1.11}$$

Studies of nonlinear systems is often conveniently conducted in terms of the so-called *action–angle variables.* These variable ($\boldsymbol{J}$, $\boldsymbol{\Theta}$) are determined by a certain canonical transformation[2] $\boldsymbol{J} = \boldsymbol{J}(\boldsymbol{q}, \boldsymbol{p})$, $\boldsymbol{\Theta} = \boldsymbol{\Theta}(\boldsymbol{q}, \boldsymbol{p})$. In the case when the system is completely integrable (see Ch. 3), in these variables the Hamiltonian appears to be independent of $\boldsymbol{\Theta}$ and takes the form $H(\boldsymbol{J})$. For a conservative system with one degree of freedom the action J is

$$J = \frac{1}{2\pi}\oint p\,\mathrm{d}q\,, \tag{1.12}$$

where the integral is taken along a cycle on the phase plane q, p corresponding to periodic motion. The canonical variables J and ϑ are associated by the canonical equations

$$\dot{\vartheta} = \frac{\mathrm{d}H(J)}{\mathrm{d}J}\,, \quad \dot{J} = -\frac{\mathrm{d}H(J)}{\mathrm{d}\vartheta} \equiv 0\,. \tag{1.13}$$

Because $J = \text{const}$ we have $\dot{\vartheta} \equiv \omega(J) = \text{const}$. It can be easily shown that $\omega(J) = 2\pi/T(J)$, where $T(J)$ is the period of the oscillations. Hence, $\vartheta = \omega(J)t + \varphi$.

Dissipative systems are very important and the most abundant class of dynamical systems. In nature all systems are dissipative, but if dissipation is sufficiently small then for a limited time the dissipation effects have no time to show themselves, and such systems behave as conservative ones. This situation occurs in the motion of celestial bodies, for example. Dissipative systems are subdivided into passive and active ones. The systems are called *passive* if they do not contain any energy sources. Owing to dissipation the total energy of such systems decreases. The systems are called *active* if they contain constant or alternating energy sources. *Self-oscillatory systems*, or

[2] A transformation is said to be canonical if its Jacobian is equal to zero.

generators, form a large and all-important subclass of active systems. Self-oscillatory systems are defined as those active systems in which undamped oscillations can exist in the absence of any external action. An amplifier is a necessary but insufficient component of any self-oscillatory system. To become a generator, an amplifier must be included in a feedback loop, so that a portion of the signal from the amplifier output is supplied to its input. Thus, a feedback loop must be an integral part of any self-oscillatory system.

1.3 Different types of mathematical models and their functions in studies of concrete systems

There is much speculation that the generality of the laws of oscillation theory is based on the similarity of the equations describing oscillatory processes in different systems. Every so often this is indeed so. However, the general laws of oscillation theory, as a rule, are independent of the concrete form of the equations. It is only sufficient that the equations belong to a certain class, which in common is moderately broad. Therefore, equations of the model do not need to be similar to the equations of the original system.

All of the models for concrete systems can be conventionally separated into four categories: models–'portraits' of investigated systems, 'black box' models, so-called aggregated models, and models for the study of a certain phenomenon which we are interested in.

Models of the first type are constructed on the basis of the description, in as much detail as possible, of all elements of the system studied. Such models are usually studied in physics. An example is the Navier–Stokes equations and their finite-difference approximations describing fluid motion with a high degree of accuracy. As another example we can point to certain models of heart activity or breathing [242]. These models, as a rule, are very complicated and cumbersome for detailed analysis and revealing the principal features of the system's behavior.

When models of the second type are constructed, the system studied is considered as 'black box' with given inputs and outputs which can be measured. Further, a model as simpler as possible is chosen. The free parameters of this model are determined from the conditions of the minimum, according to a given criterion, of the difference between the model's and system's outputs for identical inputs. Such an approach is conceptually used for reconstructing a model from experimental data (see, e.g, [109, 200]).

Models of the third type are constructed on the basis of the analysis of aggregate behavior of the individual elements of the system studied. A classical example of a model of this type is the 'prey–predator' model of Lotka–Volterra [215, 360]. Other interesting examples are the model of an immune reaction illustrating the oscillatory course of some chronic diseases [311, 189] and the model of economic progress of human society [254, 189].

Finally, models of the fourth type are constructed for the analysis of a certain phenomenon no matter in what system it occurs. For example, the phenomenon of self-excitation of self-oscillations can be modeled by the van der Pol equation; the phenomenon of synchronization of periodic self-oscillations can also be modeled by the van der Pol equation but with an external force; oscillatory character of certain chemical reactions is clearly demonstrated by the 'brusselator' and 'oregonator' models [189]; the different transitions to chaos can be modeled by maps of a certain form [85, 253]. Similar examples can be given in abundance. It should be particularly emphasized that the mere possibility and utility of such models are based on the universality of the laws of oscillation theory.

1.4 Phase space of autonomous dynamical systems and the number of degrees of freedom

For autonomous dynamical systems we can introduce the notion of *state* determined by a set of quantities called *dynamical variables*. The space of dynamical variables is said to be the *phase space*. It follows from the definition of a dynamical system that its state at each instant t must be uniquely determined by its state at some earlier instant t_0. Obviously all real systems are not dynamical, on two counts: for one thing, these systems are always acted upon by uncontrolled random forces (fluctuations); for another, the initial state of these systems cannot be determined with pinpoint accuracy. However, if the system motion is stable with respect to small perturbations then these circumstances are not essential, and the state of the real system at each instant t can be uniquely predicted, starting from its initial state. It is known, for example, that the motion of the planets and their satellites can be predicted with very high precision. But if the system motion is unstable with respect to small perturbations, then the behavior of such a system is unpredictable. One such example is the motion of air in the atmosphere. Systems of such a type belong to the class of so-called *dynamical chaotic systems*.

An important characteristic of an autonomous dynamical system is its *number of degrees of freedom*. It is customary to set this number equal to one half the number of independent variables that completely determine the state of the system, i.e., to one half the dimension of the system phase space. Such a definition of the number of degrees of freedom arose because this notion first appeared in mechanics where the one-dimensional motion of a material point is completely determined by two variables: coordinate and velocity. According to this definition the number of degrees of freedom may not be an integer. For example, if a system is described by one differential equation of third order or by three first-order differential equations, then its number of degrees of freedom equals one and a half. We note that if a system is nonautonomous, i.e., the algorithm for its transition from one state to another is explicitly dependent on time, then, formally, it may be considered as autonomous by means of the incorporation of time as one of the coordinates of its phase space.

1.5 The subject matter of the book

The book contains an introduction and two parts. In the introduction we explain the importance of oscillation theory for engineering mechanics, give the definition of dynamical systems and carry out their classification, and describe different types of mathematical models and their functions in studies of concrete systems. The first part deals with oscillatory processes in autonomous systems. The multitude of such processes can be divided into two large classes: natural oscillations and self-oscillations. Both of them are considered. In the second part we present the results of studies of oscillations in nonautonomous systems; in particular, the problems of forced oscillations and of synchronization of self-oscillations are fully discussed. Special attention is given to changes in the dynamical behavior and in the properties of nonlinear systems induced by external actions, both periodic and random.

The book concludes with a comprehensive list of references.

2. The main analytical methods of studies of nonlinear oscillations in near-conservative systems

The analytical methods of the present-day theory of nonlinear oscillations originated in the investigations by Poincaré [272, 273], Birkhoff [42] and Lyapunov [216, 217], laying the mathematical foundations for this theory. Although, it should be noted that a direct application of these mathematical methods to oscillation theory as such occurred much later, primarily owing to the works of Andronov [5, 7, 8].

A great contribution to the development of the quantitative theory of nonlinear oscillations, especially of the applied part of this theory, was made by van der Pol [350, 352, 353] who studied the operation of an electronic generator and proposed his own investigative method, namely, *the method of slowly time-varying amplitudes*. Later a rigorous justification of this method was given by Mandelshtam and Papaleksi [226].

Almost independently of Mandelshtam, Andronov and other physicists the mathematical groundwork for nonlinear oscillation theory was laid by Krylov, Bogolyubov, Mitropol'sky [159, 52, 53, 238, 239, 240] and their disciples. They worked out the most important methods for the analysis of slightly nonlinear oscillations: the asymptotic method, the averaging method and the method of equivalent linearization. These methods were developed further trough works of Moiseev [241], Volosov and Morgunov [359], Nayfeh [248], Vasilyeva [354, 355], Mischenko [237], O'Malley [259], Eckhaus [79], Sanders and Verhulst [295], and other scientists. It should be noted that the averaging method is applicable not only to slightly nonlinear systems but also to essentially nonlinear ones provided that these systems are near-conservative. Another variant of the averaging method, as applied to linear and nonlinear waves in conservative systems, was suggested by Whitham [363]. This method can be easily generalized to near-conservative systems [168].

One of the first methods for approximate calculations of stationary periodic oscillatory regimes was suggested by Poincaré [272, 273]. Currently, it is variously known as *the method of a small parameter*, or *the Poincaré method* [5, 7, 8, 62, 46, 223]. We will not describe this method because it is not used in the present book.

2.1 The van der Pol method

The van der Pol method, or the method of slowly time-varying amplitudes, is applicable to near-linear, near-conservative self-oscillatory systems. The method was suggested by van der Pol as applied to self-oscillatory systems with one degree of freedom. However, the method may be easily generalized to self-oscillatory systems with n degrees of freedom which are described by equations of the form

$$\ddot{y}_k + \omega_k^2 y_k = \mu Y_k(t, y, \dot{y}) \quad (k = 1, 2, \ldots, n), \tag{2.1}$$

where μ is a small parameter, $y = \{y_1, \ldots, y_n\}$, $\dot{y} = \{\dot{y}_1, \ldots, \dot{y}_n\}$.

For $\mu = 0$ Eqs. (2.1) are the equations of a linear oscillatory system with n degrees of freedom written in terms of normal coordinates. They have a solution

$$y_k = A_k \cos\psi_k, \tag{2.2}$$

where $\psi_k = \omega_k t + \varphi_k$, A_k and φ_k are arbitrary constants having a physical meaning of the corresponding amplitudes and phases.

Using the van der Pol method we assume that, for $\mu \neq 0$, the solution of Eqs. (2.1) has the same form (2.2), but only the amplitudes A_k and phases φ_k are slowly time-varying functions. The conditions of the slowness, which follow from the nearness of the system to conservative one, are

$$\frac{\dot{A}_k}{\omega_k A_k} \ll 1, \quad \frac{\dot{\varphi}_k}{\omega_k} \ll 1. \tag{2.3}$$

Owing to conditions (2.3) we can ignore the second derivatives of the amplitudes and phases. In so doing we find

$$\ddot{y}_k = -A_k\omega_k^2 \cos\psi_k - 2\dot{A}_k\omega_k \sin\psi_k - 2A_k\dot{\varphi}_k\omega_k \cos\psi_k. \tag{2.4}$$

Substituting (2.4) and (2.2) into (2.1) we obtain

$$\begin{aligned} &-2\dot{A}_k\omega_k \sin\psi_k - 2A_k\dot{\varphi}_k\omega_k \cos\psi_k \\ &= \mu Y_k(t, A_1\cos\psi_1, \ldots, A_n\cos\psi_n, -A_1\omega_1\sin\psi_1, \ldots, -A_n\omega_n\sin\psi_n). \end{aligned} \tag{2.5}$$

We note that A_k and φ_k on the right-hand sides of Eqs. (2.5) are considered as constant values.

Let us represent the right-hand sides of Eqs. (2.5) as

$$Y_k = f_k(\mu t, A, \varphi)\sin\psi_k + F_k(\mu t, A, \varphi)\cos\psi_k + \ldots, \tag{2.6}$$

where $A = \{A_1, \ldots, A_n\}$, $\varphi = \{\varphi_1, \ldots, \varphi_n\}$; $f_k(\mu t, A, \varphi)$ and $F_k(\mu t, A, \varphi)$ are certain functions of the oscillation amplitude and phases, and of 'slow time' μt. The explicit dependence of f_k and F_k on μt appears either when the system is acted upon by an external force at a frequency close to one of the fundamental frequencies ω_k or when any of the frequencies ω_k are close to each other, so that the difference between them is of the order of the small parameter μ.

Equating in (2.5) the coefficients of $\sin\psi_k$ and of $\cos\psi_k$, in view of (2.6), we obtain the following equations for the amplitudes and phases:

$$\dot{A}_k = -\frac{\mu}{2\omega_k} f_k(\mu t, A, \varphi), \quad \dot{\varphi}_k = -\frac{\mu}{2\omega_k A_k} F_k(\mu t, A, \varphi). \tag{2.7}$$

According to the van der Pol method, the terms in (2.6) involving higher harmonics are merely dropped.

Equations (2.7) allow us to consider periodic and quasi-periodic oscillatory regimes, to determine the stability of these regimes and find the transient processes. However, it should be noted that determination of the stability from Eqs. (2.7) gives only the necessary stability conditions which are optionally sufficient. This results from the fact that Eqs. (2.7) are derived by using a certain form of the solution of the original equations. Therefore the stability conditions found from these equations relate only to disturbances of the same form, whereas the instability with respect to disturbances of other forms is possible. A full answer to the stability problem can be obtained only in the framework of the original equations.

Of special interest is a particular case where all fundamental frequencies ω_k are closely related, so that $\omega_j - \omega_k \ll \omega_l$ for any j, k and l. In this case a solution of Eqs. (2.1) is conveniently sought as

$$y_k = A_k \cos(\omega t + \varphi_k), \tag{2.8}$$

where $\omega = (\omega_1 + \ldots + \omega_n)/n$. In so doing the equations for the amplitudes and phases become

$$\dot{A}_k = -\frac{\mu}{2\omega} f_k(\mu t, A, \varphi), \quad \dot{\varphi}_k = \Delta_k - \frac{\mu}{2\omega A_k} F_k(\mu t, A, \varphi), \tag{2.9}$$

where $\Delta_k = (\omega_k^2 - \omega^2)/(2\omega) \approx \omega_k - \omega$ is the frequency mistuning.

As mentioned above, a rigorous justification of the van der Pol method was given by Mandelshtam and Papaleksi [226]. They considered an equation describing self-oscillations in a system with one degree of freedom:

$$\ddot{y} + \omega^2 y = \mu Y(y, \dot{y}). \tag{2.10}$$

A solution of Eq. (2.10) was set in the form

$$y = A\cos(\omega t + \varphi), \tag{2.11}$$

where A and φ were considered as new variables defined by Eq. (2.11) and

$$\dot{y} = -A\omega \sin(\omega t + \varphi). \tag{2.12}$$

One of the equations for the variables A and φ follows from Eqs. (2.11) and (2.12):

$$\dot{A}\cos(\omega t + \varphi) - A\dot{\varphi}\sin(\omega t + \varphi) = 0. \tag{2.13}$$

Another equation is obtained by the substitution of Eqs. (2.11) and (2.12) into Eq. (2.10):

$$\dot{A}\sin(\omega t+\varphi)+A\dot{\varphi}\cos(\omega t+\varphi)=-\frac{\mu}{\omega}Y\Big(A\cos(\omega t+\varphi),-A\omega\sin(\omega t+\varphi)\Big). \tag{2.14}$$

Resolving Eqs. (2.13), (2.14) with respect to $\dot{A}$ and $\dot{\varphi}$ we obtain

$$\begin{aligned}\dot{A}&=-\frac{\mu}{\omega}Y\Big(A\cos(\omega t+\varphi),-A\omega\sin(\omega t+\varphi)\Big)\sin(\omega t+\varphi),\\ \dot{\varphi}&=-\frac{\mu}{\omega A}Y\Big(A\cos(\omega t+\varphi),-A\omega\sin(\omega t+\varphi)\Big)\cos(\omega t+\varphi).\end{aligned} \tag{2.15}$$

Equations (2.15) are exact equations for the amplitude and phase, because they are derived from Eq. (2.10) by a change of variables. Since the right-hand sides of Eqs. (2.15) are proportional to the small parameter μ, the variables A and φ are slow time-varying variables, justifying the assumption by van der Pol.

Mandelshtam and Papaleksi showed that the corresponding van der Pol equations can be found by averaging the right-hand sides of Eqs. (2.15) over 'fast time' for a period $T=2\pi/\omega$. By doing so, Mandelshtam and Papaleksi pioneered the averaging method, which then was developed by Bogolyubov and Mitropol'sky.

2.2 The asymptotic Krylov–Bogolyubov method

Because the van der Pol method cannot be used to find higher approximations with respect to small parameters, it is very suitable when the first approximation turns out to be sufficient. The asymptotic Krylov–Bogolyubov method is conceptually a generalizalition of the van der Pol method which allows us to calculate the higher approximations.

The method has two modifications, depending on the form of the original equations and on the problem in question. If we are interested in the calculation of multi-frequency oscillations, we conveniently set the equations in the form of (2.1). Provided we are interested in the calculation of single-frequency oscillations, we can set the original equations in the following form:

$$\dot{\boldsymbol{y}}+\mathcal{B}\boldsymbol{y}=\mu\boldsymbol{Y}(t,\boldsymbol{y}), \tag{2.16}$$

where $\boldsymbol{y}$ is a vector with n components, $\mathcal{B}$ is a quadratic matrix with elements b_{jk}, μ is a small parameter, and $\boldsymbol{Y}(t,\boldsymbol{y})$ is a nonlinear vector-function of time and of all components of the vector $\boldsymbol{y}$. We assume that for $\mu=0$ Eqs. (2.16) describe a linear conservative system.

(i) First we consider the case when the original equations of a system are set in the form of (2.1). To zero approximation, these equations have a solution (2.2). For $\mu\neq 0$, let us represent a solution of Eqs. (2.1) as a power series in μ:

$$y_k=y_k^{(0)}+\mu u_{1k}(A,\psi,\mu t)+\mu^2 u_{2k}(A,\psi,\mu t)+\dots, \tag{2.17}$$

where $y_k^{(0)} = A_k \cos\psi_k$, $\psi_k = \omega_k t + \varphi_k$, A_k and φ_k are slowly time-varying functions obeying the equations

$$\frac{\mathrm{d}A_k}{\mathrm{d}t} = \mu f_{1k}(A,\varphi,\mu t) + \mu^2 f_{1k}(A,\varphi,\mu t) + \dots ,$$
$$\frac{\mathrm{d}\varphi_k}{\mathrm{d}t} = \mu F_{1k}(A,\varphi,\mu t) + \mu^2 F_{1k}(A,\varphi,\mu t) + \dots . \tag{2.18}$$

Here $u_{1k}(A,\psi,\mu t)$, $u_{2k}(A,\psi,\mu t)$, ... and $f_{1k}(A,\psi,\mu t)$, $f_{2k}(A,\psi,\mu t)$, ..., $F_{1k}(A,\psi,\mu t)$, $F_{2k}(A,\psi,\mu t)$, ... are unknown functions, which should be found.

Taking account of (2.17) and (2.18) we obtain for the derivatives $\dot{y}_k$ and $\ddot{y}_k$ the following expressions:

$$\dot{y}_k = -A_k\omega_k \sin\psi_k + \mu\left(f_{1k}\cos\psi_k - A_k F_{1k}\sin\psi_k + \sum_{j=1}^{n}\omega_j \frac{\partial u_{1k}}{\partial\psi_j}\right) + \dots ,$$
$$\ddot{y}_k = -A_k\omega_k^2 \cos\psi_k + \mu\Bigg(-2\omega_k f_{1k}\sin\psi_k - 2\omega_k A_k F_{1k}\cos\psi_k + \sum_{j=1}^{n}\sum_{l=1}^{n}\omega_j\omega_l \frac{\partial^2 u_{1k}}{\partial\psi_j\partial\psi_l}\Bigg) + \dots . \tag{2.19}$$

We also expand the right-hand side of Eqs. (2.1) as a power series in μ:

$$\mu Y_k(t,y,\dot{y}) = \mu Y_{1k} + \dots , \tag{2.20}$$

where

$$Y_{1k} = Y_k(t,y_0,\dot{y}_0) = X_{1k}(\mu t,A,\varphi)\sin\psi_k + Z_{1k}(\mu t,A,\varphi)\cos\psi_k + \dots . \tag{2.21}$$

Substituting (2.17), (2.19) and (2.20) into Eqs. (2.1) and equating the coefficients of the same powers of μ, we obtain the equations for the unknown functions u_{jk}:

$$\sum_{j=1}^{n}\sum_{l=1}^{n}\omega_j\omega_l \frac{\partial^2 u_{1k}}{\partial\psi_j\partial\psi_l} + \omega_k^2 u_{1k} = 2\omega_k\Big(f_{1k}\sin\psi_k + A_k F_{1k}\cos\psi_k\Big) + Y_{1k}, \tag{2.22}$$

$$\dots$$

Substituting further (2.21) into (2.22) and demanding the absence of resonant constituents in functions u_{1k}, we find the unknown functions f_{1k} and F_{1k}:

$$f_{1k} = -\frac{1}{2\omega_k}X_{1k}(\mu t,A,\varphi), \quad F_{1k} = -\frac{1}{2\omega_k A_k}Z_{1k}(\mu t,A,\varphi). \tag{2.23}$$

Thus we obtain the equations of the first approximation for the amplitudes and phases:

$$\frac{\mathrm{d}A_k}{\mathrm{d}t} = -\frac{\mu}{2\omega_k} X_{1k}(\mu t, A, \varphi), \quad \frac{\mathrm{d}\varphi_k}{\mathrm{d}t} = -\frac{\mu}{2\omega_k A_k} Z_{1k}(\mu t, A, \varphi). \tag{2.24}$$

These equations coincide with (2.7) obtained by using the van der Pol method. The functions u_{1k}, which can be found from Eqs. (2.22), describe the higher harmonics and combination frequencies in the solution of the first approximation.

Using the next terms in the expansions (2.18), (2.19), and (2.20), we can find the equations of the second and higher approximations.

(ii) Let us consider further the case when the original equations of a system are set in the form of (2.16) and we are interested in finding single-frequency oscillations.

One of the periodic solutions of the generative equations (for $\mu = 0$) that is of interest to us can be written as

$$\boldsymbol{y} = A\Big(\boldsymbol{V}\mathrm{e}^{\mathrm{i}(\omega t+\varphi)} + \text{c.c.}\Big), \tag{2.25}$$

where ω is one of the system's fundamental frequencies, A and φ are arbitrary constants, $\boldsymbol{V}$ is the eigenvector of the matrix $\mathcal{B}$ corresponding to the frequency ω, and c.c. means the complex conjugate value.

For $\mu \neq 0$ a solution of Eqs. (2.16) is sought as a power series in μ:

$$\boldsymbol{y} = \boldsymbol{y}_0 + \mu \boldsymbol{u}_1(A, \psi, \mu t) + \mu^2 \boldsymbol{u}_2(A, \psi, \mu t) + \dots, \tag{2.26}$$

where $\boldsymbol{y}_0 = A\Big(\boldsymbol{V}\mathrm{e}^{\mathrm{i}\psi} + \text{c.c.}\Big)$, $\psi = \omega t + \varphi$, $\boldsymbol{u}_1$, $\boldsymbol{u}_2$, ... are unknown vector-functions, A and φ are slowly time-varying functions obeying the equations

$$\begin{aligned} \frac{\mathrm{d}A}{\mathrm{d}t} &= \mu f_1(A, \varphi, \mu t) + \mu^2 f_2(A, \varphi, \mu t) + \dots, \\ \frac{\mathrm{d}\varphi}{\mathrm{d}t} &= \mu f_1(A, \varphi, \mu t) + \mu^2 F_2(A, \varphi, \mu t) + \dots, \end{aligned} \tag{2.27}$$

Using (2.26), (2.27), let us calculate the derivative $\dot{\boldsymbol{y}}$ and expand it as a power series in μ:

$$\begin{aligned} \dot{\boldsymbol{y}} = \mathrm{i}\omega A\Big(\boldsymbol{V}\mathrm{e}^{\mathrm{i}\psi} - \text{c.c.}\Big) + \mu\Bigg(f_1\Big(\boldsymbol{V}\mathrm{e}^{\mathrm{i}\psi} + \text{c.c.}\Big) \\ + \mathrm{i}AF_1\Big(\boldsymbol{V}\mathrm{e}^{\mathrm{i}\psi} - \text{c.c.}\Big) + \omega\frac{\partial \boldsymbol{u}_1}{\partial \psi}\Bigg) + \dots. \end{aligned} \tag{2.28}$$

The right-hand side of Eqs. (2.16) is also expanded as a power series in μ:

$$\mu \boldsymbol{Y}(t, \boldsymbol{y}) = \mu \boldsymbol{Y}_1 + \mu^2 \boldsymbol{Y}_2 + \dots, \tag{2.29}$$

where

$$\boldsymbol{Y}_1 = \boldsymbol{Y}(t, \boldsymbol{y}_0), \quad \boldsymbol{Y}_2 = \frac{\partial \boldsymbol{Y}}{\partial \boldsymbol{y}}(t, \boldsymbol{y}_0)\boldsymbol{u}_1 + \frac{\partial \boldsymbol{Y}}{\partial \mu}(t, \boldsymbol{y}_0), \quad \dots.$$

Equating in Eqs. (2.16) the coefficients of the same powers of μ, we obtain the equations for the unknown vector-functions $\boldsymbol{u}_j$:

$$\omega\frac{\partial \boldsymbol{u}_1}{\partial\psi} + \mathcal{B}\boldsymbol{u}_1 = -f_1\left(\boldsymbol{V}e^{i\psi} + \text{c.c.}\right) - iAF_1\left(\boldsymbol{V}e^{i\psi} - \text{c.c.}\right) + \boldsymbol{Y}_1, \tag{2.30a}$$

$$\omega\frac{\partial \boldsymbol{u}_2}{\partial\psi} + \mathcal{B}\boldsymbol{u}_2 = -f_2\left(\boldsymbol{V}e^{i\psi} + \text{c.c.}\right) - iAF_2\left(\boldsymbol{V}e^{i\psi} - \text{c.c.}\right)$$
$$-\frac{\partial \boldsymbol{u}_1}{\partial(\mu t)} - f_1\frac{\partial \boldsymbol{u}_1}{\partial A} - F_1\frac{\partial \boldsymbol{u}_1}{\partial\psi} + \boldsymbol{Y}_2, \tag{2.30b}$$

...

Let us further expand the vector-functions $\boldsymbol{u}_1, \boldsymbol{u}_2, \ldots, \boldsymbol{Y}_1, \boldsymbol{Y}_2, \ldots$ into the Fourier series with slowly time-varying coefficients:

$$\boldsymbol{u}_j(A,\psi,\mu t) = \sum_{k=-\infty}^{\infty} \boldsymbol{U}_j^{(k)}(A,\varphi,\mu t)e^{ik\psi},$$
$$\boldsymbol{Y}_j(A,\psi,\mu t) = \sum_{k=-\infty}^{\infty} \boldsymbol{Y}_j^{(k)}(A,\varphi,\mu t)e^{ik\psi}. \tag{2.31}$$

Substituting (2.31) into (2.31) and equating the coefficients of the same harmonics, we obtain, for each j, a system of nonuniform equations for the components of the vector-functions $\boldsymbol{U}_j^{(k)}$. For $k \neq 1$ the determinant of this system is nonzero, and, hence, all of $\boldsymbol{U}_j^{(k\neq 1)}$ can be determined uniquely. For $k = 1$ the system determinant is zero. In this case we should require, for all j, the fulfillment of the compatibility conditions

$$\mathcal{A}\boldsymbol{R}_j = 0, \tag{2.32}$$

where $\boldsymbol{R}_j$ is the right-hand side of the jth system and $\mathcal{A}$ is the adjoint matrix. These conditions allow us to find $f_1(A,\varphi,\mu t)$, $F_1(A,\varphi,\mu t)$, $f_2(A,\varphi,\mu t)$, $F_2(A,\varphi,\mu t)$, ...

The first approximation. From (2.30a) and (2.31) we obtain the following equation for the vector-function $\boldsymbol{U}_1^{(k)}$:

$$(ik\omega\mathcal{E} + \mathcal{B})\boldsymbol{U}_1^{(k)} = -(f_1 + iAF_1)\delta_{k1}\boldsymbol{V} + \boldsymbol{Y}_1^{(k)}, \tag{2.33}$$

where $\mathcal{E}$ is an identity matrix and δ_{k1} is the Kronecker delta. For $k = 1$ the compatibility condition of the system (2.33) is

$$-f_1\mathcal{A}\boldsymbol{V} - iAF_1\mathcal{A}\boldsymbol{V} + \mathcal{A}\boldsymbol{Y}_1^{(1)} = 0. \tag{2.34}$$

Splitting the real part and the imaginary part in Eq. (2.34) we find

$$f_1(A, \varphi, \mu t) = \mathrm{Re}\left\{\frac{1}{\mathrm{Sp}\,\mathcal{A}} \sum_j \frac{A_{jj}}{V_j} Y_{1j}^{(1)}(A, \varphi, \mu t)\right\},$$

$$F_1(A, \varphi, \mu t) = \frac{1}{A}\,\mathrm{Im}\left\{\frac{1}{\mathrm{Sp}\,\mathcal{A}} \sum_j \frac{A_{jj}}{V_j} Y_{1j}^{(1)}(A, \varphi, \mu t)\right\}, \tag{2.35}$$

where $\mathrm{Sp}\,\mathcal{A}$ is the spur of the matrix $\mathcal{A}$.

The second approximation. To calculate the second approximation, we find the vector-functions $\boldsymbol{U}_1^{(k)}$ from Eqs. 2.34. Taking into account that $\boldsymbol{U}_1^{(1)}$ is determined to an arbitrary constant, we can put

$$\boldsymbol{U}_1^{(1)} = A\boldsymbol{V}^{(1)}, \tag{2.36}$$

where $\boldsymbol{V}^{(1)}$ is the correction to the eigenvector $\boldsymbol{V}$ in the first approximation. From (2.30b) and (2.31), in view of (2.36), we obtain the following equation for the vector-function $\boldsymbol{U}_2^{(k)}$:

$$\begin{aligned}(\mathrm{i}k\omega\mathcal{E} + \mathcal{B})\boldsymbol{U}_2^{(k)} = &-\Bigg[(f_2 + \mathrm{i}AF_2)\boldsymbol{V} + (f_1 + \mathrm{i}AF_1)\boldsymbol{V}^{(1)} \\ &+ A\left(\frac{\partial \boldsymbol{V}^{(1)}}{\partial(\mu t)} + f_1\frac{\partial \boldsymbol{V}^{(1)}}{\partial A} + F_1\frac{\partial \boldsymbol{V}^{(1)}}{\partial \varphi}\right)\Bigg]\delta_{k1} - \Bigg[\frac{\partial \boldsymbol{U}_1^{(k)}}{\partial(\mu t)} + f_1\frac{\partial \boldsymbol{U}_1^{(k)}}{\partial A} \\ &+ F_1\frac{\partial \boldsymbol{U}_1^{(k)}}{\partial \varphi} + \mathrm{i}kF_1\boldsymbol{U}_1^{(k)}\Bigg](1 - \delta_{k1}) + \boldsymbol{Y}_2^{(k)}.\end{aligned} \tag{2.37}$$

From the compatibility condition of the system (2.37) for $k = 1$ we find $f_2(A, \varphi, \mu t)$ and $F_2(A, \varphi, \mu t)$.

The third and higher approximations can be found in much the same way.

2.3 The averaging method

The rigorous theory of the averaging method was worked out by Bogolyubov [52] and developed by Mitropol'sky [239]. This theory concerns so-called equations in a standard form. By means of a certain change of variables any equations describing oscillations in near-conservative systems can be reduced to equations in a standard form.

Let us consider, for example, a near-linear, near-conservative systems described by Eqs. (2.1). In place of the variables y_k and $\dot{y}_k$ we introduce the variables A_k and φ_k which are related to y_k and $\dot{y}_k$ by

$$y_k = A_k \cos\psi_k, \quad \dot{y}_k = -A_k\omega_k \sin\psi_k, \tag{2.38}$$

where $\psi_k = \omega_k t + \varphi_k$.

It follows from (2.38) that the variables A_k and φ_k have to obey the equations

$$\dot{A}_k \cos\psi_k - A_k\dot{\varphi}_k \sin\psi_k = 0. \tag{2.39}$$

Another equation is found by the substitution of (2.38) into Eqs. (2.1):

$$\dot{A}_k \sin\psi_k + A_k\dot{\varphi}_k \cos\psi_k = -\frac{\mu}{\omega_k} Y_k(t, A\cos\psi, -A\omega\sin\psi). \tag{2.40}$$

Resolving Eqs. (2.39) and (2.40) with respect to $\dot{A}_k$ and $\dot{\varphi}_k$, we obtain

$$\begin{aligned} \dot{A}_k &= -\frac{\mu}{\omega_k} Y_k(t, A\cos\psi, -A\omega\sin\psi)\sin\psi_k, \\ \dot{\varphi}_k &= -\frac{\mu}{\omega_k A} Y_k(t, A\cos\psi, -A\omega\sin\psi)\cos\psi_k. \end{aligned} \tag{2.41}$$

Equations (2.41) are exact equations for the amplitude and phase, because they are derived from Eqs. (2.1) by the change of variables. Since the right-hand sides of Eqs. (2.41) are proportional to the small parameter μ, the variables A_k and φ_k are slow time-varying variables. According to Bogolyubov, such a system is called *equations in a standard form.*

So, let us consider the averaging method as applied to the following equations in a standard form:

$$\frac{\mathrm{d}\boldsymbol{x}}{\mathrm{d}t} = \mu\boldsymbol{X}(t, \mu t, \boldsymbol{x}). \tag{2.42}$$

We assume that the vector-function $\boldsymbol{X}(t, \mu t, \boldsymbol{x})$ can be represented as a sum of harmonic constituents with slowly time-varying amplitudes:

$$\boldsymbol{X}(t, \mu t, \boldsymbol{x}) = \sum_{\nu} e^{\mathrm{i}\nu t}\boldsymbol{X}_\nu(\mu t, \boldsymbol{x}). \tag{2.43}$$

The first approximation. Substitute into Eq. (2.42)

$$\boldsymbol{x} = \boldsymbol{\xi} + \mu\hat{\boldsymbol{X}}(t, \mu t, \boldsymbol{\xi}), \tag{2.44}$$

where the symbol $\hat{\boldsymbol{X}}$ means integrating the vector $\boldsymbol{X} - \boldsymbol{X}_0$ with respect to explicit time, namely

$$\hat{\boldsymbol{X}}(t, \mu t, \boldsymbol{\xi}) = \sum_{\nu\neq 0} \frac{\mathrm{e}^{\mathrm{i}\nu t}}{\mathrm{i}\nu} \boldsymbol{X}_\nu(\mu t, \boldsymbol{\xi}). \tag{2.45}$$

The expression (2.44) may be treated as an approximate solution of Eq. (2.42) that can be obtained by the substitution $\boldsymbol{x} = \boldsymbol{\xi}$, where $\boldsymbol{\xi}$ is a slowly time-varying function free from vibrational constituents, into the right-hand side of this equation.

Differentiating (2.44) with respect to time we find

$$\frac{\mathrm{d}\boldsymbol{x}}{\mathrm{d}t} = \frac{\mathrm{d}\boldsymbol{\xi}}{\mathrm{d}t} + \mu\frac{\partial\hat{\boldsymbol{X}}}{\partial t} + \mu^2\frac{\partial\hat{\boldsymbol{X}}}{\partial(\mu t)} + \mu\frac{\partial\hat{\boldsymbol{X}}}{\partial\boldsymbol{\xi}}\frac{\mathrm{d}\boldsymbol{\xi}}{\mathrm{d}t}. \tag{2.46}$$

It follows from (2.45) and (2.43) that

$$\frac{\partial \hat{\boldsymbol{X}}}{\partial t} = \sum_{\nu \neq 0} \mathrm{e}^{\mathrm{i}\nu t} \boldsymbol{X}_\nu(\mu t, \boldsymbol{\xi}) = \boldsymbol{X}(t, \mu t, \boldsymbol{\xi}) - \boldsymbol{X}_0(\mu t, \boldsymbol{\xi}). \tag{2.47}$$

Substituting (2.44) and (2.46) into Eq. (2.42), in view of (2.47), we obtain the equation for the vector $\boldsymbol{\xi}$:

$$\mathcal{A} \frac{\mathrm{d}\boldsymbol{\xi}}{\mathrm{d}t} = \mu\Big(\boldsymbol{X}_0(\mu t, \boldsymbol{\xi}) + \boldsymbol{X}(t, \mu t, \boldsymbol{\xi} + \mu \hat{\boldsymbol{X}}) - \boldsymbol{X}(t, \mu t, \boldsymbol{\xi})\Big) - \mu^2 \frac{\partial \hat{\boldsymbol{X}}}{\partial(\mu t)}, \tag{2.48}$$

where $\mathcal{A} = \mathcal{E} + \mu\, \partial \hat{\boldsymbol{X}} / \partial \boldsymbol{\xi}$ is quadratic matrix with elements $a_{jk} = \delta_{jk} + \mu\, \partial \hat{X}_j / \partial \xi_k$. Multiplying Eq. (2.48) from the left by $\mathcal{A}^{-1}$, we have

$$\frac{\mathrm{d}\boldsymbol{\xi}}{\mathrm{d}t} = \mu \mathcal{A}^{-1} \left(\boldsymbol{X}_0(\mu t, \boldsymbol{\xi}) + \boldsymbol{X}(t, \mu t, \boldsymbol{\xi} + \mu \hat{\boldsymbol{X}}) - \boldsymbol{X}(t, \mu t, \boldsymbol{\xi}) - \mu \frac{\partial \hat{\boldsymbol{X}}}{\partial(\mu t)} \right). \tag{2.49}$$

Taking into account that $\mathcal{A}^{-1} = \mathcal{E} - \mu \dfrac{\partial \hat{\boldsymbol{X}}}{\partial \boldsymbol{\xi}} + \ldots$, we expand the right-hand side of Eq. (2.49) as a power series in μ:

$$\frac{\mathrm{d}\boldsymbol{\xi}}{\mathrm{d}t} = \mu \boldsymbol{X}_0(\mu t, \boldsymbol{\xi}) - \mu^2 \left(\frac{\partial \hat{\boldsymbol{X}}}{\partial \boldsymbol{\xi}} \boldsymbol{X}_0 - \frac{\partial \boldsymbol{X}}{\partial \boldsymbol{\xi}} \hat{\boldsymbol{X}} + \frac{\partial \hat{\boldsymbol{X}}}{\partial(\mu t)} \right) + \ldots. \tag{2.50}$$

Ignoring in (2.50) all terms of order μ^2 and higher and putting $\boldsymbol{\xi} = \boldsymbol{x}$, we find as a first approximation

$$\frac{\mathrm{d}\boldsymbol{x}}{\mathrm{d}t} = \mu \boldsymbol{X}_0(\mu t, \boldsymbol{x}). \tag{2.51}$$

Thus, the equation of the first approximation is obtained from the exact equation (2.42) by averaging its right-hand side over explicit time. In so doing the components of the vector $\boldsymbol{x}$ are considered as constant quantities.

Returning to Eqs. (2.41) for the amplitudes and phases, we obtain as a first approximation

$$\begin{aligned} \dot{A}_k &= -\frac{\mu}{\omega_k} \overline{Y_k(t, A\cos\psi, -A\omega\sin\psi)\sin\psi_k}, \\ \dot{\varphi}_k &= -\frac{\mu}{\omega_k A} \overline{Y_k(t, A\cos\psi, -A\omega\sin\psi)\cos\psi_k}, \end{aligned} \tag{2.52}$$

where the overbar means averaging over explicit time t. Equations (2.52) are often called *the truncated equations.*

The second approximation. We give only a technique for the calculation of the second approximation and omit its justification. The justification of this technique is presented in detail in [239]. To calculate the second approximation, we substitute (2.44) into Eq. (2.42) taking the vector $\boldsymbol{\xi}$ in the argument of the vector-function $\hat{\boldsymbol{X}}(t, \mu t, \boldsymbol{\xi})$ to be constant. Then, taking account of (2.47), we obtain the equation

$$\frac{\mathrm{d}\boldsymbol{\xi}}{\mathrm{d}t} = \mu \boldsymbol{X}_0(\mu t, \boldsymbol{\xi}) + \mu\Big(\boldsymbol{X}\big(t, \mu t, \boldsymbol{\xi} + \mu\hat{\boldsymbol{X}}\big) - \boldsymbol{X}(t, \mu t, \boldsymbol{\xi})\Big) - \mu^2 \frac{\partial \hat{\boldsymbol{X}}}{\partial(\mu t)}. \tag{2.53}$$

Expanding the right-hand side of Eq. (2.53) as a power series in μ, restricting ourselves to terms of order μ^2 and ignoring vibrational terms, we obtain the following equation of the second approximation:

$$\frac{\mathrm{d}\boldsymbol{\xi}}{\mathrm{d}t} = \mu \boldsymbol{X}_0(\mu t, \boldsymbol{\xi}) + \mu^2 \left(\overline{\frac{\partial \boldsymbol{X}}{\partial \boldsymbol{\xi}}(t, \mu t, \boldsymbol{\xi}) \hat{\boldsymbol{X}}(t, \mu t, \boldsymbol{\xi})} - \frac{\partial \hat{\boldsymbol{X}}}{\partial(\mu t)} \right). \tag{2.54}$$

We note that the same equations can be obtained also from Eq. (2.50) by neglecting terms of order μ^3 and averaging over time.

The sense of the procedure described consists in the following. It can be seen from (2.44) that the vector $\boldsymbol{x}$ involves both a slow constituent and a fast one. The equations of the first approximation are obtained by substituting into the right-hand side of Eq. (2.42) only the slow constituent of $\boldsymbol{x}$, whereas the fast constituent is not taken into account. In the second approximation we take account of the effect of the fast constituent on the behavior of the slow constituent.

It is easy to verify that in the first approximation the equations for amplitudes and phases found by the averaging method coincide with the corresponding equations obtained by both the van der Pol method and the Krylov–Bogolyubov method.

It should be stressed that the averaging method, along with the Poincaré method, can be applied not only to near-linear systems but to essentially nonlinear systems as well [62, 223, 296, 297]. An example of such a calculation will be given in Ch. 6.

2.4 The averaging method in systems incorporating fast and slow variables

In the preceding section the averaging method was considered as applied to equations in a standard form (2.42), where all of the components of the vector $\boldsymbol{x}$ were slowly time-varying variables. However, many important problems demand solving equations in which only a portion of the variables are slowly time-varying functions [336, 374, 354, 237]. For example, in Eqs. (2.1) the functions Y_k can depend not only on t, y and $\dot{y}$, but also on a certain set of variables z_l described by the equations

$$\dot{z}_l = Z_l(t, y, \dot{y}, z) \quad l = 1, 2, \dots, m. \tag{2.55}$$

Performing the change of variables (2.38) in Eqs. (2.1), in view of (2.55), we obtain the following equations for A_k, φ_k and z_l:

$$\begin{aligned} \dot{A}_k &= -\frac{\mu}{\omega_k} Y_k(t, A\cos\psi, -A\omega\sin\psi, z)\sin\psi_k, \\ \dot{\varphi}_k &= -\frac{\mu}{\omega_k A} Y_k(t, A\cos\psi, -A\omega\sin\psi, z)\cos\psi_k, \\ \dot{z}_l &= Z_l(t, A\cos\psi, -A\omega\sin\psi, z). \end{aligned} \tag{2.56}$$

In the more general form these equations can be written as

$$\frac{\mathrm{d}\boldsymbol{x}}{\mathrm{d}t} = \mu\boldsymbol{X}(t, \mu t, \boldsymbol{x}, \boldsymbol{y}), \tag{2.57a}$$

$$\frac{\mathrm{d}\boldsymbol{y}}{\mathrm{d}t} = \boldsymbol{Y}(t, \mu t, \boldsymbol{x}, \boldsymbol{y}). \tag{2.57b}$$

The averaging method as applied to equations of the form (2.57a), (2.57b) was suggested by Volosov and Morgunov [358, 359].

In Eqs. (2.57a) and (2.57b) the components of the vector $\boldsymbol{x}$ are slowly time-varying variables, whereas the components of the vector $\boldsymbol{y}$ are fast time-varying variables. This means that the characteristic time of the variation of the components of the vector $\boldsymbol{y}$ is so short that the components of the vector $\boldsymbol{x}$ cannot significantly change. With this supposition, Eq. (2.57b) can be solved, and its partial solution can be written as

$$\boldsymbol{y} = \boldsymbol{y}(t, \mu t, \boldsymbol{x}, \boldsymbol{y}_0), \tag{2.58}$$

where $\boldsymbol{y}_0$ is the initial vector.

As a first approximation, in Eq. (2.57a) we can average $\boldsymbol{X}(t, \mu t, \boldsymbol{x}, \boldsymbol{y})$ over t along the integral curve (2.58), i.e., substitute

$$\overline{\boldsymbol{X}(t, \mu t, \boldsymbol{x}, \boldsymbol{y})} = \lim_{T\to\infty} \frac{1}{T} \int\limits_0^T \boldsymbol{X}\Big(t, \mu t, \boldsymbol{x}, \boldsymbol{y}(t, \mu t, \boldsymbol{x}, \boldsymbol{y}_0)\Big)\,\mathrm{d}t \tag{2.59}$$

in place of $\boldsymbol{X}(t, \mu t, \boldsymbol{x}, \boldsymbol{y})$. Since the duration of the transient process for the components of the vector $\boldsymbol{y}$ is sufficiently small, we can take as lower limit of the integral in (2.59) not zero but a certain instant t_0 when the solution (2.58) becomes stationary. Therefore $\overline{\boldsymbol{X}(t, \mu t, \boldsymbol{x}, \boldsymbol{y})}$ has to be independent of the initial vector $\boldsymbol{y}_0$, and that is why we can take as the solution (2.58) a stationary solution $\boldsymbol{y}(t, \mu t, \boldsymbol{x})$.

The derivation of higher approximations is set out in [358, 359]. This derivation is moderately complicated. In actual practice, the first approximation, as a rule, turns out to be sufficient.

An example of using the averaging method described here will be given in Ch. 6.

2.5 The Whitham method

The method suggested by Whitham [361, 362, 363] for approximate calculations of the evolution of dispersive waves is based on the variational principle of mechanics and on the theory of adiabatic invariants. This method is also appropriate in studies of lumped, essentially nonlinear, quasi-conservative systems.

Let us write the equations for a studied system in the Lagrange form:

$$\frac{\mathrm{d}}{\mathrm{d}t}\frac{\partial L}{\partial \dot{\boldsymbol{u}}} - \frac{\partial L}{\partial \boldsymbol{u}} = \mu \boldsymbol{F}(\boldsymbol{u}, \dot{\boldsymbol{u}}), \tag{2.60}$$

where $L(\boldsymbol{u}, \dot{\boldsymbol{u}})$ is the Lagrangian, $\boldsymbol{u}(t)$ is the vector with components $u_i(t)$, $\partial L/\partial \dot{\boldsymbol{u}}$ and $\partial L/\partial \boldsymbol{u}$ are the vectors with components $\partial L/\partial \dot{u}_i$ and $\partial L/\partial u$ respectively, $\mu \boldsymbol{F}(\boldsymbol{u}, \dot{\boldsymbol{u}})$ is the vector of generalized dissipative forces, and μ is a small parameter.

Like the averaging method, the Whitham method involves the transition from 'fast' variables to 'slow' ones describing the amplitude and phase of quasi-stationary oscillations. The truncated equation for the amplitude is found from the condition that the energy conservation law must be valid in the average over a period of oscillations.

Let a periodic solution of Eqs. (2.60) for $\mu = 0$, depending on an arbitrary constant A, be

$$\boldsymbol{u}(t) = \boldsymbol{f}(A, \psi), \tag{2.61}$$

where A is the amplitude of oscillations, $\psi = \omega(A)t$ is the oscillation phase, and $\boldsymbol{f}(A, \psi)$ is a periodic function of ψ of period 2π. The function $\boldsymbol{f}(A, \psi)$ is described by the equation

$$\frac{\partial}{\partial \psi}\frac{\partial L_s}{\partial \boldsymbol{f}_\psi} - \frac{\partial L_s}{\partial \boldsymbol{f}} = 0, \tag{2.62}$$

where $\boldsymbol{f}_\psi = d\boldsymbol{f}/d\psi$,

$$L_s(A, \psi) = L\Big(\boldsymbol{f}(A, \psi), \omega \boldsymbol{f}_\psi(A, \psi)\Big). \tag{2.63}$$

We seek a solution of Eqs. (2.60) for $\mu \neq 0$ in the same form as (2.61), only A and $\omega \equiv \mathrm{d}\psi/\mathrm{d}t$ are assumed to be slow functions of time. Let us show that $A(t)$ is described by the equation

$$\frac{\mathrm{d}}{\mathrm{d}t}\frac{\partial \overline{L}}{\partial \omega} = \mu \mathcal{F}(A), \tag{2.64}$$

where

$$\overline{L}(A, \omega) = \frac{1}{2\pi}\int_0^{2\pi} L_s(A, \psi)\,\mathrm{d}\psi, \tag{2.65a}$$

$$\mathcal{F}(A) = \overline{\boldsymbol{f}_\psi(A,\psi)\boldsymbol{F}\Big(\boldsymbol{f}(A,\psi),\omega\boldsymbol{f}_\psi(A,\psi)\Big)}$$
$$= \frac{1}{2\pi}\int\limits_0^{2\pi} \boldsymbol{f}_\psi(A,\psi)\boldsymbol{F}\Big(\boldsymbol{f}(A,\psi),\omega\boldsymbol{f}_\psi(A,\psi)\Big)\,\mathrm{d}\psi. \tag{2.65b}$$

Preparatory to this, we will first show that

$$\frac{\partial\overline{L}}{\partial\omega} = \frac{1}{\omega}\overline{\boldsymbol{f}_\psi\frac{\partial L}{\partial\boldsymbol{f}_\psi}}. \tag{2.66}$$

It follows from (2.63) and (2.65a) that

$$\frac{\partial\overline{L}}{\partial\omega} = \frac{1}{2\pi}\int\limits_0^{2\pi}\left(\frac{\partial L}{\partial\boldsymbol{f}}\frac{\partial\boldsymbol{f}}{\partial\omega} + \frac{\partial L}{\partial(\omega\boldsymbol{f}_\psi)}\frac{\partial(\omega\boldsymbol{f}_\psi)}{\partial\omega}\right)d\psi.$$

Taking account of Eq. (2.62) and

$$\frac{\partial(\omega\boldsymbol{f}_\psi)}{\partial\omega} = \boldsymbol{f}_\psi + \omega\frac{\partial^2\boldsymbol{f}}{\partial\psi\partial\omega},$$

we find

$$\int\limits_0^{2\pi}\left(\frac{\partial L}{\partial\boldsymbol{f}}\frac{\partial\boldsymbol{f}}{\partial\omega} + \frac{\partial L}{\partial(\omega\boldsymbol{f}_\psi)}\frac{\partial(\omega\boldsymbol{f}_\psi)}{\partial\omega}\right)\mathrm{d}\psi$$
$$= \int\limits_0^{2\pi}\left[\frac{\partial L}{\partial(\omega\boldsymbol{f}_\psi)}\boldsymbol{f}_\psi + \omega\frac{\partial}{\partial\psi}\left(\frac{\partial L}{\partial(\omega\boldsymbol{f}_\psi)}\frac{\partial\boldsymbol{f}}{\partial\omega}\right)\right]\mathrm{d}\psi.$$

Using the periodicity condition for the vector-function $\boldsymbol{f}$, we obtain (2.66).

The averaged energy conservation law, in view of (2.66), can be written as

$$\frac{\mathrm{d}\overline{E}}{\mathrm{d}t} = \overline{W}, \tag{2.67}$$

where

$$\overline{E} = \overline{\boldsymbol{f}_\psi\frac{\partial L}{\partial\boldsymbol{f}_\psi}} - \overline{L} = \omega\frac{\partial\overline{L}}{\partial\omega} - \overline{L}, \quad \overline{W} = \mu\omega\overline{\boldsymbol{f}_\psi\boldsymbol{F}} = \mu\omega\mathcal{F}, \tag{2.68}$$

$\overline{E}$ is the mean oscillation energy, and $\overline{W}$ is the work of the generalized dissipative forces in a unit of time.

Substituting (2.68) into Eq. (2.67) we obtain

$$\omega\frac{\mathrm{d}}{\mathrm{d}t}\frac{\partial\overline{L}}{\partial\omega} + \frac{\partial\overline{L}}{\partial\omega}\frac{\mathrm{d}\omega}{\mathrm{d}t} - \frac{\mathrm{d}\overline{L}}{\mathrm{d}t} = \overline{W}. \tag{2.69}$$

Since $\dfrac{\mathrm{d}\overline{L}}{\mathrm{d}t} = \dfrac{\partial\overline{L}}{\partial A}\dfrac{\mathrm{d}A}{\mathrm{d}t} + \dfrac{\partial\overline{L}}{\partial\omega}\dfrac{\mathrm{d}\omega}{\mathrm{d}t}$, we find from (2.69)

$$\omega \frac{\mathrm{d}}{\mathrm{d}t} \frac{\partial \overline{L}}{\partial \omega} - \frac{\partial \overline{L}}{\partial A} \frac{\mathrm{d}A}{\mathrm{d}t} = \overline{W}. \tag{2.70}$$

As shown below

$$\frac{\partial \overline{L}}{\partial A} = -\mu \overline{\boldsymbol{f}_A \boldsymbol{F}}, \tag{2.71}$$

where

$$\overline{\boldsymbol{f}_A \boldsymbol{F}} = \frac{1}{2\pi} \int_0^{2\pi} \frac{\partial \boldsymbol{f}}{\partial A} \boldsymbol{F} \, \mathrm{d}\psi,$$

i.e., $\partial \overline{L} / \partial A$ is of order μ. Since $\partial A / \partial t$ is also of order μ, in the first approximation with respect to μ Eq. (2.70) reduces to (2.64).

We show further that Eq. (2.71) follows from the periodicity condition of the generative solution $\boldsymbol{f}(A, \psi)$. For this we differentiate (2.65a) with respect to A and take account of (2.63). As a result we obtain

$$\begin{aligned} \frac{\partial \overline{L}}{\partial A} &= \frac{1}{2\pi} \int_0^{2\pi} \left(\frac{\partial L}{\partial \boldsymbol{f}} \frac{\partial \boldsymbol{f}}{\partial A} + \omega \frac{\partial L}{\partial (\omega \boldsymbol{f}_\psi)} \frac{\partial \boldsymbol{f}_\psi}{\partial A} \right) \mathrm{d}\psi \\ &= \frac{1}{2\pi} \int_0^{2\pi} \left[\frac{\partial L}{\partial \boldsymbol{f}} \frac{\partial \boldsymbol{f}}{\partial A} + \omega \frac{\partial}{\partial \psi} \left(\frac{\partial L}{\partial (\omega \boldsymbol{f}_\psi)} \frac{\partial \boldsymbol{f}}{\partial A} \right) - \frac{\partial \boldsymbol{f}}{\partial A} \frac{d}{dt} \frac{\partial L}{\partial (\omega \boldsymbol{f}_\psi)} \right] \mathrm{d}\psi. \end{aligned} \tag{2.72}$$

Taking account of (2.61) and the periodicity condition for the vector-function $\boldsymbol{f}$, we can rewrite (2.72) as

$$\frac{\partial \overline{L}}{\partial A} = \frac{1}{2\pi} \int_0^{2\pi} \frac{\partial \boldsymbol{f}}{\partial A} \left(\frac{\partial L}{\partial \boldsymbol{u}} - \frac{\mathrm{d}}{\mathrm{d}t} \frac{\partial L}{\partial \dot{\boldsymbol{u}}} \right) \mathrm{d}\psi. \tag{2.73}$$

Substituting Eq. (2.60) into (2.73) we obtain Eq. (2.71). This equation gives the relation between $A(t)$ and $\omega(t)$.

Part I

OSCILLATIONS IN AUTONOMOUS DYNAMICAL SYSTEMS

3. General properties of autonomous dynamical systems

3.1 Phase space of autonomous dynamical systems and its structure. Singular points and limit sets

The main elements of the phase space of a dynamical system are *singular points* and *limits sets*.[1] Singular points and limit sets can both attract all adjacent phase trajectories and repel them. The former are often spoken of as *attractors* and the latter are said to be *repellers*.

3.1.1 Singular points and their classification

Singular points in the phase space of a dynamical system determine the equilibrium states of this system. If a system is described by the equation

$$\dot{\boldsymbol{x}} = \boldsymbol{F}(\boldsymbol{x})\,, \tag{3.1}$$

its equilibrium state is determined by

$$\boldsymbol{F}(\boldsymbol{x}) = \boldsymbol{0}\,. \tag{3.2}$$

Equation(3.2) gives the coordinates of singular points in the system phase space $(x_1, x_2, \ldots, x_n)$, where $(x_1, x_2, \ldots, x_n)$ are the coordinates of the vector $\boldsymbol{x}$. A singular point $\boldsymbol{x}^*$ is said to be stable in the sense of Lyapunov if any small deviation from it ($\boldsymbol{y} = \boldsymbol{x} - \boldsymbol{x}^*$) in an initial instant $t = 0$ remains small in any succeeding instant $0 < t < \infty$; otherwise it is unstable. The more rigorous definition is: a singular point is stable if for any $\varepsilon > 0$ there exist $\delta > 0$ such that $|\boldsymbol{y}(t > 0)| < \varepsilon$ for $|\boldsymbol{y}(0)| < \delta$. If, in addition, $\lim\limits_{t\to\infty} |\boldsymbol{y}(t)| = 0$ then the corresponding singular point is said to be *stable asymptotically*. It should be noted that only an asymptotically stable singular point is an attractor.

To determine the character of a singular point, it is sufficient to consider the motion of the system in the vicinity of this singular point, where Eq. (3.1) can be linearized. The linearized equation is

$$\dot{\boldsymbol{y}} = \left.\frac{\partial \boldsymbol{F}}{\partial \boldsymbol{x}}\right|_{\boldsymbol{x}=\boldsymbol{x}^*} \boldsymbol{y}\,. \tag{3.3}$$

[1] It should be noted that singular points are also limits sets but involving only one point; however, it is convenient to talk about them separately.

A partial solution of Eq. (3.3) can be sought as

$$y = C\exp(pt)\,. \tag{3.4}$$

Substituting (3.4) into Eq. (3.3) and equating the system determinant to zero we obtain an algebraic equation of the nth degree allowing us to find n values of p. This equation is called the *characteristic equation*. The classification of singular points is performed on the basis of the characteristic equation roots $p_1, p_2, \ldots, p_n$. We consider this classification for the cases of two-dimensional and three-dimensional phase spaces.

Two-dimensional phase space. Four types of singular points are recognized in a phase plane:

- *A node* (p_1 and p_2 are real and of the same sign). If p_1 and p_2 are negative then the node is stable; otherwise it is unstable.
- *A focus* (p_1 and p_2 are complex conjugate). If the real part of p_1 and p_2 is negative then the focus is stable; otherwise it is unstable.
- *A center* (p_1 and p_2 are imaginary). To find whether such a point is stable or not, we have take into account nonlinear terms. If these terms have no effect on the behavior of phase trajectories in the vicinity of this point then the point is deemed stable (but not asymptotically).
- *A saddle* (p_1 and p_2 are real and of opposite sign). A saddle is always an unstable singular point.

It can be shown that in the case when p_1 and p_2 are real, there exist two phase trajectories which go through the singular point and in the vicinity of this point take the form of straight lines described by the equations

$$y_2 = k_1 y_1\,, \tag{3.5a}$$

$$y_2 = k_2 y_1\,, \tag{3.5b}$$

where

$$\begin{aligned} k_{1,2} &= \left(\left.\frac{\partial F_1}{\partial x_2}\right|_{x_1=x_1^*,x_2=x_2^*}\right)^{-1}\left(p_{1,2} - \left.\frac{\partial F_1}{\partial x_1}\right|_{x_1=x_1^*,x_2=x_2^*}\right) \\ &= \left.\frac{\partial F_2}{\partial x_1}\right|_{x_1=x_1^*,x_2=x_2^*}\left(p_{1,2} - \left.\frac{\partial F_2}{\partial x_2}\right|_{x_1=x_1^*,x_2=x_2^*}\right)^{-1}. \end{aligned} \tag{3.6}$$

It should be noted that an infinity of phase trajectories goes through a singular point of node or focus type, whereas only two phase trajectories go through a singular point of saddle type.

The types of singular points described above and the behavior of phase trajectories in the vicinity of these points are illustrated in Fig. 3.1 by the example of the equations

$$\dot{y}_1 = y_2 - y_1\,, \quad \dot{y}_2 = a y_1 + b y_2\,, \tag{3.7}$$

where a and b are the parameters which determine the type of the singular point $y_1 = 0$, $y_2 = 0$. It can be seen from this figure that in the case of the

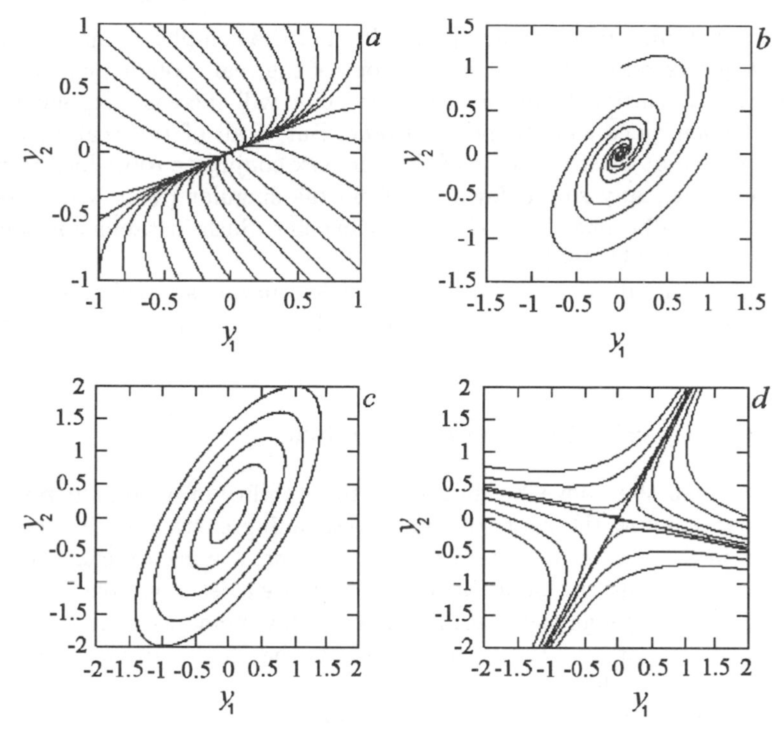

Fig. 3.1. The types of singular points and the behavior of phase trajectories in the vicinity of these points for Eqs. (3.7) for (a) $a = 0.4$, $b = -1.4$ (the node, $k_1 \approx 0.463$, $k_2 \approx -0.863$), (b) $a = -1.6$, $b = 0.6$ (the focus), (c) $a = -2$, $b = 1$ (the center), and (d) $a = 0.4$, $b = 0.6$ (the saddle, $k_1 \approx 1.8198$, $k_2 \approx -0.2198$)

node all phase trajectories, except that described by Eq. (3.5b), are tangent to the phase trajectory described by Eq. (3.5a). In the case of the saddle the phase trajectories described by Eqs. (3.5a), (3.5b) are asymptotes to which other phase trajectories tend as $t \to \infty$.

Multi-dimensional phase space. Classification of singular points in an n-dimensional phase space, where $n \geq 3$, is more complicated than in a phase plane. But the number of the types of singular points is the same as for a phase plane. They are [115, 64]:

- *A node* (all roots of the characteristic equation are real and of the same sign). If all of them are negative then the node is stable; otherwise it is unstable.
- *A focus* (the characteristic equation has both complex conjugate and real roots and all of them are of the same sign). If the real parts of all of the roots are negative then the focus is stable; otherwise it is unstable.
- *A saddle-focus* (the characteristic equation has both complex conjugate and real roots but their real parts are of different signs). A saddle-focus

is always an unstable singular point. If all of the complex conjugate roots have negative real parts and at least one of the real roots is positive then any small deviation from the singular point will increase exponentially. Such instability is said to be *aperiodic*. But if all of the real roots are negative and at least one pair of the complex conjugate roots has positive real part then any small deviation from the singular point will increase executing oscillations about this singular point. Such instability is said to be *oscillatory* [334, 166].

- *A saddle-node* (all roots of the characteristic equation are real and of different signs). Like a saddle-focus, a saddle-node is always an unstable singular point.

3.1.2 Stability criterion of singular points

As follows from the above, to determine the stability of a singular point, it is sufficient to know the roots of the corresponding characteristic equation. However, in the case of a system of a high order, solving its characteristic equation is a difficult problem. Nevertheless, there exist a number of criteria which allow us to judge the stability of a singular point without solving its characteristic equation. One such criterion is the Routh–Hurwitz criterion [127, 290]. We consider this criterion.

Let the characteristic equation for a system be

$$a_0 p^n + a_1 p^{n-1} + a_2 p^{n-2} + \ldots + a_{n-1} p + a_n = 0. \tag{3.8}$$

A quadratic matrix is set up from the coefficients of Eq. (3.8) as shown below:

$$\mathcal{M} = \begin{pmatrix} a_1 & a_0 & 0 & \ldots & 0 & 0 & 0 \\ a_3 & a_2 & a_1 & \ldots & 0 & 0 & 0 \\ a_5 & a_4 & a_3 & \ldots & 0 & 0 & 0 \\ \vdots & \vdots & \vdots & \vdots & \vdots & \vdots & \vdots \\ 0 & 0 & 0 & \ldots & a_{n-2} & a_{n-3} & a_{n-4} \\ 0 & 0 & 0 & \ldots & a_n & a_{n-1} & a_{n-2} \\ 0 & 0 & 0 & \ldots & 0 & 0 & a_n \end{pmatrix} \tag{3.9}$$

Further, n determinants are set up from the matrix $\mathcal{M}$:

$$\Delta_1 = a_1, \quad \Delta_2 = \begin{vmatrix} a_1 & a_0 \\ a_3 & a_2 \end{vmatrix}, \quad \Delta_3 = \begin{vmatrix} a_1 & a_0 & 0 \\ a_3 & a_2 & a_1 \\ a_5 & a_4 & a_3 \end{vmatrix},$$

$$\Delta_{n-1} = \begin{vmatrix} a_1 & a_0 & 0 & \dots & 0 & 0 & 0 \\ a_3 & a_2 & a_1 & \dots & 0 & 0 & 0 \\ \vdots & \vdots & \vdots & \vdots & \vdots & \vdots & \vdots \\ 0 & 0 & 0 & \dots & a_{n-1} & a_{n-2} & a_{n-3} \\ 0 & 0 & 0 & \dots & 0 & a_n & a_{n-1} \end{vmatrix}, \tag{3.10}$$

$$\Delta_n = a_n \Delta_{n-1}.$$

The Routh–Hurwitz criterion states: in order for a singular point to be stable, all of the determinants (3.10) must be greater than or equal to zero. As can be shown, it follows from here that a necessary condition of stability is $a_n/a_0 > 0$ for any n. For equations of first and second order this condition is also sufficient.

It should be noted that the condition $\Delta_n \geq 0$ is a consequence of the conditions $a_n \geq 0$ and $\Delta_{n-1} \geq 0$. Since $a_n = (-1)^n p_1 p_2 \dots p_n$, violation of the condition $a_n \geq 0$ (for $\Delta_{n-1} \geq 0$) means that one of real roots of the characteristic equation (3.8) becomes positive, i.e., aperiodic instability appears. With violation of the condition $\Delta_{n-1} \geq 0$ (for $a_n \geq 0$) a pair of complex conjugate roots crosses the imaginary axis and passes into the right half-plane, i.e., oscillatory instability appears. The latter is conveniently illustrated by the example of a characteristic equation of third order

$$a_0 p^3 + a_1 p^2 + a_2 p + a_3 = 0. \tag{3.11}$$

We assume that all of the coefficients of this equation are positive. Then the condition $\Delta_{n-1} \geq 0$ takes the form

$$\Delta_2 = a_1 a_2 - a_0 a_3 \geq 0. \tag{3.12}$$

Let Δ_2 be equal to zero. In this case Eq. (3.11) can be rewritten as

$$a_0 \left(p + \frac{a_1}{a_0} \right) \left(p^2 + \frac{a_3}{a_1} \right) = 0. \tag{3.13}$$

The roots of Eq. (3.13) are

$$p_1 = -\frac{a_1}{a_0} < 0, \quad p_{2,3} = \pm \mathrm{i} \sqrt{\frac{a_3}{a_1}},$$

i.e., Eq. (3.13) has two pure imaginary roots. Thus, we find that the change of the sign of Δ_2 means the transition of the system through the boundary of oscillatory instability.

Because the appearance of a positive root of the characteristic equation (3.8) results in violation of the condition $a_n \geq 0$, and the appearance of a pair of complex conjugate roots with positive real part results in violation of the condition $\Delta_{n-1} \geq 0$, the occurrence of instability is always associated with the violation of one of the last two Routh–Hurwitz conditions. Although the stability domain in the space of the parameters a_0, a_1, a_2, ..., a_n, which is determined by all of the Routh–Hurwitz conditions, is only a part of the wider area enclosed by the hypersurfaces $a_n = 0$ and $\Delta_{n-1} = 0$, escape from the stability domain is nevertheless possible only through these hypersurfaces.

3.2 Attractors and repellers

As mentioned above, a limit set in the phase space attracting all neighboring phase trajectories from some domain called the *attraction basin* is said to be an attractor. The simplest attractors are a stable singular point (Fig. 3.1 a and b), a stable limit cycle (Fig. 3.2 a) and a stable torus (Fig. 3.2 b). However, a system with n degrees of freedom for $n \geq 1.5$ may have not only

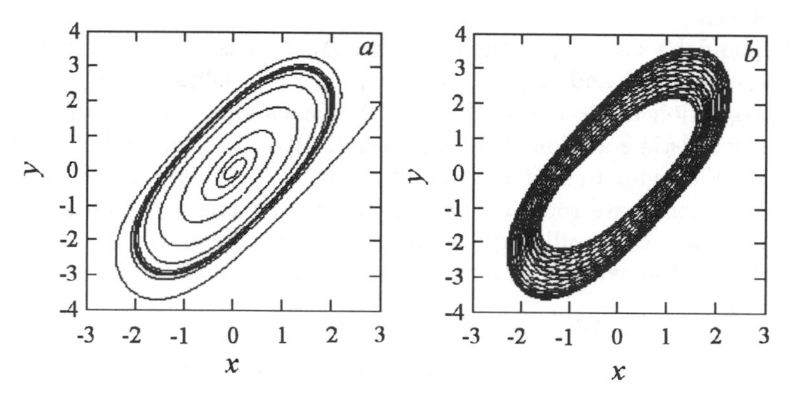

Fig. 3.2. The simplest attractors: (a) a stable limit cycle and (b) the projection of a stable two-dimensional torus on the plane xy

such simple attractors but complexly structured attractors as well. The latter are often spoken of as *strange attractors*.

Much like attractors, repellers can also be simply and complexly structured. Unstable singular points, unstable limit cycles and unstable tori are simple repellers. Complex repellers possess a structure that is similar to that for strange attractors. If there are several attractors in a phase space then their attraction basins are separated by repellers. The simplest example of such a separation is illustrated in Fig. 3.3, where the phase portrait of a system described by the equation

$$\ddot{x} + 2\beta(1 - \alpha x^2 + \gamma x^4)\dot{x} + x = 0, \tag{3.14}$$

where $\beta = 0.05$, $\alpha = 5$ and $\gamma = 2.5$, is given.

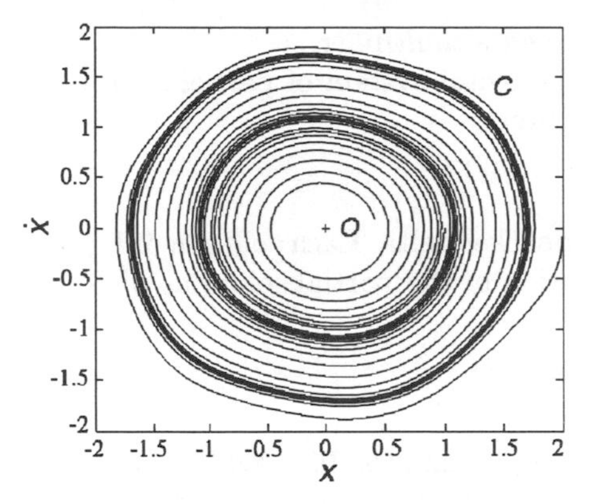

Fig. 3.3. The unstable limit cycle separating the attraction basins of the stable singular point O and the stable limit cycle C

3.3 The stability of limit cycles and their classification

In addition to the notion of stability in the sense of Lyapunov which is similar to that for singular points, the notion of the so-called *orbital stability* is important for limit cycles. The definition of orbital stability of a limit cycle Γ is as follows: Γ is orbitally stable if, with the proviso that $\rho(\gamma(0), \Gamma) < \delta$, for any $\varepsilon > 0$ there exist $\delta > 0$ such that $\rho(\gamma(t), \Gamma) < \varepsilon$ for all $t > 0$, where $\gamma(t)$ is a phase trajectory issuing from the point $\gamma(0)$, and $\rho(\gamma(t), \Gamma)$ is the distance between the point $\gamma(t)$ and the closed curve Γ.

Like singular points, to find whether or not a limit cycle is stable in the sense of Lyapunov we can linearize the original equations with respect to small deviations from the solution associated with this limit cycle. Since the solution corresponding to a limit cycle is a periodic function of period T, for these deviations we obtain linear differential equations with periodic coefficients. They can be written as

$$\dot{\boldsymbol{y}} = B(t)\boldsymbol{y}, \tag{3.15}$$

where $B(t)$ is a quadratic matrix. Each of the elements of the matrix $B(t)$ is a periodic function of period T. According to the Floquet theory [91], a partial solution of Eq. (3.15) is

$$\boldsymbol{y}(t) = F(t)\mathrm{e}^{\lambda t}, \tag{3.16}$$

where $F(t)$ is a column matrix with periodic elements, and λ is a *characteristic exponent*. It should be noted that the characteristic exponent λ is a partial case of a more general exponent called the *Lyapunov exponent*, the set of which determines the stability of any phase trajectory, and not just of a limit cycle. Because of the periodicity of $F(t)$ we obtain from (3.16)

$$\boldsymbol{y}(t + T) = \mu\boldsymbol{y}(t), \tag{3.17}$$

where $\mu = e^{pT}$ is called a *multiplier*.

Consider a fundamental system of partial solutions $\boldsymbol{y}^{(j)}(t)$ of Eq. (3.15) satisfying the condition

$$y_k^{(j)}(0) = \delta_{jk}, \tag{3.18}$$

where δ_{jk} is the Kronecker delta. Each solution of Eq. (3.15) can be presented as a linear combination of the partial solutions $\boldsymbol{y}^{(j)}(t)$:

$$\boldsymbol{y}(t) = \sum_{j=1}^{n} C_j \boldsymbol{y}^{(j)}(t), \tag{3.19}$$

where C_j are arbitrary constants. It is evident that the partial solution satisfying condition (3.17) can also be presented in the form (3.19). It follows from (3.19) and (3.18) that

$$y_k(0) = C_k. \tag{3.20}$$

Setting $t = T$ in (3.19) and taking into account (3.17) and (3.20) we obtain the following system of equations for the constants C_j:

$$\sum_{j=1}^{n} C_j \left(y_k^{(j)}(T) - \delta_{jk}\mu \right) = 0 \quad (k = 1, 2, \dots, n). \tag{3.21}$$

Equating the determinant of this system to zero we find the equation for the multiplier μ:

$$\begin{vmatrix} y_1^{(1)}(T) - \mu & y_1^{(2)}(T) & \dots & y_1^{(n-1)}(T) & y_1^{(n)}(T) \\ y_2^{(1)}(T) & y_2^{(2)}(T) - \mu & \dots & y_2^{(n-1)}(T) & y_2^{(n)}(T) \\ \vdots & \vdots & \vdots & \vdots & \vdots \\ y_{n-1}^{(1)}(T) & y_{n-1}^{(2)}(T) & \dots & y_{n-1}^{(n-1)}(T) - \mu & y_{n-1}^{(n)}(T) \\ y_n^{(1)}(T) & y_n^{(2)}(T) & \dots & y_n^{(n-1)}(T) & y_n^{(n)}(T) - \mu \end{vmatrix} = 0. \tag{3.22}$$

By analogy with the characteristic equation for a singular point, Eq. (3.22) is also said to be a characteristic equation.

It can be shown that one of the roots of Eq. (3.22) is always equal to unity. This is due to the fact that the magnitude of a deviation given along the cycle does not vary with time. Depending on the values of the other roots, the limit cycle is stable or unstable: if the absolute values of all roots are less than unity then the cycle is stable; otherwise, it is unstable.

In three-dimensional phase space limit cycles can be classified as follows [64]:

- If two characteristic exponents corresponding to two roots of Eq. (3.22) different from unity are real and of different signs then such a cycle is referred to as *a saddle cycle*. Two surfaces S^+ and S^- pass through such a cycle (see Fig. 3.4 a).[2] These surfaces, called *integral manifolds*, consist of phase trajectories making a close approach to the corresponding limit set either for $t \to \infty$ (S^+) or for $t \to -\infty$ (S^-).
- If the characteristic exponents are real and of the same sign then such a cycle is referred to as *a nodal cycle*. It can be stable (for $\lambda_{1,2} < 0$) and unstable (for $\lambda_{1,2} > 0$). Stable and unstable nodal cycles are shown in Fig. 3.4 b and c, respectively.
- A limit cycle for which the characteristic exponents are complex conjugate has no special name. In the vicinity of such a cycle phase trajectories resemble spiral lines with the cycle as their axis (Fig. 3.4 d and e).

3.4 Strange attractors: stochastic and chaotic attractors

As mentioned above, a system with n degrees of freedom for $n \geq 1.5$ may have not only simple but also complexly structured attractors which are called strange. According to Neimark, strange attractors can be separated into two categories: *stochastic attractors* and *chaotic attractors* [253]. Attractors involving only a finite or an infinite number of saddle cycles and their integral manifolds are referred to as stochastic. Attractors involving both saddle and stable cycles with small attraction basins are referred to as chaotic. All phase trajectories forming a stochastic attractor are exponentially unstable. A chaotic attractor holds at least one trajectory that is stable in the Lyapunov sense. In particular, chaotic attractors can consist either of one stable multi-revolution limit cycle with closely spaced coils or of a denumerable set of stable limit cycles with sufficiently small attraction basins (the number of cycles can be infinite).

It should be emphasized that there is no way of distinguishing between stochastic and chaotic attractors in real experiments and numerical simulations. This is associated with the inevitable presence of small uncontrollable disturbances.

Although all phase trajectories forming a stochastic attractor are unstable in the Lyapunov sense, such an attractor is a stable formation, but this stability is in the Poisson sense [42]. The term *Poisson-stable motion* means that the motion $\boldsymbol{x}(t)$ is limited and, for any t_0 and $\varepsilon > 0$, there exist such increasing indefinitely instants of time $t_0 < t_1 < t_2 < \ldots < t_\infty$ that $\left|\boldsymbol{x}(t_n) - \boldsymbol{x}(t_0)\right| < \varepsilon$ ($n = 1, 2, \ldots, \infty$).

[2] It should be noted that saddle cycles are possible only in a phase space with dimension greater than or equal to three.

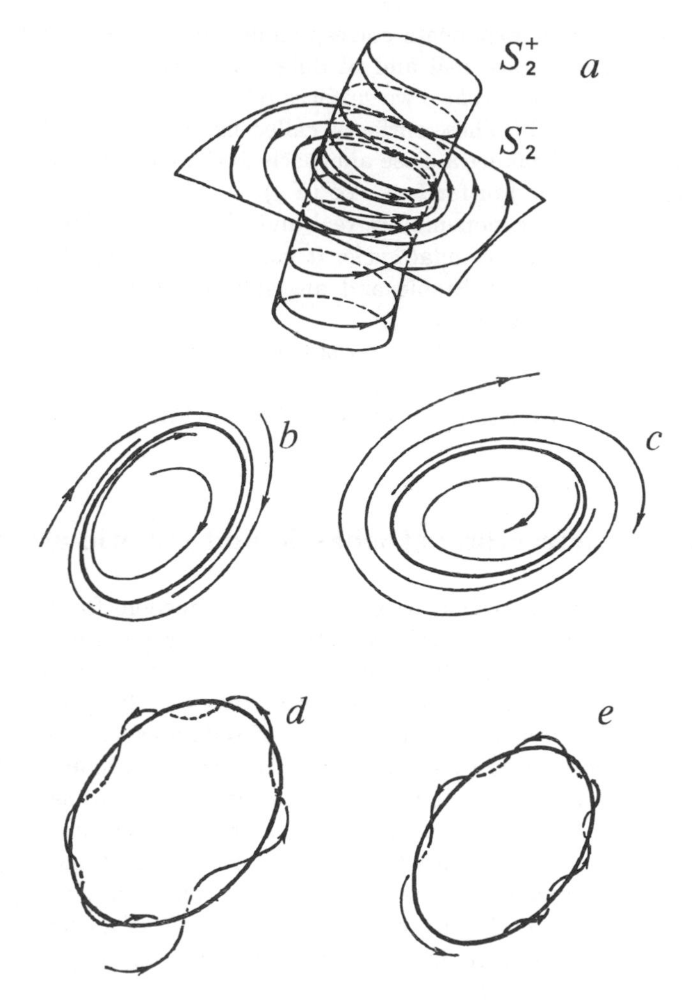

Fig. 3.4. The different types of limit cycles: (a) a saddle limit cycle and its integral manifolds, (b) and (c) stable and unstable nodal limit cycles, respectively, (d) and (e) limit cycles corresponding to complex conjugate characteristic exponents with negative and positive real part, respectively

3.4.1 Quantitative characteristics of attractors

Attractors and repellers can exist only in dissipative systems. The phase volume occupied by trajectories forming attractors and repellers is zero. Hence the dimension of an attractor and a repeller has to be less (often much less) than the dimension of the original phase space. For example, the dimension of a limit cycle is equal to one without regard to the dimension of the phase space in which it lies. This fact is very important in studies of systems with many degrees of freedom and particularly of continuous systems; these latter have an infinite-dimensional phase space, but the dimension of an attractor for such systems, as a rule, is finite. The fact that the dimension of an attractor of a continuous system is finite allows us to use a finite-dimensional phase space in studies of stationary motions in such systems.

In contrast to simple attractors whose dimension is an integer, strange attractors, as a rule, have fractional dimension. Hausdorff was the first to define the notion of fractional dimension [114]. More recently, this dimension has been referred to by his name. The Hausdorff dimension is defined as follows. Let a set of points be given in an n-dimensional space. Cover this set with n-dimensional cubes of edges $\varepsilon_i \leq \varepsilon$ and introduce the quantity $l_d = \lim_{\varepsilon \to 0} l_d(\varepsilon)$, where $l_d(\varepsilon) = \inf \sum_i \varepsilon_i^d$, where inf signifies the greatest lower bound (infimum). The quantity l_d is called the *Hausdorff measure*. Hausdorff showed that there exists a value of d ($d = d_{\mathrm{H}}$) for which l_d is finite; whereas $l_d = 0$ for $d > d_{\mathrm{H}}$ and $l_d = \infty$ for $d < d_{\mathrm{H}}$. The value d_{H} is said to be *the Hausdorff dimension.*

Undoubtedly, the Hausdorff dimension is the main quantitative characteristic of an attractor. It determines the minimal dimension of a phase space in which the attractor can be embedded. Indeed, it is known [230] that any compact set given in a space of some dimension and having finite Hausdorff dimension d_{H} can be projected one-to-one into a hyperplane of dimension not exceeding $2d_{\mathrm{H}} + 1$.

Apparently, the necessity of using the notions of Hausdorff dimension and Hausdorff measure in real studies was first perceived by Richardson in his attempt to measure the length of the fiorded coast of England [282]. This fact is described by Mandelbrot [225]. Approximating the coastline by a broken line with each unit ε in length, Richardson obtained that the length of this broken line

$$L_\varepsilon = N\varepsilon \tag{3.23}$$

increases indefinitely as ε decreases, but in so doing the quantity $\lambda = N\varepsilon^d$, where d is a certain value greater than unity, remains constant. Substituting $N = \lambda\varepsilon^{-d}$ into (3.23) we find $L_\varepsilon = \lambda\varepsilon^{1-d}$. Thus, if we take two segments of the coastline approximated by broken lines $L_{1\varepsilon}$ and $L_{2\varepsilon}$ in length then the ratio $L_{1\varepsilon}/L_{2\varepsilon} = \lambda_1/\lambda_2$, i.e., it is independent of ε. Hence, the lengths of different segments of the fiorded coast can be related to one another by using the quantity λ. The quantities λ and d can be considered as the Hausdorff measure and dimension, respectively.

We note that the calculation of the Hausdorff dimension directly is a complicated problem. Therefore, it is often preferred to calculate other dimensions which are close to the Hausdorff dimension. Such dimensions are the so-called *fractal dimension d* (or *attractor capacity*) and *correlation dimension ν*. The attractor capacity is defined by the formula

$$d = \lim_{\varepsilon \to 0} \frac{\log N(\varepsilon)}{\log \epsilon^{-1}}, \tag{3.24}$$

where $N(\varepsilon)$ is the number of n-dimensional cubes of edge ε that completely cover the attractor. It is easily shown that $d_{\mathrm{H}} \leq d$. Indeed, if for the calculation of the Hausdorff dimension we take the cubes with the same edges ε then $\tilde{l}_d(\varepsilon) = N(\varepsilon)\varepsilon^d \geq l_d(\varepsilon)$, because of the infimum operation. As seen from (3.24), $\lim_{\varepsilon \to 0} \tilde{l}_d(\varepsilon) = 1$, therefore $l_d \leq 1$. It follows from this that $d \geq d_{\mathrm{H}}$.

The correlation dimension can be calculated by using the correlation integral (see, e.g., [105, 106])

$$C(\varepsilon) = \lim_{N\to\infty} \frac{1}{N^2} \sum_{i,j=1}^{N} \vartheta\Big(\varepsilon - |\boldsymbol{y}_i - \boldsymbol{y}_j|\Big), \tag{3.25}$$

where $\vartheta(z)$ is the Heaviside step function, $\boldsymbol{y}_i$ is the vector describing the position of a representative point in the phase space at the instant $t_i = t_0 + i\tau$, and N is the number of samples. The quantity $C(\varepsilon)$ determines the relative number of pairs of points which are spaced at no more than ε. It turns out that $C(\varepsilon) \sim \varepsilon^{\nu}$ for small ε. Hence,

$$\nu = \lim_{\varepsilon\to 0} \frac{\log C(\varepsilon)}{\log \varepsilon}. \tag{3.26}$$

It was shown by Grassberger et al. [107, 119] that the dimensions d and ν are particular values of the so-called *generalized dimension*

$$D_q = \lim_{\varepsilon\to 0} \frac{\log I_q(\varepsilon)}{\log \varepsilon^{-1}}, \tag{3.27}$$

where $I_q(\varepsilon) = \dfrac{1}{1-q} \log \sum_{i=1}^{N(\varepsilon)} p_i^q$ is the Renyi entropy of order q [280], and p_i is the probability that the representative point will be in the ith cube. It is easily seen that $I_0(\varepsilon) = N(\varepsilon)$ and the formula (3.27) for $q = 0$ turns into (3.24), i.e., $d = D_0$. For $q = 2$ we have $I_q(\varepsilon) = -C(\varepsilon)$, i.e., $\nu = D_2$. Similar to the correlation dimension, the generalized dimension of any order q can be calculated by means of the generalized correlation integrals [108]:

$$C_q(\varepsilon) = \lim_{N\to\infty} \left(\frac{1}{n} \sum_{i=1}^{n} \left(\frac{1}{n} \sum_{i=1}^{n} \vartheta(\varepsilon - |\boldsymbol{y}_i - \boldsymbol{y}_j|) \right)^{q-1} \right)^{1/(q-1)}. \tag{3.28}$$

For $q = 2$ the expression (3.28) turns into (3.25). Like the ordinary correlation integral, the generalized correlation integral $C_q(\varepsilon) \sim \varepsilon^{D_q}$ for sufficiently small ε. Hence,

$$D_q = \lim_{\varepsilon\to 0} \frac{\log C_q(\varepsilon)}{\log \varepsilon}. \tag{3.29}$$

It can be shown that $D_q \geq D_{q'}$ for $q < q'$.

The dependence of the generalized dimension D_q on q characterizes the extent to which the attractor fractal properties are not uniform.[3] For uniform attractors all values of D_q are equal to each other and to the Hausdorff dimension.

Another quantitative characteristic of strange attractors, both stochastic and chaotic, is the set of Lyapunov exponents. Lyapunov exponents are defined as follows [217]. Let a dynamical system be described by Eq. (3.1). We consider two adjacent trajectories $\boldsymbol{x}(t)$ and $\boldsymbol{x}_1(t)$ starting from the points $\boldsymbol{x}_0$ and $\boldsymbol{x}_{10}$, respectively, and denote $\boldsymbol{y}(t) = \boldsymbol{x}_1(t) - \boldsymbol{x}(t)$, $\boldsymbol{y}(0) = \boldsymbol{x}_{10} - \boldsymbol{x}_0$. If the trajectories are nearby then the evolution of the vector $\boldsymbol{y}(t)$ can be described by the linearized equation

$$\dot{\boldsymbol{y}} = \left.\frac{\partial \boldsymbol{F}}{\partial \boldsymbol{x}}\right|_{\boldsymbol{x}=\boldsymbol{x}(t)} \boldsymbol{y}, \tag{3.30}$$

where $\partial \boldsymbol{F}/\partial \boldsymbol{x}$ is the matrix with elements $\partial F_i/\partial x_j$. Equation (3.30) is known to have n fundamental partial solutions $\boldsymbol{y}^{(i)}(t) = \boldsymbol{e}_i(t)$, where $\boldsymbol{e}_i(0)$ are unit orthogonal vectors, such that the quantity

$$\lambda_i(\boldsymbol{x}_0) = \lim_{t\to\infty} \frac{1}{t} \ln \frac{\left|\boldsymbol{e}_i(t)\right|}{\left|\boldsymbol{e}_i(0)\right|} \tag{3.31}$$

exists for each i and, in general, all $\lambda_i(\boldsymbol{x}_0)$ are different. The quantities $\lambda_i(\boldsymbol{x}_0)$ are called *the Lyapunov exponents*. It is evident that if the point $\boldsymbol{x}_0$ belongs to an attractor, the values of $\lambda_i(\boldsymbol{x}_0)$ are independent of $\boldsymbol{x}_0$. The number of positive Lyapunov exponents determines the number of directions of instability, and therefore it is an important characteristic of stochastic and chaotic attractors.

Another important characteristic of stochastic and chaotic attractors is the maximal Lyapunov exponent. It can be shown that in the case of an arbitrary initial vector $\boldsymbol{y}(0)$ the value

$$\lambda(\boldsymbol{x}_0) = \lim_{t\to\infty} \frac{1}{t} \ln \frac{|\boldsymbol{y}(t)|}{|\boldsymbol{y}(0)|} \tag{3.32}$$

is equal to just the maximal Lyapunov exponent. The latter characterizes the extent to which adjacent phase trajectories are divergent.

Direct computation of the Lyapunov exponents by (3.31), (3.32) for systems with exponentially unstable trajectories is in practice impossible, because, even for a very small initial distance between adjacent phase trajectories, the distance between them will increase indefinitely as t increases,

[3] Stochastic attractors, as a rule, possess fractal, i.e., strongly jagged, structure. This is the reason why these attractors have fractional dimension. The Cantor set is an example of a fractal set. Chaotic attractors, in principle, do not possess such a structure but, because of different perturbations which are always present in any numerical calculations, as in real experiments, one can be under the impression that such a structure exists.

resulting in an overflow of computer registers and substantial errors. To overcome these difficulties, Benettin et al. [33, 34] offered another algorithm for the computation of the Lyapunov exponents and, in particular, of the maximal Lyapunov exponent. According to this algorithm, the total computation time T is divided into m intervals of duration τ. The identical initial distances between adjacent phase trajectories are given on each interval (in so doing one has to take into account a turn of the initial vector). Next the local Lyapunov exponents $\lambda^{(i)}$ are calculated by the formula $\lambda^{(i)} = (1/\tau)\ln d_i$, where d_i is the ratio of the distance between phase trajectories at the end of the ith step to the initial distance. Finally, all values of $\lambda^{(i)}$ are averaged. As a result, we find the maximal Lyapunov exponent:

$$\lambda = \frac{1}{m}\sum_{i=1}^{m}\lambda^{(i)} = \frac{1}{m\tau}\sum_{i=1}^{m}\ln d_i\,. \tag{3.33}$$

The algorithm for the calculation of the other Lyapunov exponents is similar to that for the maximal Lyapunov exponent, but it incorporates the necessary Gram–Schmidt orthogonalization procedure.

By analogy with generalized dimensions generalized Lyapunov exponents were introduced by Fujisaka [93] by the formula

$$\lambda_q = \frac{1}{\tau}\ln\left(\frac{1}{m}\sum_{i=1}^{m}d_i^q\right)^{1/q}. \tag{3.34}$$

It can be shown that for $q \to 0$ the expression (3.34) turns into (3.33), i.e., $\lambda = \lambda_0$. Furthermore, if all d_i are equal ($d_i = d$) then λ_q are also equal ($\lambda_q = \lambda = (1/\tau)\ln d$). Thus, distinctions between the generalized Lyapunov exponents characterize the nonuniformity of the attractor with respect to the divergence of phase trajectories in different areas of this attractor.

It should be noted that in the case of chaotic attractors the divergence of neighboring phase trajectories, as well as fractal properties of the attractors, depends upon the fact that there are disturbances. However, an important point is that the extent to which the neighboring phase trajectories are divergent, which is described by the Lyapunov exponents, is only slightly dependent, within certain limits, on the value and the character of the disturbances; it is largely determined by the dynamical system in itself. This is also true in respect to the dimensions of chaotic attractors characterizing their illusory fractal properties.

In addition to the above, we note that dimensions calculated in terms of the Lyapunov exponents $\lambda_1, \lambda_2, \ldots, \lambda_n$ are often used. The Kaplan–Yorke formula [143] determining the so-called *Lyapunov dimension* is

$$d_{\mathrm{L}} = j + \sum_{i=1}^{j}\frac{\lambda_i}{|\lambda_{j+1}|}\,, \tag{3.35}$$

where all λ_i are in descending order ($\lambda_1 \geq \lambda_2 \geq \ldots \geq \lambda_n$), and j is determined by the conditions $\lambda_1 + \lambda_2 + \ldots + \lambda_j \geq 0$, $\lambda_1 + \lambda_2 + \ldots + \lambda_j + \lambda_{j+1} < 0$.

3.4.2 Reconstruction of attractors from experimental data

The fact mentioned above that the dimension of an attractor is less (often much less) than the dimension of the original phase space allows us to study stationary self-oscillations of a system in a phase space of small dimension. How can one construct such a phase space? One way of doing this is to use the Takens theorem [326]. According to this theorem, the attractor of a system under consideration can be reconstructed from a time series for one of the coordinates of the original phase space of the system. Let us denote this time series by $x(t)$. Starting from $x(t)$ we can construct a new dynamical system of arbitrary dimension m taking as the vector describing the position of the representative point on the attractor in the phase space of the system constructed, the m-dimensional vector $\mathbf{y}(t) = \{x(t), x(t+\tau), \dots, x(t+(m-1)\tau)\}$. The Takens theorem states that for almost any time series $x(t)$ (which must be generic) and almost any time delay τ the attractor of the m-dimensional dynamical system constructed will be equivalent topologically to the attractor of the original system if $m \geq 2d_{\mathrm{H}} + 1$, where d_{H} is the Hausdorff dimension of the original attractor. Since the Hausdorff dimension, as a rule, is not known in advance, the number m must be taken large enough. To find the minimal value of m, called *the embedding dimension*, one can use different methods of a phase space transformation. One such method was proposed by Broomhead and King [60]. It is based on the Karhunen–Loeve theorem about the expansion of a multi-variable process in uncorrelated components, known in pattern recognition theory [95]. Another method proposed by Landa and Rosenblum [178, 181] is based on the algorithm for expansion of a process to a *well adapted basis*, created by Neimark for optimal coding of biomedical information [251]. Analysis of these methods, a comparison of them, and a variety of examples are given in [120, 178, 181]. These methods can be applied both to the original phase space and to Takens' phase space. It is evident that Takens' technique of attractor reconstruction is suitable for the analysis of experimental data as well. True, in this case difficulties associated with the presence of uncontrollable noise often emerge. The procedures for transformation of the phase space coordinates described above allow us to resolve the problem of partial filtering noise at the same time. As shown in [177], several consecutive applications of one of these procedures yields a large dividend in filtering quality. Other difficulties, which in some instances generate essential errors, are associated with appropriate choice of time series used and with the finite length of this series [185].

Using the reconstructed attractor we can determine all quantitative characteristics of the original attractor. However, for the computation of Lyapunov exponents some difficulties appear. By reconstructing the attractor from experimental data a single phase trajectory is only known at discrete points. Therefore, in this case specialized algorithms are necessary. Such an algorithm was devised by Wolf et al. [367], and a modified Wolf algorithm was suggested by Landa and Chetverikov [175]. Both of these algorithms are also appropriate for computation of generalized Lyapunov exponents.

3.5 Poincaré cutting surface and point maps

The notions of the *point map* and *cutting surface* were first introduced by Poincaré [273]. Later these notions were extended and generalized by Andronov [5, 8] and Neimark [250].

To construct a point map we must choose in the phase space a cutting surface, i.e., a surface that is intersected by all phase trajectories a sufficiently large number of times. Consider now a phase trajectory which at the initial instant of time intersects the cutting surface at a point M_0 with coordinates $x_0^{(1)}, x_0^{(2)}, \ldots, x_0^{(m)}$, where m is the dimension of the cutting surface. Let the following intersection occur at a point M_1 with coordinates $x_1^{(1)}, x_1^{(2)}, \ldots, x_1^{(m)}$. In its turn the point M_1 is mapped into a point M_2 and so on. The coordinates $x_n^{(1)}, x_n^{(2)}, \ldots, x_n^{(m)}$ and $x_{n+1}^{(1)}, x_{n+1}^{(2)}, \ldots, x_{n+1}^{(m)}$ are related by the following equations:

$$x_{n+1}^{(j)} = f_j(x_n^{(1)}, x_n^{(2)}, \ldots, x_n^{(m)}) \quad (j = 1, 2, \ldots, m). \tag{3.36}$$

Equations (3.36) can be rewritten in the form of an operator equation

$$M_{n+1} = TM_n, \tag{3.37}$$

where T is an operator mapping the point M_n into the point M_{n+1}. The point map described by Eq. (3.36) (or (3.37)) is called *the succession map*, and T is called *the succession operator*.

Besides the succession map, the so-called *translation map* plays an important part in studies of oscillatory systems, mainly of nonautonomous systems. A point map describing the transition of a point $\boldsymbol{x}(t)$ to a point $\boldsymbol{x}(t+\tau)$ that is related to the point $\boldsymbol{x}(t)$ by differential equations of the dynamical system under consideration, is called the translation map of this system. In terms of *the translation operator* T_τ this map can be written as

$$\boldsymbol{x}(t+\tau) = T_\tau \boldsymbol{x}(t). \tag{3.38}$$

It follows from the definition of a point map that singular points and limit cycles have to be mapped into themselves giving so-called *fixed points* of the map. Like singular points and limit cycles in a phase space fixed points can also be both stable and unstable. To determine the stability of a fixed point, it is sufficient to consider a linearized map in the vicinity of this point. Let the map under consideration be described by Eq. (3.36) and let the coordinates of the fixed point considered be $x_1^*, x_2^*, \ldots, x_m^*$. The linearized map is

$$\xi_{n+1}^{(j)} = a_{j1}\xi_n^{(1)} + aj2\xi_n^{(2)} + \ldots + a_{jm}\xi_n^{(m)}, \quad j = 1, 2, \ldots, m, \tag{3.39}$$

where

$$a_{jk} = \left| \frac{\partial f_j}{\partial x^{(1)}} \right|_{x^{(1)}=x_1^*,\dots,x^{(m)}=x_m^*};$$

$\xi^{(j)} = x^{(j)} - x_j^*$ is the difference between the jth coordinate of a certain point of the map close to the fixed point considered and the jth coordinate of this point. Substituting into (3.39) $\xi_{n+1}^{(j)} = \mu \xi_n^{(j)}$, where μ is called a *multiplier of the map fixed point*, and equating the determinant of the system to zero, we obtain the characteristic equation for the multiplier μ:

$$\begin{vmatrix} a_{11}-\mu & a_{12} & \dots & a_{1m} \\ a_{21} & a_{22}-\mu & \dots & a_{2m} \\ \vdots & \vdots & \vdots & \vdots \\ a_{m1} & a_{m2} & \dots & a_{mm}-\mu \end{vmatrix} = 0. \tag{3.40}$$

The fixed point is stable if the absolute values of all roots of Eq. (3.40) are less than one; otherwise, it is unstable. This statement as applied to a one-dimensional map is known as *the Koenigs theorem* [156, 157].

One-dimensional maps are the most vivid representatives of point maps. Such a map can be written as

$$x_{n+1} = f(x_n), \tag{3.41}$$

where $f(x)$ is said to be *the succession function*. If x_n is a point on the interval I then the map (3.41) is spoken of as the map of I into itself. The plot of x_{n+1} versus x_n allows us to find fixed points of the map and to determine their stability. An example of such a plot is shown in Fig. 3.5. In order to find fixed points of the map we have to draw the bisectrix. The points at which the succession function intersects the bisectrix are just the fixed points. If x_0 is the initial point then the successive point x_1 can be found as follows: from the

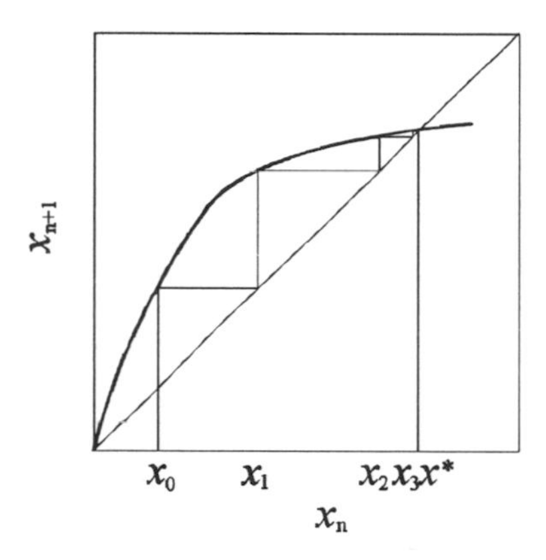

Fig. 3.5. An example of the Lamerey diagram

point x_0 we erect a perpendicular to the abscissa and from the intersection point of this perpendicular with the plot $f(x)$ we draw a straight line parallel the abscissa to the intersection with the bisectrix. The x-coordinate of this point of intersection is just x_1. The point x_2 and all successive points can be found in a similar way. The diagram shown in Fig. 3.5 is often called the *Lamerey diagram.*

At the present time point maps play an increasingly important part in the theory of nonlinear oscillations and in its branch—nonlinear dynamics [253]. Firstly, this is due to the universality of the laws of oscillation theory [191]: many important phenomena were discovered with point maps. Secondly, the study of even very complicated systems can often be reduced to the study of one-dimensional maps. Many similar examples will be demonstrated in this book.

3.6 Some routes for the loss of stability of simple attractors and the appearance of strange attractors

The best known routes for the loss of stability of a singular point are (1) the birth of a stable limit cycle from this point, and (2) its fusion with an unstable limit cycle. In the western literature these bifurcations are known as *the direct and reverse Hopf bifurcations*, respectively. However, Hopf classified these bifurcations in 1942 [123], whereas Andronov and Leontovich discovered them (though for systems with one degree of freedom only) in 1939 [6].

The most widespread route for the loss of stability of a limit cycle resulting in the transition to more complicated periodic and chaotic motions is the so-called *Feigenbaum scenario* consisting in the appearance of an infinite period-doubling bifurcation sequence [85]. It is pertinent to note that the existence of such sequences was proved by Sharkovsky [305] well before the work by Feigenbaum; however, Feigenbaum was the first to establish universal quantitative laws for this bifurcation sequence. He considered a one-dimensional quadratic map, often called the logistic map, of the form

$$x_{n+1} = 1 - \mu x_n^2, \tag{3.42}$$

where μ is the bifurcation parameter and showed that the values of the parameter μ, for which the sequential period-doubling bifurcations occur, are related by the law

$$\lim_{n\to\infty} \frac{\mu_n - \mu_{n-1}}{\mu_{n+1} - \mu_n} = \delta, \tag{3.43}$$

where μ_n is the value of μ such that the nth period-doubling bifurcation occurs, and $\delta = 4.6692\ldots$ is the universal constant, which came to be known as *the Feigenbaum constant.* It follows from (3.43) that the differences of two adjacent bifurcation values of μ $(\mu_n - \mu_{n-1})$ for sufficiently large n form a convergent geometric progression with the denominator $q = \delta^{-1}$. For

$\mu = \mu_\infty = \lim_{n\to\infty} \mu_n$ all stable cycles previously existing in the system disappear and stochasticity appears. With further increase of the parameter μ stable periodic motions with different periods again arise, but their attraction basins are sufficiently small. Therefore the oscillations in the system for these values of μ are of a chaotic character and a chaotic attractor corresponds to these oscillations. Only over moderately narrow ranges of μ the stable cycles have considerable attraction basins, which allows them to be observed. These ranges of μ are called *the windows of periodicity*. As a rule, it is possible to observe cycles with triple and quintuple periods. We note that the transition to chaos in accordance with the Feigenbaum scenario manifests itself in characteristic changes in the shape of the power spectra of self-oscillations [86, 247], which allow us to observe this transition very easily.

Another possible route for the appearance of a chaotic attractor is the fusion of a stable cycle with an unstable one and the subsequent disappearance of both of these cycles. As a result of such a bifurcation chaotic motion, possessing the property of so-called *intermittency*, may arise. This property implies that in the system phase space the representative point moves for a long time close to the vanished cycles. This motion manifests itself in the near-regular oscillations of the system (so-called 'laminar' phases). These long parts of almost regular motion alternate with short irregular splashes ('turbulent' phases). The average duration of laminar phases can be evaluated by using a model map (see, e.g., [122, 125, 172, 173, 198]).

As a result of the loss of stability of a limit cycle a stable two-dimensional torus may come into being. Further this two-dimensional torus may become unstable and give rise to a stable three-dimensional torus. In its turn, the loss of stability of the three-dimensional torus, as shown by Ruelle and Takens [292], should cause, as a rule, the birth of a stochastic (or chaotic) attractor, since in the general case a four-dimensional torus is unstable. Moreover, a stochastic or chaotic attractor can be born on the surface of the three-dimensional torus itself, without initiating its destruction [253].

3.7 Integrable and nonintegrable systems. Action–angle variables

Systems described by differential equations may be separated into two classes: integrable and nonintegrable systems. A system described by differential equations of the Nth order is referred to as completely integrable if it has N independent integrals of motion. According to the Liouville theorem, for Hamiltonian systems the existence of only $n = N/2$ integrals of motion is, as a rule, sufficient for integrability [364, 17]. It is only necessary for these integrals to be in involution, i.e., the Poisson brackets for any pair of them would be equal to zero.

If a Hamiltonian system is completely integrable, its Hamiltonian $H(\boldsymbol{q}, \boldsymbol{p})$ can be reduced to the form $H(\boldsymbol{P})$, where $\boldsymbol{P}$ is the generalized momentum vector, by means of a canonical transformation of the variables $\boldsymbol{Q} = \boldsymbol{Q}(\boldsymbol{q}, \boldsymbol{p})$, $\boldsymbol{P} = \boldsymbol{P}(\boldsymbol{q}, \boldsymbol{p})$.[4] As follows from canonical equations, in this case all of the components of the vector $\boldsymbol{P}$ are constants. So-called *action variables* are well suited as the generalized momenta $\boldsymbol{P}$.

A system with a Hamiltonian of the form

$$H(\boldsymbol{q}, \boldsymbol{p}) = \sum_{s=1}^{n} \left(\frac{p_s^2}{2} + U_s(q_s) \right), \tag{3.44}$$

where q_s and p_s $(s = 1 \dots n)$ are the generalized coordinates and momenta, respectively, and $U_s(q_s)$ are arbitrary functions, is an example of a completely integrable system. For such a system the equations of motion are

$$\dot{q}_s = p_s, \quad \dot{p}_s = -\frac{\mathrm{d}U_s}{\mathrm{d}q_s} \equiv -g_s(q_s). \tag{3.45}$$

Hence, the system of the equations of motion decompose into n independent sets of equations, each describing a conservative nonlinear oscillator with one degree of freedom. In this case the energy conservation equation (1.8) also decomposes into n independent equations of the form

$$\frac{p_s^2}{2} + U_s(q_s) = E_s. \tag{3.46}$$

The action variables J_s are defined as

$$J_s = \frac{1}{2\pi} \oint p_s \, \mathrm{d}q_s = \frac{1}{\pi} \int_{q_{s\min}}^{q_{s\max}} \sqrt{2\big(E_s - U_s(q_s)\big)} \, \mathrm{d}q_s, \tag{3.47}$$

where $q_{s_{\min}}$ and $q_{s_{\max}}$ are the roots of the equation $U_s(q_s) = E_s$.

Canonical variables ϑ_s related to the action variables J_s by the canonical equations

$$\dot{\vartheta}_s = \frac{\mathrm{d}H(J_s)}{\mathrm{d}J_s} \tag{3.48}$$

are called *angle variables*. Because $J_s = \text{const}$, we find from (3.48) that $\dot{\vartheta}_s \equiv \omega_s(J_s) = \text{const}$. It can be shown that $\omega_s(J_s) = 2\pi/T_s(J_s)$, where $T_s(J_s)$ is the oscillation period of the sth oscillator.

Owing to the existence of n conservation laws for completely integrable systems, each trajectory in the system phase space has to lie in an n-dimensional surface. If the trajectories lie within a limited region of the phase space, this surface is a torus described by the following equations in parametric form:

$$J_s = \text{const}, \quad \vartheta_s = \omega_s t + \varphi_s \quad (s = 1, 2, \dots, n). \tag{3.49}$$

An example of a two-dimensional torus is given in Fig. 3.6. The ratio of the frequencies $\rho = \omega_1/\omega_2$ is called *the Poincaré rotation number*. If for

[4] A transformation is said to be canonical if its Jacobian is equal to zero.

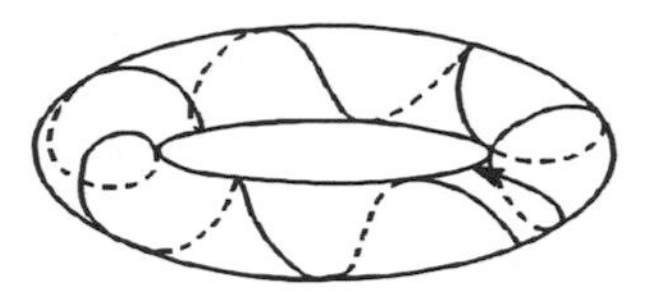

Fig. 3.6. An example of a two-dimensional torus

certain J_1 and J_2 the Poincaré rotation number is equal to k/m, where k and m are integers, the corresponding trajectory lying on the torus closes after k revolutions in the direction of the angle ϑ_1 and m revolutions in the direction of the angle ϑ_2. If ρ is irrational then the corresponding trajectory covers the torus surface entirely. This may result in the ergodic property. Such a trajectory corresponds to quasi-periodic oscillations.

It is evident from the foregoing that any motion of an integrable Hamiltonian system is either periodic or quasi-periodic, i.e., it is regular. By contrast, motions of nonintegrable Hamiltonian systems can be stochastic, i.e., irregular. Examples of such systems are given in Ch. 5.

4. Examples of natural oscillations in systems with one degree of freedom

Oscillations are called *natural*, or *free*, if they are executed in systems free from energy sources, i.e., in conservative or in passive dissipative systems. In addition, oscillations executed in systems with slowly time-varying parameters we call natural too. In principle, such systems contain energy sources but these sources are not responsible for the excitation of oscillations.

4.1 Oscillator with nonlinear restoring force

We consider here natural oscillations in certain systems described by equations of the form

$$\ddot{x} + f(x) = 0, \tag{4.1}$$

where $f(x)$ is a nonlinear function describing the restoring force.

4.1.1 Pendulum oscillations

Let us consider a pendulum represented schematically in Fig. 4.1 and assume that it oscillates in a single plane and friction forces are negligibly small. The equation for the angle of rotation φ is

Fig. 4.1. Schematic image of a pendulum

$$ml^2\ddot{\varphi} + mgl\sin\varphi = 0, \tag{4.2}$$

where m is the mass of the ball, and l is the length of the thread.

Using the energy conservation law we obtain the equation for trajectories in the phase plane φ, $\dot{\varphi}$:

$$\frac{ml^2\dot{\varphi}^2}{2} + mgl(1 - \cos\varphi) = E, \tag{4.3}$$

where E is the energy of the pendulum. For $E < 2mgl$ Eq. (4.3) describes closed trajectories corresponding to oscillations of the pendulum about its stable equilibrium state, whereas for $E > 2mgl$ this equation describes non-closed trajectories corresponding to rotation of the pendulum. The phase trajectories described by Eq. (4.3) are illustrated in Fig. 4.2 a. These two kinds of trajectories are separated by peculiar trajectories passing through

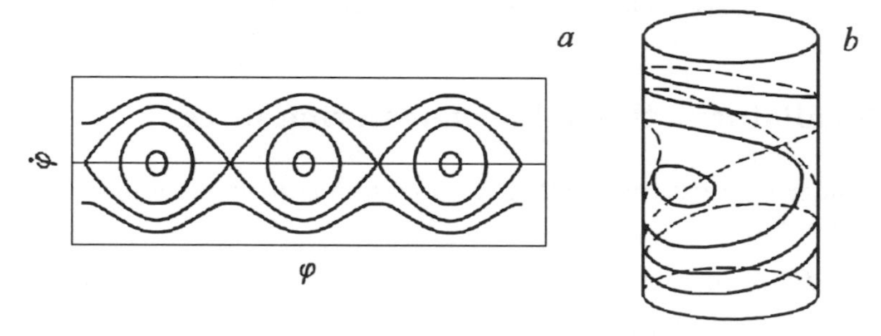

Fig. 4.2. Phase portrait of the pendulum oscillations (a) in the phase plane and (b) in the cylindrical phase space

the singular saddle points ($\dot{\varphi} = 0$, $\varphi = \pm\pi$, $\pm 3\pi$, ...). Such trajectories are called *separatrices*. Since the values of φ different from each other by 2π are physically indistinguishable, it is convenient to roll up the phase plane φ, $\dot{\varphi}$ into a cylinder. The phase portrait of the pendulum oscillations in the resulting cylindrical phase space is shown in Fig. 4.2 b.

Generally Eqs. (4.2) and (4.3) cannot be solved analytically. Analytical solutions can be obtained only in the case of small oscillations ($E \ll mgl$), when they are near-sinusoidal, and in the case when $E = 2mgl$ (this value of E corresponds to the motion of a representative point along a separatrix). If the latter is the case, Eq. (4.3) is split into two equations:

$$\dot{\varphi}_{\pm}(t) = \pm 2\omega_0 \cos\frac{\varphi}{2}, \tag{4.4}$$

where $\omega_0 = \sqrt{g/l}$ is the natural frequency of small oscillations of the pendulum. By integrating (4.4) we find

$$\varphi_{\pm}(t) = \pm\Big(4\arctan \mathrm{e}^{\omega_0 t} - \pi\Big). \tag{4.5}$$

The time dependencies of $\varphi = \varphi_+$ and $\dot{\varphi}/\omega_0 = \dot{\varphi}_+/\omega_0$ are depicted in Fig. 4.3. These solutions play an important part in soliton theory.

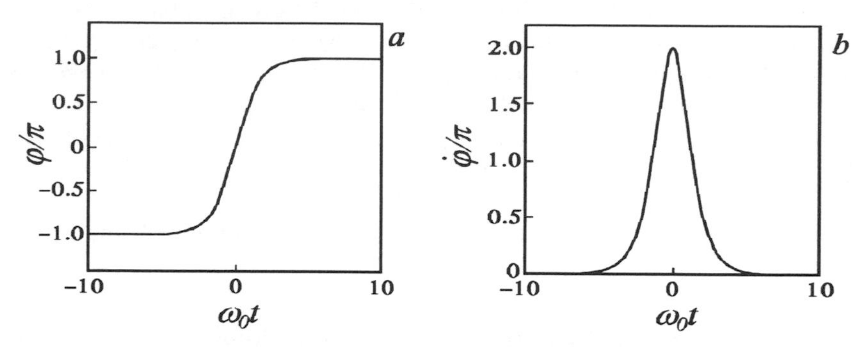

Fig. 4.3. The time dependencies of $\varphi = \varphi_+$ and $\dot{\varphi}/\omega_0 = \dot{\varphi}_+/\omega_0$ corresponding to the motion of a representative point along a separatrix

4.1.2 Oscillations of a pendulum placed between the opposite poles of a magnet

Let the ball of the pendulum shown in Fig. 4.1 be made of iron and placed between the opposite poles of a magnet (Fig. 4.4). If we approximate the mag-

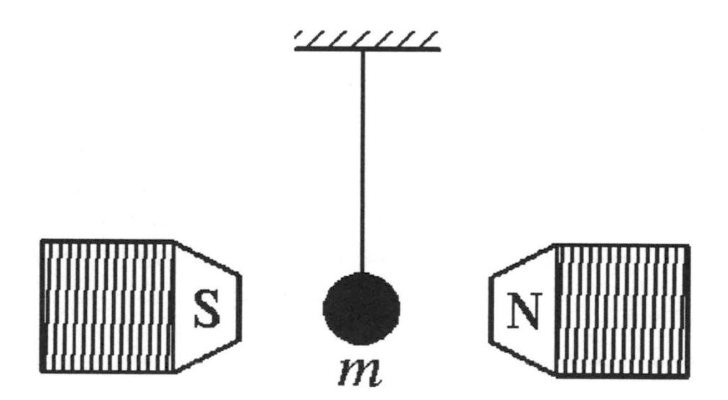

Fig. 4.4. Schematic image of a pendulum placed between the opposite poles of a magnet

netic force moment acting on the ball by $M(\varphi) = ml(a_1\varphi - b_1\varphi^3)$ and restrict ourselves to the case of small oscillations of the pendulum, then Eq. (4.2) is reworked as follows:

$$ml^2\ddot{\varphi} + ml\left[(g - a_1) - \left(\frac{g}{6} - b_1\right)\varphi^2\right]\varphi = 0. \tag{4.6}$$

This equation can be rewritten in the form

$$\ddot{\varphi} - a\varphi + b\varphi^3 = 0, \tag{4.7}$$

where $a = a_1/l - \omega_0^2$, $b = b_1/l - \omega_0^2/6$. In the case when $a_1 > l\omega_0^2$ ($a > 0$) the equilibrium position $\varphi = 0$ becomes aperiodically unstable (the singular point $\varphi = 0$, $\dot{\varphi} = 0$ on the phase plane φ, $\dot{\varphi}$ becomes of saddle type). If, in addition to the above, the inequality $b_1 > l\omega_0^2/6$ holds then there exist two stable equilibrium positions about which the pendulum can oscillate. These equilibrium states correspond to singular points on the phase plane of center type. But if $b_1 l < \omega_0^2/6$ and $a > 0$ then the ball sticks to one of the magnet poles.

Equations of the form (4.7) were first studied by Duffing [76] and are now known as Duffing equations.

4.1.3 Oscillations described by Duffing equations

We consider here different types of natural oscillations in an oscillator described by the Duffing equation

$$\ddot{x} + ax + bx^3 = 0. \tag{4.8}$$

These types of oscillations are determined by the signs of the coefficients a and b.

(i) $a > 0$, $b > 0$. In this case Eq. (4.8) has a single singular point of centre type and all phase trajectories are closed (Fig. 4.5 a). A solution of Eq. (4.8) can be expressed in terms of the Jacobi elliptic cosine [134] as

$$x = A\,\mathrm{cn}(\Omega t, k), \tag{4.9}$$

where A is the amplitude of the oscillations, $\Omega = \sqrt{a + bA^2} = 4\mathbf{K}(k)/T$, T is the period of the oscillations, $\mathbf{K}(k)$ is the full elliptic integral of the first kind, and $k = \sqrt{b/2}A/\Omega$ is the modulus of the Jacobi elliptic function. For small A this solution has a near-harmonic form with frequency $\omega = \sqrt{a}(1 + 3bA^2/8a)$. For large A the form of oscillations is essentially different from the harmonic one and the oscillation frequency ω is directly proportional to A, i.e.,

$$\omega = \frac{2\pi\sqrt{b}\,A}{4\mathbf{K}(1/\sqrt{2})} \approx \frac{2\pi\sqrt{b}\,A}{6.42}.$$

(ii) $a > 0$, $b < 0$. In this case Eq. (4.8) has three singular points: one point with coordinate $x = 0$ is of center type and the two points with coordinates $x_{1,2} = \pm\sqrt{a/|b|}$ are of saddle type. The equation of the phase trajectories, being equivalent to the energy conservation equation, is

$$\frac{\dot{x}^2}{2} + \frac{ax^2}{2} - \frac{|b|x^4}{4} = E. \tag{4.10}$$

For $E < a^2/4|b|$ the phase trajectories are closed, and for $E > a^2/4|b|$ they are nonclosed (Fig. 4.5 b). The motion along the separatrix between the points with coordinates x_1 and x_2 (curve 2) is described by the following equation:

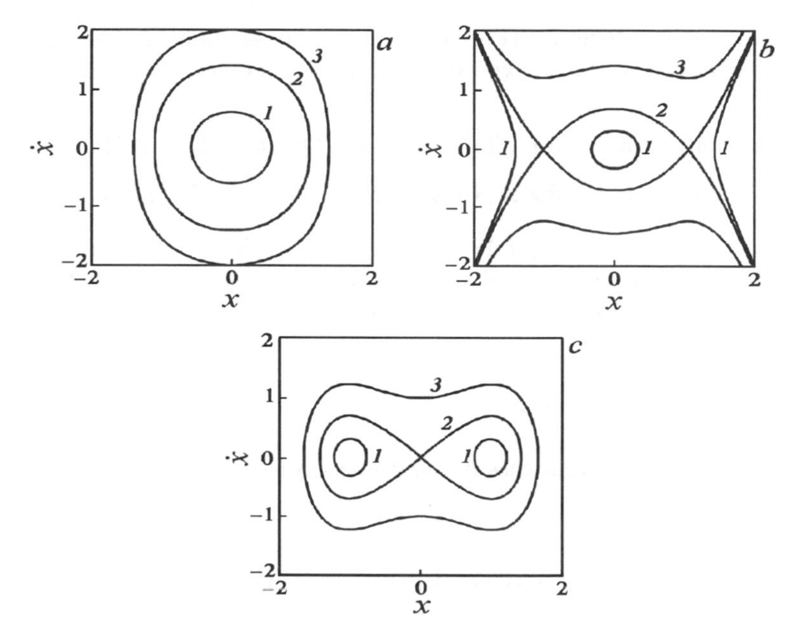

Fig. 4.5. The phase portrait of the Duffing oscillator for (a) $a = 1$, $b = 1$; (b) $a = 1$, $b = -10$; (c) $a = -1$, $b = 1$. Curves labeled 1 correspond to (a) $E = 0.2$, (b) $E = 0.05$, (c) $E = -0.2$; curves 2 correspond to (a) $E = 1$, (b) $E = 0.25$, (c) $E = 0$; curves 3 correspond to (a) $E = 2$, (b) $E = 1$, (c) $E = 0.5$

$$x = \pm\sqrt{\frac{a}{|b|}}\tanh\sqrt{\frac{a}{2}}\,t. \tag{4.11}$$

The general solution of Eq. (4.8) for $a > 0$ and $b < 0$, associated with closed phase trajectories, is expressed in terms of the Jacobi elliptic sine as

$$x = A\,\mathrm{sn}(\Omega t, k), \tag{4.12}$$

where $\Omega = \sqrt{a - |b|A^2/2}$, and $k = \sqrt{|b|/2}\,A/\Omega$ is the modulus of the elliptic sine. The expression (4.12) is valid for $A \le \sqrt{a/|b|}$, i.e., $k \le 1$. It describes periodic oscillations with period $4\mathbf{K}(k)/\Omega$. For $k \to 1$ the oscillation period tends to infinity and $x(t)$ tends to that described by the expression (4.11).

(iii) $a < 0$, $b > 0$. As in the previous case, Eq. (4.8) has three singular points, but the point with coordinate $x = 0$ is of saddle type and the points with coordinates $x_{1,2} = \pm\sqrt{a/|b|}$ are of center type. The equation of the phase trajectories in this case is conveniently written as

$$\frac{\dot{x}^2}{2} - \frac{|a|x^2}{2} + \frac{bx^4}{4} = E - \frac{a^2}{4b}. \tag{4.13}$$

It follows from this that, for $E < a^2/4b$, there are two sets of closed phase trajectories, each enclosing one of the singular points with coordinates $x_{1,2}$. For $E > a^2/4b$, the phase trajectories are also closed, but they surround all singular points together (Fig. 4.5 c). In the case under consideration the solution of Eq. (4.8) is also expressed in terms of the Jacobi elliptic functions as

$$x = \begin{cases} \pm A\,\mathrm{dn}(\omega_1 t, k_1) & \text{for } \sqrt{|a|/b} < A < \sqrt{2|a|/b} \quad (E < a^2/4b) \\ A\,\mathrm{cn}(\omega_2 t, k_2) & \text{for } A > \sqrt{2|a|/b} \quad (E > a^2/4b)\,, \end{cases} \tag{4.14}$$

where $\omega_1 = \sqrt{b/2}A$, $\omega_2 = \sqrt{bA^2 - |a|}$, $k_1 = \omega_2/\omega_1$, $k_2 = \omega_1^2/\omega_2^2$. The solution associated with the motion of a representative point along the separatrix follows from (4.14) for $k_1 \to k_2 \to 1$. It is

$$x = \pm\sqrt{\frac{2|a|}{b}}\,\frac{1}{\cosh\sqrt{|a|}\,t}\,. \tag{4.15}$$

(iv) $a < 0$, $b < 0$. In this case all solutions go to infinity; hence it is of no interest.

4.1.4 Oscillations of a material point in a force field with the Toda potential

In the theory of nonlinear oscillations we often deal with an equation of the form

$$\ddot{x} + \mathrm{e}^x - 1 = 0. \tag{4.16}$$

This equation can be considered as the equation of motion of a material point in a force field with the potential

$$U(x) = \mathrm{e}^x - x \tag{4.17}$$

known as the Toda potential.

From Eq. (4.16) we find the following equation for the phase trajectories:

$$\frac{\dot{x}^2}{2} + \mathrm{e}^x - x - 1 = \frac{A^2}{2}\,, \tag{4.18}$$

where A is the amplitude of oscillations of the variable $\dot{x}$. The phase portrait corresponding to Eq. (4.18) is shown in Fig. 4.6 a. For any values of amplitude A the oscillations are periodic with period $T(A)$ increasing monotonically as A increases (Fig. 4.6 b) [168].

The shape of the oscillations can be analytically calculated in two limiting cases:

(i) $A < 1$. In this case e^x can be expanded as a power series in x up to the quadratic term of the expansion. As a result Eq. (4.16) becomes

$$\ddot{x} + x\left(1 + \frac{x}{2}\right) = 0. \tag{4.19}$$

A solution of Eq. (4.19) is expressed in terms of the Jacobi elliptic cosine as

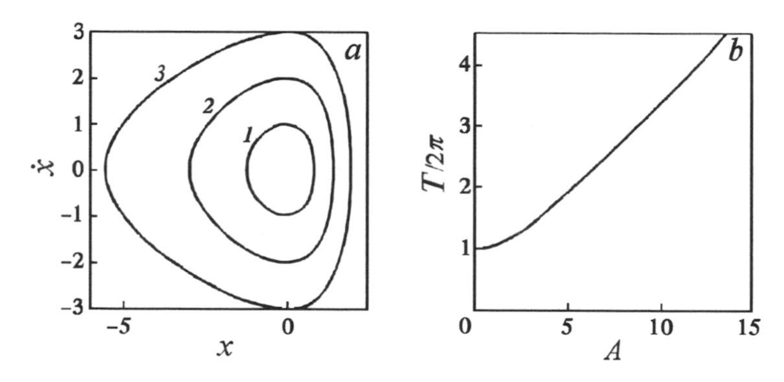

Fig. 4.6. (a) The phase portrait of the motion of a material point in a force field with the Toda potential. The curves 1, 2 and 3 correspond to $A = 0.5$, $A = 2$ and $A = 4.5$, respectively. (b) The amplitude dependence of the oscillation period

$$x = 12k^2\omega^2\mathrm{cn}^2(\omega t, k) - 1 - 4\omega^2(2k^2 - 1), \tag{4.20}$$

where $\omega = (1 + k^4 - k^2)^{-1/4}/2$, and the modulus of the elliptic function k is related to the amplitude A by

$$3A^2 = 2 + \frac{(2k^2 - 1)(2 + k^2 - k^4)}{(1 - k^2 + k^4)^{3/2}}.$$

It can be seen from this that $k < 1$ for $A < 1$.

(ii) $A \gg 1$. In this case the condition $\mathrm{e}^x \gg 1$ holds in the neighborhood of maxima of the function $x(t)$, and an approximate solution of Eq. (4.16) in this neighborhood is

$$x = \ln\left(\frac{A^2}{2\cosh^2\Big(A(t - t_{\max})/2\Big)}\right), \tag{4.21}$$

where $t_{\max}$ is the instant at which the function $x(t)$ peaks. The truth of the solution (4.21) can be verified by direct substitution. In the neighborhood of minima of the function $x(t)$, and vice versa, $\mathrm{e}^x \ll 1$, and the approximate solution of Eq. (4.16) is

$$x = -\frac{A^2 - (t - t_{\min})^2}{2}, \tag{4.22}$$

where $t_{\min}$ is the instant at which the function $x(t)$ is minimal. For $A \gg 1$ the oscillation period is proportional to A, namely, $T(A) \approx 2A$. Starting from the formulas (4.21) and (4.22) one can construct the time dependencies $x(t)$ for $A \gg 1$. An example of such a dependence is presented in Fig. 4.7 a. For comparison, the time dependencies $x(t)$ calculated by computer are shown in Fig. 4.7 b.

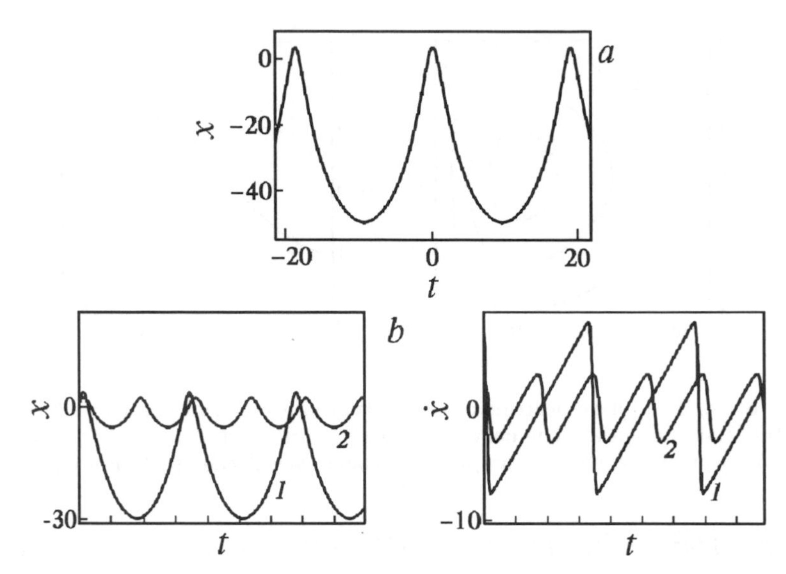

Fig. 4.7. The shape of oscillations of a material point moving in a force field with the Toda potential, for large amplitudes. It is both constructed from (4.21) and (4.22) (a) and calculated numerically (b) for $A = 7.565$ (curve 1) and $A = 2.963$ (curve 2)

4.2 Oscillations of a bubble in fluid

As early as 1917, in connection with the cavitation problem Rayleigh published a work [278] in which he derived the equation of oscillations of a spherical gas bubble in an ideal incompressible fluid. Below we give this derivation. We start from the one-dimensional Euler equation written in spherical coordinates

$$\frac{\partial u}{\partial t} + u\frac{\partial u}{\partial r} = -\frac{1}{\rho}\frac{\partial p}{\partial r} \tag{4.23}$$

and the continuity equation

$$\frac{\partial (r^2 u)}{\partial r} = 0, \tag{4.24}$$

where u and ρ are the fluid velocity and density, respectively, and p is the pressure. Equations (4.23) and (4.24) are valid for $r \geq R$, where R is the radius of the bubble.

Introducing the velocity potential φ, so that $u = -\partial\varphi/\partial r$, we integrate Eq. (4.23) over r from r to ∞. Taking into account that u and φ vanish for $r \to \infty$, we obtain

$$\frac{\partial \varphi}{\partial t} - \frac{u^2}{2} + \frac{1}{\rho}\Big(p(\infty) - p(r)\Big) = 0, \tag{4.25}$$

where $p(\infty) = p_0$ is the hydrostatic pressure in the fluid. It follows from Eq. (4.24) that $u(r) = C/r^2$, where the constant C is determined from the boundary condition $u(R) = \dot{R}$. The latter gives $C = R^2\dot{R}$. Hence

$$u(r) = \frac{R^2}{r^2}\dot{R}, \quad \varphi = \frac{R^2}{r}\dot{R}. \tag{4.26}$$

Substituting, further, (4.26) into Eq. (4.25) and putting $r = R$, we obtain the equation of bubble oscillations derived by Rayleigh:

$$R\ddot{R} + \frac{3}{2}\dot{R}^2 - \frac{1}{\rho}\Big(p(R) - p_0\Big) = 0. \tag{4.27}$$

In order to calculate $p(R)$ we take account of the condition that pressures in and outside the bubble must be the same. Outside the bubble the pressure is equal to the sum of the fluid pressure $p(R)$ and the pressure $2\sigma/R$ created by the surface tension forces. Inside the bubble the pressure is equal to the gas pressure $p_g(R)$. Considering all processes in the gas as polytropic, one can write the following expression for $p_g(R)$:

$$p_g(R) = \left(p_0 + \frac{2\sigma}{R_0}\right)\left(\frac{R_0}{R}\right)^{3\kappa}, \tag{4.28}$$

where κ is the polytropic exponent, and R_0 is the stationary value of the bubble radius determined by the equation $p(R)\big|_{R=R_0} = p_0$. Equating $p_g(R)$ to the pressure outside the bubble, i.e., to $p(R) + 2\sigma/R$, we find

$$p(R) = \left(p_0 + \frac{2\sigma}{R_0}\right)\left(\frac{R_0}{R}\right)^{3\kappa} - \frac{2\sigma}{R}. \tag{4.29}$$

Equation (4.27), in view of (4.29), can be rewritten in the form of an energy conservation law:

$$\frac{R^3\dot{R}^2}{2} + \frac{1}{\rho}\left[\frac{R_0^3}{3\kappa - 1}\left(p_0 + \frac{2\sigma}{R_0}\right)\left(\left(\frac{R_0}{R}\right)^{3\kappa-1} - 1\right) + \sigma(R^2 - R_0^2) + p_0\frac{R^3 - R_0^3}{3}\right] = E. \tag{4.30}$$

This equation can also be written in the Lagrange form with the Lagrangian L described by the following expression:

$$L = \frac{R^3\dot{R}^2}{2} - \frac{1}{\rho}\left[\frac{R_0^3}{3\kappa - 1}\left(p_0 + \frac{2\sigma}{R_0}\right)\left(\left(\frac{R_0}{R}\right)^{3\kappa-1} - 1\right) + \sigma(R^2 - R_0^2) + p_0\frac{R^3 - R_0^3}{3}\right]. \tag{4.31}$$

It follows from (4.31) and (4.30) that kinetic and potential energies of the bubble are described by the first and second terms of the expression (4.30), respectively. We call attention to the fact that the bubble's kinetic energy depends not only on the rate of change of the radius, but also on the value of the radius itself.

The form of the phase trajectories determined by Eq. (4.30) is shown in Fig. 4.8.

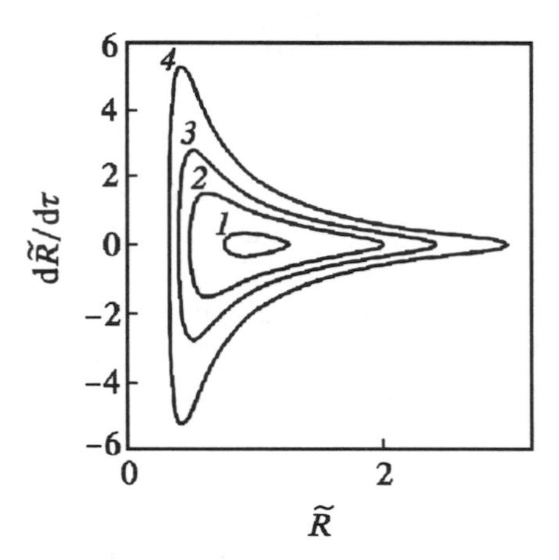

Fig. 4.8. The phase portrait of the oscillations of a bubble in a fluid, in the coordinates $\tilde{R} = R/R_0$, $d\tilde{R}/d\tau$, where $\tau = \omega_0 t$, for $\sigma/(R_0 p_0) = 0.0725$ and $\kappa = 4/3$

In the case of small oscillations, when $|R - R_0| \ll R_0$, Eq. (4.27), in view of (4.29), takes the form:

$$\ddot{\xi} + \frac{3\kappa}{\rho R_0^2}\left(p_0 + \frac{3\kappa - 1}{3\kappa}\frac{2\sigma}{R_0}\right)\xi = 0, \tag{4.32}$$

where $\xi = R - R_0$. It follows from this equation that the frequency of small oscillations ω_0 is equal to

$$\omega_0 = \frac{1}{R_0}\sqrt{\frac{3\kappa}{\rho}\left(p_0 + \frac{3\kappa - 1}{3\kappa}\frac{2\sigma}{R_0}\right)}. \tag{4.33}$$

As one would expect, the value of ω_0 is the greater, the less is R_0.

When the amplitude of the oscillations increases the oscillation shape differs more and more from the harmonic one, and the oscillation period increases (see Fig. 4.9).

4.3 Oscillations of species populations described by the Lotka–Volterra equations

All the examples considered previously are described by equations containing no terms with first derivatives to odd power, and, therefore, these equations

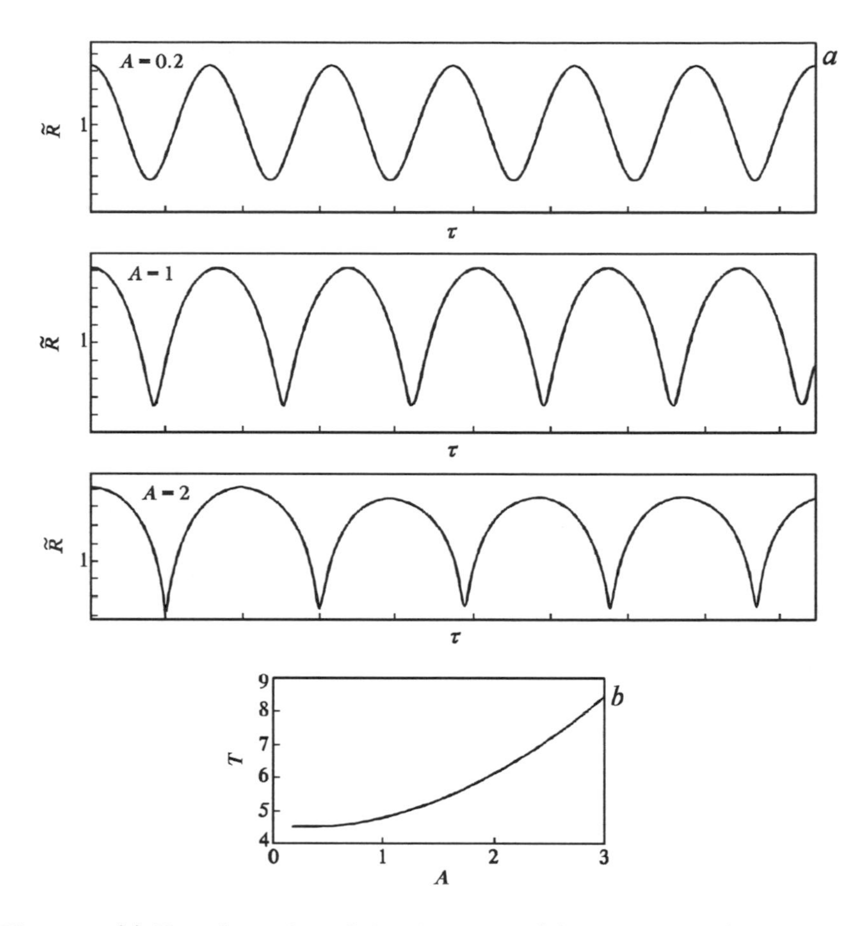

Fig. 4.9. (a) Transformation of the shape and (b) the change of the period of the bubble oscillations with increase of the amplitude $A = 2(R_{\max}/R_0 - 1)$ for $\sigma/(R_0 p_0) = 0.0725$ and $\kappa = 4/3$

are not changed when time becomes of opposite sign. Phase trajectories determined by these equations are symmetric about the x-axis. The Lotka–Volterra equations do not possess this property. These equations are

$$\dot{x} = k_1 x - a_1 xy, \quad \dot{y} = -k_2 y + a_2 xy, \tag{4.34}$$

where x is the number of prey and y is the number of predators. In the first equation the term $k_1 x$ describes the natural increase in the number of prey, and the term $a_1 xy$ describes the decrease in the number of prey at the cost of being eaten by predators. In regard to predators, it is suggested that their number can only decrease by itself (this decrease is described by the term $-k_2 y$), and an increase in the number of predators can be accounted for by eating the prey (this increase is described by the term $a_2 xy$).

It follows from (4.34) that the equation of integral curves on the phase plane x, y is

$$\frac{\mathrm{d}y}{\mathrm{d}x} = \frac{y}{x}\frac{a_2 x - k_2}{k_1 - a_1 y}. \tag{4.35}$$

This equation is easily integrated, and its solution is conveniently written as

$$\frac{k_2}{k_1}\left(\frac{x}{x_0} - 1 - \ln\frac{x}{x_0}\right) + \frac{y}{y_0} - 1 - \ln\frac{y}{y_0} = E, \tag{4.36}$$

where $x_0 = k_2/a_2$, $y_0 = k_1/a_1$ are coordinates of the singular point corresponding to the steady state of the system. Phase trajectories described by Eq. (4.36) are depicted in Fig. 4.10 for several values of the ratio k_2/k_1. Equa-

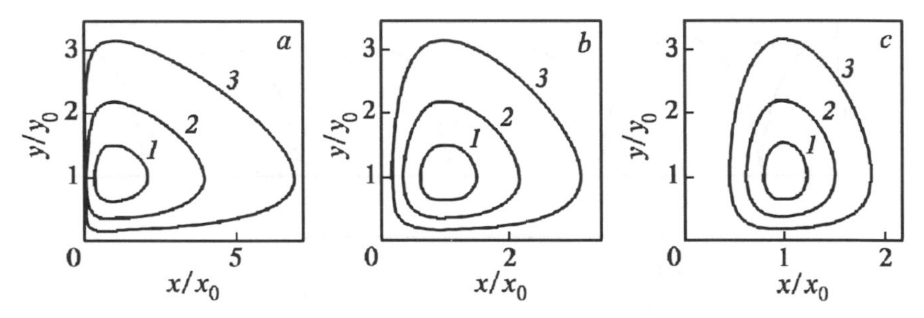

Fig. 4.10. The phase trajectories in the plane with the coordinates x/x_0, y/y_0 for the Lotka–Volterra equations for (a) $k_2/k_1 = 0.25$, (b) $k_2/k_1 = 1$, (c) $k_2/k_1 = 4$. Curves 1, 2 and 3 correspond to $E = 0.1$, $E = 0.4$ and $E = 1$, respectively

tion (4.36) can be considered as the energy integral for the Lotka–Volterra equations.

By excluding the variable y Eqs. (4.34) are transformed into a single equation of the form

$$\ddot{x} - \frac{\dot{x}^2}{x} + (k_1 x - \dot{x})(a_2 x - k_2) = 0. \tag{4.37}$$

Equation (4.37), like Eq. (4.1), describes a nonlinear oscillator, but with the nonlinear restoring force dependent not only on x, but on $\dot{x}$ as well. The phase portrait, corresponding to Eq. (4.37), in the $x\,\dot{x}$ plane for $k_1 = k_2 = 1$ is presented in Fig. 4.11. It is seen from this figure that the phase trajectories are asymmetric about the x-axis. This is because Eq. (4.37) changes when t is replaced by $-t$. Figure 4.12 depicts the time dependence of x for a sufficiently large oscillation amplitude. It is seen that in this case the oscillations present a periodic sequence of nonsymmetric pulses.

Let us note that the Lotka–Volterra equations can be rewritten in canonical variables $q = \ln x$, $p = \ln y$. In terms of these variables the Hamiltonian of the system and the corresponding canonical equations are

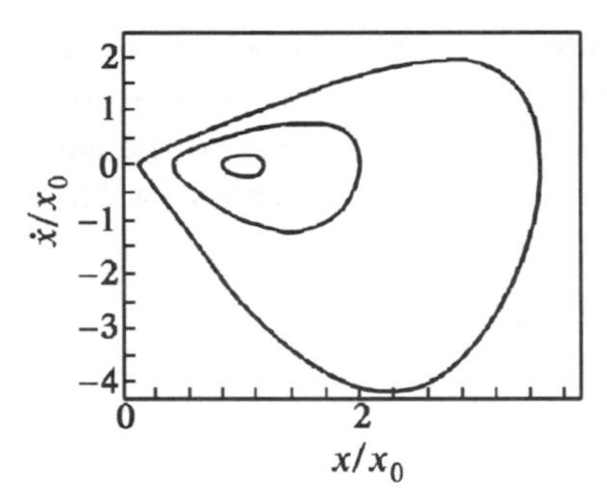

Fig. 4.11. The phase portrait of the Lotka–Volterra equations on the plane x/x_0, $\dot{x}/x_0$ for $k_1 = k_2 = 1$

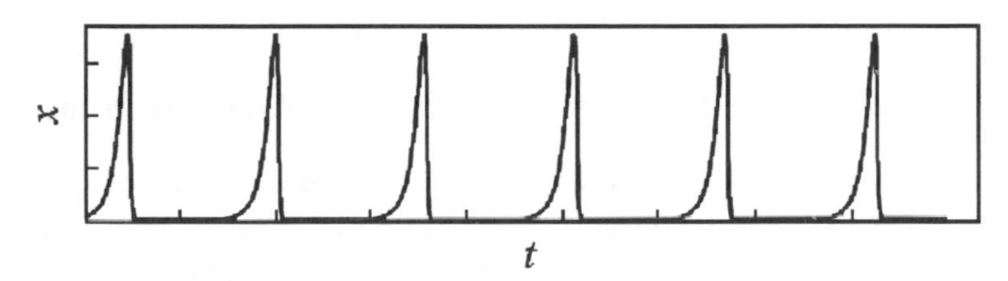

Fig. 4.12. An example of the time dependence of x for a sufficiently large oscillation amplitude

$$H = k_1 p - a_1 \mathrm{e}^p + k_2 q - a_2 \mathrm{e}^q, \tag{4.38}$$

$$\dot{q} = k_1 - a_1 \mathrm{e}^p, \quad \dot{p} = -k_2 + a_2 \mathrm{e}^q. \tag{4.39}$$

4.4 Natural oscillations in a system with slowly time-varying natural frequency

For simplicity's sake we consider a linear oscillator described by the equation

$$\ddot{x} + \omega^2(\epsilon t)x = 0, \tag{4.40}$$

where $\omega(\epsilon t)$ is a slowly varying function in time, and ϵ is a small parameter.

Let us go in Eq. (4.40) from t to slow time $\tau = \epsilon t$. As a result we obtain an equation with the small parameter ϵ^2 at higher derivative. Introducing the large parameter $\lambda = \epsilon^{-1}$, we rewrite this equation as

$$\frac{\mathrm{d}^2 x}{\mathrm{d}\tau^2} + \lambda^2 \omega^2(\tau)x = 0. \tag{4.41}$$

For the approximate solution of similar equations there exist a number of methods, such as the JWKB method [92, 248], the asymptotic method for large eigenvalues [41, 327], and methods based on such transformations of dependent and independent variables for which the original equation for sufficiently large values of the parameter λ is reduced to the simplest form [248].

In the first approximation with respect to the small parameter $\epsilon = \lambda^{-1}$, all these methods lead to the same expression for $x(t)$, which can be written as

$$x(t) = A(\epsilon t) \sin \psi(t, \epsilon t), \tag{4.42}$$

where

$$A(\epsilon t) = A_0 \sqrt{\frac{\omega_0}{\omega(\epsilon t)}}, \tag{4.43}$$

$$\psi(t, \epsilon t) = \int \omega(\epsilon t)\, \mathrm{d}t. \tag{4.44}$$

In (4.43) A_0 and ω_0 are the values of the amplitude A and the frequency ω, respectively, for $\epsilon = 0$.

Let us note that the approximate solution (4.42) is valid if the sign of $\omega^2(\epsilon t)$ does not change with time, i.e., the solution is always of an oscillatory character. Otherwise all methods mentioned above, except for a method based on the Langer transformation [248], cease to function. The Langer transformation results in a solution expressed via the Airy special functions, the behavior of which changes its character from oscillatory to exponential as their argument reverses sign.

It follows from (4.40), (4.42)–(4.44) that the energy of the oscillations, as a first approximation with respect to ϵ, is

$$E = \frac{\dot{x}^2}{2} + \frac{\omega^2(\tau)x^2}{2} = \frac{A_0^2 \omega_0 \omega}{2} - \epsilon \frac{A_0^2}{2} \frac{\mathrm{d}\omega}{\mathrm{d}\tau} \sin 2\psi. \tag{4.45}$$

From this it is seen that the energy E not only varies slowly with time, but also contains a small (of the order of ϵ) fast oscillating part. Averaging (4.45) over the phase ψ, we obtain the following expression for the averaged energy:

$$\overline{E} = \frac{A_0^2 \omega_0 \omega(\tau)}{2}. \tag{4.46}$$

So, the averaged energy is a function varying slowly with time. It is known (see, e.g., [189]) that the averaged energy of a linear oscillator is $\overline{E} = J\omega$, where J is the action. It follows from this and (4.46) that $J = A_0^2 \omega_0 / 2$. Hence we have directly shown that, in a first approximation with respect to the small parameter ϵ, the action J is independent of ϵ, i.e., it is indeed an adiabatic invariant [189].

5. Natural oscillations in systems with many degrees of freedom. Normal oscillations

Each dynamical system can be characterized by a set of natural oscillations called *normal oscillations*. The number of normal oscillations depends on the number of degrees of freedom of the system. For linear systems the number of normal oscillations is exactly equal to the number of degrees of freedom, whereas for nonlinear systems this is not necessarily so. For continuous systems having an infinitely large number of degrees of freedom the number of normal oscillations is infinite, but in the case of bounded systems it is denumerable.

Of special interest are normal oscillations in conservative systems, where they are undamped. It is apparent that when the damping is small the normal oscillations will be only slightly changed in their shape. However, it is important that different normal oscillations can damp with different rates, often resulting in a number of intriguing effects.

5.1 Normal oscillations in linear conservative systems

The general theory of normal oscillations in linear systems with fixed parameters is adequately covered in many books (see, e.g., [62, 322]). Therefore here we focus our attention on the fundamental tenets of this theory and on some concrete results only.

In the general case the equation of oscillations in a linear conservative system can be written in the complex form

$$\mathcal{M}\ddot{\boldsymbol{x}} + \mathcal{H}\dot{\boldsymbol{x}} + \mathcal{K}\boldsymbol{x} = 0, \tag{5.1}$$

where $\mathcal{M}$ and $\mathcal{K}$ are symmetric matrices designating the mass matrix and the stiffness matrix, respectively, $\mathcal{H}$ is an antisymmetric matrix designating the matrix of gyroscopic forces, e.g., of the Coriolis forces. It follows from the antisymmetry of the matrix $\mathcal{H}$ that its diagonal elements must be equal to zero.

The general solution of Eq. (5.1) can be written in the form of a linear combination of the so-called normal coordinates ξ_s:

$$\boldsymbol{x} = \frac{1}{2}\sum_{s=1}^{n}(\boldsymbol{V}_s\xi_s + \text{c.c.}), \tag{5.2}$$

where $\boldsymbol{V}_s$ is the sth complex eigenvector, and n is the number of degrees of freedom of the system.

By definition, the coordinates ξ_s are normal coordinates if they are described by decoupled equations of the form

$$\ddot{\xi}_s + \omega_s^2 \xi_s = 0, \tag{5.3}$$

where $s = 1, 2, \ldots, n$, and ω_s is the sth natural frequency of the oscillations of the system which is the sth root of the characteristic equation

$$\det \| -\mathcal{M}\omega^2 + \mathrm{i}\mathcal{H}\omega + \mathcal{K} \| = 0. \tag{5.4}$$

We note that because of the properties of the matrices $\mathcal{M}$, $\mathcal{H}$ and $\mathcal{K}$ indicated above, Eq. (5.4) contains only even powers of ω, and the squares of all roots of this equation are real numbers.

The eigenvectors $\boldsymbol{V}_s$ are determined by the systems of homogeneous equations

$$(-\mathcal{M}\omega_s^2 + \mathrm{i}\mathcal{H}\omega_s + \mathcal{K})\boldsymbol{V}_s = 0 \tag{5.5}$$

for which the determinant is equal to zero. Hence all components of the vector $\boldsymbol{V}_s$ are determined to within a constant multiplier.

5.2 Normal oscillations in nonlinear conservative systems

In contrast to linear systems, for nonlinear systems the superposition principle does not hold. Therefore, the general solution of a nonlinear system of equations cannot be represented as a sum of normal oscillations. However, the notion of normal oscillations exists for nonlinear systems as well as linear ones. This notion is often found to be very useful, in particular for the construction of generating solutions for the purpose of using different perturbation methods.

By normal oscillations of nonlinear systems are meant such periodic motions for which all the system's variables are expressed in terms of one of them (e.g., x_1) by means of algebraic relations, i.e.,

$$x_i = f_i(x_1) \quad (i = 2, 3, \ldots, N). \tag{5.6}$$

In the regime of normal oscillations a system behaves like one with one degree of freedom. It is interesting to note that the definition of normal oscillations for nonlinear systems is similar to the definition of Riemann simple waves (see, e.g., [189]).

An interesting particular case, when the relations (5.6) are linear, i.e.,

$$x_i = K_i x_1 \quad (K_1 = 1), \tag{5.7}$$

was studied in detail by Rosenberg [286, 287]. It is an easy matter to find the conditions under which this case can take place. For this purpose let us consider a system described by the equations

$$\ddot{x}_i + \frac{\partial U(x_1, x_2, \dots, x_N)}{\partial x_i} = 0 \quad (i = 1, 2, \dots, N), \tag{5.8}$$

where $U(x_1, x_2, \dots, x_N)$ is the potential energy. Substituting (5.7) into (5.8) we obtain

$$\begin{aligned} &\ddot{x}_1 + \left.\frac{\partial U(x_1, x_2, \dots, x_N)}{\partial x_1}\right|_{x_i = K_i x_1} = 0, \\ &K_2 \ddot{x}_1 + \left.\frac{\partial U(x_1, x_2, \dots, x_N)}{\partial x_2}\right|_{x_i = K_i x_1} = 0, \\ &\vdots \\ &K_N \ddot{x}_1 + \left.\frac{\partial U(x_1, x_2, \dots, x_N)}{\partial x_N}\right|_{x_i = K_i x_1} = 0 \quad (i = 2, 3, \dots, N). \end{aligned}$$

It follows from the compatibility requirements for this system of equations that $N - 1$ conditions must be satisfied for the function U:

$$\frac{1}{K_i} \left.\frac{\partial U(x_1, x_2, \dots, x_N)}{\partial x_i}\right|_{x_i = K_i x_1} = \left.\frac{\partial U(x_1, x_2, \dots, x_N)}{\partial x_1}\right|_{x_i = K_i x_1} \quad (i = 2, 3, \dots, N). \tag{5.9}$$

The conditions (5.9) can be satisfied only if the function U has a specified symmetry.

5.3 Examples of normal oscillations in linear and nonlinear conservative systems

5.3.1 Two coupled linear oscillators with gyroscopic forces

As an example, we consider a system of two coupled linear oscillators described by the equations

$$m_1 \ddot{x} + h\dot{y} + k_1 x + k_{12} y = 0, \quad m_2 \ddot{y} - h\dot{x} + k_2 y + k_{12} x = 0. \tag{5.10}$$

The characteristic equation for the system of Eqs. (5.10) is

$$m_1 m_2 \omega^4 - (m_1 k_2 + m_2 k_1 + h^2)\omega^2 + k_1 k_2 - k_{12}^2 = 0.$$

The solution of this equation gives the frequencies of the normal oscillations:

$$\begin{aligned} \omega_{1,2} = \frac{1}{2m_1 m_2} \Big(& m_1 k_2 + m_2 k_1 + h^2 \\ & \pm \sqrt{(m_1 k_2 - m_2 k_1 + h^2)^2 + 4m_2(k_{12}^2 m_1 + k_1 h^2)} \Big). \end{aligned} \tag{5.11}$$

Putting

$$x = \frac{\xi_1 + \xi_2 + \text{c.c.}}{2}, \quad y = \frac{V_1\xi_1 + V_2\xi_2 + \text{c.c.}}{2},$$

where $\xi_{1,2} = C_{1,2}\exp(\mathrm{i}\omega_{1,2}t)$ are normal coordinates, we find

$$V_{1,2} = \frac{m_1k_2 - m_2k_1 + h \pm \sqrt{(m_1k_2 - m_2k_1 + h^2)^2 + 4m_2(k_{12}^2 m_1 + k_1h^2)}}{2m_2(k_{12} + \mathrm{i}h\omega_{1,2})}. \tag{5.12}$$

It is seen from (5.12) that for $h \neq 0$ the values of $V_{1,2}$ are complex, which is to say that there is a phase shift between oscillations of x and y. For $h = 0$ the values $V_{1,2}$ are real numbers of opposite sign. The positive value of V corresponds to in-phase oscillations of x and y, and the negative value of V corresponds to anti-phase oscillations of x and y.

5.3.2 Examples of normal oscillations in two coupled nonlinear oscillators

As an example of normal oscillations in nonlinear systems we consider a system with two degrees of freedom described by the equations [231]

$$\ddot{x} + x^3 + \gamma(x-y)^3 = 0, \quad \ddot{y} + y^3 - \gamma(x-y)^3 = 0. \tag{5.13}$$

Substituting $y = Kx$ into (5.13) we obtain, in view of (5.9), the following equation for K:

$$(K^2 - 1)\big(\gamma(K-1)^2 + K\big) = 0. \tag{5.14}$$

Equation (5.14) possesses four solutions:

$$K_{1,2} = \pm 1, \quad K_{3,4} = 1 - \frac{1}{2\gamma}\big(1 \pm \sqrt{1-4\gamma}\big).$$

It should be noted that the solutions $K_{3,4}$ are real if $\gamma \leq 1/4$.

The solutions K_1 and K_2 correspond to in-phase and anti-phase oscillations of the dynamical variables x and y with identical amplitudes; the solutions K_3 and K_4 correspond to anti-phase oscillations of the dynamical variables x and y with different amplitudes. These four types of normal oscillations are described by a common equation of the form $\ddot{x} + \alpha x^3 = 0$ but with different values of the coefficient α:

$$\alpha = \begin{cases} 1 & \text{for } K = 1, \\ 1 + 8\gamma & \text{for } K = -1, \\ 1 + (1/\sqrt{\gamma})(-K_{3,4})^{3/2} & \text{for } K = K_{3,4}. \end{cases}$$

It follows from the example cited that the number of types of normal oscillations for nonlinear systems, in contrast to linear ones, can be more than the number of degrees of freedom.

As another example of normal oscillations in nonlinear systems, let us consider the so-called Yang–Mills equations [369], which are well known in the theory of elementary particles. These equations are

$$\ddot{x} + xy^2 = 0, \quad \ddot{y} + x^2 y = 0. \tag{5.15}$$

For Eqs. (5.15) we immediately obtain two types of normal oscillations: in-phase and anti-phase with $y = \pm x$. Both types are described by the equation

$$\ddot{x} + x^3 = 0.$$

5.3.3 An example of normal oscillations in three coupled nonlinear oscillators

An interesting example of normal oscillations in three coupled nonlinear oscillators concerns a device proposed by Lavrov in the 'Mekhanobr' Institute in St Petersburg, Russia [188]. The schematic image of this device is shown

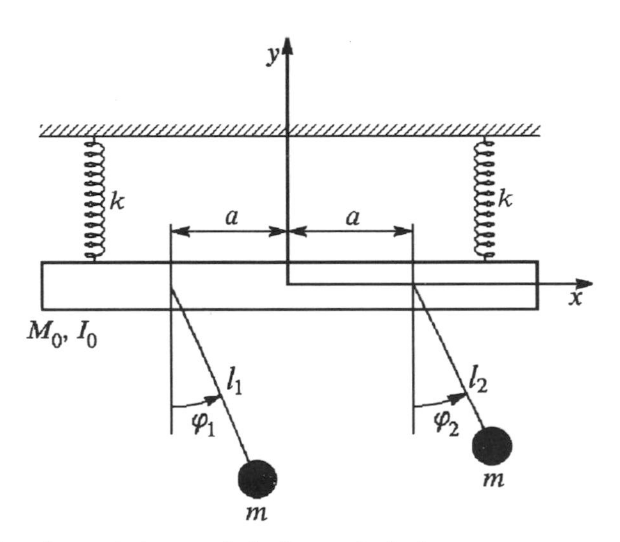

Fig. 5.1. The schematic image of the Lavrov's device

in Fig. 5.1. Two pendulums of identical masses m and different lengths (l_1 and l_2) are suspended on a plank of mass M_0 symmetrically about its center of gravity. The plank, in turn, is suspended by two sufficiently soft springs of stiffness c on a solid bracket. It was found experimentally that the pendulums, being initially deflected through some angles in the same plane, switch, after a short complexly shaped transient process, to anti-phase oscillations with the same frequencies but with different amplitudes. These oscillations persist over a long period of time.

Let us derive the equations of oscillations of these pendulums by means of the Lagrange function L having the following form:

$$L = \frac{M}{2}(\dot{x}^2 + \dot{y}^2) + \frac{I}{2}\dot{\varphi}^2 + m\left(\frac{l_1^2\dot{\varphi}_1^2}{2} + \frac{l_2^2\dot{\varphi}_2^2}{2}\right) + m\dot{x}(l_1\dot{\varphi}_1\cos\varphi_1$$
$$+ l_2\dot{\varphi}_2\cos\varphi_2) + m\dot{y}(l_1\dot{\varphi}_1\sin\varphi_1 + l_2\dot{\varphi}_2\sin\varphi_2)$$
$$+ ma\dot{\varphi}(l_1\dot{\varphi}_1\sin(\varphi - \varphi_1) - l_2\dot{\varphi}_2\sin(\varphi - \varphi_2))$$
$$+ mg(l_1\cos\varphi_1 + l_2\cos\varphi_2), \tag{5.16}$$

where x and y are the deviations of the center of gravity of the plank from its equilibrium position in the horizontal and vertical directions, respectively, φ is the angle of rotation of the plank about the axis passing through its center of gravity (this angle is measured from the horizontal in the anti-clockwise direction), φ_1 and φ_2 are the angles of deviation of the pendulums measured from the vertical in the anti-clockwise direction, $M = M_0 + 2m$, $I = I_0 + 2ma^2$, I_0 is the moment of inertia of the plank about the axis passing through its center of gravity, and a is the distance from the center of gravity of the plank to the suspension axis of one of the pendulums. In the expression (5.16) we ignore the potential energy of the springs with stiffness c, assuming it to be small.

The Lagrange equations for the system under consideration are

$$\ddot{\varphi}_1 + \omega_1^2\sin\varphi_1 + \frac{\ddot{x}}{l_1}\cos\varphi_1 + \frac{\ddot{y}}{l_1}\sin\varphi_1$$
$$+\frac{a}{l_1}\Big(\ddot{\varphi}\sin(\varphi - \varphi_1) + \dot{\varphi}^2\cos(\varphi - \varphi_1)\Big) = 0,$$
$$\tag{5.17}$$
$$\ddot{\varphi}_2 + \omega_2^2\sin\varphi_2 + \frac{\ddot{x}}{l_2}\cos\varphi_2 + \frac{\ddot{y}}{l_2}\sin\varphi_2$$
$$+\frac{a}{l_2}\Big(\ddot{\varphi}\sin(\varphi - \varphi_2) + \dot{\varphi}^2\cos(\varphi - \varphi_2)\Big) = 0,$$

$$\ddot{x} = -\frac{ml_1}{M}(\ddot{\varphi}_1\cos\varphi_1 - \dot{\varphi}_1^2\sin\varphi_1) - \frac{ml_2}{M}(\ddot{\varphi}_2\cos\varphi_2 - \dot{\varphi}_2^2\sin\varphi_2),$$
$$\tag{5.18}$$
$$\ddot{y} = -\frac{ml_1}{M}(\ddot{\varphi}_1\sin\varphi_1 + \dot{\varphi}_1^2\cos\varphi_1) - \frac{ml_2}{M}(\ddot{\varphi}_2\sin\varphi_2 + \dot{\varphi}_2^2\cos\varphi_2),$$

$$\ddot{\varphi} = -\frac{mal_1}{I}\Big(\ddot{\varphi}_1\sin(\varphi - \varphi_1) - \dot{\varphi}_1^2\cos(\varphi - \varphi_1)\Big)$$
$$+\frac{mal_2}{I}\Big(\ddot{\varphi}_2\sin(\varphi - \varphi_2) - \dot{\varphi}_2^2\cos(\varphi - \varphi_2)\Big), \tag{5.19}$$

where $\omega_{1,2} = \sqrt{g/l_{1,2}}$. Substituting (5.18) into Eqs. (5.17) we obtain

$$\begin{aligned}
&\left(1-\frac{m}{M}\right)\ddot{\varphi}_1+\omega_1^2\sin\varphi_1-\frac{ml_2}{Ml_1}\Big(\ddot{\varphi}_2\cos(\varphi_1-\varphi_2)+\dot{\varphi}_2^2\sin(\varphi_1-\varphi_2)\Big)\\
&\qquad+\frac{a}{l_1}\Big(\ddot{\varphi}\sin(\varphi-\varphi_1)+\dot{\varphi}^2\cos(\varphi-\varphi_1)\Big)=0,\\
&\left(1-\frac{m}{M}\right)\ddot{\varphi}_2+\omega_2^2\sin\varphi_2-\frac{ml_1}{Ml_2}\Big(\ddot{\varphi}_1\cos(\varphi_1-\varphi_2)+\dot{\varphi}_1^2\sin(\varphi_1-\varphi_2)\Big)\\
&\qquad-\frac{a}{l_2}\Big(\ddot{\varphi}\sin(\varphi-\varphi_2)+\dot{\varphi}^2\cos(\varphi-\varphi_2)\Big)=0.
\end{aligned}\tag{5.20}$$

Equations (5.20) in combination with Eq. (5.19) form a closed system. The forms of normal oscillations for this system can be analytically found in a special case only, as the pendulums are identical, i.e., $l_1 = l_2 = l$. These forms are $\varphi_1 = \varphi_2 = \psi$, $\varphi = 0$, and $\varphi_1 = -\varphi_2 = \vartheta$, $\varphi = 0$, where ψ and ϑ are described by the following equations:

$$\begin{aligned}
&\left(1-\frac{2m}{M}\right)\ddot{\psi}+\omega_0^2\sin\psi=0,\\
&\left(1-\frac{2m}{M}\sin^2\vartheta\right)\ddot{\vartheta}+\omega_0^2\sin\vartheta-\frac{m}{M}\dot{\vartheta}^2\sin 2\vartheta=0.
\end{aligned}\tag{5.21}$$

Here $\omega_0 = \sqrt{g/l}$ is the natural frequency of small oscillations of the pendulums. The first shape corresponds to in-phase oscillations of the pendulums, and the second one to anti-phase oscillations of the pendulums. It follows from Eqs. (5.21) that the frequency of small anti-phase oscillations is equal to ω_0, whereas the frequency of small in-phase oscillations is more than ω_0. From Eqs. (5.18) we can find the displacements x and y of the plank:

$$x=\begin{cases}-2(ml/M)\sin\psi & \text{for in-phase oscillations}\\ 0 & \text{for anti-phase oscillations,}\end{cases}$$

$$y=\begin{cases}2(ml/M)\cos\psi & \text{for in-phase oscillations}\\ 2(ml/M)\cos\vartheta & \text{for anti-phase oscillations.}\end{cases}$$

In order to give at least a qualitative explanation for the results of Lavrov's experiments, we consider the approximation of small oscillations of the pendulums, when Eqs. (5.20) take the form

$$\begin{aligned}
&\left(1-\frac{m}{M}\right)\ddot{\varphi}_1+\omega_1^2\varphi_1-\frac{m\omega_1^2}{M\omega_2^2}\ddot{\varphi}_2=0,\\
&\left(1-\frac{m}{M}\right)\ddot{\varphi}_2+\omega_2^2\varphi_2-\frac{m\omega_2^2}{M\omega_1^2}\ddot{\varphi}_1=0.
\end{aligned}\tag{5.22}$$

We see that in this approximation Eqs. (5.22) are independent of φ. Putting $\varphi_2 = k\varphi_1$, from the condition for the compatibility of Eqs. (5.22) we obtain the following equation for k:

$$\left(1-\frac{m}{M}\right)k\omega_1^2 - \frac{m}{M}\omega_2^2 = k\left(\omega_2^2\left(1-\frac{m}{M}\right) - \frac{m}{M}k\omega_1^2\right). \tag{5.23}$$

Eq. (5.23) possesses two roots:

$$k_{1,2} = \frac{1}{2}\left(\frac{\omega_2^2}{\omega_1^2}-1\right)\left(\frac{M}{m}-1\right) \pm \sqrt{\frac{1}{4}\left(\frac{\omega_2^2}{\omega_1^2}-1\right)^2\left(\frac{M}{m}-1\right)^2 + \frac{\omega_2^2}{\omega_1^2}}.$$

The first root corresponds to in-phase normal oscillations, and the second root corresponds to anti-phase normal oscillations (we suppose that l_1 is greater than l_2, i.e., $\omega_1 < \omega_2$). The frequencies $\tilde{\omega}_{1,2}$ of the in-phase and the anti-phase oscillations are equal, respectively, to

$$\tilde{\omega}_{1,2} = \left(\frac{(1-\mu)(\omega_1^2+\omega_2^2) \pm \sqrt{(1-2\mu)(\omega_1^2-\omega_2^2)^2 + \mu^2(\omega_1^2+\omega_2^2)^2}}{2(1-2\mu)}\right)^{1/2}, \tag{5.24}$$

where $\mu = m/M$. It is seen from (5.24) that the frequency of the in-phase oscillations, as for the case considered above, is greater than the frequency of the anti-phase oscillations.

Let us show that the in-phase oscillations of the pendulums die down significantly more rapidly than the anti-phase oscillations if damping of the pendulums and the plank is accounted for. In the linear approximation Eqs. (5.17)–(5.19) with regard to damping become:

$$\begin{aligned} &\ddot{\varphi}_1 + 2\alpha\dot{\varphi}_1 + \omega_1^2\varphi_1 + \frac{\ddot{x}}{l_1} = 0, \\ &\ddot{\varphi}_2 + 2\alpha\dot{\varphi}_2 + \omega_2^2\varphi_2 + \frac{\ddot{x}}{l_2} = 0, \\ &\ddot{x} + 2\beta\dot{x} + \mu(l_1\ddot{\varphi}_1 + l_2\ddot{\varphi}_2) = 0, \end{aligned} \tag{5.25}$$

where α and β are damping factors of free oscillations of the pendulums and the plank, respectively (for the sake of simplicity we assume the damping factors for both pendulums to be the same). The characteristic equation for the system (5.25) is

$$(p+2\beta)(p^2+2\alpha p+\omega_1^2)(p^2+2\alpha p+\omega_2^2) - \mu p^3(2p^2+4\alpha p+\omega_1^2+\omega_2^2) = 0. \tag{5.26}$$

Assuming damping factors to be sufficiently small, we seek two roots of Eq. (5.26) in the form $p_{1,2} = \mathrm{i}\tilde{\omega}_{1,2} + \delta_{1,2}$, where $\tilde{\omega}_{1,2}$ are the frequencies of the normal oscillations determined by the formula (5.24) and $\delta_{1,2} \ll \omega_{1,2}$ are the damping factors. As a first approximation we find

$$\delta_{1,2} = -\frac{\alpha(1-\mu)+\beta\mu}{1-2\mu} - \frac{\mu^2}{1-2\mu}\frac{(\alpha+\beta)(\omega_1^2+\omega_2^2)\tilde{\omega}_{1,2}^2}{(1-\mu)(\omega_1^2+\omega_2^2)\tilde{\omega}_{1,2}^2 - 2\omega_1^2\omega_2^2}. \tag{5.27}$$

Based on (5.27) and (5.24) we can calculate the difference of the damping factors for the in-phase and anti-phase normal oscillations:

$$|\delta_1| - |\delta_2| = \frac{2\mu^2(\alpha+\beta)(\omega_1^2+\omega_2^2)}{(1-\mu)^2(\omega_1^2-\omega_2^2)^2 + 4\mu^2\omega_1^2\omega_2^2}(\tilde{\omega}_1^2 - \tilde{\omega}_2^2). \tag{5.28}$$

Because $\tilde{\omega}_1 > \tilde{\omega}_2$, it follows from (5.28) that the damping factor for the in-phase oscillations is always more than that for the anti-phase oscillations, i.e., in-phase oscillations do die down more rapidly than anti-phase ones. With this we can explain the results of Lavrov's experiment described above.

We emphasize that the difference between the damping factors for in-phase and anti-phase oscillations is the greater, the larger is the ratio of the pendulum mass m to the plank mass M_0. This is seen from Fig. 5.2 in which

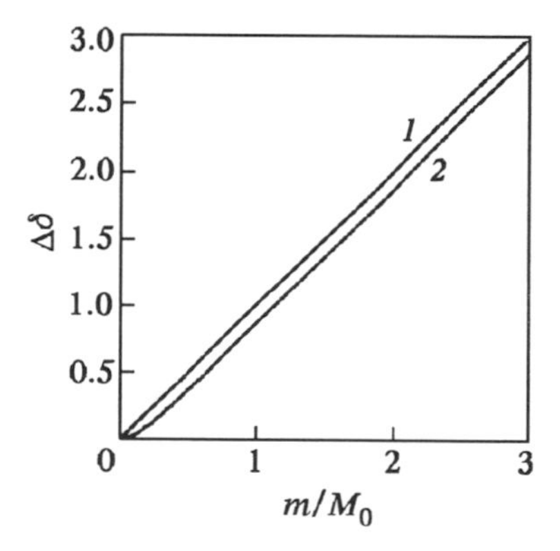

Fig. 5.2. The dependencies of $\Delta\delta = (|\delta_1| - |\delta_2|)/2(\alpha+\beta)$ on the ratio m/M_0 for pendulums with much the same length (curve 1 is constructed for $\omega_1 = \omega_2$) and for pendulums significantly different in length (curve 2 is constructed for $\omega_2^2 = 2\omega_1^2$)

the dependencies of $|\delta_1| - |\delta_2|$ on m/M_0 are depicted for pendulums with much the same lengths and for pendulums significantly different in length. We see that the difference $|\delta_1| - |\delta_2|$ depends only slightly on the ratio ω_1/ω_2.

5.3.4 Normal oscillations in linear homogeneous and periodically inhomogeneous chains

Let us consider chains consisting of identical balls and springs. Such chains play a great role both as objects in their own right and as finite-difference models of homogeneous continuous media.

We consider two types of chains. The first type is when the balls of mass m are connected to one another by springs with stiffness k (see Fig. 5.3 a), and the second type is when the balls are mounted rigidly to hard weightless pivots connected to one another by means of hinges suspended by springs (Fig. 5.3 b).

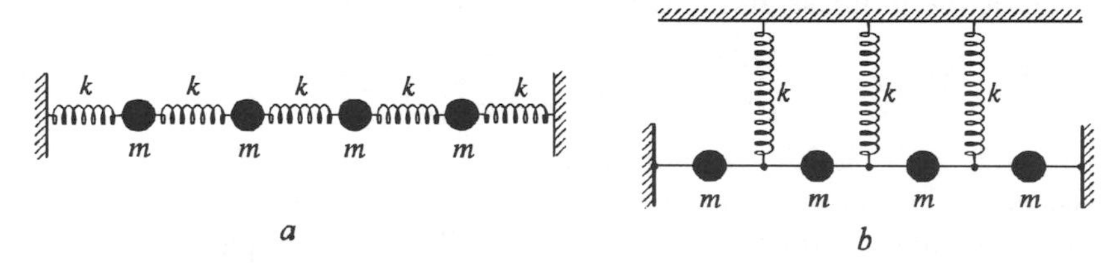

Fig. 5.3. The chains with (a) spring-connected and (b) inertially coupled balls

Let us derive the equations of oscillations in such chains using the Lagrange function. Denoting a displacement of the sth ball shown in Fig. 5.3 a by x_s, we write the expression for the Lagrange function:

$$L = \sum_s \left(\frac{m\dot{x}_s^2}{2} - \frac{k(x_{s+1} - x_s)^2}{2} \right). \tag{5.29}$$

For the chain shown in Fig. 5.3 b, we denote the vertical displacement of the sth hinge by y_s. In this case

$$L = \sum_s \left(\frac{m(\dot{y}_{s+1} - \dot{y}_s)^2}{8} - \frac{ky_s^2}{2} \right). \tag{5.30}$$

From (5.29) and (5.30) we obtain the following equations of the chains under consideration:

$$m\ddot{x}_s + k(2x_s - x_{s+1} - x_{s-1}) = 0 \qquad \text{for the first chain,} \tag{5.31}$$

$$\frac{m(2\ddot{y}_s - \ddot{y}_{s+1} - \ddot{y}_{s-1})}{4} + ky_s = 0 \quad \text{for the second chain,} \tag{5.32}$$

$$(s = 1, 2, \ldots, n).$$

For solving Eqs. (5.31), (5.32) we must add to these equations the boundary conditions at the ends of the chains. So, for the chains shown in Fig. 5.3 we have $x_0 = 0$, $x_{n+1} = 0$, $y_0 = 0$, $y_{n+1} = 0$. Let us seek a partial solution of Eqs. (5.31) (or (5.32)) in the form of one of normal oscillations

$$x_s = a_s \cos \omega t, \tag{5.33}$$

where ω is one of the frequencies of normal oscillations yet to be determined, and a_s is the oscillation amplitude of the sth ball, i.e., the sth component of the eigenvector. Substituting (5.33) into Eqs. (5.31), we obtain the following system of homogeneous equations for the amplitudes a_s:

$$(-m\omega^2 + 2k)a_s - ka_{s+1} - ka_{s-1} = 0. \tag{5.34}$$

A partial solution of Eqs. (5.34) can be sought in the form

$$a_s = A\mathrm{e}^{is\beta}\,, \tag{5.35}$$

where β is an unknown value. Substituting (5.35) into (5.34) we obtain the following relation between ω and β:

$$\cos\beta = 1 - \frac{m\omega^2}{2k}\,. \tag{5.36}$$

It follows from (5.36) that

$$\beta = \pm 2 \arcsin\frac{\omega}{\omega_0} \equiv \pm\gamma, \tag{5.37}$$

where $\omega_0 = 2\sqrt{k/m}$. From this we see that β is a real value for the frequencies $\omega \leq \omega_0$ and complex for $\omega > \omega_0$. The dependence of γ on ω/ω_0 is depicted in Fig. 5.4 a.

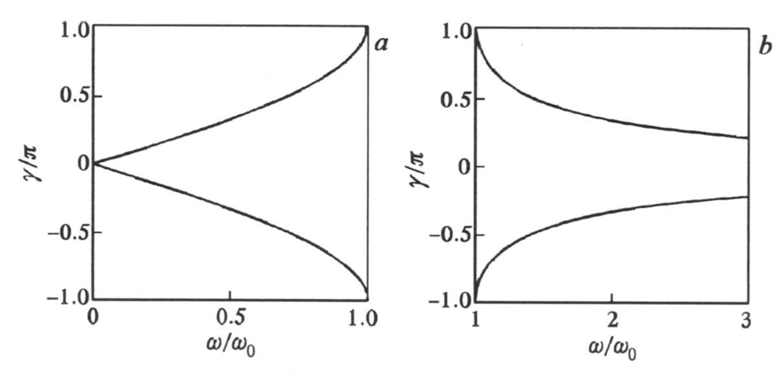

Fig. 5.4. Dispersion dependencies for the chains with (a) spring-connected and (b) inertially coupled balls

The general solution of Eqs. (5.34) is

$$a_s = A_1\mathrm{e}^{is\gamma} + A_2\mathrm{e}^{-is\gamma}, \tag{5.38}$$

where γ is associated with ω by the relation (5.37). Substituting (5.38) into two boundary conditions we obtain two homogeneous equations for A_1 and A_2. The condition for the system determinant to be equal to zero gives the characteristic equation for the determination of the unknown frequency ω. In the general case, the solution of this characteristic equation is complicated because it is transcendental. Therefore, we restrict ourselves to the simplest example when $x_0 = 0$, $x_{n+1} = 0$. In this case $A_1 = -A_2$, and the characteristic equation is

$$\sin(n+1)\gamma = 0. \tag{5.39}$$

It follows from (5.39) that $\gamma = \gamma_q = q\pi/(n+1)$, where q is an integer. Substituting these values of γ into (5.37) we find the frequencies of normal oscillations ω_q:

$$\omega_q = 2\sqrt{\frac{k}{m}} \sin \frac{q\pi}{2(n+1)} . \tag{5.40}$$

From this we see that the values of the frequencies are different for $q = 0, 1, 2, \ldots, n+1$. But, as follows from (5.38), $a_s = 0$ for $q = 0$ and $q = n+1$. Thus, nontrivial solutions are obtained, as they must be, only for n frequencies with $q = 1, 2, \ldots, n$. Let us note that all natural frequencies found are less than ω_0; hence the values β associated with them are real numbers. The oscillation modes corresponding to each of the natural frequencies ω_q for $n = 6$ are demonstrated in Fig. 5.5.

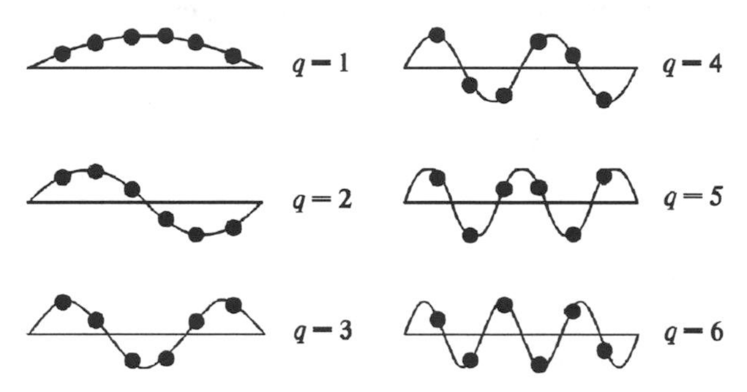

Fig. 5.5. The oscillation modes for the chain containing six spring-connected balls

In a similar manner, we can find the natural frequencies and oscillation modes for the chain shown in Fig. 5.3 b. In so doing we obtain the following relation between β and ω:

$$\beta = \pm 2 \arcsin \frac{\omega_0}{\omega} = \pm\gamma, \tag{5.41}$$

where $\omega_0 = \sqrt{k/m}$. The dependence of γ on ω/ω_0 determined by the formula (5.41) is depicted in Fig. 5.4 b. Let us call attention to the fact that in this case γ decreases as ω increases. If the boundary conditions are $y_0 = 0$, $y_{n+1} = 0$ then the natural frequencies of the chain are equal to

$$\omega_q = \sqrt{\frac{k}{m}} \left(\sin \frac{q\pi}{2(n+1)} \right)^{-1} , \tag{5.42}$$

where $q = 1, 2, \ldots, n$. It is seen from (5.42) that all of the frequencies ω_q lie beyond ω_0. The oscillation modes associated with each of the natural frequencies ω_q for $n = 6$ are identical to those shown in Fig. 5.5, the only difference being that the smallest frequency corresponds to the greatest q.

Let us consider further natural oscillations in the chain shown in Fig. 5.6.

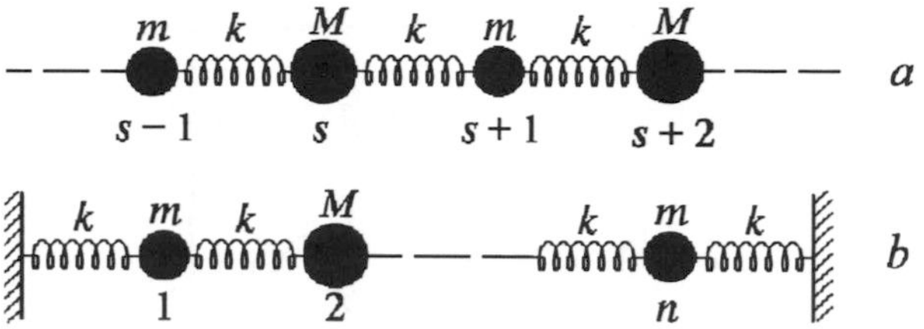

Fig. 5.6. The chains of spring-connected balls with alternating masses: (a) the boundless chain, and (b) the chain of n coupled balls

This chain is different from the chain considered above in that the masses of the balls are not identical, and the heavy balls of mass M alternate with light balls of mass m. We enumerate all balls in succession and assume that the heavy balls correspond to even numbers s and light ones correspond to odd s. In this case the equations of oscillations of the balls are

$$\begin{aligned} &m\ddot{x}_s + k(2x_s - x_{s+1} - x_{s-1}) = 0 \quad \text{for } s = 1, 3, 5, \dots, \\ &M\ddot{x}_s + k(2x_s - x_{s+1} - x_{s-1}) = 0 \quad \text{for } s = 2, 4, 6, \dots . \end{aligned} \tag{5.43}$$

Since the masses of the balls are different, the amplitudes of oscillations of the balls are also different. Because of this, we seek a solution of Eqs. (5.43) in the form

$$x_s = \begin{cases} A\mathrm{e}^{\mathrm{i}s\beta}\cos\omega t & \text{for } s = 1, 3, 5, \dots \\ B\mathrm{e}^{\mathrm{i}s\beta}\cos\omega t & \text{for } s = 2, 4, 6, \dots . \end{cases} \tag{5.44}$$

Substituting (5.44) into (5.43), we obtain the following equations for the amplitudes A and B:

$$\begin{aligned} &\left(-m\omega^2 + 2k\right) A - k\left(\mathrm{e}^{\mathrm{i}\beta} + \mathrm{e}^{-\mathrm{i}\beta}\right) B = 0, \\ &-k\left(\mathrm{e}^{\mathrm{i}\beta} + \mathrm{e}^{-\mathrm{i}\beta}\right) A + \left(-M\omega^2 + 2k\right) B = 0. \end{aligned} \tag{5.45}$$

For the solution (5.44) to be nontrivial, the determinant of the system of Eqs. (5.45) must be equal to zero. This condition gives the relation between β and ω:

$$\omega^4 - (\omega_1^2 + \omega_2^2)\omega^2 + \omega_1^2\omega_2^2(\sin\beta)^2 = 0, \tag{5.46}$$

where $\omega_1 = \sqrt{2k/M}$, $\omega_2 = \sqrt{2k/m}$. It follows from (5.46) that to every value of β in the range from $-\pi$ to π there correspond two values of the frequency ω (see Fig. 5.7).

Solving Eqs. (5.45), in view of (5.46), we find the relation between the amplitudes of the oscillations of heavy and light balls:

$$B = \sqrt{\left(1 - \frac{\omega^2}{\omega_2^2}\right)\left(1 - \frac{\omega^2}{\omega_1^2}\right)^{-1}} A \operatorname{sign}\left(\left(1 - \frac{\omega^2}{\omega_2^2}\right)\cos\beta\right). \tag{5.47}$$

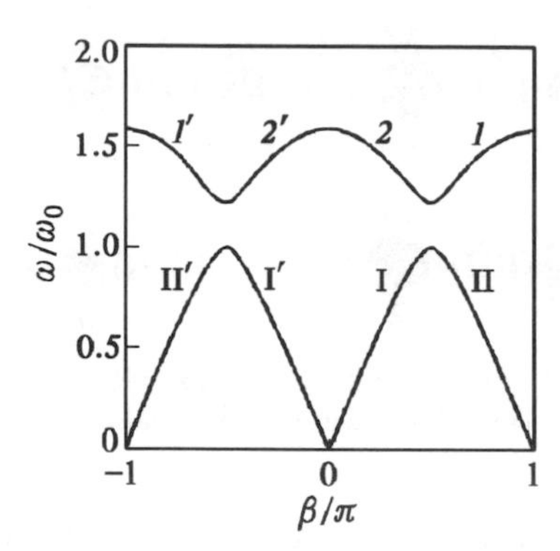

Fig. 5.7. Dispersion dependencies for the chains of spring-connected balls with alternating masses

From this it can be seen that for $\omega \leq \omega_1$ the amplitude of the oscillations of the heavy balls (B) exceeds the amplitude of the oscillations of the light balls (A). This difference in amplitudes is the greater, the closer is the frequency ω to the boundary value ω_1 (see Fig. 5.8 a). In the case of $\omega_2 \leq \omega \leq \sqrt{\omega_1^2 + \omega_2^2}$ we have, in contrast, $B < A$ (Fig. 5.8 b).

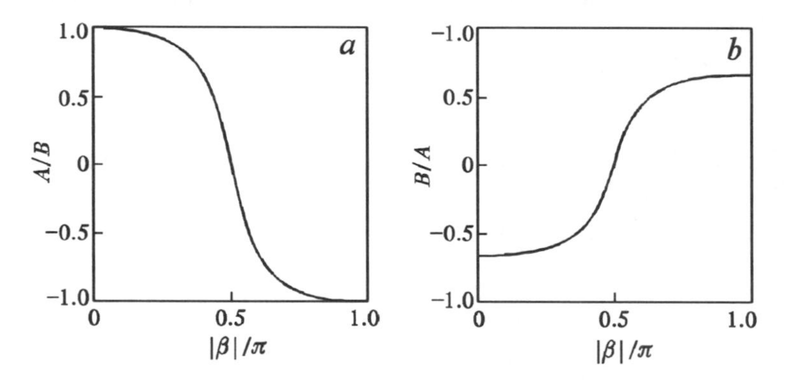

Fig. 5.8. Dependencies of the ratio between the amplitude of the oscillations of the heavy balls (B) and the amplitude of the oscillations of the light balls (A) on the 'wave' number β for (a) $\omega \leq \omega_1$, and (b) $\omega \geq \omega_2$

In view of (5.47), we can rewrite the solution (5.44) as

$$x_s = \frac{A+B}{2}(1 + \mu \cos \pi s)\mathrm{e}^{\mathrm{i}s\beta} \cos \omega t, \tag{5.48}$$

where $\mu = (1 - A/B)/(1 + A/B)$. The dependencies of $\mu_1(\beta) = \mu$ for $\omega \leq \omega_1$ and $\mu_2(\beta) = -\mu$ for $\omega \geq \omega_2$ are shown in Fig. 5.9. Taking account of the fact that $\exp\big(\mathrm{i}s(\beta + \pi)\big) = \exp\big(\mathrm{i}s(\beta - \pi)\big)$, the expression (5.48) can be rewritten

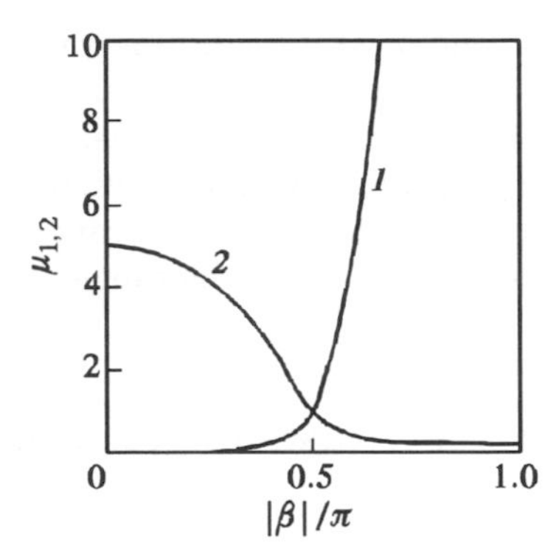

Fig. 5.9. Dependencies of $\mu_1(\beta) = \mu$ for $\omega \leq \omega_1$ (curve 1) and of $\mu_2(\beta) = -\mu$ for $\omega \geq \omega_2$ (curve 2)

in the following form:

$$x_s = \frac{A+B}{2}\left(\mathrm{e}^{\mathrm{i}s\beta} + \mu\mathrm{e}^{\mathrm{i}s(\beta-\pi)}\right)\cos\omega t.$$

It follows from here that four 'waves' with different values of β are possible for each ω. This corresponds to four dispersion branches in Fig. 5.7 labeled I,I′; II,II′; 1, 1′; and 2, 2′. It is significant that the wave amplitudes associated with the branches I and II′ , I′ and II, 1 and 2′ , 1′ and 2 are not independent. The ratio of the wave amplitudes associated with the branches II,II′ to the wave amplitudes associated with the branches I′,I is $(1-|A/B|)/(1+|A/B|) \leq 1$. Similarly, the ratio of the wave amplitudes associated with the branches 2′, 2 to the wave amplitudes associated with the branches 1, 1′ is $(1-|B/A|)/(1+|B/A|) \leq 1$.

The branches I,I′, 1 and 1′ correspond to waves with normal dispersion for which the frequency ω increases with increase of the 'wave' number β. In this case the 'phase' velocity ω/β and the 'group' velocity $\mathrm{d}\omega/\mathrm{d}\beta$ are equally directed. The branches II,II′,2 and 2′ correspond to waves with anomalous dispersion for which the frequency ω decreases with increase of the 'wave' number β. In this case the 'phase' velocity ω/β and the 'group' velocity $\mathrm{d}\omega/\mathrm{d}\beta$ are oppositely directed. It follows from these results that excitation of a normally dispersive wave inevitably causes excitation of a corresponding anomalously dispersive wave with smaller amplitude. The group velocities of these waves are equally directed, and the phase velocities are oppositely directed. Conversely, excitation of an anomalously dispersive wave entails excitation of a corresponding normally dispersive wave with larger amplitude.

If $m \to M$, then the amplitudes of anomalously dispersive waves approach zero, and ω_1 approaches ω_2. As a result, the dependence of ω on β takes the same form as for the chain with balls of identical masses (Fig. 5.4 a).

If the chain contains n balls and is fixed at the ends, i.e., $x_0 = x_{n+1} = 0$, then the eigenvalues of β are equal to $\beta_q = \pm q\pi/(n+1)$, where $q = 1, 2, \ldots, n$.

This is further proof that one must consider the variation of β in the range $-\pi$ to π rather than in the range $-\pi/2$ to $\pi/2$, as is done in many books.

It should be noted that the chain described can be considered as an one-dimensional model of a crystal lattice with two different types of atoms. With this model in mind, frequencies that lie in the range $\omega \leq \omega_1$ are spoken of as *acoustical*, and frequencies that lie in the range $\omega_2 \leq \omega \leq \sqrt{\omega_1^2 + \omega_2^2}$ are spoken of as *optical*.

5.3.5 Examples of natural oscillations in nonlinear homogeneous chains

Let us consider a chain of identical balls of mass m connected by identical nonlinear springs (Fig. 5.10). The equation of motion of the jth ball is

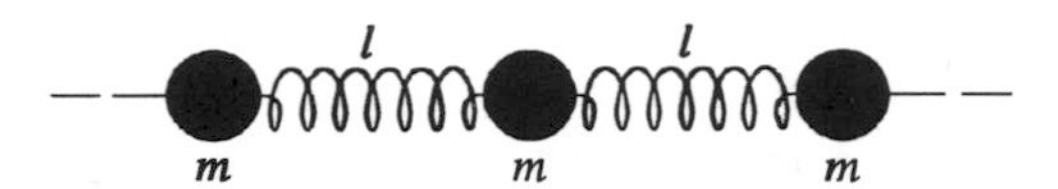

Fig. 5.10. The chain of identical balls connected by nonlinear springs

$$m\ddot{x}_j = f(x_j - x_{j-1}) - f(x_{j+1} - x_j), \tag{5.49}$$

where $f(z)$ is a nonlinear function of the spring strain z proportional to the spring elastic force, and x_j is the displacement of the jth ball from its equilibrium state. If the elastic force depends exponentially on strain, i.e.,

$$f(z) = -\alpha\left(1 - \mathrm{e}^{-z}\right), \tag{5.50}$$

then the chain is spoken of as a Toda chain [339, 340, 341, 342]. The Hamiltonian and equations of motion for the Toda chain are

$$H = \sum_j \left(\frac{p_j^2}{2} + \alpha\left(\mathrm{e}^{-(x_{j+1}-x_j)} - 1\right)\right), \tag{5.51}$$

$$\dot{p}_j = \alpha\left(\mathrm{e}^{-(x_j - x_{j-1})} - \mathrm{e}^{-(x_{j+1}-x_j)}\right), \tag{5.52}$$

where $p_j = m\dot{x}_j$ is the momentum of the jth ball.

Let us consider a particular case when the Toda chain with N elements is closed in a ring, i.e., $x_{j+N} = x_j$. In this case the law of total momentum conservation follows from Eq. (5.52). It is

$$p = \sum_{j=1}^{N} p_j = \text{const.} \tag{5.53}$$

Thus, we have two integrals of motion: the conservation laws of energy and momentum. The rest of the integrals of motion were found by Henon [118]. For example, the third integral of motion is

$$\sum_{j=1}^{N}\left(\frac{1}{3}p_j^3+\alpha\big(p_j+p_{j+1}\big)\mathrm{e}^{-(x_{j+1}-x_j)}\right)=\text{const.} \tag{5.54}$$

It should be noted that the physical meaning of this integral, as well as of all the rest, remains incomprehensible. The existence of N integrals of motion for the Toda chain with N elements testifies to the integrability of the equations of this chain. The rigorous proof of integrability for the Toda chain was given by Manakov [224] and Flashka [90] not only for a ring but for the linear chain as well.

Let us show that in the Toda chain stationary 'waves' akin to solitons in continuous media are possible. With this in mind let us consider the chain shown in Fig. 5.10. The equation of motion of the jth ball in this chain has the form of (5.49), where the elastic force $f(z)$ is determined by the expression (5.50). Toda [340] succeeded in solving Eqs. (5.49) by substituting a new variable y_j determined by the equation

$$\dot{y}_j=f(z_j), \tag{5.55}$$

where

$$z_j=x_j-x_{j-1} \tag{5.56}$$

is the strain of the jth spring (see [189]); $f(z)$ is defined by (5.50). With this change of variables Eqs. (5.49) become

$$m\dot{x}_j=y_j-y_{j+1}+mC, \tag{5.57}$$

where C is an arbitrary constant. Eliminating the variables x_j and z_j from Eqs. (5.56), (5.55) and (5.57), in view of (5.50), we find the following equations for y_j:

$$\ddot{y}_j=\frac{\alpha+\dot{y}_j}{m}\left(y_{j-1}-2y_j+y_{j+1}\right). \tag{5.58}$$

A partial solution of Eqs. (5.58) can be sought in the form of a 'running wave':

$$y_j(t)=\varphi(\xi_j), \tag{5.59}$$

where $\xi_j=\omega t-\beta j$, and φ is a periodic function of ξ_j with period 2π. Substituting (5.59) into (5.58) we obtain the following equation for the function $\varphi(\xi_j)$:

$$m\omega^2\varphi''=(\alpha+\omega\varphi')\Big(\varphi(\xi_j-\beta)-2\varphi(\xi_j)+\varphi(\xi_j+\beta)\Big), \tag{5.60}$$

where the prime indicates differentiation with respect to ξ_j.

Toda showed that a solution of Eq. (5.60) can be expressed in terms of the Jacobi elliptic zeta function [134] as

$$\varphi(\xi_j) = A\,\mathrm{zn}\left(\frac{\mathbf{K}(k)}{\pi}\,\xi_j, k\right), \tag{5.61}$$

where

$$\mathrm{zn}(\vartheta, k) = \int\limits_0^{\vartheta} (\mathrm{dn}(x,k))^2\,\mathrm{d}x - \frac{\mathbf{E}(k)}{\mathbf{K}(k)}\,\vartheta$$

is the Jacobi elliptic zeta function, and $\mathbf{K}(k)$ and $\mathbf{E}(k)$ are the full elliptic integrals of the first kind and of the second kind, respectively. Substituting (5.61) into Eq. (5.60) we obtain the following equations relating the amplitude A, the modulus of the elliptic function k, the frequency ω and the phase shift β:

$$A = \frac{m\omega\mathbf{K}(k)}{\pi}, \tag{5.62}$$

$$\omega = \frac{\pi\omega_0}{2\mathbf{K}(k)}\left[1 - \left(1 - \frac{\mathbf{E}(k)}{\mathbf{K}(k)}\right)\mathrm{sn}^2\left(\frac{\mathbf{K}(k)}{\pi}\,\beta, k\right)\right]^{-1/2}\mathrm{sn}\left(\frac{\mathbf{K}(k)}{\pi}\,\beta, k\right), \tag{5.63}$$

where $\omega_0 = 2\sqrt{\alpha/m}$. It is easy to verify that Eq. (5.63) for small k reduces to the dispersion equation for the corresponding linear chain (Eq. (5.37)). The change of dispersion dependencies with increase of the modulus k is shown in

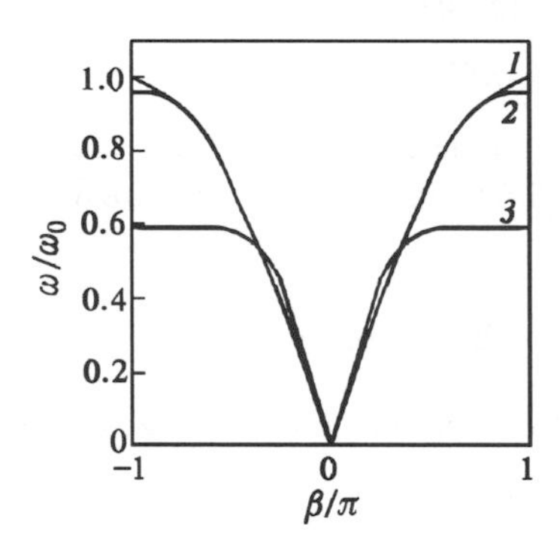

Fig. 5.11. The change of dispersion dependencies with increase of the modulus k for the Toda chain: the curves labeled 1, 2, and 3 correspond to $k^2 \leq 0.6$, $k^2 = 0.8$ and $k^2 = 1 - 10^{-5}$, respectively

Fig. 5.11. For $k < 0.8$ the dispersion dependencies are practically unchanged. If k increases further then the maximum value of ω decreases and a small near-horizontal portion of the dispersion curve appears. When k approaches unity the length of this portion increases and the value of ω associated with this portion decreases according to the formula

$$\omega = \frac{\omega_0 \pi}{2\sqrt{\mathbf{K}(k)}} . \tag{5.64}$$

It follows from (5.57) and (5.61) that the velocity of the jth ball is

$$\dot{x}_j(\xi_j) = \frac{A}{m}\left[\operatorname{zn}\left(\frac{\mathbf{K}(k)}{\pi}\,\xi_j, k\right) - \operatorname{zn}\left(\frac{\mathbf{K}(k)}{\pi}(\xi_j - \beta), k\right)\right] + C. \tag{5.65}$$

It is easily seen that $\dot{x}_j$ is a periodic function of ξ_j with period 2π.

The strain of the jth spring can be found from Eq. (5.55) and the expression (5.61). As a result we obtain the following equation:

$$\alpha(1 - \mathrm{e}^{-z_j}) = \frac{A\omega}{\pi}\left[\mathbf{E}(k) - \mathbf{K}(k)\operatorname{dn}^2\left(\frac{\mathbf{K}(k)}{\pi}\xi, k\right)\right]. \tag{5.66}$$

Taking account of Eq. (5.62), we find from (5.66):

$$z_j(\xi_j) = -\ln\left\{1 - \frac{4\omega^2\mathbf{K}^2(k)}{\omega_0^2\pi^2}\left[\frac{\mathbf{E}(k)}{\mathbf{K}(k)} - \operatorname{dn}^2\left(\frac{\mathbf{K}(k)}{\pi}\xi_j, k\right)\right]\right\}. \tag{5.67}$$

It is evident that z_j is also a periodic function of ξ_j with period 2π.

The dependencies of $\dot{x}_j$ and z_j on $\xi_j/(2\pi)$ constructed by the formulas (5.65) and (5.67) for two values of the modulus k are shown in Fig. 5.12. It is seen from Fig. 5.12 that for k close to unity, short pulses of compression of the springs alternate with long periods corresponding to expansion of the springs. The pulses of the spring compression are reminiscent, with respect to their shape, of solitons in the Korteweg–de Vries equation (see, e.g., [257, 189]). As k approaches unity the duration of the pulses decreases indefinitely and their period increases, but these changes follow a logarithmic law. It is interesting that the change of the ball velocity is similar to 'light' or 'dark' solitons, depending on β.

We note that $\mathbf{E}(k) \to 1$, $\mathbf{K}(k) \to \ln\big(4/\sqrt{1-k^2}\,\big)$, $\operatorname{dn}\vartheta \to 1/\cosh\vartheta$ and $\operatorname{zn}\vartheta \to \tanh\vartheta - \vartheta/\mathbf{K}(k)$ as $k \to 1$. From this it follows that, for $k \to 1$ and for the frequencies ω determined by the formula (5.64),

$$\dot{x}_j(\xi_j) \approx \frac{\omega_0\sqrt{\mathbf{K}(k)}}{2}\left[\sum_{n=-\infty}^{\infty} \tanh\left(\frac{\mathbf{K}(k)}{\pi}(\xi_j + 2n\pi)\right)\right.$$
$$\left. - \tanh\left(\frac{\mathbf{K}(k)}{\pi}(\xi_j + 2n\pi - \beta)\right) - \frac{\beta}{\pi}\right] + C. \tag{5.68}$$

The formula (5.68) allows us to calculate $x_j(t)$ analytically for k close to 1. Integrating (5.68) over t we obtain

$$x_j(\xi_j) \approx \sum_{n=-\infty}^{\infty} \ln\left[\frac{\cosh\big(\mathbf{K}(k)(\xi_j/\pi + 2n)\big)}{\cosh\big(\mathbf{K}(k)(\xi_j/\pi + 2n - \beta/\pi)\big)}\right] - \frac{\beta\mathbf{K}(k)}{\pi^2}\,\xi_j - x_0 + Ct,$$

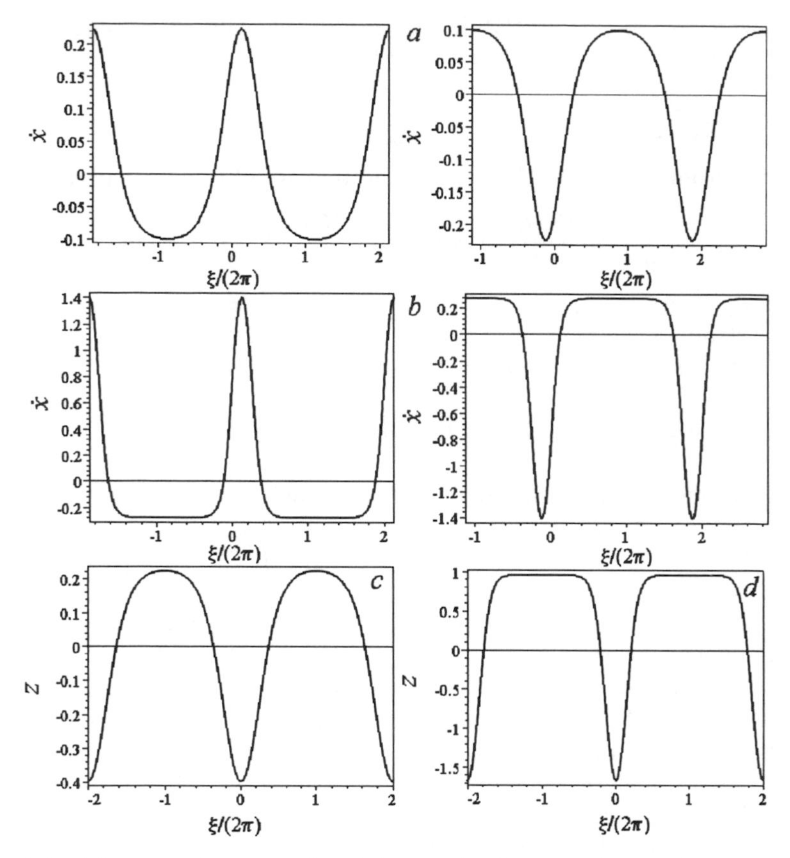

Fig. 5.12. The dependencies of (a, b) $\dot{x}_j$ and (c, d) z_j on ξ/π for $C = 0$ and (a, c) $k = 0.99$ and (b, d) $k = 1 - 10^{-7}$. Shown in (a, b) are 'light' solitons for $\beta = \pi/4$ (on the left) and 'dark' solitons for $\beta = 7\pi/4$ (on the right)

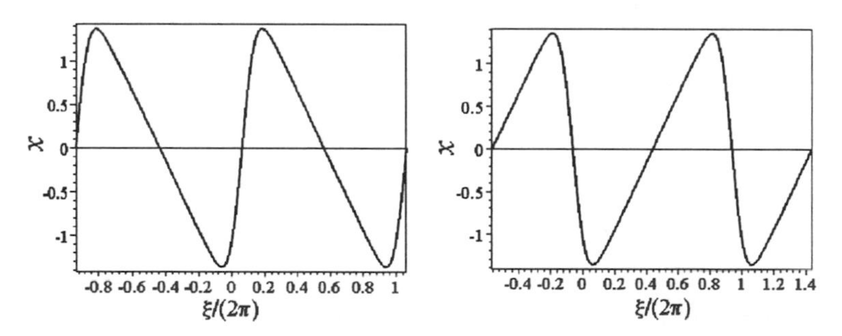

Fig. 5.13. The plot of x_j versus ξ_j for $C = 0$, $k = 1 - 10^{-7}$, $\beta = \pi/4$ (on the left) and $\beta = 7\pi/4$ (on the right)

(5.69)

where

$$x_0 = \frac{1}{2\pi}\int_0^{2\pi}\sum_{n=-\infty}^{\infty}\ln\left[\frac{\cosh\big(\mathbf{K}(k)(\xi_j/\pi+2n)\big)}{\cosh\big(\mathbf{K}(k)(\xi_j/\pi+2n-\beta/\pi)\big)}\right] - \frac{\beta\mathbf{K}(k)}{\pi^2}\,\xi_j\,\mathrm{d}\xi_j.$$

The plot of $x_j(\xi_j)$ is illustrated in Fig. 5.13 for $C=0$ and $k=1-10^{-7}$. We see that oscillations of x_j have a saw-tooth form.

As already noted, the Toda chain equations are completely integrable. However, the property of the integrability is nonrough and a small change in the elastic forces can lead immediately to nonintegrability of the system. So, Fermi, Pasta and Ulam [87] considered a chain of oscillators with quadratic nonlinearity, the equations for which were obtained from Eqs. (5.52) by expanding the exponents into a series taking account of linear and squared terms only and putting $\beta=-1$, $\alpha=1$. These equations are

$$\ddot{x}_j = x_{j+1} - 2x_j + x_{j-1} + \frac{1}{2}(x_{j+1}-x_j)^2 - \frac{1}{2}(x_j - x_{j-1})^2 \quad (j=1,2,\ldots). \tag{5.70}$$

Another chain simulated by Fermi, Pasta and Ulam [87] contains cubic nonlinearities. Its equations are

$$\ddot{x}_j = x_{j+1} - 2x_j + x_{j-1} + \alpha\left((x_{j+1}-x_j)^3 - (x_j-x_{j-1})^3\right) \quad (j=1,2,\ldots). \tag{5.71}$$

The systems (5.70) and (5.71) were found to be nonintegrable although they are very close to integrable ones [224]. It is precisely this fact that explains the failure of the well-known Fermi–Pasta–Ulam numerical experiment, by means of which they tried to demonstrate the fundamental theorem of thermodynamics, namely, the theorem concerning the uniform distribution of energy throughout all degrees of freedom. By presetting a perturbation approximating one of the modes of oscillations for the generating linear system, Fermi, Pasta and Ulam expected that, because of nonlinearity, this perturbation would be uniformly distributed throughout all modes with time, i.e., stochastization of oscillations would occur. At first the perturbation indeed did transform to the other modes, but in due course it again focused almost entirely on the initial mode. As was shown in succeeding works, the Fermi–Pasta–Ulam experiment fails for the reason that, because Eqs. (5.70) and (5.71) are sufficiently close to integrable ones, a moderately prolonged time and moderately large energy are necessary for stochastization of oscillations in these chains.

5.4 Stochasticity in Hamiltonian systems close to integrable ones

If a Hamiltonian system is nonintegrable, stochasization of oscillations is possible. We consider several examples of this phenomenon.

First of all, we continue consideration of the Fermi–Pasta–Ulam chains and dwell on the results of the paper [61]. In this paper some particular solutions of Eqs. (5.70) and (5.71) are obtained and quantitative assessments of oscillation energy needed for stochastization are given. These particular solutions are associated with the highest modes of ring chains from N cells. For these modes $\beta = \pi$, i.e., $x_j(t) = -x_{j+1}(t) \equiv x(t)$. In the case of a chain described by Eq. (5.70) $x(t)$ must obey the equation $\ddot{x} + 4x = 0$. It follows from this equation that $x = A\cos 2t$. Investigation of the stability of this solution revealed that it becomes unstable as $A^2 > E_{\mathrm{cr}}/2N$, where $E_{\mathrm{cr}} = \pi(2-\sqrt{3})$ is the critical value of the oscillation energy E; this latter is $E = 2NA^2$. In the case of the chain described by Eq. (5.71) $x(t)$ must obey the Duffing equation

$$\ddot{x} + 4x + 16\alpha x^3 = 0. \tag{5.72}$$

A solution of Eq. (5.72) for $\alpha > 0$ is expressed in terms of the Jacobi elliptic cosine (see (4.9)) as $x = B\mathrm{cn}(\Omega t, k)$, where $\Omega = 2\sqrt{1+4\alpha B^2}$, $k = B\sqrt{2\alpha}/\sqrt{1+4\alpha B^2}$. The energy associated with this solution is $E = 2NB^2(1+2\alpha B^2) = Nk^2(1-k^2)/\alpha(1-2k^2)^2$. This solution loses stability as $k > k_{\mathrm{cr}}$, where

$$k_{\mathrm{cr}}^2 = \frac{2\pi^3}{\left(\sqrt{96}+3\pi\right)N^2} \approx \frac{3.226}{N^2}.$$

It follows from this that $k_{\mathrm{cr}} \ll 1$ for large N. In this case $E_{\mathrm{cr}} \approx 3.226/\alpha N$. A periodic solution of Eq. (5.72) for $\alpha < 0$ is expressed in terms of the Jacobi elliptic sine (see (4.12)). It can be shown that for this solution $E_{\mathrm{cr}} \approx 0.214N/\alpha$. Hence, for sufficiently large N, in the case $\alpha > 0$ chaos has to appear for significantly smaller values of the nonlinear parameter than for $\alpha < 0$.

Below we give a number of examples of the appearance of stochasticity in Hamiltonian systems close to integrable ones.

5.4.1 The ring Toda chain and the Henon–Heiles system

That even a very small departure of a Hamiltonian system from integrable one can cause its stochastic behavior was first justified numerically by Henon and Heiles [117]. The system studied by them can be obtained by using certain approximations in the equations of a Toda ring chain consisting of three elements.

Let us consider a chain involving three identical balls of mass m located on the circumference and connected by springs (Fig. 5.14). If the elastic forces

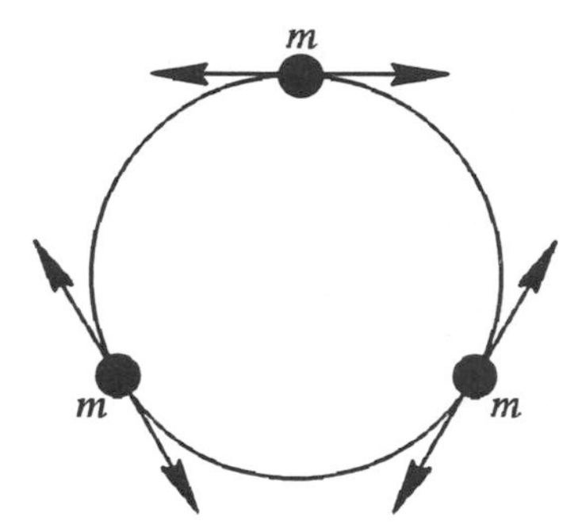

Fig. 5.14. The Toda ring chain of three elements

acting between the balls are supposedly dependent exponentially on distance, such a chain is a Toda chain. In this case the equations of motion of the balls can be written as

$$\begin{aligned} \ddot{x}_1 &= f(x_1 - x_3) - f(x_2 - x_1), \\ \ddot{x}_2 &= f(x_2 - x_1) - f(x_3 - x_2), \\ \ddot{x}_3 &= f(x_3 - x_2) - f(x_1 - x_3), \end{aligned} \tag{5.73}$$

where x_j is the deviation of the jth ball from its equilibrium position,

$$f(z) = -\alpha \left(1 - \mathrm{e}^{-\beta z}\right). \tag{5.74}$$

The Hamiltonian of the system (5.73) has the form

$$H = \sum_{j=1}^{3} \left(\frac{p_j^2}{2} + \frac{\alpha}{\beta} \left(\mathrm{e}^{-\beta(x_{j+1} - x_j)} - 1 \right) \right), \tag{5.75}$$

where $p_j = \dot{x}_j$, $\alpha = k/\beta$, $x_4 = x_1$.

From Eqs. (5.73) we obtain the following conservation laws of energy and total momentum:

$$H = E = \text{const}, \quad p = \sum_{j=1}^{3} p_j = \text{const}. \tag{5.76}$$

The third integral of motion found by Henon [118] is

$$\sum_{j=1}^{3} \left(\frac{p_j^3}{3} + \alpha(p_j + p_{j+1}) \mathrm{e}^{-\beta(x_{j+1} - x_j)} \right) = \text{const}. \tag{5.77}$$

The existence of three integrals of motion for the Toda chain involving three elements is evidence of its integrability.

To obtain the Henon–Heiles system we introduce new generalized coordinates x, y and momenta p_x, p_y by the formulas

$$x = -\frac{\sqrt{3}\beta}{4}(x_1 + x_2), \quad y = -\frac{\beta}{4}(x_1 - x_2), \quad p_x = \frac{\mathrm{d}x}{\mathrm{d}\tau}, \quad p_y = \frac{\mathrm{d}y}{\mathrm{d}\tau},$$

where $\tau = \beta\sqrt{3\alpha}t$. Assuming that the total momentum is zero, i.e., $p_3 = -(p_1 + p_2)$, we can put $x_3 = -(x_1 + x_2)$. It immediately follows from this that

$$\beta(x_3 - x_1) = 2(y + \sqrt{3}x), \quad \beta(x_2 - x_3) = 2(y - \sqrt{3}x),$$
$$\frac{1}{2}(p_1^2 + p_2^2 + p_3^2) = 12\alpha(p_x^2 + p_y^2).$$

Substituting these expressions into (5.75), we obtain the Hamiltonian of the

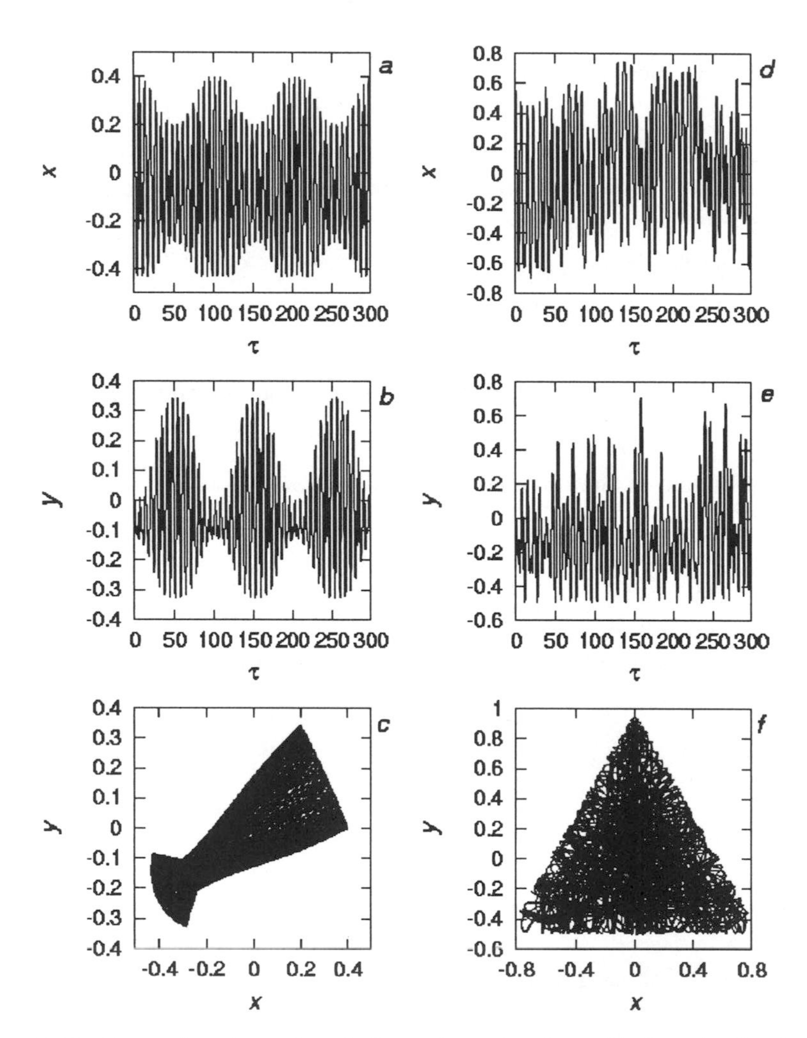

Fig. 5.15. Examples of quasi-periodic and stochastic oscillations in the Henon–Heiles system (5.80) for (a, b, c) $x(0) = 0.4$ and (d, e, f) $x(0) = 0.575$: the plots are (a, d) $x(\tau)$, (b, e) $y(\tau)$ and (c, f) the projections of the phase portraits on the plane xy

system in new variables. It is conveniently written as

$$H = \frac{p_x^2 + p_y^2}{2} + \frac{1}{24}\Bigg(\exp\Big(2\big(y + \sqrt{3}\,x\big)\Big) + \exp\Big(2\big(y - \sqrt{3}\,x\big)\Big) + \exp\Big(-4y\Big) - 3\Bigg). \tag{5.78}$$

Expanding the exponents in expression (5.78) as a power series and restricting ourselves to cubic terms of the expansion, we obtain the Henon–Heiles Hamiltonian:

$$H = \frac{p_x^2 + p_y^2}{2} + \frac{x^2 + y^2}{2} + x^2 y - \frac{y^3}{3}. \tag{5.79}$$

The equations of motion associated with this Hamiltonian are

$$\ddot{x} = -x(1 + 2y), \quad \ddot{y} = -(y + x^2 - y^2), \tag{5.80}$$

where the dots denote differentiation with respect to time τ.

The simulation of Eqs. (5.80), performed first by Henon and Heiles [117], showed that for sufficiently large values of energy the solution of these equations is of a chaotic character. This means that the system is nonintegrable. The results of a recent replication of the simulation of Eqs. (5.80) made by the author are shown in Fig. 5.15 for two initial values of the variable x: for $x = 0.4$ (a) and $x = 0.575$ (b). We see that in the first case the oscillations are quasi-periodic, corresponding to a two-dimensional torus in the system phase space. In the second case the oscillations are stochastic with a corresponding stochastic set in the system phase space.

Computation of Eqs. (5.49) performed by the author for $N = 4$ showed that these equations have stochastic solutions in a wide range of initial conditions. This means that even small errors arising from the fact that the solution is found numerically make the system to be nonintegrable. Examples of such solutions are illustrated in Fig. 5.16.

5.4.2 Stochastization of oscillations in the Yang–Mills equations

Stochastic oscillations in the Yang–Mills system (5.15) were first detected numerically by Chirikov and Shepelyansky [66] and later by Nicolaevsky and Shchur [256]. This suggests that this system is nonintegrable.

From the results presented here it may be deduced that the Hamiltonian models of certain systems, as a rule, are nonrough and the behavior of the corresponding real system can differ essentially from the behavior of the model.

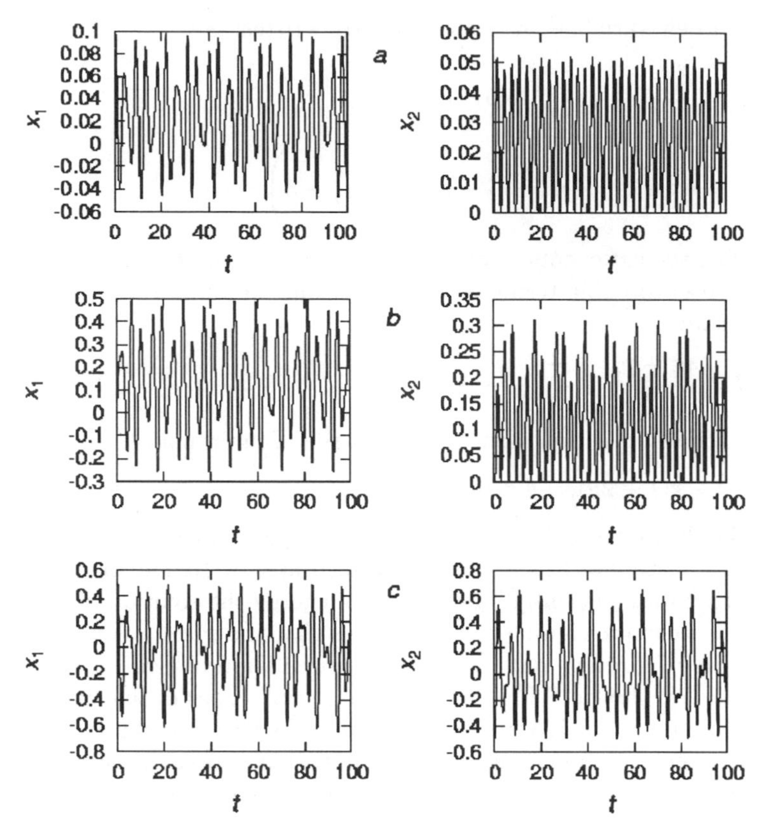

Fig. 5.16. Examples of stochastic solutions of Eqs. (5.49) for (a) $x_1(0) = 0.1$, $x_2(0) = x_3(0) = x_4(0) = \dot{x}_1(0) = \dot{x}_2(0) = \dot{x}_3(0) = \dot{x}_4(0) = 0$, (b) $x_1(0) = 0.5$, $x_2(0) = x_3(0) = x_4(0) = \dot{x}_1(0) = \dot{x}_2(0) = \dot{x}_3(0) = \dot{x}_4(0) = 0$, and (c) $x_1(0) = 0.5$, $x_2(0) = -0.5$, $x_3(0) = x_4(0) = \dot{x}_1(0) = \dot{x}_2(0) = \dot{x}_3(0) = \dot{x}_4(0) = 0$

6. Self-oscillatory systems with one degree of freedom

Studies of self-oscillatory systems with one degree of freedom are markedly simpler than these of systems with two or more degrees of freedom. Therefore, whenever feasible, investigators endeavor to describe any real system by a model with one degree of freedom. In many cases such a model adequately depicts the most essential features of the processes studied.

Classical examples of self-oscillatory systems described adequately by models with one degree of freedom are simple electronic generators [5, 8, 334, 166], the Froude pendulum [5, 8, 321, 322, 166] and some types of clock [5, 8]. However, a plethora of other, more complicated, systems have been successfully described by models with one degree of freedom as well.

6.1 The van der Pol, Rayleigh and Bautin equations

Vacuum tube generators are classical examples of self-oscillatory systems which in certain simple cases can be described by the van der Pol equation or by the Rayleigh equation. These are precisely the systems which were studied in the first works of van der Pol [350, 353]. In more recent times vacuum tubes have been replaced by transistors, but in essence the operation of these electronic generators remains the same.

The van der Pol and Rayleigh equations are respectively

$$\ddot{x} - \mu(1 - \alpha x^2)\dot{x} + x = 0 \tag{6.1}$$

and

$$\ddot{y} - \mu(1 - \alpha \dot{y}^2)\dot{y} + y = 0. \tag{6.2}$$

It should be noted that the Rayleigh equation can be obtained from the van der Pol equation by the substitution $x = \sqrt{3}\,\dot{y}$. Equations (6.1) and (6.2) describe the processes both of oscillation self-excitation (for $\mu > 0$) and of limitation of their amplitude (for $\alpha > 0$).

The phase portraits for Eqs. (6.1) and (6.2) are much the same for $\mu \ll 1$ (see Fig. 6.1 a) and essentially different for $\mu > 1$ (Fig. 6.1 b). In the first case the equilibrium state is associated with an unstable focus and the limit cycle has a near-elliptic shape; before it approaches the representative point executes many revolutions around the focus. In the second case the

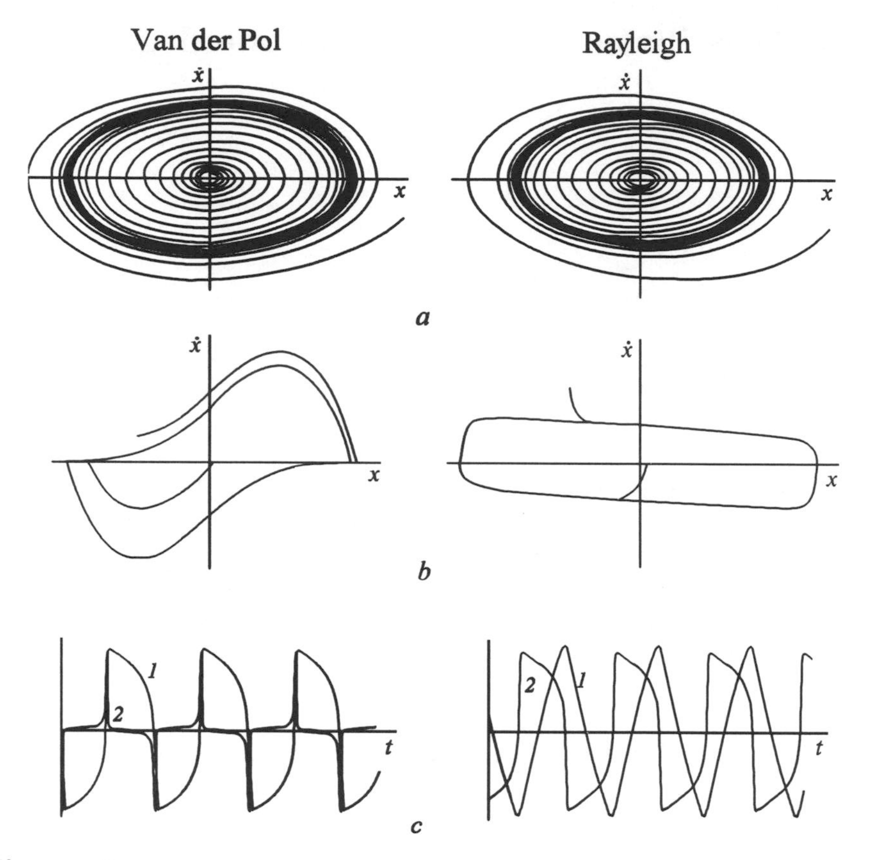

Fig. 6.1. The phase portraits for the van der Pol and Rayleigh equations for (a) $\mu \ll 1$ and (b) $\mu = 5$; (c) the solutions of the van der Pol and Rayleigh equations for $\mu = 5$: the curves 1 and 2 show $x(t)$ and $\dot{x}(t)$, respectively

equilibrium state is associated with an unstable node and the limit cycle has a complicated shape; the time it takes for the representative point to approach the limit cycle is very short. For example, the solutions of the van der Pol and the Rayleigh equations for $\mu = 5$ are shown in Fig. 6.1 c. We see that in the case of $\mu \gg 1$ the oscillations closely resemble discontinuous ones in shape. Whereas for $\mu \ll 1$ the self-oscillation period is practically independent of μ and approximately equal to the period of natural oscillations of the oscillatory circuit $T_0 = 2\pi$, for $\mu \gg 1$ the self-oscillation period is completely determined by the value of the parameter μ, namely $T \approx 8\mu/3\sqrt{3}$ (see below). It can be shown by considering the process of relaxation to stationary self-oscillations that the relaxation time is of the order of μ^{-1}. Thus, for $\mu \gg 1$ the self-oscillation period is inversely proportional to the relaxation time. Owing to such a dependence of the self-oscillation period on the relaxation time the term 'relaxation self-oscillations' was coined.

A less known model of the simplest self-oscillatory system is that described by the Bautin equations [30, 7]:

$$\dot{x} = \omega y + a^2 x - x(x^2 + y^2), \quad \dot{y} = -\omega x + a^2 y - y(x^2 + y^2). \tag{6.3}$$

These equations are of interest because they have an exact solution. Indeed, introducing the amplitude A and phase φ by the formulas $x = A\cos\varphi$, $y = -A\sin\varphi$, we obtain for A and φ the following equations:

$$\dot{A} = (a^2 - A^2)A, \quad \dot{\varphi} = \omega. \tag{6.4}$$

These equations are easily integrated. Their solution is

$$A = a\Big(1 + C\exp(-2a^2 t)\Big)^{-1/2}, \quad \varphi = \omega t, \tag{6.5}$$

where $C = (a^2 - A_0^2)/A_0^2$, and A_0 is the value of A at the initial instant. It follows from (6.5) and from the expressions for x and y that the stationary solution of Eqs. (6.3) is harmonic oscillations with frequency ω and amplitude a.

The van der Pol and Rayleigh equations are of primary importance in the theory of self-oscillations because they reflect the most fundamental features of all self-oscillatory systems with soft excitation, namely the possibility of self-excitation of oscillations and the possibility of limitation of the amplitude of oscillations. Furthermore, these equations are capable of describing self-oscillatory systems of both Thomsonian (for $\mu \ll 1$) and relaxation (for $\mu \gg 1$) types. It follows from the above that in Thomsonian systems the energy change during an oscillation 'period' is small in comparison with the energy stored in the system. This fact is reflected in the smallness of the parameter μ and makes itself evident in the phase portrait (see Fig. 6.1 a) as well as in the plot of the Poincaré point map for $\mu \ll 1$ (see Fig. 6.2 a). This plot was constructed in the following manner: the line $x = 0$ was chosen as the Poincaré cutting line and the dependence of the nth intersection point coordinate $\dot{x}_n$ on the $(n-1)$th intersection point coordinate $\dot{x}_{n-1}$ was plotted. It is seen from Fig. 6.2 a that the plot of the point map is close to the bisectrix, and the transient process covers a great many 'periods' of

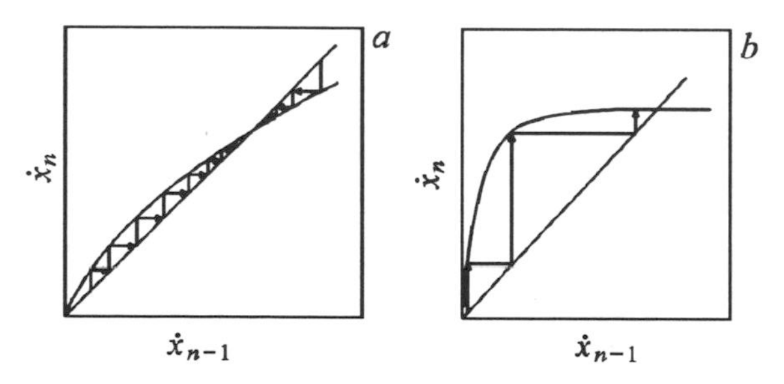

Fig. 6.2. The qualitative plots of the point maps for the van der Pol and Rayleigh equations in the cases of (a) $\mu \ll 1$ and (b) $\mu \sim 1$

oscillations. For $\mu \sim 1$ the qualitative plots of the point map are shown in Fig. 6.2 b.

6.1.1 The Kaidanovsky–Khaikin frictional generator and the Froude pendulum

A mechanical self-oscillatory system of frictional type was proposed by Kaidanovsky and Khaikin as early as 1933 [136]. It consists of a mass m fixed to a wall by a spring of stiffness k and lying over a transmission belt moving with a certain velocity v (see Fig. 6.3 a). The equation of the oscillations of the

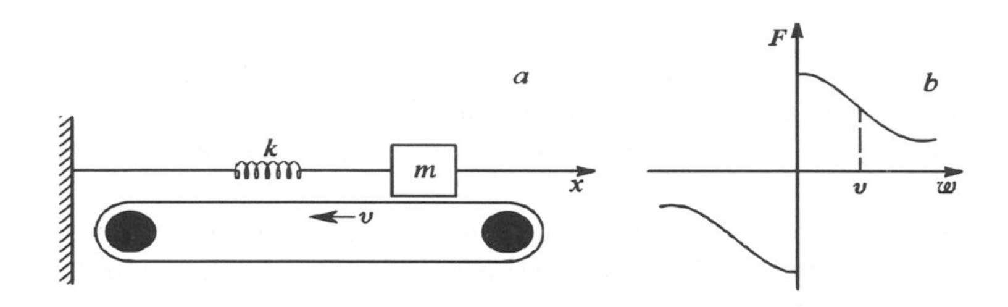

Fig. 6.3. (a) The Kaidanovsky–Khaikin frictional self-oscillatory system; (b) the dependence of the dry friction force F on the relative velocity $w = \dot{x} - v$

mass is

$$m\ddot{x} + h\dot{x} + F(\dot{x} - v) + kx = 0, \tag{6.6}$$

where $-h\dot{x}$ is the friction force of the mass m on the air, and $-F(\dot{x} - v)$ is the dry friction force of the mass m on the belt, depending on the relative velocity $w = \dot{x} - v$. The dependence $F(w)$ can, under certain conditions, have the form shown in Fig. 6.3 b, i.e., it can contain a dropping part associated with negative friction. Representing $F(\dot{x} - v)$ for small $\dot{x}$ as

$$F(\dot{x} - v) = F(v) - \gamma(v)\dot{x} + \alpha(v)\dot{x}^3, \tag{6.7}$$

we can reduce Eq. (6.6) to the Rayleigh equation.

Another frictional self-oscillatory system is the so-called *Froude pendulum* described even by Rayleigh [279]. Its arrangement is shown in Fig. 6.4. The clutch B is fitted on an uniformly rotating shaft A. The moment of the friction force of the shaft on the clutch depends on the angular velocity Ω of the shaft rotation relative to the clutch. Usually this dependence has a form which is similar to that shown in Fig. 6.3 b. By approximating it much like (6.7) and scaling φ and $\dot{\varphi}$ as needed, we obtain the Rayleigh equation for the pendulum angular deviation φ.

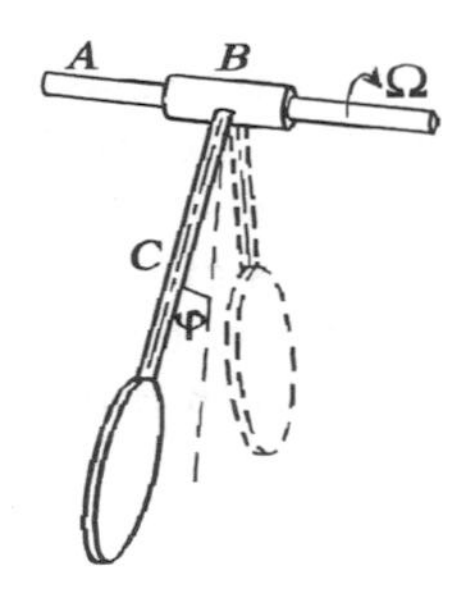

Fig. 6.4. Schematic image of the Froude pendulum

6.2 Soft and hard excitation of self-oscillations

After transition through the self-excitation threshold the amplitude of self-oscillations described by both the van der Pol equation and the Rayleigh equation increases smoothly (Fig. 6.5). This means that the excitation of self-oscillations is *soft*.

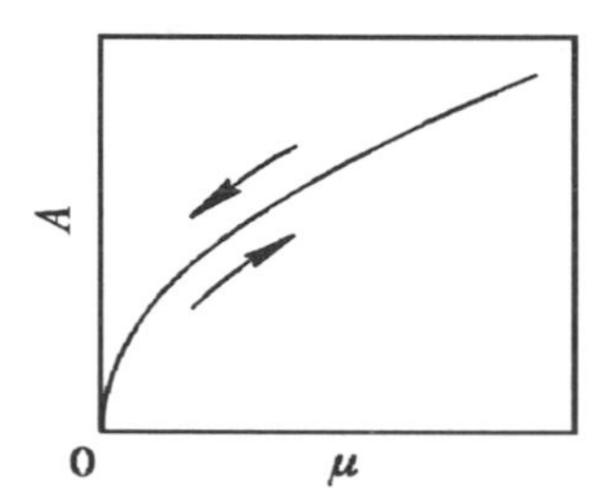

Fig. 6.5. The qualitative dependence of the self-oscillation amplitude A on exceeding the self-excitation threshold μ for the van der Pol and Rayleigh equations

If the coefficient α in the van der Pol equation is negative then this equation cannot describe the process of limitation of the self-oscillation amplitude: as the amplitude increases, the increment increases too. To describe the amplitude limitation in this case we consider the equation of the more general form:

$$\ddot{x} - (\eta - \alpha x^2 - \beta x^4)\dot{x} + x = 0. \tag{6.8}$$

The phase portrait of a system described by Eq. (6.8) for $\alpha < 0$, $\beta > 0$, is shown in Fig. 6.6 a. If $\eta^* < \eta < 0$, where η^* is a certain critical value of the parameter η determined by parameters α and β, then the system equilibrium state is stable and the phase portrait contains both a stable and unstable limit cycle. For small deviations from the equilibrium state the oscillations are damped and the system returns to its equilibrium state. For sufficiently large initial deviations the representative point goes into the stable limit cycle in the course of time, i.e., self-oscillations are excited. For $\eta = 0$ the unstable limit cycle fuses with the stable equilibrium state and transmits its instability to it. As a result only the stable limit cycle remains (Fig. 6.6 b). The dependencies of the amplitudes of the stable and unstable limit cycles on the parameter η are shown in Fig. 6.6 c. We see that beyond transition through the self-excitation threshold ($\eta = 0$) the amplitude of the stationary

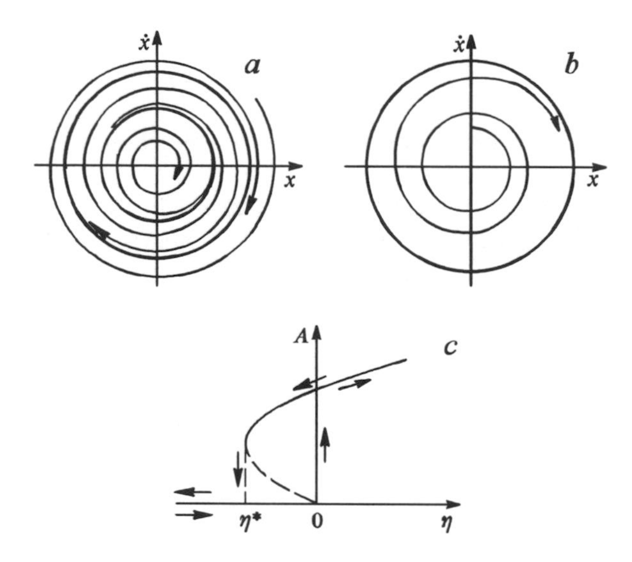

Fig. 6.6. The phase portrait of the system described by Eq. (6.8) for (a) $\alpha < 0$, $\beta > 0$, $\eta^* < \eta < 0$, and (b) for $\alpha < 0$, $\beta > 0$, $\eta > 0$; (c) the dependencies of the amplitudes of the stable (solid line) and unstable (dashed line) limit cycles on the parameter η

self-oscillations increases by a jump, if prior to the transition the system has been in the equilibrium state. Hence, Eq. (6.8) for $\alpha < 0$, $\beta > 0$ describes a system with *hard* excitation. In the case of hard excitation the self-oscillations can also exist when there is no self-excitation (for $\mu < 0$), but for their excitation an initial perturbation of a certain finite value must be applied to the system. The possibility of hard excitation is caused by the presence of nonlinear negative friction in the system described by the term αx^2. So, for $\mu < 0$ the self-excitation of oscillations is absent, but a stable limit cycle exists in the system phase space. For $\mu = 0$ the so-called *reverse Andronov bifurcation* occurs (see Ch. 3) and the system becomes a self-excited one.

6.3 Truncated equations for the oscillation amplitude and phase

6.3.1 Quasi-linear systems

Equations of the form (6.1), (6.2) and (6.8) are special cases of a more general equation of a self-oscillatory system with one degree of freedom which can be written as

$$\ddot{x} - \mu\Psi(x,\dot{x})\dot{x} + \omega_0^2\Big(1 + \epsilon F(x)\Big)x = 0, \tag{6.9}$$

where $\Psi(x,\dot{x})$ and $F(x)$ are nonlinear functions. For small μ and ϵ this equation describes a quasi-linear and quasi-conservative system, i.e., a system of Thomsonian type. First we consider just this case.

For small μ and ϵ we can use for finding an approximate solution of Eq. (6.9) one of the methods described in Ch. 2, e.g., the averaging method. To use this method we substitute into Eq. (6.9) new variables A and φ in place of x and $\dot{x}$ according to the formulas

$$x = A\cos(\omega_0 t + \varphi), \quad \dot{x} = -A\omega_0\sin(\omega_0 t + \varphi). \tag{6.10}$$

For the variables A and φ we obtain the following exact equations:

$$\begin{aligned} \dot{A} &= \mu A\Psi(A\cos\psi, -A\omega_0\sin\psi)\sin^2\psi + \epsilon\omega_0^2 AF(A\cos\psi)\sin\psi\cos\psi, \\ \dot{\varphi} &= \mu\Psi(A\cos\psi, -A\omega_0\sin\psi)\sin\psi\,\cos\psi + \epsilon\omega_0^2 AF(A\cos\psi)\cos^2\psi, \end{aligned} \tag{6.11}$$

where $\psi = \omega_0 t + \varphi$.

In the first approximation the averaged equations are

$$\begin{aligned} \dot{A} &= \frac{\mu A}{T}\int_0^T \Psi(A\cos\psi, -A\omega_0\sin\psi)\sin^2\psi\,\mathrm{d}t \\ &\quad + \frac{\epsilon\omega_0^2 A}{T}\int_0^T F(A\cos\psi)\sin\psi\cos\psi\,\mathrm{d}t, \\ \dot{\varphi} &= \frac{\mu}{T}\int_0^T \Psi(A\cos\psi, -A\omega_0\sin\psi)\sin\psi\,\cos\psi\,\mathrm{d}t \\ &\quad + \frac{\epsilon\omega_0^2}{T}\int_0^T F(A\cos\psi)\cos^2\psi\,\mathrm{d}t, \end{aligned} \tag{6.12}$$

where $T = 2\pi/\omega_0$.

If $\Psi(x,\dot{x})$ and $F(x)$ are polynomials then, as follows from (6.12), only those terms of these polynomials enter into Eqs. (6.12) for which the summary exponent is even. The terms with odd exponents influence only higher approximations.

Let the function $F(x)$ be of parabolic form and $\Psi(x,\dot{x})$ depend only on x and be a polynomial of degree 4, i.e.,

$$F(x) = ax^2, \quad \Psi(x) = 1 + \beta x - \alpha x^2 + \delta x^3 + \gamma x^4. \tag{6.13}$$

In this case Eqs. (6.12) become

$$\dot{A} = \frac{\mu}{2}\left(1 - \frac{\alpha A^2}{4} + \frac{\gamma A^4}{8}\right) A, \tag{6.14a}$$

$$\dot{\varphi} = \frac{3\epsilon\omega_0^2 a A^2}{8}. \tag{6.14b}$$

It is seen that, indeed, the terms βx and δx^3 do not enter into the equations of the first approximation.

It follows from Eq. (6.14b) that for $\epsilon = 0$ the correction to the frequency, $\dot{\varphi}$, is equal to zero, i.e., the oscillation frequency is independent of the oscillation amplitude. Such oscillations are spoken of as *isochronous*. For $\epsilon \neq 0$ the oscillations become nonisochronous.

Equation (6.14a) possesses three steady-state solutions:

$$A_1 = 0, \tag{6.15a}$$

$$A_2 = \sqrt{\frac{\alpha}{\gamma}\left(1 - \sqrt{1 - \frac{8\gamma}{\alpha^2}}\right)}, \tag{6.15b}$$

$$A_3 = \sqrt{\frac{\alpha}{\gamma}\left(1 + \sqrt{1 - \frac{8\gamma}{\alpha^2}}\right)}, \tag{6.15c}$$

The solution (6.15a) corresponds to the system equilibrium state $x = 0$, $\dot{x} = 0$. It is stable for $\mu \leq 0$ and unstable for $\mu > 0$. The solutions (6.15b) and (6.15c) describe possible stationary oscillation amplitudes. These solutions must be real. A necessary condition for this is $\gamma < \alpha^2/8$. For $\alpha > 0$ and $\gamma > 0$ this condition is also sufficient. For $\mu > 0$, the solution (6.15b) determines the stable limit cycle amplitude, and the solution (6.15c) determines the unstable limit cycle amplitude. In this case the amplitude of the unstable limit cycle is greater than the amplitude of the stable limit cycle. This means that for initial conditions outside the unstable limit cycle the oscillations increase indefinitely. The latter points to the fact that nonlinear terms in the original equations are taken incompletely into account. If we add the necessary terms, we find another stable limit cycle. Then two modes of self-oscillations with different amplitudes become possible. One of these modes is excited softly, whereas the other is excited hardly.

For $\mu < 0$, the solution (6.15b) determines the unstable limit cycle amplitude, and the solution (6.15c) determines the stable limit cycle amplitude, i.e., in this case only hard excitation of oscillations is possible.

For $\alpha > 0$, $\gamma < 0$ only the solution (6.15b) is real; for $\mu > 0$ it determines the stable cycle amplitude, and for $\mu < 0$ it determines the unstable cycle amplitude.

For $\alpha < 0$, $\gamma < 0$ only the solution (6.15c) is real; for $\mu > 0$ it determines the stable cycle amplitude, and for $\mu < 0$ it determines the unstable cycle amplitude.

For $\alpha < 0$, $\gamma > 0$ real values of the stationary amplitudes are absent.

Let us further derive the equations for the oscillation amplitude and phase in the second approximation. For the sake of simplicity, in (6.13) the terms δx^3 and γx^4 will not be taken into account. Substituting (6.13) into Eqs. (6.11), we obtain

$$\dot{A} = \mu\left(1 + \beta A \cos\psi - \alpha A^2 \cos^2\psi\right) A \sin^2\psi + \epsilon\omega_0^2 a A^3 \sin\psi \cos^3\psi,$$

$$\dot{\varphi} = \mu\left(1 + \beta A \cos\psi - \alpha A^2 \cos^2\psi\right) \sin\psi \cos\psi + \epsilon\omega_0^2 a A^2 \cos^4\psi. \tag{6.16}$$

To find the equations of the second approximation, we integrate Eqs. (6.16) over 'fast' time, ignoring slowly varying constituents in the right-hand sides of these equations. As a result, we obtain the following expressions for the fast varying constituents of A and φ:

$$\tilde{A} = -\frac{\mu A}{4\omega_0}\left(\sin 2\psi - \beta A \sin\psi + \frac{\beta A}{3}\sin 3\psi - \frac{\alpha A^2}{8}\sin 4\psi\right) - \frac{\epsilon\omega_0 a A^3}{8}\left(\cos 2\psi + \frac{1}{4}\cos 4\psi\right),$$

$$\tilde{\varphi} = -\frac{\mu}{4\omega_0}\left[\left(1 - \frac{\alpha A^2}{2}\right)\cos 2\psi + \beta A \cos\psi + \frac{\beta A}{3}\cos 3\psi - \frac{\alpha A^2}{8}\cos 4\psi\right] + \frac{\epsilon\omega_0 a A^2}{4}\left(\sin 2\psi + \frac{1}{8}\sin 4\psi\right). \tag{6.17}$$

Further we substitute into the right-hand sides of Eqs. (6.16) $A + \tilde{A}$ and $\varphi + \tilde{\varphi}$ in place of A and φ and average over time. As a result, we obtain the following equations of the second approximation:

$$\dot{A} = \frac{\mu}{2}\left(1 - \frac{\alpha A^2}{4} + \frac{\epsilon\omega_0^2 a\alpha A^4}{16}\right) A,$$

$$\dot{\varphi} = -\frac{\mu^2}{8\omega_0}\left(1 - \frac{3}{2}\alpha A^2 + \frac{11}{32}\alpha^2 A^4 + \frac{\beta^2 A^2}{3}\right) + \frac{3\epsilon\omega_0 a A^2}{8}\left(1 - \frac{17\epsilon\omega_0^2 a A^2}{32}\right). \tag{6.18}$$

It follows from (6.18) that, for $\epsilon = 0$, the equation for the oscillation amplitude in the second approximation is the same as before, whereas in the equation for the phase additional terms describing a correction to the oscillation frequency appear. Since these terms depend on the amplitude, the oscillations become nonisochronous even for $\epsilon = 0$.

We note that the corresponding equations of the second approximation obtained by the Krylov–Bogolyubov asymptotic method (for $\beta = 0$ and $\epsilon = 0$) [53] differ from (6.18) by the coefficients (1 in place of 3/2 and 7/32 in place of 11/32).

6.3.2 Transient processes in the van der Pol generator

It follows from Eq. (6.14a) that for the van der Pol generator the truncated equation for the oscillation amplitude can be written as

$$\dot{A} = \frac{\mu}{2}\left(1 - \frac{A^2}{A_{\mathrm{st}}^2}\right)A, \tag{6.19}$$

where $A_{\mathrm{st}} = 2/\sqrt{\alpha}$ is the stationary value of the oscillation amplitude.

Let us denote the initial value of the oscillation amplitude by A_0 and integrate Eq. (6.19). As a result, we obtain

$$A(t) = A_0 \mathrm{e}^{\mu t/2}\left(1 + \frac{A_0^2}{A_{\mathrm{st}}^2}(\mathrm{e}^{\mu t} - 1)\right)^{-1/2}. \tag{6.20}$$

It is seen from (6.20) that the excitation of self-oscillations is possible only if there is a certain initial deviation from the equilibrium state. Because there are fluctuations in any real system, this condition is always fulfilled. The time needed for the amplitude to be close to its stationary value, which we will call the *relaxation time*, depends on A_0 only slightly.

Let us define the relaxation time as the time the amplitude squared takes to attain 0.9 from its stationary value. As follows from (6.20), this time is

$$t_{\mathrm{r}} = \frac{1}{\mu}\ln\left[9\left(\frac{A_{\mathrm{st}}^2}{A_0^2} - 1\right)\right]. \tag{6.21}$$

It is seen from (6.21) that for $A_0 \ll A_{\mathrm{st}}$ the relaxation time indeed does depend on A_0 only slightly and it is of order $1/\mu$.

6.3.3 Essentially nonlinear quasi-conservative systems

As an example, we consider a so-called van der Pol–Duffing generator described by the equation

$$\ddot{x} + ax + bx^3 = \mu(1 - \alpha x^2)\dot{x}, \tag{6.22}$$

where μ is a small parameter. As shown in Ch. 4, a solution of the generative equation ($\mu = 0$) for $a > 0$ and $b > 0$ can be expressed in terms of the Jacobi elliptic cosine as

$$x_0(t) = A\,\mathrm{cn}(\psi(t), k), \tag{6.23}$$

where $\psi(t) = \Omega t + \varphi$, $\Omega = \sqrt{a + bA^2} = 4\mathbf{K}(k)/T$, T is the period of the oscillations, $\mathbf{K}(k)$ is the full elliptic integral of the first kind, $k = \sqrt{b/2}A/\Omega$ is the modulus of the Jacobi elliptic function, and A and φ are arbitrary constants.

Let us consider $A(t)$ and $\psi(t)$ as new variables which are related to the initial variable $x(t)$ and $\dot{x}(t)$ by

$$x(t) = A(t)\,\mathrm{cn}(\psi(t), k), \quad \dot{x}(t) = -A\Omega\,\mathrm{sn}(\psi(t), k)\,\mathrm{dn}(\psi(t), k) \tag{6.24}$$

The following exact equations for $A(t)$ and $\psi(t)$ can be found from (6.22) and (6.24):

$$\left(\mathrm{cn}(\psi, k) + \frac{\partial \mathrm{cn}(\psi, k)}{\partial k}\frac{\mathrm{d}k}{\mathrm{d}A}A\right)\dot{A} - A\mathrm{sn}(\psi, k)\mathrm{dn}(\psi, k)(\dot{\psi} - \Omega) = 0,$$

$$\left[\left(1 + \frac{A}{\Omega}\frac{\mathrm{d}\Omega}{\mathrm{d}A}\right)\mathrm{sn}(\psi, k)\mathrm{dn}(\psi, k) + \frac{\partial\Big(\mathrm{sn}(\psi, k)\mathrm{dn}(\psi, k)\Big)}{\partial k}\frac{dk}{dA}A\right]\dot{A}$$
$$+ A\mathrm{cn}(\psi, k)\Big(\mathrm{dn}^2(\psi, k) - k^2\mathrm{sn}^2(\psi, k)\Big)(\dot{\psi} - \Omega)$$
$$= \mu\Big(1 - \alpha A^2\mathrm{cn}^2(\psi, k)\Big)A\mathrm{sn}(\psi, k)\mathrm{dn}(\psi, k).$$

Resolving these equations with respect to $\dot{A}$ and $\dot{\psi}$ we obtain

$$\dot{A} = \mu\Big(1 - \alpha A^2\mathrm{cn}^2(\psi, k)\Big)A\mathrm{sn}^2(\psi, k)\mathrm{dn}^2(\psi, k)$$
$$\dot{\psi} = \Omega + \mu\Big(1 - \alpha A^2\mathrm{cn}^2(\psi, k)\Big)\mathrm{sn}(\psi, k)\mathrm{cn}(\psi, k)\mathrm{dn}(\psi, k) \tag{6.25}$$
$$\times\left(1 + k(1 - 2k^2)\frac{\partial \ln \mathrm{cn}(\psi, k)}{\partial k}\right).$$

Equations (6.25) are the so-called standard equations with rapidly rotating phase [239]. The corresponding averaged equations are

$$\dot{A} = \frac{\mu A}{4\mathbf{K}(k)}\int_0^{4\mathbf{K}(k)}\Big(1 - \alpha A^2\mathrm{cn}^2(\psi, k)\Big)\mathrm{sn}^2(\psi, k)\mathrm{dn}^2(\psi, k)\,\mathrm{d}\psi, \tag{6.26a}$$

$$\dot{\psi} = \Omega. \tag{6.26b}$$

A steady-state solution of Eq. (6.26a) is described by the following transcendental equation:

$$\alpha A^2 = \frac{\int_0^{4\mathbf{K}(k)} \mathrm{sn}^2(\psi, k)\mathrm{dn}^2(\psi, k)\,\mathrm{d}\psi}{\int_0^{4\mathbf{K}(k)} \mathrm{sn}^2(\psi, k)\mathrm{cn}^2(\psi, k)\mathrm{dn}^2(\psi, k)\,\mathrm{d}\psi}. \tag{6.27}$$

The right-hand side of Eq. (6.27) can be expressed in terms of the full elliptic integrals of the first and second kinds.

6.4 The Rayleigh relaxation generator

We consider a relaxation generator described by the Rayleigh equation

$$\ddot{x} - \epsilon(1 - \dot{x}^2)\dot{x} + x = 0, \tag{6.28}$$

where $\epsilon \gg 1$. Introducing the small parameter $\mu = 1/\epsilon$ we rewrite Eq. (6.28) in the form of two equations of first order as

$$\dot{x} = y, \tag{6.29a}$$

$$\mu\dot{y} = (1 - y^2)y - \mu x. \tag{6.29b}$$

For finding an approximate solution of Eqs. (6.29a), (6.29b) we apply the averaging method in systems incorporating fast and slow variables. It follows from Eqs. (6.29a), (6.29b) that the variable x is slow and the variable y is fast. Consequently, in solving Eq. (6.29b) the variable x can be considered as a constant. Stationary solutions of Eq. (6.29b)under this condition are real roots of the cubic equation

$$y^3 - y + \mu x = 0. \tag{6.30}$$

An example of the plot of y versus x for $\mu = 0.05$ is shown in Fig. 6.7 by the heavy line. Over the range

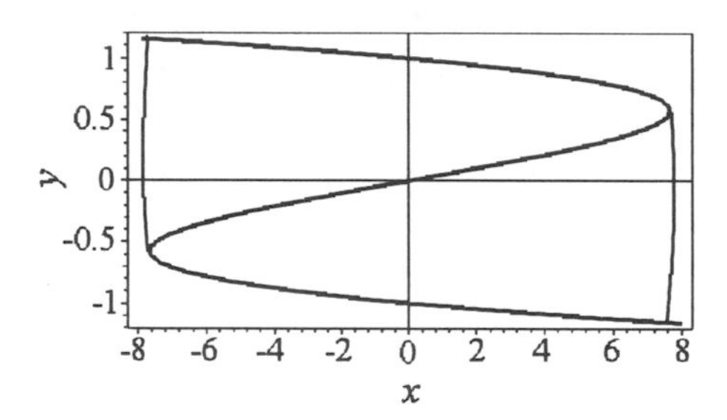

Fig. 6.7. An example of the phase portrait for a Rayleigh relaxation generator

$$-\frac{2}{3\mu\sqrt{3}} < x < \frac{2}{3\mu\sqrt{3}} \tag{6.31}$$

Eq. (6.30) has three real roots, whereas outside this range it has only one real root. For $x = \pm 2/(3\mu\sqrt{3})$ Eq. (6.30) has two real roots, $y_1 = \pm 1/\sqrt{3}$, $y2 = \mp 2/\sqrt{3}$.

Approximate solutions of Eq. (6.30) can be found analytically under the condition $\mu x \ll 1$. Expanding y sequentially about $y_{10} = 1$, $y_{20} = -1$ and $y_{30} = 0$ we find three solutions:

$$y_1 = 1 - \frac{\mu}{2}x, \quad y_2 = -1 - \frac{\mu}{2}x, \quad y_3 = \mu x. \tag{6.32}$$

It is seen that the approximate solutions (6.32) are close to the exact ones over the range (6.31).

Let us consider the stability of the solutions found. Linearizing Eq. (6.29b) with respect to small deviations $\xi_i = y - y_i$ $(i = 1, 2, 3)$ we obtain the following equations:

$$\mu \dot{\xi}_i = (1 - 3y_i^2)\xi_i \quad (i = 1, 2, 3). \tag{6.33}$$

It follows from (6.33) that only those solutions are stable which satisfy the constraint $|y_i \geq 1/\sqrt{3}$. It is easily seen that this constrain is fulfilled for the upper and the lower parts of the dependence $y(x)$ and is not fulfilled for its middle part.

Substituting $y_{1,2}$ from (6.32) into Eq. (6.29a) and integrating the latter we find solutions of this equation under the initial condition $x(0) = x_0$:

$$x_{1,2}(t) = x_0 e^{-\mu t/2} \pm \frac{2}{\mu}\left(1 - e^{-\mu t/2}\right). \tag{6.34}$$

For $\mu t \ll 1$ the solutions (6.34) can be simplified as

$$x_{1,2}(t) = (x_0 \pm t)\left(1 - \frac{\mu t}{2}\right). \tag{6.35}$$

Let the initial conditions be $x_0 = -2/(3\mu\sqrt{3})$, $y_0 = y_1(x_0)$. Then, as follows from (6.35), x will increase at an approximately constant rate and at the instant $t = 4/(3\mu\sqrt{3})$ it will attain the value $2/(3\mu\sqrt{3})$. At this instant the solution $y_1(x)$ becomes unstable because it merges with the unstable solution $y_3(x)$. As a result y will jump to take on the value $y_2\left(2/(3\mu\sqrt{3})\right)$. After this x will decrease to the value $-2/(3\mu\sqrt{3})$ and then a jump will occur again, i.e., y will take on the value $y_1\left(-2/(3\mu\sqrt{3})\right)$. Further on the process will be repeated. Thus, we obtain a limit cycle on the phase plane x, y (see Fig. 6.7).

If $y_0 \neq y_{1,2}(x_0)$ then y takes on one of these values very rapidly, practically for a constant x. This means that the corresponding phase trajectories are near vertical.

An approximate expression for the oscillation period T can be found on the assumption that the jumps occur instantly and x varies with the constant velocity ± 1. As a result we find

$$T \approx \frac{8}{3\mu\sqrt{3}} = \frac{4}{3\pi\sqrt{3}}\,\epsilon T_0, \tag{6.36}$$

where $T_0 = 2\pi$ is the period of natural oscillations for $\epsilon = 0$. It is seen from (6.36) that in the relaxation regime the oscillation period is approximately $\epsilon/4$ times as much than the period of natural oscillations.

The shape of relaxation oscillations for the Rayleigh equation is shown in Fig. 6.1.

6.5 Clock movement mechanisms and the Neimark pendulum. The energetic criterion of chaotization of self-oscillations

A clock movement mechanism is a self-oscillatory system which may be roughly divided into three basic parts: (1) an oscillatory element, e.g., a pendulum, a balance-wheel and the like; (2) a wind-up mechanism, e.g., a weight or a spring; (3) an escapement mechanism serving as a coupling agent between the oscillatory element and the wind-up mechanism. Clock movement mechanisms are so constructed that for certain positions of the oscillatory element the escapement mechanism gives a certain impulse to it at the expense of the energy of the wind-up mechanism. The duration of this impulse is different for different clock movement constructions, but, as a rule, it is very short. Usually the escapement mechanism functions twice in the oscillation period in the vicinity of the equilibrium position of the oscillatory element.

There are many models of clock movement mechanisms, distinguished by the escapement mechanisms (see [31]). We consider only the simplest model, namely the model with impacts and linear friction in the oscillatory element [8]. Let the clock pendulum acquire a certain constant momentum of value p as it passes through its equilibrium position from left to right. We assume that the pendulum is subject in its motion to a friction force which is proportional to its velocity with coefficient 2β. With these assumptions the equation of small oscillations of the pendulum is

$$\ddot{x} + 2\beta\dot{x} + \omega_0^2 x = p\sum_s \delta(t - t_s), \tag{6.37}$$

where t_s $(s = 1, 2, 3, \ldots)$ are consecutive instants of time at which impacts occur. The phase portrait of the solution of Eq. (6.37) is shown in Fig. 6.8 a.

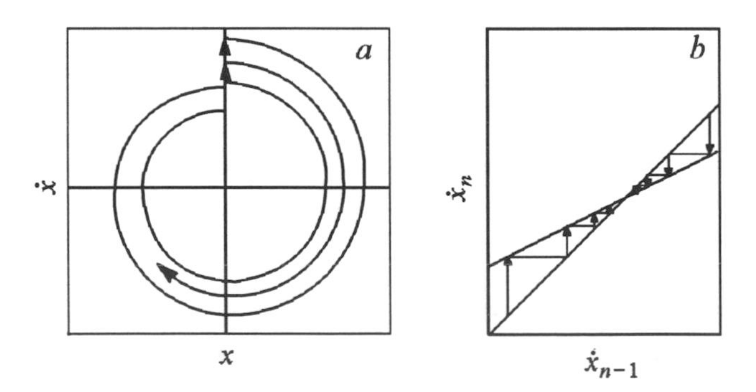

Fig. 6.8. (a) The phase portrait of the solution of Eq. (6.37), and (b) the plot $\dot{x}_n$ versus $\dot{x}_{n-1}$

Because Eq. (6.37) involves the function $\delta(t - t_s)$, the method of point maps [8, 250] is convenient to study the qualitative behavior of the solution of this equation. To find the point map associated with Eq. (6.37) we denote the value of the pendulum velocity immediately after the nth impact by $\dot{x}_n$. It is evident that the velocity value just before this impact is $\gamma\dot{x}_{n-1}$, where $\gamma = \exp(2\pi\beta/\omega)$, $\omega = \sqrt{\omega_0^2 - \beta^2}$. It follows from this that

$$\dot{x}_n = \gamma\dot{x}_{n-1} + p. \tag{6.38}$$

Equation (6.38) describes a mapping from the point $\dot{x}_{n-1}$ into the point $\dot{x}_n$. This map is plotted in Fig. 6.8 b. The stable fixed point of the map corresponds to the stable limit cycle on the phase plane.

A model proposed by Neimark [252, 253] to demonstrate the possibility of stochastic oscillations in dynamical systems is 'the reverse of the clock' in some sense. The model consists of a pendulum with negative friction (e.g., the Froude pendulum) whose oscillations are limited owing to the impacts which reduce the pendulum velocity by p at the instants when the pendulum passes through the equilibrium position with velocity $\dot{x} \geq a$, where a is a certain given positive value. The equation of small oscillations of such a pendulum is

$$\ddot{x} - 2\beta\dot{x} + \omega_0^2 x = -p\sum_s \delta(t - t_s), \tag{6.39}$$

where t_s are the instants at which $x = 0$ and $\dot{x} \geq a$. It is easy to obtain from this equation that the consecutive values of the velocity immediately after passing the equilibrium position from left to right are related by the equation

$$\dot{x}_n = \begin{cases} \gamma\dot{x}_{n-1} & \text{for } \gamma\dot{x}_{n-1} < a, \\ \gamma\dot{x}_{n-1} - p & \text{for } \gamma\dot{x}_{n-1} \geq a, \end{cases} \tag{6.40}$$

where $\gamma = \exp(2\pi\beta/\omega)$. The phase portrait[1] of the solution of Eq. (6.39) and plots of $\dot{x}_n$ versus $\dot{x}_{n-1}$ in two essentially different cases are shown in Fig. 6.9 a, b, c. In both cases the map has two unstable fixed points with coordinates $\dot{x} = 0$ and $\dot{x} = p/(\gamma - 1)$. In the first case, which takes place for sufficiently small values of p (when $p < (\gamma - 1)a$), the amplitude of the pendulum velocity $\dot{x}$, from a certain instant onwards, turns out to be greater than $p/(\gamma - 1)$ and then increases indefinitely. In the second case, when $(\gamma - 1)a < p < a$, two completely different types of behavior of the pendulum are possible, depending on the initial conditions: if $\dot{x}_0 > p/(\gamma - 1)$ then oscillations, as in the first case, increase indefinitely, but if $\dot{x}_0 < p/(\gamma - 1)$ then stochastic oscillations arise. In the latter case the consecutive mappings of the point $\dot{x}_0$ (i.e., $\dot{x}_1$, $\dot{x}_2$, ...) are everywhere dense in the interval J with ends at $a - p$ and a. An example of the time dependence of x in the stochastic regime [253] is depicted in Fig. 6.9 e.

[1] We note that the phase space for the Neimark pendulum is not a plane, but a two-sheeted surface.

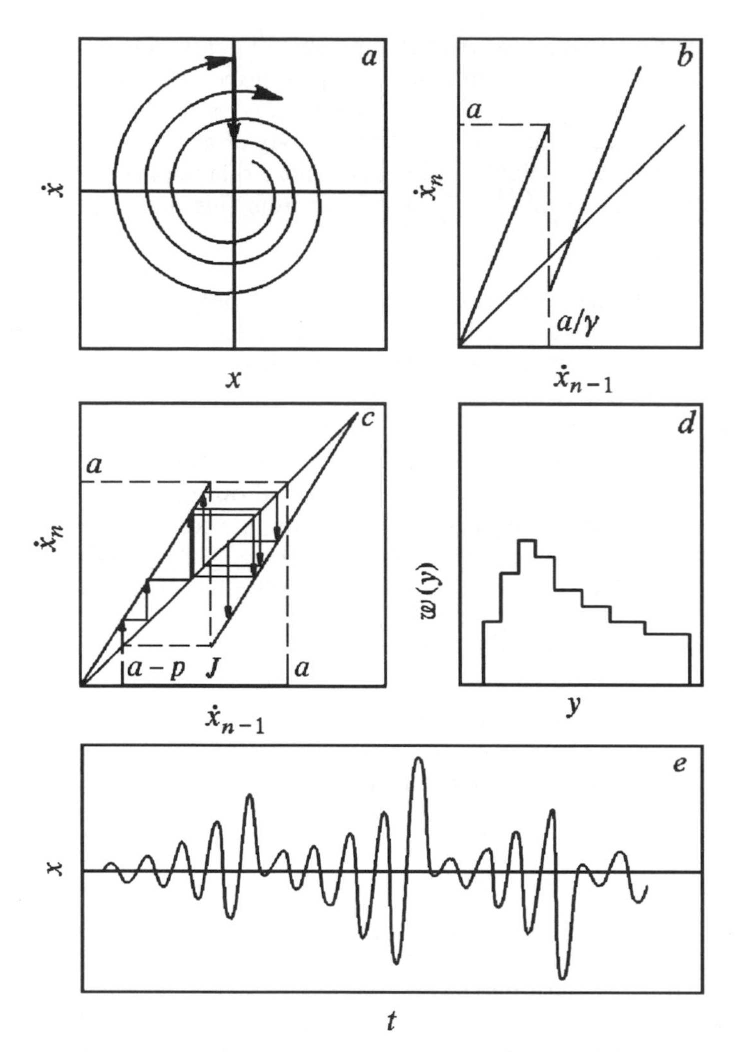

Fig. 6.9. (a) The phase portrait of the solution of Eq. (6.39) and (b, c) plots of $\dot{x}_n$ versus $\dot{x}_{n-1}$ for (b) $p < (\gamma - 1)a$ and (c) $(\gamma - 1)a < p < a$. (d) The stationary probability distribution for the values of $y \equiv \dot{x}_n$ for $\gamma = 1.171$, $p/a = 0.717$. (e) An example of the time dependence of x obtained by computation of Eq. (6.39)

By virtue of the simplicity of Eq. (6.39) and of the point map (6.40) associated with it we have succeeded in calculating the stationary probability distribution for the values of $y \equiv \dot{x}_n$ [167, 170]. An example of such a distribution is given in Fig. 6.9 d.

A reasonable extension of Eq. (6.39) to the case of weakly nonlinear friction of the pendulum is the following equation [75]:

$$\ddot{x} - 2\delta(1 - \alpha\dot{x}^2)\dot{x} + \omega_0^2 x = -p\sum_s \delta(t - t_s). \tag{6.41}$$

Owing to the nonlinearity the point map for Eq. (6.41) cannot be exactly calculated analytically. However, its qualitative behavior can easily be imagined. Depending upon the parameters, the map can be one of those illustrated by

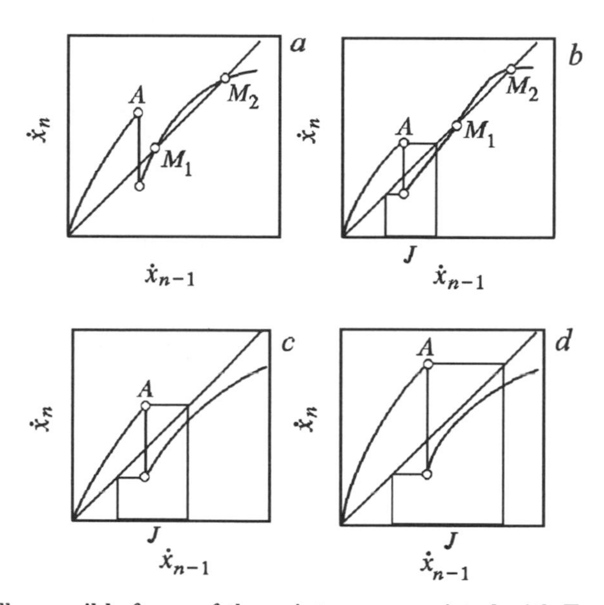

Fig. 6.10. The possible forms of the point map associated with Eq. (6.41)

Fig. 6.10. Unlike (6.40), the point map associated with Eq. (6.41) can have either three (a and b) or one (c and d) fixed point. In the case a, when the fixed point M_1 lies below the point A, for any initial conditions periodic oscillations are excited that correspond to the stable fixed point M_2. In the case b, which differs from a in that the fixed point M_1 lies above the point A, depending upon the initial conditions, either periodic or stochastic oscillations are excited, i.e., in the system phase space there are two attractors: a stable limit cycle, and a strange attractor. The domains of attraction of these attractors are separated by an unstable limit cycle corresponding to the unstable fixed point M_1. Finally, in the cases c and d for any initial conditions only chaotic oscillations are possible. The ranges of $\dot{x}_{n-1}$ corresponding to stationary stochastic oscillations are marked off in Fig. 6.10 by heavy lines and labeled J. The transition from a and b to c and d, respectively, occurs when the stable fixed point M_2 fuses with the unstable fixed point M_1 and both of them disappear. If the point M_0, at which the fusion occurs, lies below the point A (see d), then there is no strange attractor before the transition, and only a stable limit cycle and a non-attracting homoclinic structure exist.

The stable and unstable limit cycles merge just in the region of this structure which becomes attracting in the process and forms a chaotic attractor. As a result the appearance of chaos is accompanied afterwards by intermittency (see Ch. 3). But if the point M_0 lies above the point A (see c), then a strange attractor exists before the transition along with a stable limit cycle. Since the stable cycle merges with the unstable one outside the attractor region, intermittency does not arise. As the parameter changes in the reverse direction, hysteresis, which is typical of hard transitions, is observed.

In energy terms, the Neimark pendulum differs from a clock movement in that in the first case energy slowly increases owing to the negative friction and decreases rapidly as a result of the impact, whereas in the second case energy increases rapidly owing to the interaction with the escapement mechanism and slowly dissipates. Starting from this, Neimark hypothesized the energetic criterion of the chaotization of self-oscillations in dynamical systems [253]. However, we emphasize that for the majority of systems using this criterion is very difficult, if it is at all feasible, if for no other reason than the indefiniteness of the notion of energy for systems of non-mechanical origin.

7. Self-oscillatory systems with one and a half degrees of freedom

7.1 Self-oscillatory systems with inertial excitation

7.1.1 The model equations of self-oscillatory systems with inertial excitation

In systems with one degree of freedom no mechanism for the excitation of self-oscillations other than negative friction (or negative resistance) can exist. However, in systems with one and a half degrees of freedom, along with negative friction, another, less known mechanism is widespread. The excitation of oscillations in such systems is conditioned by the so-called *inertial interaction* between dynamical variables occurring as a result of inertiality in the negative feedback loop. Self-oscillatory systems with such mechanism of excitation were called by Babitsky and Landa *systems with inertial excitation* [20, 21]. They are widely met in studies of many physical and engineering problems.

The block diagram of the simplest self-oscillatory system with inertial excitation is shown in Fig. 7.1. The inertial interaction between dynamical

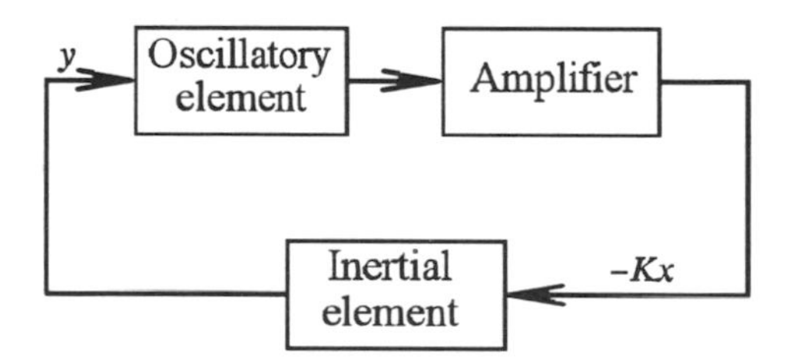

Fig. 7.1. The block diagram of the simplest self-oscillatory system with inertial excitation

variables, like negative friction, can be both linear and nonlinear. A linear interaction can, under certain conditions, result in the self-excitation of oscillations, whereas a nonlinear interaction can result in the hard excitation of self-oscillations.

The simplest model equations of systems with linear inertial interaction are

$$\ddot{x} + 2\delta\dot{x} + \omega_0^2 x = -ky + f(x, \dot{x}, y), \quad \dot{y} + \gamma y = ax + \varphi(x, \dot{x}, y), \tag{7.1}$$

where $f(x, \dot{x}, y)$ and $\varphi(x, \dot{x}, y)$ are nonlinear functions free from linear terms, and a is proportional to the gain factor K of the amplifier. The inertiality of the feedback loop is characterized by the parameter γ. The condition of self-excitation of oscillations can be shown to be $\gamma \leq \gamma_{\mathrm{cr}}$, where $\gamma_{\mathrm{cr}} = -\delta + \sqrt{\delta^2 + ak/2\delta - \omega_0^2}$. Since γ_{cr} should be positive, self-excitation can occur only for $ak \geq 2\delta\omega_0^2$. The condition $\gamma \leq \gamma_{\mathrm{cr}}$ signifies that the feedback loop must be sufficiently inertial (an inertialess feedback loop corresponds to $\gamma \to \infty$).

Systems with inertial excitation do not necessarily need to be self-excited ones. A swing is an example of a system that is not self-excited. Notice that the oscillations of a swing are often considered in text-books on the theory of oscillations as an example of parametrically excited oscillations. In fact, this approach is inaccurate. If oscillations of a swing are excited by a man who stands on it and lifts his body's center of gravity up and down at the proper instants (Fig. 7.2), then such a system is not parametrically excited. The

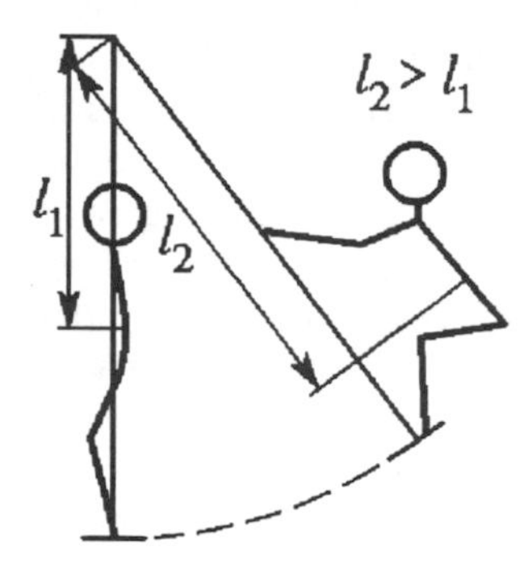

Fig. 7.2. The man–swing system; it is self-oscillatory, not parametrically excited, because its oscillations are maintained by the position-dependent forces

frequency of the oscillations of the man's center of gravity is not constant and it is tuned according to variations of the frequency of the oscillations of the swing. Thus a swing represents a control system with feedback. This system, undeniably, is a self-oscillatory one. At the same time it is not a self-excited system. For the excitation of self-oscillations some finite initial perturbation is necessary, i.e., the excitation of oscillations is hard. The simplest equations by which the excitation of oscillations of the swing can be described are

$$\ddot{x} + 2\delta\dot{x} + \omega_0^2 x = -bxy, \quad \dot{y} + \gamma y = ax^2, \tag{7.2}$$

where variable y describes the position of the man's center of gravity and γ is the inertiality factor of the control circuit. If the feedback is slightly inertial, i.e., γ is too large to put $y \approx (a/\gamma)x^2$, then for the variable x we obtain the Duffing equation having no self-oscillatory solution. Self-oscillations can exist only for not very large value of γ, i.e., when the feedback loop is sufficiently inertial. As this takes place, the excitation of self-oscillations, as seen from Eqs. (7.2), can occur only in a hard manner.

If the parameter γ in Eqs. (7.1) is close to γ_{cr}, i.e., the system is nearby its threshold of self-excitation, then we can use the Krylov–Bogolyubov asymptotic method for finding a truncated equation for the oscillation amplitude. For this purpose we rewrite Eqs. (7.1) as

$$\begin{aligned} &\ddot{x} + 2\delta\dot{x} + \omega_0^2 x + ky = \epsilon f(x, \dot{x}, y), \\ &\dot{y} + \gamma_{cr} y - ax = -\epsilon(\gamma - \gamma_{cr})y + \epsilon\varphi(x, \dot{x}, y), \end{aligned} \tag{7.3}$$

where ϵ is a small parameter which in the final results should be put equal to unity. A solution of Eqs. (7.3) for $\epsilon = 0$ can be represented as

$$\begin{aligned} &x(t) = x_0(t) = \tilde{A}e^{i\omega t} + \text{ c.c.}, \\ &y(t) = y_0(t) = \frac{2\delta}{k}(\gamma_{cr} - i\omega)\tilde{A}e^{i\omega t} + \text{c.c.}, \end{aligned} \tag{7.4}$$

where $\tilde{A}$ is the complex amplitude,

$$\omega = \sqrt{\omega_0^2 + 2\delta\gamma_{cr}} \tag{7.5}$$

is a real positive root of the characteristic equation

$$\omega^3 - i(\gamma_{cr} + 2\delta)\omega^2 - (\omega_0^2 + 2\delta\gamma_{cr})\omega + i(\omega_0^2\gamma_{cr} + ak) = 0.$$

As a first approximation, a solution of Eqs. (7.3) is sought in the form

$$x(t) = x_0(t) + \epsilon u(t), \quad y(t) = y_0(t) + \epsilon v(t), \tag{7.6}$$

where $x_0(t)$, $y_0(t)$ are determined by the expressions (7.4), $u(t)$ and $v(t)$ are unknown functions. Operating in accordance with the Krylov–Bogolyubov method (see Ch. 2), we obtain for $\tilde{A}$ the following truncated equation:

$$\frac{d\tilde{A}}{dt} = \frac{\delta}{\omega}\frac{(\gamma_{cr} - \gamma)(\omega + i\gamma)}{2\delta + \gamma + i\omega}\tilde{A} + \frac{\omega - i\gamma}{2(2\delta + \gamma + i\omega)}\overline{f} + \frac{k}{2(2\delta + \gamma + i\omega)}\overline{\varphi}, \tag{7.7}$$

where

$$\overline{f} = \frac{1}{2\pi}\int_0^{2\pi/\omega} f(x_0, \dot{x}_0, y_0)e^{-i\omega t}\,dt, \quad \overline{\varphi} = \frac{1}{2\pi}\int_0^{2\pi/\omega} \varphi(x_0, \dot{x}_0, y_0)e^{-i\omega t}dt.$$

Setting $\tilde{A} = Ae^{i\psi}$ in Eq. (7.7) and splitting this equation into real and imaginary parts, we obtain the equations for the real amplitude A and the phase ψ:

$$\begin{aligned}
\dot{A} = \frac{1}{\omega^2 + (2\delta + \gamma)^2} \Big\{ & 2\delta(\delta + \gamma)(\gamma_{\mathrm{cr}} - \gamma)A \\
& + \frac{1}{2}\Big[2\delta\omega \operatorname{Re}\left(\overline{f}\mathrm{e}^{-\mathrm{i}\psi}\right) + \left(\omega^2 + \gamma(2\delta + \gamma)\right)\operatorname{Im}\left(\overline{f}\mathrm{e}^{-\mathrm{i}\psi}\right) \\
& + k(2\delta + \gamma)\operatorname{Re}\left(\overline{\varphi}\mathrm{e}^{-\mathrm{i}\psi}\right) + k\omega \operatorname{Im}\left(\overline{\varphi}\mathrm{e}^{-\mathrm{i}\psi}\right)\Big]\Big\}, \\
\dot{\psi} = \frac{1}{\omega^2 + (2\delta + \gamma)^2} \Big\{ & \frac{\delta}{\omega}(\gamma^2 - \omega_0^2)(\gamma_{\mathrm{cr}} - \gamma) \\
& - \frac{1}{2A}\Big[\left(\omega^2 + \gamma(2\delta + \gamma)\right)\operatorname{Re}\left(\overline{f}\mathrm{e}^{-\mathrm{i}\psi}\right) - 2\delta\omega \operatorname{Im}\left(\overline{f}\mathrm{e}^{-\mathrm{i}\psi}\right) \\
& + k\omega \operatorname{Re}\left(\overline{\varphi}\mathrm{e}^{-\mathrm{i}\psi}\right) - k(2\delta + \omega)\operatorname{Im}\left(\overline{\varphi}\mathrm{e}^{-\mathrm{i}\psi}\right)\Big]\Big\}.
\end{aligned} \tag{7.8}$$

Away from the threshold of self-excitation, when γ is much less than γ_{cr}, the solution (7.6) is not valid. In this case chaotic oscillations can appear. For functions f and φ of certain special forms such oscillations were studied numerically in [169, 22].

In the case when γ is close to γ_{cr} chaotic or stochastic oscillations can arise by a jump. This can occur when Eqs. (7.8) have no stationary solution associated with a stable limit cycle. It is precisely such a situation that takes place for the Lorenz equations (7.9) (see below) for $\sigma = 10$, $b = 8/3$ and a certain critical value of the parameter r.

7.1.2 Examples of self-oscillatory systems with inertial excitation

The Lorenz system and its mechanical analog. The Lorenz equations, having the form

$$\dot{X} = -\sigma(X - Y), \quad \dot{Y} = rX - Y - XZ, \quad \dot{Z} = -bZ + XY, \tag{7.9}$$

are well known as a three-mode model of thermoconvection [214]. It is less known that the Lorenz system belongs to the class of self-oscillatory systems with inertial excitation [20]. Indeed, Eqs. (7.9) can be reduced to the system of equations (7.1) by excluding Y and a change of variables

$$x = X - \sqrt{b(r-1)}, \quad y = \sigma\left(Z - X^2/2\sigma - (r-1)(1 - b/2\sigma)\right).$$

In so doing one should put $2\delta = \sigma + 1$, $\omega_0^2 = b(r-1)$, $k = \sqrt{b(r-1)}$, $f(x, y) = -(3kx/2 + x^2/2 + y)x$, $a = (2\sigma - b)k$, $\gamma = b$, $\varphi(x) = (\sigma - b/2)x^2$.

As mentioned above, in the case of the Lorenz equations, for $\sigma = 10$, $b = 8/3$, stochasticity appears by a jump, when r rises to the point 24.74.

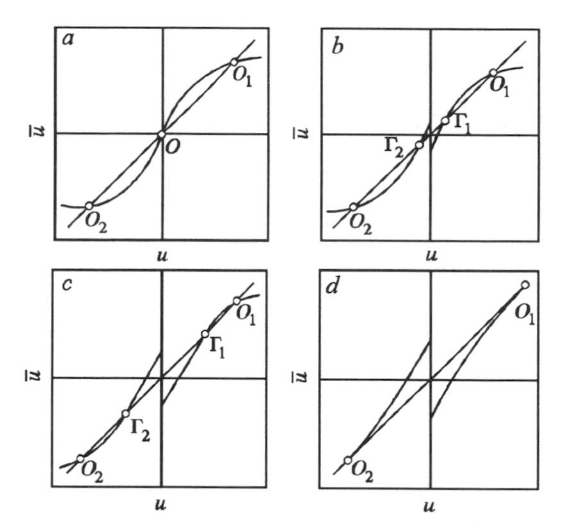

Fig. 7.3. Evolution, with increasing parameter r, of the one-dimensional point map associated with the Lorenz system for $\sigma = 10$, $b = 8/3$

The scenario for the appearance of stochasticity can be conveniently traced by considering the evolution of the approximate one-dimensional point map shown in Fig. 7.3 [253]. Figure 7.3 a corresponds to the presence of two stable equilibrium states O_1 and O_2; Fig. 7.3 b corresponds to the presence of two stable equilibrium states O_1 and O_2 and of two unstable limit cycles Γ_1 and Γ_2; Fig. 7.3 c corresponds to the presence of two stable equilibrium states O_1 and O_2 and of a stochastic attractor; and Fig. 7.3 d corresponds to the loss of stability of the equilibrium states O_1 and O_2 and the presence of a stochastic attractor only. The shape of oscillations of the variables X, Y, Z for $r = 28$ is illustrated in Fig. 7.4.

For sufficiently large r oscillations in a Lorenz system are always periodic but far from harmonic. In the limiting case of very large r their shape and quantitative characteristics can be calculated approximately in an analytic way [306, 20, 21]. As r decreases, the transition to chaos occurs in accordance with the Feigenbaum scenario. Inside the region of chaos there are several more periodicity windows. We emphasize that nowadays the overall picture of the behavior of the solution for the Lorenz system is completely understood.

A mechanical system that is described by equations similar to the Lorenz equations was considered by Zlochevsky. It consists of two solid bodies, a carrying body B and a carried axially symmetric rotor R (Fig. 7.5). This system possesses the property that, as the rotor moves relative to the carrying body, the spatial distribution of masses remains unchanged. Therefore, the tensor for the system is constant. Such a system is called a *gyrostat*. It is assumed that its center of mass (the point O) is fixed, and that its ellipsoid of inertia is an ellipsoid of revolution, i.e., $I_y = I_z = I$, where I_y and I_z are the

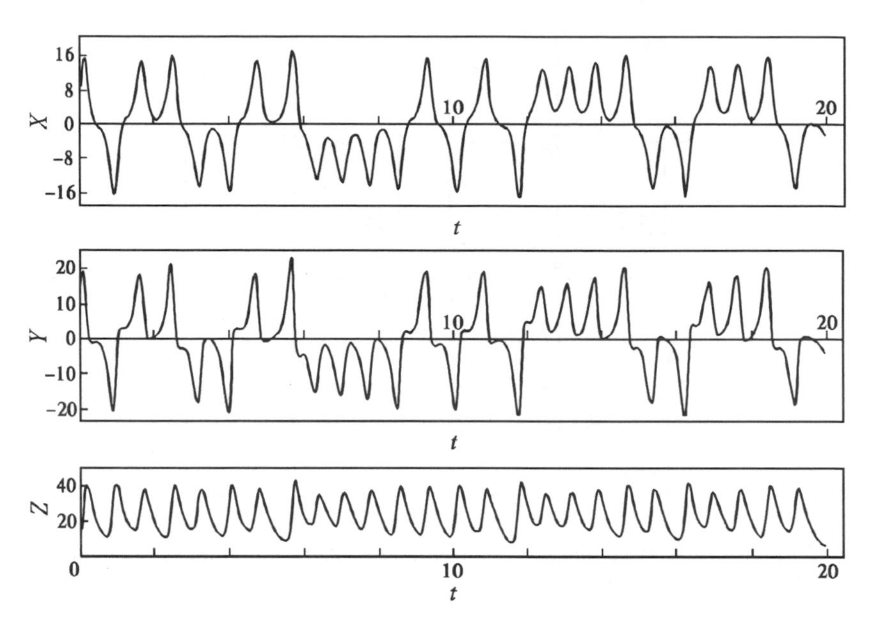

Fig. 7.4. The shape of oscillations of the variables X, Y, Z for $\sigma = 10$, $b = 8/3$, $r = 28$

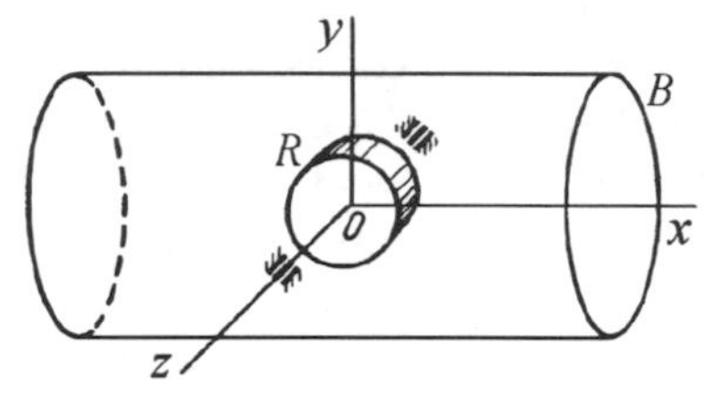

Fig. 7.5. The schematic image of a gyrostat: the carrying body B can rotate about the fixed center of mass O with a certain angular velocity; the rotor R rotates with a constant velocity about its symmetry axis (the z-axis) fixed on the carrying body B transversely to its symmetry axis (the x-axis)

moments of inertia of the gyrostat about the y-axis and z-axis, respectively. The rotor is supposedly rotating with a constant angular velocity Ω about its symmetry axis z, which is fixed on the carrying body transversely to its symmetry axis x. The moment $\boldsymbol{M}$ of external forces applied to the gyrostat depends on the angular velocity $\boldsymbol{\omega}$ of the carrying body rotating about the fixed center of mass O.

In the coordinates bound to the carrying body, the equations of motion of the gyrostat are

$$\begin{aligned}
&I_x\dot{\omega}_x + N\omega_y = M_x,\\
&I\dot{\omega}_y + (I_x - I)\omega_x\omega_z - N\omega_x = M_y,\\
&I\dot{\omega}_z + (I - I_x)\omega_y\omega_x = M_z,
\end{aligned} \tag{7.10}$$

where ω_x, ω_y, ω_z are the projections of the angular velocity $\boldsymbol{\omega}$ on the x-, y-, z-axes, I_x is the moment of inertia of the gyrostat about x-axis, $N = J\Omega$ is the relative moment of momentum of the rotor, and M_x, M_y, M_z are the projections of the moment $\boldsymbol{M}$ of external forces on the x-, y-, z-axes.

As is known [366], for $M_x = M_y = M_z = 0$, $\boldsymbol{\omega} \neq 0$ and $\Omega \neq 0$ the gyrostat either vibrates periodically or rotates permanently. Such motions are associated with the trajectories in the system phase space resting on the energy ellipsoid of the carrying body described by the equation

$$\frac{I_x \omega_x^2}{2} + \frac{I(\omega_y^2 + \omega_z^2)}{2} = T = \text{const}, \tag{7.11}$$

where T is the kinetic energy of the gyrostat.

If M_x, M_y and M_z are different from zero and equal to

$$M_x = (N_1 + N)\omega_y - N_1\omega_x, \quad M_y = -N_2\omega_y, \quad M_z = -N_3\omega_z,$$

where N_1, N_2 and N_3 are certain coefficients, then Eqs. (7.10) become

$$\begin{aligned} I_x\dot{\omega}_x &= N_1(\omega_y - \omega_x), \\ I\dot{\omega}_y &= N\omega_x - (I_x - I)\omega_x\omega_z - N_2\omega_y, \\ I\dot{\omega}_z &= (I_x - I)\omega_y\omega_x - N_3\omega_z. \end{aligned} \tag{7.12}$$

Putting in (7.12)

$$X = \frac{(I_x - I)\omega_x}{N_2}, \quad Y = \frac{(I_x - I)\omega_y}{N_2}, \quad Z = \frac{(I_x - I)\omega_z}{N_2},$$
$$\tau = \frac{N_2}{I}t, \quad \sigma = \frac{N_1 I}{N_2 I_x}, \quad r = \frac{NI}{N_2 I_x}, \quad b = \frac{N_3}{N_2},$$

we obtain the Lorenz equations in standard form (7.9).

The Helmholtz resonator with nonuniformly heated walls. The Helmholtz resonator with non-uniformly heated walls, described even in Rayleigh's treatise [279], belongs to the class of so-called *thermo-mechanical* systems. Systems are said to be thermo-mechanical if they involve an oscillatory element that interacts with a heat energy source. Rayleigh offered a qualitative explanation for the following phenomenon which was well known to glass blowers. If, at the end of a glass tube of diameter of several millimeters and length of the order of 10 cm, a globe is blown, then, while the globe is still hot, the tube emits a sound. The frequency of this sound is close to the natural frequency of the Helmholtz resonator, i.e., an acoustical resonator involving some cavity, which is often of a spherical shape, and a tube (a throat) (Fig. 7.6 at the left). It is easily shown that the natural frequency of the Helmholtz resonator is

$$\omega_0 = \sqrt{\frac{\kappa p_0 S}{\rho_0 l V}}, \tag{7.13}$$

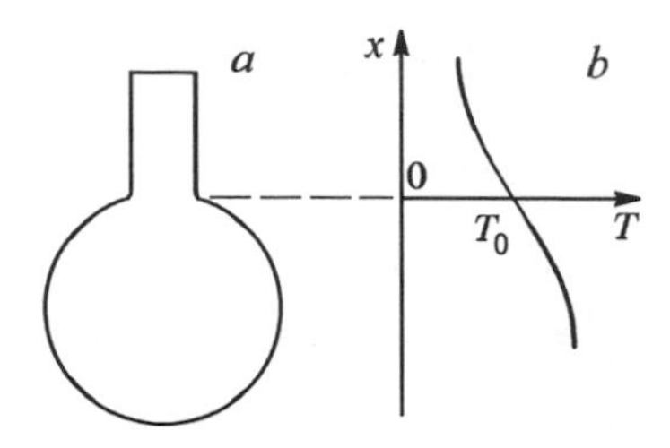

Fig. 7.6. (a) The Helmholtz resonator, and (b) the distribution of wall temperature

where p_0 is atmospheric pressure, ρ_0 is the density of the gas filling the resonator at atmospheric pressure, $\kappa = c_p/c_v$ is the isentropic exponent, which is equal to the ratio between the specific heat at a constant pressure c_p, and the specific heat at a constant volume c_v, S is the tube cross-section area, l is the tube length, and V is the volume of the resonator cavity.

A quantitative explanation for the phenomenon described by Rayleigh was given by Teodorchik [334]. The equation of motion of the mass of gas contained in the tube, without regard for compressibility, is

$$\rho_0 Sl\ddot{x} = S\Delta p(x), \tag{7.14}$$

where x is the gas displacement, $\Delta p(x) = p(x) - p_0$, and $p(x)$ is the gas pressure within the cavity. To calculate $\Delta p(x)$ we use the state equation for an ideal gas and the first principle of thermodynamics, which for sufficiently small x are conveniently written as

$$pV = mc_v(\kappa - 1)T, \quad Q(x) = mc_v\vartheta(x) + p_0 Sx, \tag{7.15}$$

where T is the mean gas temperature within the cavity, $Q(x)$ is the quantity of heat obtained by the gas in the cavity, when the gas in the tube is displaced an amount x, as a result of heat exchange with the non-uniformly heated walls, and $\vartheta(x) = T - T_0$ is the change of the gas temperature in the cavity as the gas in the tube is displaced the same amount x.[1] It follows from the first equation of (7.15) that $\Delta p(x) = -(p_0 S/V)x + (mc_v(\kappa - 1)/V)\vartheta(x)$ for small x. Substituting $\vartheta(x)$ from the second equation of (7.15) into this relationship, we find

$$\Delta p(x) = -\kappa\frac{p_0 S}{V}x + \frac{\kappa - 1}{V}Q(x). \tag{7.16}$$

Hence, Eq. (7.14) becomes

$$\rho_0 Sl\ddot{x} + \kappa\frac{p_0 S^2}{V}x = \frac{\kappa - 1}{V}SQ(x). \tag{7.17}$$

Apropos, the formula (7.13) for the natural frequency of the Helmholtz resonator follows from this for $Q(x) = 0$.

[1] We assume that heat exchange occurs so rapidly that the gas temperature can be assumed to be the same everywhere within the cavity.

We assume that the heat exchange between the moving gas and the resonator walls obeys Newton's law. According to this law the rate of heat flow is proportional to the temperature difference between an object and its environment. In the case under consideration the role of the environment is played by the resonator walls. It should be pointed out that, once the globe has been blown, the temperature of the resonator walls decreases rapidly in the direction from the globe to the tube. The temperature distribution is presented in Fig. 7.6 b. So, executing oscillations, the gas moves along the walls with a variable temperature. Therefore it is possible to introduce a certain effective wall temperature $T_{\rm w}(x)$ as the gas is displaced an amount x. This effective temperature can be set as $T_{\rm w}(x) = T_0 - \alpha x + \beta x^3$. Since an instantaneous value of the gas temperature in the resonator cavity is $T = T_0 + \vartheta$, Newton's law can be written as

$$\dot{Q} = K\Big(T_{\rm w}(x) - T\Big) = -\,K\Big(\alpha x - \beta x^3 + \vartheta\Big), \tag{7.18}$$

where K is the heat transfer factor. Substituting ϑ from the second equation of (7.15) into (7.18) we obtain

$$\dot{Q} = -K\left[\left(\alpha - \frac{p_0 S}{m c_v}\right) x - \beta x^3 + \frac{Q}{m c_v}\right]. \tag{7.19}$$

Equations (7.17), (7.19) can be rewritten as

$$\ddot{x} + 2\delta\dot{x} + \omega_0^2 x = kQ, \quad \dot{Q} + \gamma Q = -ax + bx^3, \tag{7.20}$$

where the factor δ is determined by the friction of the gas on the tube walls, ω_0 is the natural frequency of the Helmholtz resonator, $k = (\kappa - 1)/(ml)$, $\gamma = K/(mc_v)$, $a = K\big(\alpha - p_0 S/(mc_v)\big)$, and $b = \beta K$. It is easily seen that Eqs. (7.20) fall into the category of Eqs. (7.1) describing systems with inertial self-excitation. In the system under consideration the self-excitation of oscillations is caused by a sufficiently large value of the temperature gradient α. Even in the case when the term $2\delta\dot{x}$ in Eqs. (7.20) can be ignored, there exists a nonzero critical value of α beyond which the oscillations are not excited. This critical value is equal to $p_0 S/(mc_v)$. If δ is accounted for, the critical value of α depends on the inertiality parameter γ and equals

$$\alpha_{\rm cr} = \frac{p_0 S}{m c_v} + \frac{2\delta}{k}\,(\omega_0^2 + 2\delta\gamma + \gamma^2). \tag{7.21}$$

The frequency of the self-oscillations in the vicinity of the excitation threshold is equal to $\omega = \sqrt{\omega_0^2 + 2\delta\gamma}$. If δ and γ are sufficiently small, $\omega \approx \omega_0$, which is observed in experiments.

We note that the result obtained essentially differs from that found by Teodorchik [334]. This arises from the difference between the initial equations written by Teodorchik without a strict derivation and Eqs. (7.20).

Oscillations of a heated wire with a weight at its center. We now consider another example of a thermo-mechanical self-oscillatory system. Let a weightless stretched metallic wire with a weight at its center be included in an alternating current circuit of frequency ω (see Fig. 7.7). Under certain

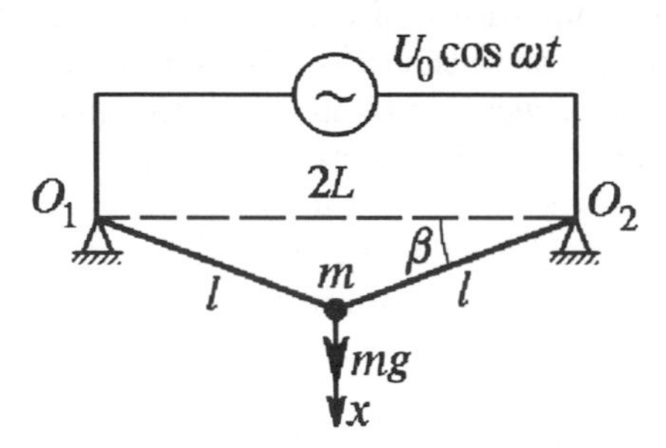

Fig. 7.7. Schematic image of a stretched metallic wire with a weight at its center, which is included in an alternating current circuit of frequency ω

conditions, such a wire can execute self-oscillations both in a vertical plane and around the O_1O_2-axis. Vertical self-oscillations of a wire with a weight at its center included in a circuit, which is similar to that pictured in Fig. 7.7 but also involving an electro-mechanical circuit breaker, were studied in detail by Teodorchik [328, 334]. Vertical self-oscillations in the circuit shown in Fig. 7.7 were first demonstrated experimentally by Dobronravov and Shalnikov, as an illustration in the A.F. Ioffe lectures, as early as 1924.[2]. An analytical consideration of the vertical oscillations of a wire in the circuit presented in Fig. 7.7 was undertaken by Vermel [356]. However, a mistake made in this work caused the results to be incorrect.

We consider below self-oscillations of the wire included in the circuit shown in Fig. 7.7, both in a vertical plane and around the O_1O_2-axis. The equation for the vertical oscillations of a weight of mass m is

$$m\ddot{x} = mg - 2F \sin\beta - h\dot{x}, \tag{7.22}$$

where F is the tension in the wire, $\beta = \arcsin x/l$, $l = \sqrt{x^2 + L^2}$ is half of the wire length at a temperature of $T_0 + \vartheta$, T_0 is the temperature of the wire environment, $2L$ is the distance between the supports, and h is the friction factor.

The tension F in a wire heated by an electric current to the temperature $T_0 + \vartheta$ is [204]

$$F = ES \, \frac{l - l_0(1 + \alpha\vartheta)}{l_0}\,, \tag{7.23}$$

[2] Apparently, these demonstrations remain unpublished, but they are described in [266] heard from A.I. Shalnikov.

where E is Young's modulus, S is the cross-section of the wire, α is the coefficient of linear thermal expansion, $l_0 = \sqrt{x_0^2 + L^2}$ is half of the length of the wire in the unstretched state at $\vartheta = 0$, and x_0 is the sag of the wire for $\vartheta = 0$, $m = 0$. Assuming $l - l_0 \ll l_0$, from (7.23) we find

$$F \sin\beta = ES\left(\frac{x^2 - x_0^2}{2l_0^2} - \alpha\vartheta\right)\frac{x}{l_0}. \tag{7.24}$$

We write an equation for the temperature deviation ϑ assuming that, firstly, the heat conductivity of the wire is so large that the temperatures of all parts of the wire and the weight have time to be equalized in one oscillation period, and, secondly, cooling of the wire occurs according to Newton's law with a heat transfer factor q which, in general, depends on the modulus of the velocity. Ignoring this dependence, we obtain the following equation for ϑ:

$$mc\dot{\vartheta} = -q\vartheta + \frac{U_0^2}{2R}(1 + \cos 2\omega t), \tag{7.25}$$

where c is the specific heat of the weight, and R is the resistance of the wire. The resistance R increases as the wire is heated and lengthens,[3] so that

$$R = R_0\left(1 + \beta_1\vartheta + \beta_2\frac{l - l_0}{l_0}\right), \tag{7.26}$$

where β_1 and β_2 are certain coefficients (the value of β_2 is approximately equal to the Poisson factor). We will assume that the temperature of the wire has no time to change considerably during the period of the current oscillations $2\pi/\omega$, and that the natural frequency ν of the wire lies outside regions of parametric excitation of the oscillations of the wire [275]. In this case the term $(U_0^2/2R)\cos 2\omega t$ in Eq. (7.25) can be neglected.

The steady-state solution x_{st}, ϑ_{st} of Eqs. (7.22) and (7.25), in view of (7.24) and (7.26), is determined by the following algebraic equations:

$$\begin{gathered} ES\frac{x_{st}}{l_0}\left(\frac{x_{st}^2 - x_0^2}{l_0^2} - 2\alpha\vartheta_{st}\right) = mg, \\ \vartheta_{st} \approx U_0^2/2R_0q. \end{gathered} \tag{7.27}$$

To investigate the stability of the steady-state solution x_{st}, ϑ_{st}, we derive linearized equations for the deviations from this solution $\xi = (x - x_{st})/l_0$, $\eta = (\vartheta - \vartheta_{st})/\vartheta_{st}$, which are conveniently written as

$$\ddot{\xi} + 2\delta\dot{\xi} + \nu^2\xi = k\eta, \quad \dot{\eta} + \gamma\eta = -a\xi, \tag{7.28}$$

where

[3] The change of resistance depending on deformation is called the *tenso-resistive effect*.

$$2\delta = \frac{h}{m}, \quad \nu^2 = \frac{g}{x_{\rm st}} + \frac{2ESx_{\rm st}^2}{ml_0^3}, \quad k = 2ES\,\frac{\alpha\vartheta_{\rm st}x_{\rm st}}{ml_0^2},$$
$$\gamma = \frac{q}{mc}\,(1+\beta_1\vartheta_{\rm st}), \quad a = \frac{qx_{\rm st}}{mcl_0}\,\beta_2.$$

We see that Eqs. (7.28), like the equations for the Helmholtz resonator with non-uniformly heated walls, fall in the class of Eqs. (7.1) describing the simplest systems with inertial self-excitation.

The condition for the self-excitation of oscillations in such a system is

$$ak \geq 2\delta(\nu^2 + 2\delta\gamma + \gamma^2). \tag{7.29}$$

From this it is seen that the dependence of the resistance of the wire on its deformation, reflected in the coefficient β_2, in common with sufficient inertiality of the change in the temperature of the wire, reflected in the factor γ, is the main cause of the self-excitation of oscillations.

Let us now consider self-oscillations of the wire about the O_1O_2-axis. It is evident that these self-oscillations can be excited only in a hard manner because of the dependence of the heat transfer factor q on the modulus of the velocity of motion of the wire. Indeed, owing to this dependence the oscillations of the wire should cause the appearance of the second harmonic in the oscillations of the temperature that should in turn cause modulation of the length of the wire at the same frequency. This can further lead to the 'parametric' excitation of oscillations of the wire.

The equation for the rotational oscillations of a wire with a weight can be written as

$$m\,\frac{\mathrm{d}(r^2\dot\varphi)}{\mathrm{d}t} + Hr\dot\varphi + mgr\sin\varphi = 0, \tag{7.30}$$

where r is the distance from the weight to the O_1O_2-axis, and H is the coefficient of friction. As a current flows through the wire, the distance r varies because of heat lengthening the wire. Taking account of the fact that $l = l_0(1+\alpha\vartheta)$, we find

$$r = \sqrt{l_0^2(1+\alpha\vartheta)^2 - L^2}\,. \tag{7.31}$$

The equation for ϑ can be written in the form of (7.25). In studies of the rotational oscillations of the wire we can ignore the tenso-resistive effect but, in return, the dependence of the heat transfer factor q on the modulus of the mean velocity of the wire must be taken into account. Since this velocity is proportional to the velocity of the weight, we can set

$$q = q_0 + q_1 f(|v|), \tag{7.32}$$

where q_0 is the heat transfer factor as the weight and the wire are at rest, q_1 is a certain coefficient, v is the total velocity of the weight whose modulus is

$$|v| = \sqrt{r^2\dot\varphi^2 + \dot r^2}\,, \tag{7.33}$$

and $f(|v|)$ is a nonlinear function, which in a certain range of $|v|$ can be approximated by the formula $f(|v|) = \sqrt{|v|}$ [23]. Equation (7.25), in view of (7.32) and (7.33), becomes

$$mc\dot{\vartheta} = -\left(q_0 + q_1 f\left(\sqrt{r^2\dot{\varphi}^2 + \dot{r}^2}\right)\right)\vartheta + \frac{U_0^2}{2R}, \tag{7.34}$$

where r and ϑ are related by (7.31).

An approximate analytical study of the phenomenon under consideration can be performed in the case when the factor H, the deviation of ϑ from its steady-state value ϑ_{st}, and the coefficient q_1 are sufficiently small (of the order of a small parameter ϵ). In this case we can put

$$r \approx r_0 + \epsilon\alpha \frac{(1+\alpha\vartheta_{st})l_0^2}{r_0}\eta, \tag{7.35}$$

where

$$r_0 = \sqrt{l_0^2(1+\alpha\vartheta_{st})^2 - L^2}, \quad \vartheta_{st} = \frac{1}{\beta_1 + 2q_0R_0/U_0^2}, \quad \eta = \vartheta - \vartheta_{st};$$

β_1 is determined by (7.26). Setting $\varphi \sim \sqrt{\epsilon}$, $q_1 f(|v|)\vartheta_{st} \sim \epsilon$ and taking account of (7.35), we rewrite Eqs. (7.30) and (7.34), with regard to terms only of order no higher than ϵ, as:

$$\ddot{\varphi} + \omega_0^2\varphi = \epsilon\left[-2\left(\delta + \alpha\frac{(1+\alpha\vartheta_{st})l_0^2}{r_0^2}\dot{\eta}\right)\dot{\varphi}\right.$$
$$\left. + \omega_0^2\varphi\left(\frac{\varphi^2}{6} + 2\alpha\frac{(1+\alpha\vartheta_{st})l_0^2}{r_0^2}\eta\right)\right], \tag{7.36a}$$

$$\dot{\eta} + \gamma\eta + \gamma_1\vartheta_{st} f\left(r_0|\dot{\varphi}|\right) = 0, \tag{7.36b}$$

where $2\delta = H/mr_0$, $\omega_0^2 = g/r_0$, $\gamma = q_0/mc$, $\gamma_1 = q_1/mc$.

It can be seen that Eqs. (7.36a) and (7.36b) describe a system with a nonlinear inertial feedback. This is why the excitation of self-oscillations can only be hard.

A solution of Eq. (7.36a) for $\epsilon = 0$ is

$$\varphi = A\cos(\omega_0 t + \psi). \tag{7.37}$$

Substituting (7.37) into Eq. (7.36b) and expanding the function $f\left(r_0A|\sin(\omega_0 t + \psi)|\right)$ in a Fourier series, we find η:

$$\eta = \sum_{n=1}^{\infty}\left(B_n(A)\cos 2n(\omega_0 t + \psi) + C_n(A)\sin 2n(\omega_0 t + \psi)\right), \tag{7.38}$$

where the coefficients B_n and C_n are determined by the coefficients of the Fourier series expansion for the function f. If we further substitute (7.37) and (7.38) into Eq. (7.36a), assuming that A and ψ are slowly time-varying functions, and take account of the terms with the fundamental frequency only, we obtain the following truncated equations for the amplitude A and the phase ψ:

$$\dot{A} = \left(-\delta + \alpha \frac{(1+\alpha\vartheta_{\text{st}})l_0^2}{r_0^2} \omega_0 C_1(A) \right) A, \tag{7.39a}$$

$$\dot{\psi} = -\frac{\omega_0}{16} A^2 - \frac{1}{2} \alpha \frac{(1+\alpha\vartheta_{\text{st}})l_0^2}{r_0^2} \omega_0 B_1(A). \tag{7.39b}$$

The steady-state solution of Eq. (7.39a) gives the value of the amplitude of the unstable cycle. To calculate the amplitude of the stable cycle we should take account of the second-order terms. In principle this is not a very complicated problem, but the calculations turn out to be rather awkward; therefore we will not give these calculations here.

We note that the wire self-oscillations described above are, apparently, one of the causes of the swing of wires in electro-transmission lines, even in still air.

Self-oscillations in the Vallis model for nonlinear interaction between ocean and atmosphere. An interesting example of a self-oscillatory system with hard inertial excitation is a model suggested by Vallis [348, 349] with the aim explaining the intriguing climatological phenomenon known as 'El Niño/Southern Oscillation' . This phenomenon consists of strong annual oscillations of the surface flow velocity and of the surface temperature of ocean water in the tropics. It exerts a pronounced effect on the climate of all our planet.

A prerequisite for the Vallis model is the availability of the movement of air upward or downward depending on whether it is over warm or cold ocean water. This movement causes a wind which, in its turn, results in a surface flow in the ocean. Therefore the surface temperatures at the western and eastern edges of the ocean basin are affected by horizontal advection and, due to continuity, by up- and down-welling. The model equations are

$$\begin{aligned} \dot{u} &= \frac{B}{2l}(T_{\text{e}} - T_{\text{w}}) - C(u + u^*), \\ \dot{T}_{\text{w}} &= \frac{u}{2l}(T_0 - T_{\text{e}}) - A(T_{\text{w}} - T^*), \\ \dot{T}_{\text{e}} &= \frac{u}{2l}(T_{\text{w}} - T_0) - A(T_{\text{e}} - T^*), \end{aligned} \tag{7.40}$$

where u is the velocity of the ocean surface flow, T_{w} and T_{e} are the relative temperatures at the western and eastern edges of the ocean basin respectively, A^{-1} is the relaxation time of the temperature, B is the coupling factor between the temperature difference and the ocean surface flow, C is a coefficient associated with internal friction in ocean water, l is the ocean basin width, T_0 is the relative temperature deep in the ocean, u^* is the velocity of the tradewind, and T^* is the steady-state ocean temperature for $u^* = 0$.

Equations (7.40) for $A = 1\,\text{year}^{-1}$, $B = 1.6\,\text{m}^2\text{s}^{-1}\text{K}^{-1}$, $C = 0.25\,\text{month}^{-1}$, $l = 7500\,\text{km}$, $T_0 = 0\,\text{K}$, $T^* = 12\,\text{K}$, $u^* = u_0(1+\sin\omega t)$, where $u_0 = 0.45\,\text{m}\,\text{s}^{-1}$,

$\omega = 2\pi\,\text{year}^{-1}$, were simulated numerically by Göber et al. [100]. It was found that the model variables executed chaotic oscillations which were caused, in the author's opinion, by the periodic variation of the tradewind velocity. However, even for $u^* = u_0$ the system described by Eqs. (7.40) possesses a chaotic attractor in addition to two stable steady-state solutions $u = u_1 \approx -3.7818$, $T_w = T_{w1} \approx 1.6726$, $T_e = T_{e1} \approx -1.2989$ and $u = u_2 \approx 3.3237$, $T_w = T_{w2} \approx -1.4420$, $T_e = T_{e2} \approx 1.9236$ and one unstable steady-state solution $u = u_3 \approx 0.0081$, $T_w = T_{w3} \approx 11.7922$, $T_e = T_{e3} \approx 12.2008$. The chaotic attractor corresponds to the chaotic self-oscillations shown in Fig. 7.8. These self-oscillations closely resemble those found in [100].

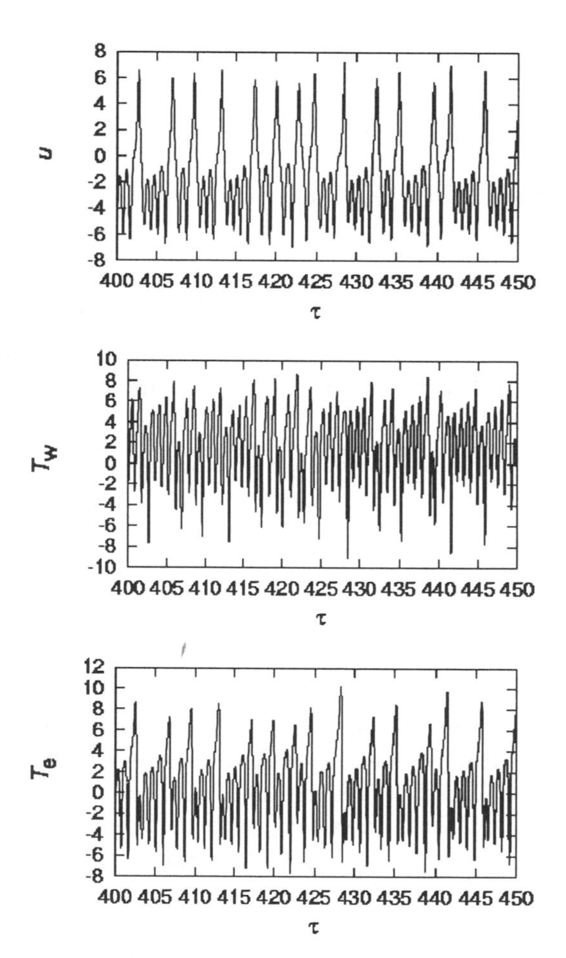

Fig. 7.8. Chaotic self-oscillations in the Vallis model for the parameters indicated in the text

We emphasize that a chaotic attractor is retained even for $u_0 = 0$ if B is sufficiently large. The availability of the chaotic attractor is caused by coupling between the temperature difference and the ocean surface flow. This coupling is determined by the factor B. For $B < B_{cr}$, where B_{cr} depends on u_0, there exists no chaotic attractor.

Let us show that the self-oscillations in the system under consideration result from hard inertial excitation. For this Eqs. (7.40) are conveniently rewritten in the variables u, $Y = T_e + T_w$, $Z = T_e - T_w$:

$$\begin{aligned}
\dot{u} &= \frac{B}{2l} Z - C(u + u_0), \\
\dot{Y} &= -\frac{u}{2l} Z - A(Y - 2T^*), \\
\dot{Z} &= \frac{u}{2l} Y - AZ.
\end{aligned} \tag{7.41}$$

Eliminating Z from (7.41) we find

$$\begin{aligned}
&\ddot{u} + (C + A)\dot{u} + ACu = \frac{B}{4l^2} Yu - ACu_0, \\
&\dot{Y} + AY = -\frac{Cu_0}{B} u - \frac{u}{B}(\dot{u} + Cu) + 2AT^*.
\end{aligned} \tag{7.42}$$

By substituting $x = u - u_s$, $y = Y - Y_s + u_s(u - u_s)/B$, where u_s, Y_s is one of stable steady-state solutions of Eqs. (7.42), we find the following equations for x and y:

$$\begin{aligned}
&\ddot{x} + (C + A)\dot{x} + \left(AC + \frac{u_s^2}{4l^2} - \frac{BY_s}{4l^2}\right) x = \frac{Bu_s}{4l^2} y + \frac{B}{4l^2} x \left(y - \frac{u_s}{B} x\right), \\
&\dot{y} + Ay = \frac{Au_s - C(u_0 + 2u_s)}{B} x - \frac{1}{B} (\dot{x} + Cx)x.
\end{aligned} \tag{7.43}$$

It is easily seen that Eqs. (7.43) belong to the class of systems with inertial excitation.

Self-oscillations of an air-cushioned body. The example considered below is concerned with a large variety of problems on the self-oscillations of a body in a fluid flow. The strict setting up of such problems needs solution of partial differential equations. However, in specific cases these problems can be reduced to solving ordinary differential equations. These equations can be treated as lumped models of the processes under consideration. Below it is shown that the problem on self-oscillations of an air-cushioned body is reduced by some approximations to a model of systems with inertial self-excitation of the form (7.1).

We consider a rigid plate of mass m suspended on an air cushion created by air coming through a number of nozzles from a chamber connected to a compressor (see Fig. 7.9 a) [180, 184]. If we ignore the friction of the plate

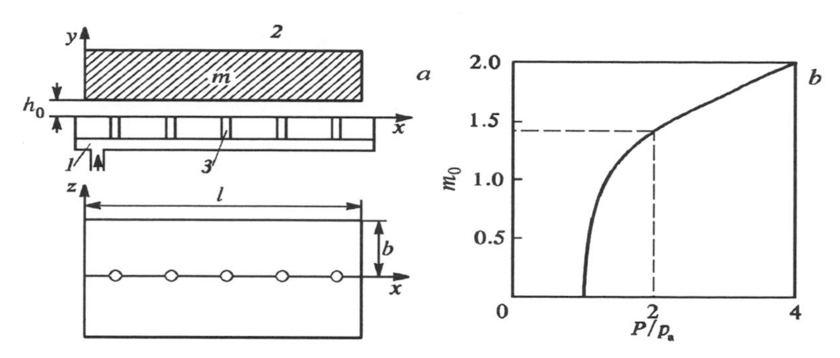

Fig. 7.9. (a) Schematic image of a rigid plate (2) suspended on an air cushion created by air coming through nozzles (3) from the chamber (1) connected to a compressor. (b) The dependence of the value $m_0 \equiv m_0(x, p_a, P)/\alpha_0(p_a) \sin \lambda x$, characterizing the mass flow rate, on P/p_a

on the air, then the equation of the motion of the plate is

$$m\ddot{h} = F - mg, \tag{7.44}$$

where h is the depth of the cushion, F is the aerodynamic force determined by the expression

$$F = 2 \int_0^b \int_0^l \Big(p(x,z) - p_a \Big) \, dx \, dz, \tag{7.45}$$

p_a is atmospheric pressure, and $p(x, z)$ is the pressure created by the air cushion.

To calculate $p(x, z)$ one may use the equation derived by Reynolds in his studies of gas lubrication [281]. This equation can be obtained from the Navier–Stokes equations and the continuity equation with the assumptions that the cushion depth is small in comparison with the linear dimensions of the plate and that the inertia of the air can be neglected. The latter is associated with the fact that for the processes under consideration the Reynolds number is much greater than unity. It follows from the first assumption that

$$\frac{\partial u}{\partial x} \sim \frac{\partial u}{\partial z} \ll \frac{\partial u}{\partial y}, \quad \frac{\partial w}{\partial x} \sim \frac{\partial w}{\partial z} \ll \frac{\partial w}{\partial y}, \quad \frac{\partial p}{\partial y} \ll \frac{\partial p}{\partial x} \sim \frac{\partial p}{\partial z}, \tag{7.46}$$

where u, v and w are the components of the vector of the air velocity $\boldsymbol{v}$. Neglecting in the Navier–Stokes equations the air's inertia and the dependence of the velocity components on x and z in accordance with (7.46), we obtain the following equations:

$$\eta \frac{\partial^2 u}{\partial y^2} = \frac{\partial p}{\partial x}, \quad \eta \frac{\partial^2 w}{\partial y^2} = \frac{\partial p}{\partial z}, \tag{7.47}$$

where η is the dynamic viscosity.

The boundary conditions for Eqs. (7.47) are

$$u(x,0,z) = u(x,h,z) = 0, \quad w(x,0,z) = w(x,h,z) = 0.$$

Taking account of these conditions and of the fact that p depends on y only slightly, we find from (7.47)

$$u = \frac{1}{2\eta}\frac{\partial p}{\partial x} y(y-h), \quad w = \frac{1}{2\eta}\frac{\partial p}{\partial z} y(y-h). \tag{7.48}$$

Further, integrating the continuity equation having the form

$$\frac{\partial \rho}{\partial t} + \operatorname{div}(\rho \boldsymbol{v}) = 0,$$

where ρ is the air density associated with the pressure p by a certain equation of state, over y from 0 to h and taking into account the facts that ρ, like p, depends on y only slightly and that $v(x,0,z) = 0$, $v(x,h,z) = \dot{h}$, we obtain

$$\frac{\partial(\rho h)}{\partial t} + \frac{\partial}{\partial x}\left(\rho \int_0^h u \, \mathrm{d}y\right) + \frac{\partial}{\partial z}\left(\rho \int_0^h w \, \mathrm{d}y\right) = 0.$$

Substitution of (7.48) into this equation gives

$$12\eta \frac{\partial(\rho h)}{\partial t} - h^3\left[\frac{\partial}{\partial x}\left(\rho\frac{\partial p}{\partial x}\right) + \frac{\partial}{\partial z}\left(\rho\frac{\partial p}{\partial z}\right)\right] = 0. \tag{7.49}$$

The simplest results appear when the processes occurring in the air are assumed to be isothermal ones, i.e., $\rho \sim p$. In this case Eq. (7.49) becomes

$$24\eta \frac{\partial(ph)}{\partial t} = h^3\left(\frac{\partial^2 p^2}{\partial x^2} + \frac{\partial^2 p^2}{\partial z^2}\right). \tag{7.50}$$

This is precisely the equation which is known as Reynolds' equation.

Ignoring the flow around the plate 2, the boundary conditions for Eq. (7.50) can be written as

$$p(x,b) = p_{\mathrm{a}}, \quad p(0,z) = p(l,z) = p_{\mathrm{a}}. \tag{7.51}$$

In addition to (7.51) we must set a boundary condition for the blow-under line (i.e., for $z = 0$). For this purpose we approximate the discrete set of nozzles by a continuous slot. Ignoring flowing air along this slot and assuming that the slot outlet ends in a pocket of a certain volume, we can write the following balance condition for the air mass issuing from the slot [270]:

$$\frac{\partial(\rho_0 V)}{\partial t} = \frac{\rho_0}{p_0} m_0(x, p_0) - 2\rho_0 \int_0^h w \, \mathrm{d}y, \tag{7.52}$$

where ρ_0 and p_0 are the air density and pressure, respectively, at the blow-under line, $V = V_0 + hd$, V_0 is the pocket volume per unit length of the blow-under line, d is the pocket width along the z-axis, and $(\rho_0/p_0)m_0(x, p_0)$

is the mass of air issuing from the slot in unit time per unit length of the blow-under line, which is called *the mass flow rate.* The mass flow rate depends on both the pressure p_0 at the blow-under line and the pressure P in the blow-under chamber 1. An empirical formula for this dependence is given in [270]. It is conveniently written as

$$m_0(x, p_0, P) = \alpha(x, p_0, P)\vartheta(p_0/P), \tag{7.53}$$

where $\alpha(x, p_0)$ is a function characterizing the distribution of the mass flow rate along the blow-under line, and $\vartheta(p_0/P)$ is the outflow function. Owing to the boundary conditions (7.51) $\alpha(x, p_0)$ should vanish for $x = 0$ and $x = l$. Hence, as a first approximation, it can be taken to be $\alpha(x, p_0, P) = \alpha_0(p_0)\sqrt{P/p_0}\sin\lambda x$, where $\lambda = \pi/l$,

$$\alpha_0(p_0) = \left(\frac{2}{\kappa+1}\right)^{(\kappa+1)/2(\kappa-1)} p_0 S_d\sqrt{\frac{\kappa p_0}{\rho_0}}, \tag{7.54}$$

κ is the isentropic exponent, and S_d is the outflow area per unit length of the blow-under line. The outflow function $\vartheta(p_0/P)$ can be approximated by the Prandtl formula

$$\vartheta(r) = \begin{cases} 1 & \text{for } r \le 0.5, \\ 2\sqrt{r(1-r)} & \text{for } 0.5 \le r \le 1. \end{cases} \tag{7.55}$$

Substituting (7.48) into (7.52), setting $\rho_0 \sim p_0$, and neglecting hd in comparison with V_0, we obtain

$$V_0\frac{\partial p_0}{\partial t} = m_0(x, p_0) + \frac{h^3}{12\eta}\frac{\partial p_0^2}{\partial z}\bigg|_{z=0}. \tag{7.56}$$

In a general way, the solution of Eqs. (7.44), (7.50) with the boundary conditions (7.51), (7.56) is complicated even for the steady-state regime. Therefore, we consider the particular case in which the air pressure p differs only slightly from atmospheric pressure [180]. In this case we can put $p_0 \approx p_a$ in the expressions (7.53), (7.54). Then, as follows from (7.53), (7.55), the value $m_0 \equiv m_0(x, p_a, P)/\alpha_0\sin\lambda x$ will depend on P/p_a, as shown in Fig. 7.9 b.

Putting $p = p_a + \tilde{p}$, where $\tilde{p} \ll p_a$, and taking into account that $\partial(ph)/\partial t \approx p_a\dot{h}$, we write Eq. (7.50) and the boundary conditions (7.51), (7.56) in the following approximate form:

$$12\eta\dot{h} = h^3\left(\frac{\partial^2\tilde{p}}{\partial x^2} + \frac{\partial^2\tilde{p}}{\partial z^2}\right), \tag{7.57}$$

$$\tilde{p}(x, b) = \tilde{p}(0, z) = \tilde{p}(l, z) = 0, \tag{7.58}$$

$$V_0\frac{\partial\tilde{p}_0}{\partial t} = \alpha_0 m_0\sin\lambda x + \frac{p_a h^3}{6\eta}\frac{\partial\tilde{p}_0}{\partial z}\bigg|_{z=0}. \tag{7.59}$$

Taking account of the approximate formula $x(x-l) \approx -(8/\pi\lambda^2)\sin\lambda x$, we can seek a solution of Eq. (7.57) with the boundary conditions (7.58), (7.59) in the form

$$\tilde{p} = \frac{6\eta}{h^3}\left(\frac{8}{\pi\lambda^2}\sin\lambda x\cosh\lambda(b-z) + x(x-l)\right)\dot{h}$$
$$+ p_{\mathrm{a}}T(t)\sin\lambda x\sinh\lambda(b-z), \tag{7.60}$$

where $T(t)$ is an unknown function. This solution satisfies Eq. (7.57) exactly and the boundary conditions (7.58) approximately. From (7.59) we obtain the equation for $T(t)$:

$$\dot{T} + \frac{p_{\mathrm{a}}\lambda h^3}{6\eta V_0\tanh\lambda b}T = \frac{\alpha_0 m_0}{p_{\mathrm{a}}V_0\sinh\lambda b} - \frac{8}{\pi\lambda V_0}\dot{h}$$
$$- \frac{48\eta(\cosh\lambda b - 1)}{\pi\lambda^2 h^3 p_{\mathrm{a}}\sinh\lambda b}\left(\ddot{h} - 3\frac{\dot{h}^2}{h}\right). \tag{7.61}$$

Let us further calculate the force F according to the formula (7.45), in view of (7.60). In doing so we obtain

$$F = \frac{4p_{\mathrm{a}}}{\lambda^2}(\cosh\lambda b - 1)T + \frac{2\eta b l^3}{h^3}\left(\frac{96}{\pi^4\lambda b}\sinh\lambda b - 1\right)\dot{h}. \tag{7.62}$$

Substitution of (7.62) into Eq. (7.44) gives the following equation:

$$m\ddot{h} = \frac{2\eta b l^3}{h^3}\left(\frac{96}{\pi^4\lambda b}\sinh\lambda b - 1\right)\dot{h} + \frac{4p_{\mathrm{a}}}{\lambda^2}(\cosh\lambda b - 1)T - mg. \tag{7.63}$$

Equations (7.63), (7.61) completely describe the motion of the plate. In particular, we can find from these equations and (7.60) the steady-state values of h, T and $\tilde{p}$:

$$h_{\mathrm{st}} = \frac{1}{\lambda}\left(\frac{24\eta\alpha_0 m_0(\cosh\lambda b - 1)}{mgp_{\mathrm{a}}\cosh\lambda b}\right)^{1/3}, \quad T_{\mathrm{st}} = \frac{mg\lambda^2}{4p_{\mathrm{a}}(\cosh\lambda b - 1)},$$
$$\tilde{p}_{\mathrm{st}} = p_{\mathrm{a}}T_{\mathrm{st}}\sin\lambda x\sinh\lambda(b-z). \tag{7.64}$$

It can be seen from this that the assumption $\tilde{p} \ll p_{\mathrm{a}}$, which has been used above, is valid if $mg\lambda^2\sinh\lambda b \ll 4p_{\mathrm{a}}(\cosh\lambda b - 1)$. This constraint can be fulfilled only in the case of a sufficiently light plate.

In order to investigate the stability of the steady-state solution (7.64), it is convenient to substitute into Eqs. (7.63), (7.61)

$$\xi = h - h_{\mathrm{st}}, \quad \zeta = T - T_{\mathrm{st}} + \frac{2mg}{\alpha_0 l\tanh\lambda b}\dot{\xi} + \frac{4}{\pi\lambda V_0\cosh^2(\lambda b/2)}\xi$$

and to write for them the following linearized equations:

$$\ddot{\xi} + 2\delta\dot{\xi} + \omega_0^2\xi = K\zeta, \quad \dot{\zeta} + \gamma\zeta = -a\xi, \tag{7.65}$$

where

$$2\delta = \frac{p_{\mathrm{a}}gl}{m_0\tanh\lambda b}\left(\frac{\pi^3 b}{12l}\coth\frac{\lambda b}{2} - \frac{16}{\pi^2}\right), \quad \omega_0^2 = \frac{32p_{\mathrm{a}}}{\pi m\lambda^3 V_0}\tanh^2\frac{\lambda b}{2},$$

$$K = \frac{4p_a}{m\lambda^2}(\cosh\lambda b - 1), \quad \gamma = \frac{4m_0}{\lambda^2 V_0 mg}\tanh\frac{\lambda b}{2},$$
$$a = \frac{3m_0}{p_a V_0 h \sinh\lambda b} - \frac{4\gamma}{\pi\lambda V_0 \cosh^2(\lambda b/2)}.$$

Thus, we have shown that the system under consideration is reduced to equations of the form (7.1), i.e., it belongs to the class of systems with inertial self-excitation.

The condition of self-excitation of the system (7.65) can easily be transformed to the form

$$\frac{3g}{h_{st}} > p_a\left(\frac{\pi^2\lambda b}{12\tanh\lambda b} - \frac{8}{\pi^2}\right)\left[\frac{4\pi}{m\lambda^3 V_0} + \frac{p_a g^2 l^2}{(\alpha_0 m_0)^2\tanh\lambda b}\coth\frac{\lambda b}{2}\left(\frac{\pi^2\lambda b}{12}\coth\frac{\lambda b}{2} - \frac{16}{\pi^2}\right)\right]. \quad (7.66)$$

Since $h_{st} \sim (\alpha_0 m_0)^{1/3}$, it is easily seen that the condition (7.66) can be fulfilled only over a certain range of m_0 ($m_{0cr} < m_0 < m_0^*$) and for $V_0 > V_{0cr}$, where V_{0cr} is a certain critical value of V_0.

The expression (7.5) for the oscillation frequency in the neighborhood of the threshold of self-excitation can be transformed to the form

$$\omega = \sqrt{\frac{3g\gamma}{h_{st}(\gamma + 2\delta)}}. \quad (7.67)$$

It follows from this that for plates of small mass, as $\gamma \gg \delta$, the frequency of self-oscillation increases with increase of the plate mass because h_{st} decreases. For plates of relatively large mass, as $\gamma \ll \delta$, the self-oscillation frequency decreases with increase of the plate mass because γ/h_{st} decreases.

Equations (7.63), (7.61) allow us to calculate the oscillation amplitude in the neighborhood of the threshold of self-excitation as well.

The system described was also studied experimentally by Bakscis. The plate had linear dimensions $l = 10$ cm, $b = 1$ cm and mass $m = 100$ g. The ratio of the pressure P in the blow-under chamber to atmospheric pressure was about 1.5. For these parameters $h_{st} \approx 0.5$ mm. Self-oscillations with frequency $\omega/2\pi = 35$ Hz and amplitude of about 0.2 mm were observed.

We note that a system which is akin to that considered above, is used in practice in the automatic assembly of machine components [24].

7.2 Self-oscillatory systems with inertial nonlinearity

A nonlinear inertial feedback can result not only in the hard excitation of self-oscillations but in the limitation of the amplitude of the self-oscillations excited. Systems in which this occurs are reffered to as *systems with inertial nonlinearity*.

Contrary to systems with inertial excitation, a notion of self-oscillatory systems with inertial nonlinearity has long existed. This notion was first introduced by Meacham in 1938 [235]. In the mid 1940s a major contribution to the investigation of systems with inertial nonlinearity was made by Teodorchik [331, 332, 334]. The simplest equations of these systems can be written in the form

$$\begin{aligned} &\ddot{x} - \Big(\mu - y + F(y)\Big)\dot{x} + \omega_0^2 x = f(x, \dot{x}, y), \\ &\dot{y} + \gamma y = \gamma\varphi(x, \dot{x}, y), \end{aligned} \tag{7.68}$$

where f, F and φ are nonlinear functions free of linear terms. It can be seen that in systems with inertial nonlinearity of the form (7.68) the self-excitation of oscillations is conditioned by negative friction ($\mu > 0$), and the limitation of their amplitude occurs as a result of nonlinear inertial interaction between dynamical variables x and y.

The generator studied by Teodorchik (see Fig. 7.10 a) differs from the

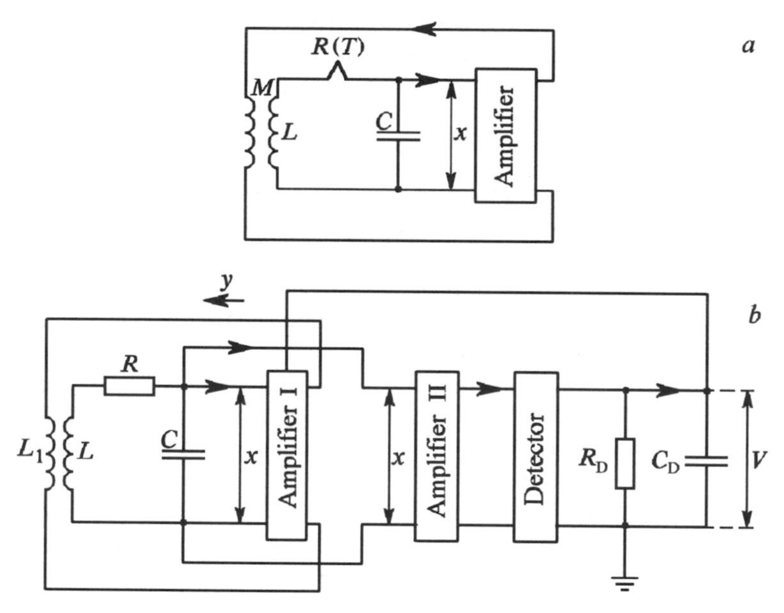

Fig. 7.10. The block diagrams of generators (a) with a thermistor and (b) with a detector

usual vacuum tube generator investigated by van der Pol [350, 353] in that in the oscillatory circuit there is a thermistor in place of an ordinary resistor R, i.e., a resistor whose value depends crucially on its temperature. Since the temperature, in its turn, depends inertially on the current through the oscillatory circuit, the availability of the thermistor causes an extra half degree of freedom. Thus, a generator with inertial nonlinearity is a self-oscillatory system with one and half degrees of freedom.

To clarify the role of just inertial nonlinearity, Teodorchik assumed that the amplifier is linear, i.e., all nonlinearity is concentrated in the thermistor $R(T)$. With this assumption the equation of oscillations in the Teodorchik generator with a thermistor is

$$L\frac{\mathrm{d}I}{\mathrm{d}t} + R(T)I + \frac{1}{C}\int\left(I - MS\frac{dI}{dt}\right)dt = 0, \tag{7.69}$$

where I is the current through the oscillatory circuit, M is the mutual induction factor, and S is the steepness of the volt-ampere characteristic of the amplifier. To derive the equation for the temperature T we suppose that heat transfer obeys Newton's law. Then we obtain

$$mc\frac{\mathrm{d}T}{\mathrm{d}t} + kT = R(T)I^2, \tag{7.70}$$

where m is the mass of the thermistor filament, c is its specific heat, and k is the heat transfer factor. Setting the thermistor resistance $R(T)$ to be $R(T) = R_0 + bLT$, we transform Eqs. (7.69), (7.70) to the form

$$\begin{aligned} &\ddot{I} - (\mu - bT)\dot{I} + \omega_0^2 I = -bI\dot{T}, \\ &\dot{T} + \gamma T = \alpha_0 I^2 + \alpha_1 T I^2, \end{aligned} \tag{7.71}$$

where $\omega_0 = 1/\sqrt{LC}$, $\mu = \omega_0^2 MS - R_0/L$, $\gamma = k/mc$, $\alpha_0 = R_0/mc$, $\alpha_1 = bL/mc$. It is easily seen that Eqs. (7.71) belong to the class of Eqs. (7.68).

For sufficiently small μ and γ ($\mu/\omega_0 \sim \gamma/\omega_0 \sim \epsilon$, where ϵ is a small parameter that in the final results should be put equal to unity) Eqs. (7.71) can be approximately solved by the Krylov–Bogolyubov method. For this we rewrite Eqs. (7.71) as[4]

$$\begin{aligned} &\ddot{I} - (\mu - bT)\dot{I} + \omega_0^2 I = \epsilon(\mu - bT)\dot{I} + \epsilon^2 bI\Big(\gamma T - (\alpha_0 + \alpha_1 T)I^2\Big), \\ &\dot{T} = -\epsilon\Big(\gamma T - (\alpha_0 + \alpha_1 T)I^2\Big). \end{aligned} \tag{7.72}$$

In accordance with the Krylov–Bogolyubov method we set

$$\begin{aligned} &I = A\cos\psi + \epsilon u_1(A, \psi, T_0) + \dots, \\ &T = T_0 + \epsilon v_1(A, \psi, T_0) + \dots, \end{aligned} \tag{7.73}$$

where $\psi = \omega_0 t + \varphi$, $u_1, \dots$ and $v_1, \dots$ are unknown functions containing no constant constituent and constituent at the frequency ω_0. In Eqs. (7.73) A, φ and T_0 are supposed to be slowly varying functions of time obeying the equations

[4] We take into account that $bT \sim \mu \sim \epsilon\omega_0$.

$$\dot{A} = \epsilon f_1(A, \varphi, T_0) + \dots, \quad \dot{\varphi} = \epsilon f_2(A, \varphi, T_0) + \dots,$$
$$\dot{T}_0 = \epsilon f_3(A, \varphi, T_0) + \dots, \tag{7.74}$$

where $f_1, \dots, f_2, \dots$ and $f_3, \dots$ are unknown functions.

In the first approximation with respect to the small parameter ϵ we obtain the following equations for the functions u_1 and v_1:

$$\frac{\partial^2 u_1}{\partial \psi^2} + u_1 = -\frac{1}{\omega_0}\Big(\big(A(\mu - bT_0) - 2f_1\big)\sin\psi - 2Af_2\cos\psi\Big),$$
$$\frac{\partial v_1}{\partial \psi} = -\frac{1}{\omega_0}\Big(\gamma T_0 - (\alpha_0 + \alpha_1 T_0)A^2\cos^2\psi + f_3\Big). \tag{7.75}$$

Equating the resonance terms in the right-hand sides of these equations to zero, we find the functions f_1, f_2 and f_3:

$$f_1 = \frac{1}{2}(\mu - bT_0)A, \quad f_2 = 0, \quad f_3 = -\gamma T_0 + \frac{(\alpha_0 + \alpha_1 T_0)A^2}{2}. \tag{7.76}$$

Further, from Eqs. (7.75) we find the functions u_1 and v_1:

$$u_1 = 0, \quad v_1 = \frac{1}{4}(\alpha_0 + \alpha_1 T_0)A^2 \sin 2\psi. \tag{7.77}$$

As follows from the fact that $u_1 = 0$, in the first approximation with respect to ϵ the oscillations excited in the generator with inertial nonlinearity are pure harmonic. This is one of the distinctions between such a generator and a van der Pol generator.

Substituting (7.76) into Eqs. (7.74), we obtain the following truncated equations for A, φ and T: [5]

$$\dot{A} = \frac{1}{2}(\mu - bT)A, \quad \dot{\varphi} = 0,$$
$$\dot{T} = -\gamma T + \frac{1}{2}(\alpha_0 + \alpha_1 T)A^2. \tag{7.78}$$

As was mentioned in [166] a generator with a thermistor is unsuitable for experimental investigations because it does not allow us to vary its parameters over a wide range. A more convenient modification of a generator with inertial nonlinearity was proposed by Kaptsov. It differs from the Teodorchik generator in that the role of the inertial element is played not by a thermistor but by a detector (see Fig. 7.10 b). The voltage across the detector reduces the steepness of the volt-ampere characteristic, and thereby limits the oscillation amplitude. The extent of inertiality is determined by the detector time constant $R_\mathrm{d}C_\mathrm{d} \equiv 1/\gamma$. The equations for such a generator are

[5] From this point on we will drop the subscript '0' on T.

$$\ddot{x} + \frac{R}{L}\dot{x} + \omega_0^2 x = M\omega_0^2 \dot{y}, \quad y = Sx, \tag{7.79}$$

where S is the steepness of the volt-ampere characteristic of the amplifier which supposedly depends on the detector voltage V, i.e.,

$$S = S_0 - \frac{b}{\omega_0^2} V. \tag{7.80}$$

Substituting (7.80) into (7.79) and eliminating y we obtain the equation for x:

$$\ddot{x} - (\mu - bV)\dot{x} + \omega_0^2 x = -b\dot{V}x, \tag{7.81}$$

where $\mu = \omega_0^2 M S_0 - R/L$. The detector voltage V is related to the voltage x by the following equation:

$$\dot{V} + \gamma V = \gamma f(x), \tag{7.82}$$

where $f(x)$ is a nonlinear function determined by the properties of the detector. If the detector is quadratic, i.e., $f(x) = \alpha x^2$, then Eqs. (7.81) and (7.82) are akin to Eqs. (7.71). In the case of a linear detector, for which $f(x) = K\vartheta(x)x$, where $\vartheta(x)$ is Heaviside's step function, Eqs. (7.81) and (7.82) are somewhat different from Eqs. (7.71). In the case when the parameters μ and γ are small we obtain for the case of a linear detector the following truncated equations for A, φ and V:

$$\begin{aligned} \dot{A} &= \frac{1}{2}(\mu - bV)A, \quad \dot{\varphi} = 0, \\ \dot{V} &= -\gamma\left(V - \frac{K}{\pi}A\right). \end{aligned} \tag{7.83}$$

Equations (7.83) only differ by the last term from Eqs. (7.78). As has been shown by analytical studies and computations, this difference is not essential and the majority of processes in all of the generators considered follow a qualitatively similar course.

Another important distinction between a generator with inertial nonlinearity and a van der Pol generator lies in the character of transient processes. For definiteness, let us investigate the stability of the steady-state solution of Eqs. (7.78) corresponding to a generator with a thermistor or with a quadratic detector. This solution is

$$T_{\text{st}} = \frac{\mu}{b}, \quad A_{\text{st}} = \sqrt{\frac{2\mu\gamma}{\alpha_0 b + \alpha_1 \mu}}. \tag{7.84}$$

The character of the stability of the steady-state solution (7.84) and of the transient process essentially depends on the relation between the parameters γ and μ. If $\gamma \gg \mu$, then in Eqs. (7.78) we can consider the variable T as 'fast' one and eliminate it by substituting a steady-state solution of the equation for T into the equation for A. This steady-state solution is

$$T = \frac{\alpha_0 A^2}{2\gamma - \alpha_1 A^2}. \tag{7.85}$$

By substituting (7.85) into the first equation of (7.78) we obtain the equation for the oscillation amplitude in a form similar to the van der Pol equation:

$$\dot{A} = \frac{1}{2}\left(\mu - \frac{b\alpha_0 A^2}{2\gamma - \alpha_1 A^2}\right) A. \tag{7.86}$$

This means that in the case of $\gamma \gg \mu$ the generator behaves as an inertialess one.

In the case of $\gamma \sim \mu$ the equations for the deviations $\xi(t)$ and $\eta(t)$ from the steady-state solution (7.84) are

$$\dot{\xi} = -\frac{bA_{\mathrm{st}}}{2}\eta, \quad \dot{\eta} = \frac{2\mu\gamma}{bA_{\mathrm{st}}}\xi - \gamma\frac{\alpha_0 b}{\alpha_0 b + \alpha_1\mu}\eta. \tag{7.87}$$

Differentiating the second equation of (7.87) with respect to t and substituting $-bA_{\mathrm{st}}/2$ in place of $\dot{\xi}$, we obtain the following equation for η:

$$\ddot{\eta} + \gamma a\dot{\eta} + \mu\gamma\eta = 0, \tag{7.88}$$

where $a = \alpha_0 b/(\alpha_0 b + \alpha_1\mu)$.

The general solution of Eq. (7.88) is

$$\eta = C_1 \exp\left\{-\frac{\gamma a}{2}\left(1 + \sqrt{1 - \frac{4\mu}{\gamma a^2}}\right)t\right\} + C_2 \exp\left\{-\frac{\gamma a}{2}\left(1 - \sqrt{1 - \frac{4\mu}{\gamma a^2}}\right)t\right\}, \tag{7.89}$$

where the constants C_1 and C_2 are found from the initial conditions.

It follows from (7.89) that, for $\gamma a^2 \le 4\mu$, the transient process is of oscillatory character with frequency $\Omega = \sqrt{\gamma(\mu - \gamma a^2/4)}$. As the inertial parameter γ decreases from $4\mu/a^2$ to zero, Ω increases initially, peaking at $\gamma = 2\mu/a^2$, and then decreases to zero. The oscillatory character of transient processes is a distinguishing feature of self-oscillatory systems with inertial nonlinearity.

If the parameters μ and γ are not small then oscillations in a generator with inertial nonlinearity can be chaotic. Such regimes for a generator with a detector were studied in detail numerically and experimentally by Anischenko et al. A review of the results obtained by them is given in [12, 13, 14, 15].

7.3 Some other systems with one and a half degrees of freedom

7.3.1 The Rössler equations

The Rössler equations were proposed as one of the models for a hypothetical chemical reaction in which oscillations of reagent concentrations can be chaotic [289]. We consider these equations because at the present time they can be compared only with the Lorenz equations in regard to popularity and widespread use in studies of dynamical chaos.

The Rössler equations are

$$\dot{X} = -Y - Z, \quad \dot{Y} = X + eY, \quad \dot{Z} = d - cZ + XZ, \tag{7.90}$$

where c, d and e are positive numbers. These equations have either no singular points or two singular points with coordinates

$$X_{1,2} = \frac{c}{2} \pm \sqrt{\frac{c^2}{4} - ed}, \quad Y_{1,2} = -Z_{1,2} = -\frac{X_{1,2}}{e}.$$

It is seen from this that the singular points exist only for $c \geq 2\sqrt{ed}$. The investigation of the stability of these singular points results in the following characteristic equation:

$$p^3 + (c - e - X_{1,2})\, p^2 + \left(1 - ce - \frac{1+e^2}{e} X_{1,2}\right) p$$
$$+ c(1 - e) - (2 - e)X_{1,2} = 0. \tag{7.91}$$

It follows from this that the first singular point is aperiodically unstable for any values of the parameters, whereas the second singular point can be both stable and unstable. After its birth the second singular point, along with the first one, is aperiodically unstable. It becomes stable for

$$c = c_0 = \sqrt{4ed\left(1 + \frac{e^2}{4(1-e)}\right)}. \tag{7.92}$$

The second singular point remains stable up to a certain critical value of $c = c_{\rm cr}$ for which the condition of oscillatory instability comes into play. This condition is

$$(c - e - X_2)\left(1 - ce - \frac{1+e^2}{e} X_2\right) - c(1 - e) + (2 - e)X_2 \leq 0. \tag{7.93}$$

Numerical study of Eqs. (7.90) was first performed by Rössler [289] for $e = d = 0.2$ and varying the parameter c. In this case $c_{\rm cr} \approx 1.233$. For $c_0 \leq c \leq c_{cr}$ both singular points are stable. But hard excitation of periodic self-oscillations occurs in the system (see Fig. 7.11 a). As condition (7.93) comes into play, the excitation of self-oscillations becomes soft. Up to $c \approx 2.8$ the self-oscillations are periodic (Fig. 7.11 b). As the parameter c increases further, these oscillations undergo a sequence of period-doubling bifurcations (Fig. 7.11 c, d, e), which comes to a close for $c \approx 4.2$. A chaotic attractor with a pronounced layered structure then appears in the system phase space (Fig. 7.11 f, g, h). This structure vanishes for $c \approx 4.6$ (Fig. 7.11 i).

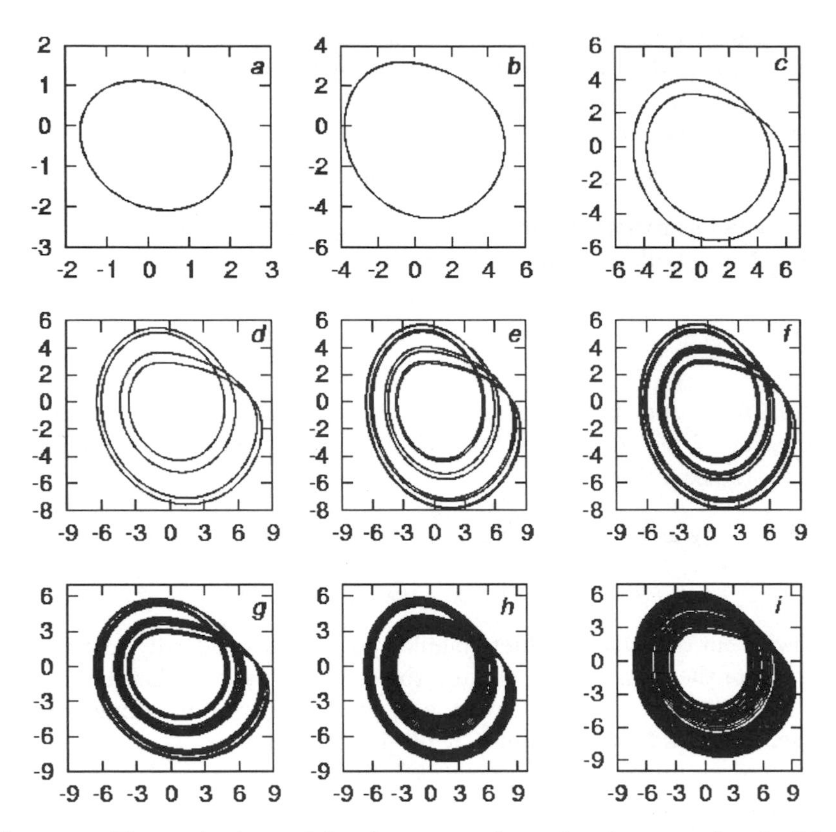

Fig. 7.11. The projections of the phase portrait on the plane x, y for the Rössler system for $e = d = 0.2$, (a) $c = 1$, (b) $c = 2.6$, (c) $c = 3$, (d) $c = 4$, (e) $c = 4.18$, (f) $c = 4.21$, (g) $c = 4.23$, (h) $c = 4.3$, and (i) $c = 4.6$

7.3.2 A three-dimensional model of an immune reaction illustrating the oscillatory course of some chronic diseases

The simplest mathematical model of an immune reaction in the human organism, suggested by Smirnova and Stepanova as early as 1971 [311, 284], describes oscillations in concentrations of mature plasmocytes (of special immune cells) (z), of antibodies produced by them (y), and of an antigene (x) caused by the presence in the organism of pathogenic bacteria or viruses. It can be represented as

$$\dot{x} = \beta x - xy, \quad \dot{y} = -y - xy + \sigma z, \quad \dot{z} = -k(z - x), \tag{7.94}$$

where kx describes the rise in the cell number z in response to the antigene, βx characterizes the birth of the antigene carriers in the organism of a sick person, the terms $-xy$ in the first and second equations of (7.94) describe the death of the antigenes and the antibodies resulting from their interaction,

σz characterizes the production of the antibodies by the cells z, and the terms $-kz$ and $-y$ describe the natural death of the plasmocytes z and the antibodies y, respectively. A certain modification of Eqs. (7.94) is given in [284, 285], differing from (7.94) in that a nonlinear function describing the limitation of the rate of rise of the plasmocyte number as the antigene number increases is substituted into Eqs. (7.94) in place of the term kx.

We note that Eqs. (7.94) can be treated as a self-oscillatory model of the interaction of two kinds of 'predator' whose numbers are determined by y and z with 'prey' whose number is determined by x. In this model the 'predator' of kind y (the antibodies) 'eats up' the 'prey' (antigene) not directly but by means of another 'predator' (the mature plasmocytes z).

Equations (7.94) have two singular points with coordinates $x = y = z = 0$ and $x = z = \beta/(\sigma - \beta) \equiv x_0$, $y = \beta$. It is seen from this that the second point has a physical meaning only if $\sigma > \beta$. For $\beta > 0$ the first singular point is always unstable, and the second singular point can be both stable and unstable. To obtain a condition for its stability we substitute the deviations $\xi = x - x_0$, $\eta = y - \beta$, $\zeta = z - x_0$ into Eqs. (7.94). As a result we obtain

$$\dot{\xi} = -x_0\eta - \xi\eta, \quad \dot{\eta} = -\beta\xi - \gamma\eta + \sigma\zeta - \xi\eta, \quad \dot{\zeta} = k(\xi - \zeta), \tag{7.95}$$

where $\gamma = 1 + x_0 = \sigma/(\sigma - \beta)$.

The characteristic equation corresponding to linearized equations (7.95) is

$$p^3 + \left(\frac{\sigma}{\sigma - \beta} + k\right) p^2 + \frac{k\sigma - \beta^2}{\sigma - \beta} p + \beta k = 0. \tag{7.96}$$

It follows from (7.96) that the singular point (x_0, β, x_0) is a saddle-node for $\beta > \sqrt{\sigma k}$, a saddle-focus for $\beta_{cr} < \beta < \sqrt{\sigma k}$, and a stable focus for $\beta \leq \beta_{cr}$, where

$$\beta_{cr} = \frac{k(\sigma + k) - \sqrt{k^2(\sigma + k)^2 - 4\sigma k(k^2 - 1)}}{2(k - 1)}. \tag{7.97}$$

So, for $\beta > \beta_{cr}$ the singular point under consideration is unstable. If $\beta_{cr} < \beta < \sqrt{\sigma k}$ then its instability is of an oscillatory character.

As the singular point (x_0, β, x_0) loses its stability a stable limit cycle appears around it. This limit cycle corresponds to the periodic course of a disease that is inherent in such chronic diseases as malaria, fever and so on. The results of the numerical simulation of Eqs. (7.95) for $k = 1.8$, $\sigma = 2.5$, $\beta = 2.4$ are illustrated in Fig. 7.12.[6] It should be noted that no chaotic solutions of Eqs. (7.95) have been found by us.

[6] For the indicated values of k and σ the critical value of β is approximately equal to 2.071.

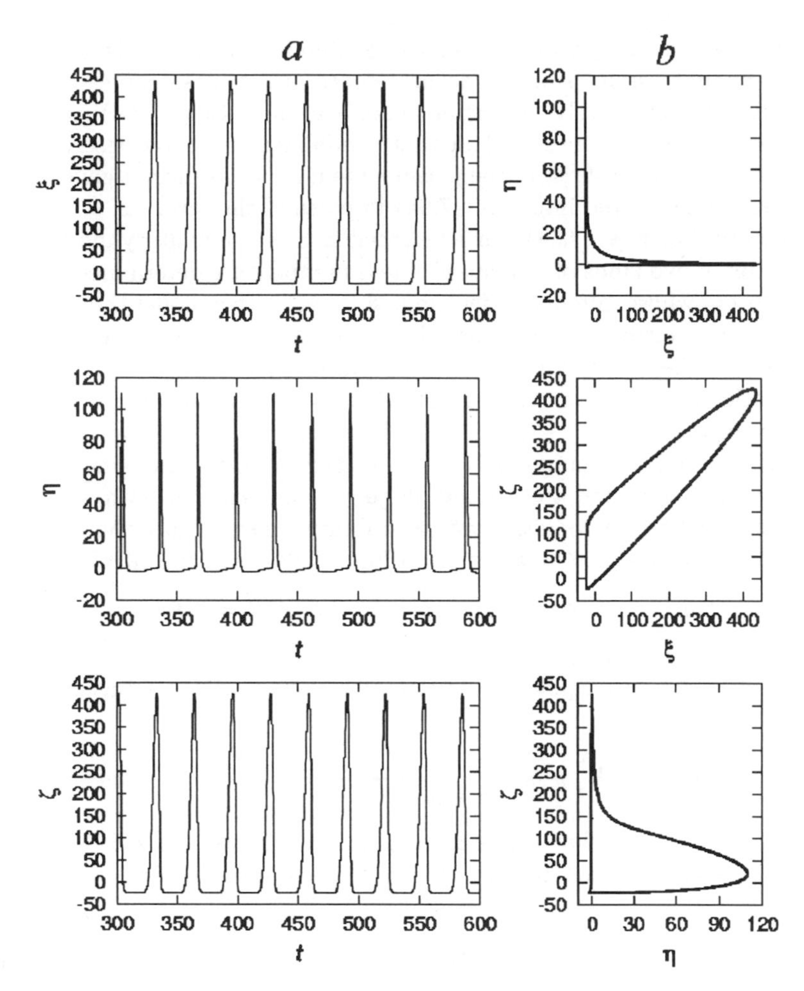

Fig. 7.12. (a) The time dependencies of ξ, η and ζ; (b) the projections of the phase portrait on the plane $\xi\,\eta$, $\xi\,\zeta$ and $\eta\,\zeta$: $k = 1.8$, $\sigma = 2.5$, $\beta = 2.4$

8. Examples of self-oscillatory systems with two or more degrees of freedom

8.1 Generator with an additional circuit

A triode generator with an additional circuit is the best known example of a self-oscillatory system with two degrees of freedom. Van der Pol the first to study such a generator [351]. Later these generators were studied in detail by Andronov and Vitt [4] and by Skibarko and Strelkov [309]. An interesting and important phenomenon, having come to be known as the frequency hysteresis, occurs in such a generator. This phenomenon consists in that the self-oscillation frequency depends on the system pre-history.

We consider the triode generator with an additional circuit studied by Andronov and Vitt [4]. Its schematic image is shown in Fig. 8.1. Setting

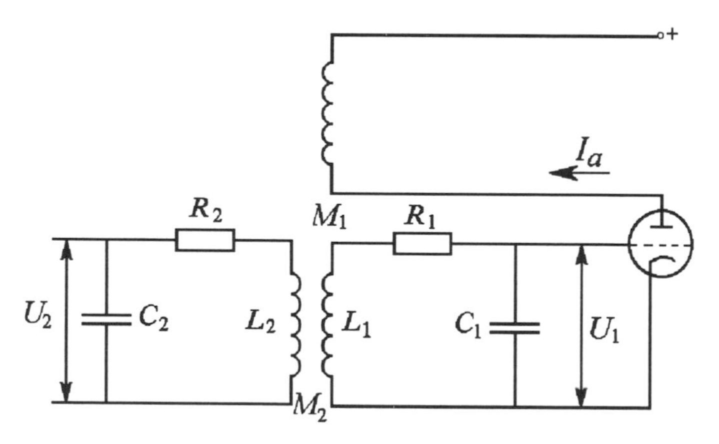

Fig. 8.1. The schematic image of the triode generator with two inductively coupled circuits

$I_a = S_0 U_1 - S_1 U_1^3/3$, we can write the equations of such a generator in the following form:

$$\ddot{x} - \mu(1 - x^2)\dot{x} + x = m_1 \ddot{y},$$
$$\ddot{y} + \epsilon\mu\dot{y} + \xi y = m_2 \ddot{x}, \tag{8.1}$$

where

$$\mu = \frac{M_1 S_0 - R_1 C_1}{\sqrt{L_1 C_1}}, \quad \epsilon = \frac{R_2 L_1 C_1}{L_2(M_1 S_0 - R_1 C_1)},$$

$$x = \sqrt{\frac{M_1 S_1}{M_1 S_0 - R_1 C_1}} U_1, \quad y = \sqrt{\frac{M_1 S_1}{M_1 S_0 - R_1 C_1}} U_2$$

are dimensionless voltage drops across the triode grid and the capacitor C_2, respectively, $m_{1,2} = M_2 C_{2,1}/(L_{1,2} C_{1,2})$ are the coupling factors, $\xi = L_1 C_1/(L_2 C_2)$ is the frequency detuning squared, $M_{1,2}$ are the factors of mutual induction between the coils L_1 and L and L_1 and L_2, respectively; the dots mean the differentiation with respect to dimensionless time $\tau = t/\sqrt{L_1 C_1}$.

Andronov and Vitt solved Eqs. (8.1) by using the Poincaré method [273] on the assumption that the parameter μ is sufficiently small. However, it is the author's opinion that using the Krylov–Bogolyubov method is more preferable. According to this method we seek a solution of Eqs. (8.1) in the form

$$\begin{aligned} x &= A\cos\psi + \mu u_1(\psi) + \mu^2 u_2(\psi) + \dots, \\ y &= kA\cos\psi + \mu v_1(\psi) + \mu^2 v_2(\psi) + \dots, \end{aligned} \tag{8.2}$$

where $\psi = \omega t + \varphi$, ω is one of the normal frequencies determined by the equation

$$(1 - m_1 m_2)\omega^4 - (1+\xi)\omega^2 + \xi = 0, \tag{8.3}$$

$$k = \frac{\omega^2 - 1}{m_1 \omega^2} = \frac{m_2 \omega^2}{\omega^2 - \xi}, \tag{8.4}$$

A and φ are determined by the equations

$$\begin{aligned} \dot{A} &= \mu f_1(A,\varphi) + \mu^2 f_2(A,\varphi) + \dots, \\ \dot{\varphi} &= \mu F_1(A,\varphi) + \mu^2 F_2(A,\varphi) + \dots, \end{aligned} \tag{8.5}$$

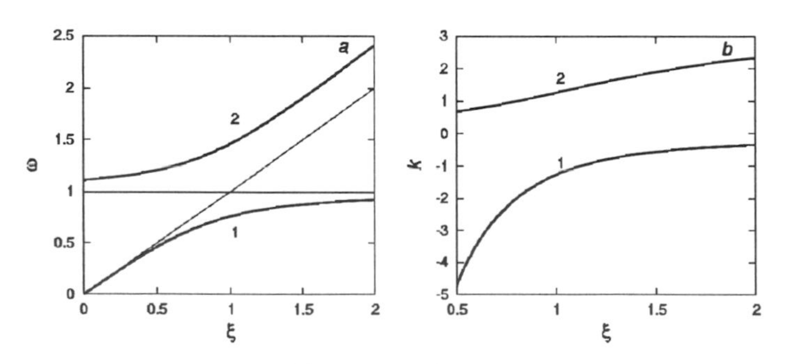

Fig. 8.2. The dependencies of (a) $\omega_{1,2}$ and (b) $k_{1,2}$ on ξ for $m_1 = 0.25$, $m_2 = 0.4$

$u_1(\psi)$, $u_2(\psi)$, ..., $v_1(\psi)$, $v_2(\psi)$, ..., $f_1(A,\varphi)$, $f_2(A,\varphi)$, ..., and $F_1(A,\varphi)$, $F_2(A,\varphi)$, ... are unknown functions.

Equation (8.3) possesses two real roots, ω_1 and ω_2. The dependencies of these roots squared on ξ are shown in Fig. 8.2 a. This plot is usually called the Vien diagram. To each $\omega_{1,2}$ there corresponds its own value of k: k_1 and k_2. The dependencies of $k_{1,2}$ on ξ are given in Fig. 8.2 b.

Substituting (8.2) into Eqs. (8.1) and equating the coefficients of μ we find the equations for u_1 and v_1:

$$\begin{aligned}
\omega^2 \frac{\mathrm{d}^2 u_1}{\mathrm{d}\psi^2} + u_1 - m_1\omega^2 \frac{\mathrm{d}^2 v_1}{\mathrm{d}\psi^2} &= -\Big(1 - A^2\cos^2\psi\Big) A\omega\sin\psi \\
&\quad + 2\omega(1 - m_1 k)\Big(f_1\sin\psi + AF_1\cos\psi\Big), \\
\omega^2 \frac{\mathrm{d}^2 v_1}{\mathrm{d}\psi^2} + \xi v_1 - m_2\omega^2 \frac{\mathrm{d}^2 u_1}{\mathrm{d}\psi^2} &= k\epsilon A\omega\sin\psi \\
&\quad + 2\omega(k - m_2)\Big(f_1\sin\psi + AF_1\cos\psi\Big).
\end{aligned} \tag{8.6}$$

A stationary solution of Eqs. (8.6) can be represented as

$$u_1 = B_{11}\sin\psi + B_{13}\sin 3\psi\,, \quad v_1 = C_{11}\sin\psi + C_{13}\sin 3\psi\,. \tag{8.7}$$

Since the determinant of the system of equations for B_{11} and C_{11} is equal to zero, the unknown functions f_1 and F_1 have to be found from the compatibility conditions. As a result, we obtain

$$f_1 = -\frac{A\omega^2\Big((\xi - \omega^2)(4 - A^2) + 4\epsilon(\omega^2 - 1)\Big)}{8\Big(\omega^2 - \xi + \xi(\omega^2 - 1)\Big)}\,, \quad F_1 = 0. \tag{8.8}$$

It follows from here that in the first approximation the correction to the frequency is equal to zero, and the steady-state value of the oscillation amplitude squared is

$$A^2 = 4\left(1 - \epsilon\frac{1 - \omega^2}{\xi - \omega^2}\right). \tag{8.9}$$

So, to each of the normal frequencies $\omega_{1,2}$ there corresponds its own value of the oscillation amplitude $A_{1,2}$. The plots of $A^2_{1,2}$ versus ξ are illustrated in Fig. 8.3 for different values of ϵ. The investigation of stability of the solutions found, performed by Andronov and Vitt, shows that the self-oscillations at the frequency ω_1 are stable for $A_1^2 \geq A_2^2/2$, and the self-oscillations at the frequency ω_2 are stable for $A_2^2 \geq A_1^2/2$. In Fig. 8.3 stable parts of the curves are shown by heavy lines. It is easily seen from Fig. 8.3 that for $\epsilon < 1$ there exists a range of ξ where the oscillations at both the frequency ω_1 and the frequency ω_2 are stable. In this range frequency hysteresis takes place. For $\epsilon = 1$ the hysteresis is absent, but for $\xi = 1$ a jump in the frequency occurs. For $\epsilon > 1$ the first approximation gives a range of ξ where the self-oscillations

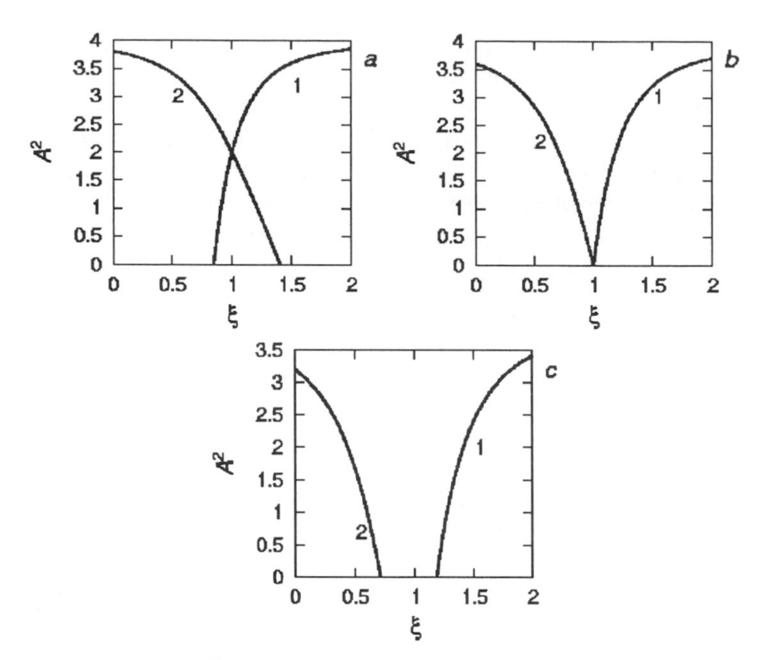

Fig. 8.3. The plots of $A^2_{1,2}$ versus ξ for $m_1m_2 = 0.1$: (a) $\epsilon = 0.5$, (b) $\epsilon = 1$ and (c) $\epsilon = 2$

do not exist. It is evident that in this range the equilibrium state of Eqs. (8.1) must be stable. By using the Routh–Hurwitz criterion we obtain the following conditions for the stability of the equilibrium state:

$$\epsilon \geq 1, \quad \xi_1 \leq \xi \leq \xi_2, \tag{8.10}$$

where

$$\xi_{1,2} = \frac{\epsilon}{\epsilon - m_1m_2}\left(1 - m_1m_2 \mp (\epsilon - 1)\sqrt{\frac{m_1m_2}{\epsilon} - \mu^2}\right). \tag{8.11}$$

The conditions (8.10) have meaning only for $\epsilon \leq m_1m_2/\mu^2$. It can be shown that the range of ξ determined by the conditions (8.10) and the range of ξ, where in the first approximation the self-oscillations do not exist, coincide only if

$$\mu^2 \ll \frac{m_1m_2}{\epsilon}. \tag{8.12}$$

Thus, we can conclude that the results found by us in the first approximation are valid only if the condition (8.12) is fulfilled.

It is very interesting that, if the condition (8.12) is not fulfilled, the true results can be obtained by using the harmonic linearization method [276], because in doing so a term of order μ^2 is taken properly into account. Setting $x = A\cos\omega t$ and linearizing the nonlinear function $x^2\dot{x}$, we obtain the following linearized equations:

$$\ddot{x} + x - \mu\left(1 - \frac{A^2}{4}\right)\dot{x} = m_1\ddot{y},$$
$$\ddot{y} + \xi y + \epsilon\mu\dot{y} = m_2\ddot{x}. \tag{8.13}$$

The characteristic equation associated with Eqs. (8.13) is

$$(1 - m_1 m_2)p^4 + \mu\left(\epsilon - 1 + \frac{A^2}{4}\right)p^3 + \left(1 + \xi - \epsilon\mu^2\left(1 - \frac{A^2}{4}\right)\right)p^2$$
$$+\mu\left(\epsilon - \xi\left(1 - \frac{A^2}{4}\right)\right)p + \xi = 0. \tag{8.14}$$

Since we are interested in a periodic solution of Eqs. (8.13), we can set $p = i\omega$ in Eq. (8.14). Equating both the real and imaginary parts of Eq. (8.14) to zero, we obtain two equations for A and ω:

$$(1 - m_1 m_2)\omega^4 - \left(1 + \xi - \epsilon\mu^2\left(1 - \frac{A^2}{4}\right)\right)\omega^2 + \xi = 0, \tag{8.15}$$

$$1 - \frac{A^2}{4} = \epsilon\frac{\omega^2 - 1}{\omega^2 - \xi}. \tag{8.16}$$

Equation (8.16) coincides with (8.9) and Eq. (8.15) differs from (8.3) by the term of order μ^2.

Substituting A^2 from Eq. (8.16) into Eq. (8.15) we obtain the equation for the frequency ω. The solution of this equation allows us to find the dependence of ω^2 on ξ. Substituting further $\omega^2(\xi)$ into Eq. (8.16), we find the dependence of A^2 on ξ. The dependencies of ω^2 and A^2 on ξ for $m_1 m_2 = 0.1$, $\mu^2 = 0.06$, and different values of ϵ are given in Fig. 8.4. We see that these dependencies essentially differ from those obtained in the first approximation with respect to μ. In particular, the effect of quenching the self-oscillations is absent.

It should be noted that the results obtained by the harmonic linearization method are in good agreement with the numerical simulation of the initial equations (8.1) [276].

8.2 A lumped model of bending-torsion flutter of an aircraft wing

We consider here the conditions for the appearance of bending-torsion flutter of a cantilever homogeneous wing represented as a thin elastic beam. It should be emphasized that this problem is presently of no practical importance, but it may be useful for many other currently available problems.

We use the known expressions for aerodynamic forces acting on the wing which follow from the so-called quasi-stationary theory [110, 44]. In this case

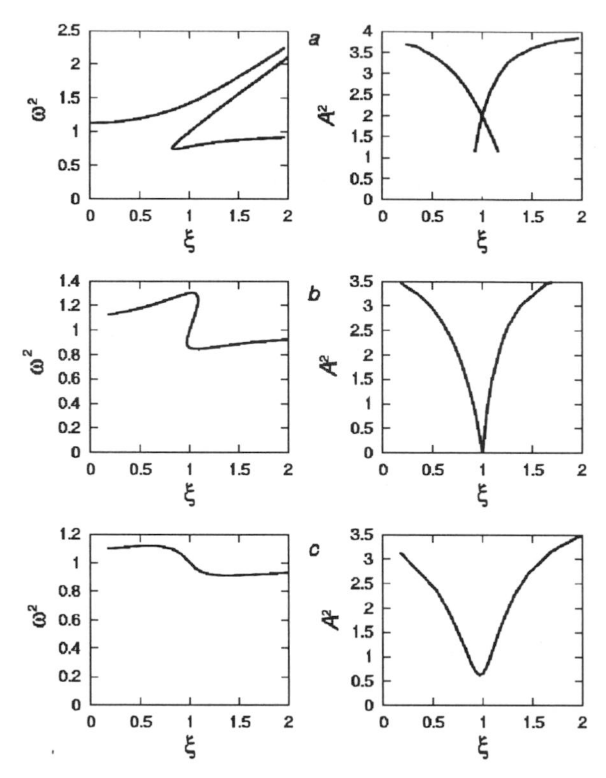

Fig. 8.4. The plots of ω^2 and A^2 versus ξ for $m_1m_2 = 0.1$, $\mu^2 = 0.06$: (a) $\epsilon = 0.5$, (b) $\epsilon = 1$ and (c) $\epsilon = 2$

the equations for bending-torsion oscillations of the wing in flight are [110, 168]:

$$
\begin{aligned}
&EI\,\frac{\partial^4 u}{\partial x^4} + m\,\frac{\partial^2 u}{\partial t^2} - m\sigma\,\frac{\partial^2 \Theta}{\partial t^2} \\
&- \frac{\partial C_y}{\partial \alpha}\,\rho V^2 b\left[\Theta + \frac{b}{V}\left(\frac{3}{4} - \frac{a}{b}\right)\frac{\partial \Theta}{\partial t} - \frac{1}{V}\frac{\partial u}{\partial t}\right] = 0, \\
&GI_p\,\frac{\partial^2 \Theta}{\partial x^2} - m\sigma\,\frac{\partial^2 u}{\partial t^2} + I_m\,\frac{\partial^2 \Theta}{\partial t^2} \\
&- \frac{\partial C_{mE}}{\partial \alpha}\,\rho V^2 b^2\left[\Theta + \frac{b}{V}\left(\frac{3}{4} - \frac{a}{b} - \frac{\pi}{16}\left(\frac{\partial C_{mE}}{\partial \alpha}\right)^{-1}\right)\frac{\partial \Theta}{\partial t} - \frac{1}{V}\frac{\partial u}{\partial t}\right] = 0,
\end{aligned}
\tag{8.17}
$$

where x is the section coordinate, u is the sag of the wing, Θ is the torsion angle, V is the flight velocity, EI is the wing stiffness with respect to bending, GI_p is the wing rigidity with respect to torsion, m is the linear mass, I_m is the linear moment of inertia, σ is the distance between the wing section center of gravity and the rigidity axis, b is the wing span, a is the distance between the wing leading edge and the rigidity axis, ρ is the air density, $\partial C_y/\partial \alpha$ is the derivative of the lift coefficient with respect to the angle of attack, $\partial C_{mE}/\partial \alpha$ is the derivative of the lift moment coefficient with respect to the angle of attack, and l is the wing length.

To find the critical value of the flight velocity corresponding to the onset of flutter, we use the Galerkin method [97]. This method is as follows. Let there be a linear distributed one-dimensional oscillatory system described by the equations

$$\frac{\partial^2 \boldsymbol{u}(x,t)}{\partial t^2} - \mathcal{L}\boldsymbol{u} = 0, \tag{8.18}$$

where $\mathcal{L}$ is a linear differential operator. The boundary conditions for this system are assumed to be given. An approximate solution of Eq. (8.18) is sought as a sum of certain given functions $\boldsymbol{f}_k(x)$ satisfying the boundary conditions with unknown time-dependent coefficients $z_k(t)$:

$$\boldsymbol{u}(x,t) = \sum_{k=1}^{n} z_k(t)\boldsymbol{f}_k(x). \tag{8.19}$$

The coefficients $z_k(t)$ are commonly called *the Galerkin coordinates*. Substituting (8.19) into the left-hand side of Eq. (8.18) we find a so-called vector $\boldsymbol{\Phi}(x,t,z,\dot{z})$ which is commonly called *the residual vector*. If the expression (8.19) were an exact solution of Eq. (8.18), the residual vector would be equal to zero. According to the Galerkin method, the equations for the Galerkin coordinates $z_k(t)$ are found from the conditions for orthogonality of the residual vector to each of the functions $\boldsymbol{f}_k(x)$:

$$\int_0^l \boldsymbol{\Phi}(x,t,z,\dot{z})\boldsymbol{f}_1(x)\,\mathrm{d}x, \ \ldots, \int_0^l \boldsymbol{\Phi}(x,t,z,\dot{z})\boldsymbol{f}_n(x)\,\mathrm{d}x. \tag{8.20}$$

The physical sense of this procedure is as follows. The representation of a solution in the form (8.18) is equivalent to imposing certain constraint forces on the system. Equations (8.20) correspond to the requirement that the work of these constraint forces in any virtual displacement must be equal to zero.

We seek a solution of Eqs. (8.17) as

$$\begin{aligned} u(x,t) &= z_1(t)y_1(x) + z_2(t)y_2(x), \\ \Theta(x,t) &= z_1(t)\vartheta_1(x) + z_2(t)\vartheta_2(x), \end{aligned} \tag{8.21}$$

where

$$\begin{aligned} &y_1(x) = \cosh kx - \cos kx - C(\sinh kx - \sin kx), \quad y_2(x) = 0, \\ &\vartheta_1(x) = 0, \quad \vartheta_2(x) = \sin\frac{\pi x}{2l}, \end{aligned} \tag{8.22}$$

and $k \approx 1.875/l$ is the least root of the equation $1 + \cosh kl \cos kl = 0$,

$$C = -\frac{\cosh kl + \cos kl}{\sinh kl + \sin kl} \approx 0.7341.$$

The functions $y_1(x)$ and $\vartheta_2(x)$ are the first eigen-functions of the equations for disconnected bending and torsion oscillations of a cantilever homogeneous beam in a vacuum.

Substituting (8.21) into Eqs. (8.17) we find the components of the residual vector:

$$\begin{aligned}
\Phi_1 &= \left[m\,\frac{\mathrm{d}^2 z_1}{\mathrm{d}t^2} + \frac{\partial C_y}{\partial \alpha}\,\rho V b\,\frac{\mathrm{d}z_1}{\mathrm{d}t} + EIk^4 z_1\right] y_1(x) \\
&\quad - \left[m\sigma\,\frac{\mathrm{d}^2 z_2}{\mathrm{d}t^2} + \frac{\partial C_y}{\partial \alpha}\,\rho V b^2\left(\frac{3}{4} - \frac{a}{b}\right)\frac{\mathrm{d}z_2}{\mathrm{d}t} + \frac{\partial C_y}{\partial \alpha}\,\rho V^2 b z_2\right]\vartheta_2(x), \\
\Phi_2 &= \left[-m\sigma\,\frac{\mathrm{d}^2 z_1}{\mathrm{d}t^2} + \frac{\partial C_{mE}}{\partial \alpha}\,\rho V b^2\,\frac{\mathrm{d}z_1}{\mathrm{d}t}\right] y_1(x) \\
&\quad + \left\{I_m\,\frac{\mathrm{d}^2 z_2}{\mathrm{d}t^2} + \left[\frac{\pi}{16}\,\rho V b^3 - \frac{\partial C_{mE}}{\partial \alpha}\,\rho V b^3\left(\frac{3}{4} - \frac{a}{b}\right)\right]\frac{\mathrm{d}z_2}{\mathrm{d}t}\right. \\
&\quad \left. + \left(GI_p\,\frac{\pi^2}{4l^2} - \frac{\partial C_{mE}}{\partial \alpha}\,\rho V^2 b^2\right) z_2\right\}\vartheta_2(x).
\end{aligned} \tag{8.23}$$

The equations for the Galerkin coordinates are found from

$$\int_0^l \Phi_1 y_1(x)\,\mathrm{d}x = 0, \qquad \int_0^l \Phi_2 \vartheta_2(x)\,\mathrm{d}x = 0. \tag{8.24}$$

Taking account of (8.23) we obtain from (8.24) equations for $z_1(t)$ and $z_2(t)$. These equations are conveniently rewritten as

$$\begin{aligned}
&\ddot{z}_1 + 2\delta_{11}\dot{z}_1 + 2\delta_{12}\dot{z}_2 + \omega_1^2 z_1 + \kappa_1 z_2 = 0, \\
&\ddot{z}_2 + 2\delta_{21}\dot{z}_1 + 2\delta_{22}\dot{z}_2 + \omega_2^2 z_2 + \kappa_2 z_1 = 0,
\end{aligned} \tag{8.25}$$

where

$$\begin{aligned}
\delta_{11} &= \frac{\rho b}{2qm}\left(\frac{\partial C_y}{\partial \alpha} + \frac{0.9246mb}{I_m}\,\frac{\partial C_{mE}}{\partial \alpha}\right) V, \\
\delta_{12} &= -\frac{0.345\rho b^2}{qm}\left\{\frac{\partial C_y}{\partial \alpha}\left(\frac{3}{4} - \frac{a}{b}\right) - \frac{mb\sigma}{I_m}\left[\frac{\pi}{16} - \frac{\partial C_{mE}}{\partial \alpha}\left(\frac{3}{4} - \frac{a}{b}\right)\right]\right\} V, \\
\delta_{21} &= \frac{0.67\rho b^2}{qI_m}\left(\frac{\partial C_{mE}}{\partial \alpha} + \frac{\sigma}{b}\,\frac{\partial C_y}{\partial \alpha}\right) V, \\
\delta_{22} &= \frac{\rho b^3}{2qI_m}\left\{\left[\frac{\pi}{16} - \frac{\partial C_{mE}}{\partial \alpha}\left(\frac{3}{4} - \frac{a}{b}\right)\right] - \frac{0.9246\sigma}{b}\,\frac{\partial C_y}{\partial \alpha}\left(\frac{3}{4} - \frac{a}{b}\right)\right\} V,
\end{aligned}$$

$$\omega_1^2 = \frac{EIk^4}{qm}, \quad \omega_2^2 = \frac{GI_p}{I_m}\frac{\pi^2}{4l^2} - \frac{\rho V^2 b^2}{I_m}\left(\frac{\partial C_{mE}}{\partial \alpha} + \frac{0.9246\sigma}{b}\frac{\partial C_y}{\partial \alpha}\right),$$
$$\kappa_1 = \frac{0.69\sigma}{q}\left[\frac{GI_p}{I_m}\frac{\pi^2}{4l^2} - , \frac{\rho V^2 b^2}{I_m}\left(\frac{\partial C_{mE}}{\partial \alpha} + \frac{I_m}{b\sigma m}\frac{\partial C_y}{\partial \alpha}\right)\right],$$
$$\kappa_2 = \frac{1.34EI\sigma k^4}{I_m q}, \quad q = 1 - \frac{0.9246m\sigma^2}{I_m}.$$

Equations (8.25) describe oscillations in a system with two degrees of freedom that is a lumped model of bending-torsion oscillations of a wing in flight.

To find the onset of flutter we can set $z_{1,2}(t) = A_{1,2}\exp(i\omega t)$, where ω is an unknown frequency. Substituting these expressions into (8.25) and equating the real and imaginary parts of the system determinant to zero, we obtain the following equations for critical values of V and ω:

$$\omega^4 - \left(\omega_1^2 + \omega_2^2 + 4(\delta_{11}\delta_{22} - \delta_{12}\delta_{21})\right)\omega^2 + \omega_1^2\omega_2^2 - \kappa_1\kappa_2 = 0,$$
$$\omega^2 = \frac{\omega_1^2\delta_{22} + \omega_2^2\delta_{11} - \kappa_1\delta_{21} - \kappa_2\delta_{12}}{\delta_{11} + \delta_{22}}. \tag{8.26}$$

8.3 A model of the vocal source

Nowadays many different models of voice production are known [89, 313]. These models are of paramount importance both for the best understanding of the mechanism of voice production and in certain practical applications, e.g., in speech synthesis and recognition, in voice pathology [121], and so on. The heart of any such model is a model of the vocal source, i.e., of vocal folds located in the human (or animal) larynx. The arrangement of the vocal folds is rather complicated. The schematic vertical cross-section of the larynx, taken from [35] and somewhat simplified, is shown in Fig. 8.5 a. The vocal folds are formed from the cords of two muscles (the vocal muscle marked 1 and the aryten-thyr-oid muscle marked 2) and conjunctive tissues. Above the vocal folds there are the so-called false vocal folds labeled 3. They are free from internal muscles, but they play an important role in the production of hissing sounds. When the genuine vocal folds are resected, their functions are sometimes taken over by the false vocal folds [36]. Between the genuine and false vocal folds there is some expansions of the larynx called Morgan's ventricles which are labeled 4. Variations of stiffness, length and shape of the vocal folds, occurring as their internal muscles are contracted, cause a certain change in the fundamental frequency of the self-oscillations excited.

Many models of vocal folds, both rather complicated [337, 338, 213] and moderately simple [131, 315, 59, 180, 316], are known. The extent to which one or other of these models is complicated is determined mainly by the techniques for the calculation of aerodynamic forces. For example, in the

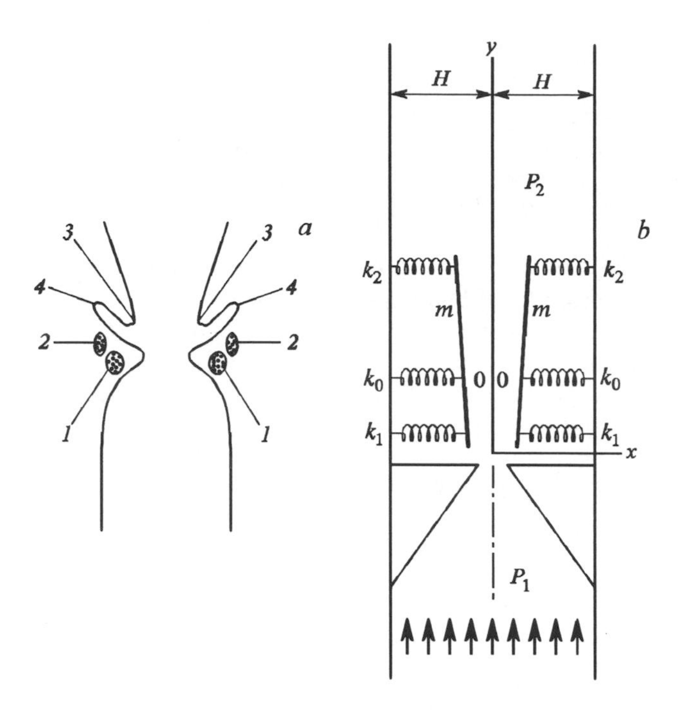

Fig. 8.5. (a) The schematic vertical cross-section of the larynx. The vocal muscles are labeled 1, the aryten-thyr-oid muscles are labeled 2, the false vocal folds are labeled 3, and Morgan's ventricles are labeled 4. (b) The model of the vocal folds

widely known two-mass model of Ishizaka and Flanagan [131] aerodynamic forces were calculated in a quasi-static approximation by using the Bernoulli law. In so doing, the velocity of motion of the glottis walls, the viscosity and inertia of air were ignored. Such a technique was also used in a simplified version of this model suggested by Herzel and Knudsen [121].

Below we consider a model of vocal folds suggested by Landa [180] and partially studied in [196]. This model involves two identical, absolutely rigid, plates attached by springs to the walls of a tube with rectangular (for simplicity) cross-section (see Fig. 8.5 b). Air enters the tube from a reservoir of sufficiently large volume V as a result of the pressure drop $\Delta P = P_1 - P_2$, where P_2 is atmospheric pressure, and can cause self-oscillations of the plates. It can be shown that the excitation of self-oscillations is possible if each plate has at least two degrees of freedom, i.e., it can both move progressively in the direction orthogonal to the air flow and turn about the axis passing through its center of mass. In this regard the excitation of self-oscillations of the plates is similar to the excitation of bending-torsion flutter of an aircraft wing.

Assuming the motion of the plates to be completely symmetric relative to the tube mid-plane, we write the equations of motion for one of the plates as

$$m\ddot{h}_a + \alpha\dot{h}_a + k(h_a - \tilde{h}_a) + \kappa(\varphi - \tilde{\varphi}) = F,$$
$$J\ddot{\varphi} + \beta\dot{\varphi} + K(\varphi - \tilde{\varphi}) + \kappa(h_a - \tilde{h}_a) = M, \tag{8.27}$$

where h_a is half of the distance between the centers of mass (O) of the plates, φ is the angle of rotation about the axis passing through the center of mass, $\tilde{h}_a$ and $\tilde{\varphi}$ are the values of h_a and φ for undeformed springs, m is the plate mass, J is the plate moment of inertia about the axis passing through the center of mass, $k = k_1 + k_2 + k_0$ is the total stiffness factor of the springs, and $K = k_1a^2 + k_2(b-a)^2$ is the total stiffness factor with respect to the rotation; a is the distance between the center of mass and the lower edge of the plate, b is the plate length along the air flow, $\kappa = k_1a - k_2(b-a)$ is the coupling factor characterizing the effect of the displacement of the plate center of mass on the plate rotation and vice versa, α and β are the damping factors,

$$F = l\int_0^b p(y)\,\mathrm{d}y, \quad M = l\int_0^b (y-a)p(y)\,\mathrm{d}y, \tag{8.28}$$

are the aerodynamic force and its moment acting on the plate which are caused by the air flow, $p(y)$ is the difference between the air pressure in the interplate slot and atmospheric pressure, and l is the plate transverse dimension. It is easily seen that Eqs. (8.27) can be obtained by the Galerkin method [97] as a lumped model of bending-torsion oscillations of a thin beam.

Equations (8.27) are conveniently written in another form:

$$\ddot{S}_0 + 2\delta_{11}\dot{S}_0 + 2\delta_{12}\dot{S}_b + \omega_1^2(S_0 - \tilde{S}_0) + \kappa_1(S_b - \tilde{S}_b) = F_0,$$
$$\ddot{S}_b + 2\delta_{21}\dot{S}_0 + 2\delta_{22}\dot{S}_b + \omega_2^2(S_b - \tilde{S}_b) + \kappa_2(S_0 - \tilde{S}_0) = F_b, \tag{8.29}$$

where

$$\delta_{11} = \frac{b-a}{b}\frac{\alpha}{2m} + \frac{a}{b}\frac{\beta}{2J}, \quad \delta_{12} = \frac{a}{b}\left(\frac{\alpha}{2m} - \frac{\beta}{2J}\right),$$
$$\delta_{21} = \frac{b-a}{a}\delta_{12}, \quad \delta_{22} = \frac{a}{b}\frac{\alpha}{2m} + \frac{b-a}{b}\frac{\beta}{2J},$$
$$\omega_1^2 = \frac{b-a}{b}\frac{k}{m} + \frac{a}{b}\frac{K}{J} + \frac{\kappa}{mJb}\Big(J + ma(b-a)\Big),$$
$$\omega_2^2 = \frac{a}{b}\frac{k}{m} + \frac{b-a}{b}\frac{K}{J} - \frac{\kappa}{mJb}\Big(J + ma(b-a)\Big),$$
$$\kappa_1 = \frac{a}{b}\left(\frac{k}{m} - \frac{K}{J}\right) - \frac{\kappa}{mJb}\Big(J - ma^2\Big),$$
$$\kappa_2 = \frac{b-a}{b}\left(\frac{k}{m} - \frac{K}{J}\right) + \frac{\kappa}{mJb}\Big(J - m(b-a)^2\Big),$$
$$\tilde{S}_0 = 2l(\tilde{h}_a + a\tilde{\varphi}), \quad \tilde{S}_b = 2l(\tilde{h}_a - (b-a)\tilde{\varphi}),$$

and

$$F_0 = 2l\left(\frac{F}{m} + \frac{aM}{J}\right), \quad F_b = 2l\left(\frac{F}{m} - \frac{(b-a)M}{J}\right). \tag{8.30}$$

It should be emphasized that the vocal folds, in executing self-oscillations, collide with one another. This process plays an important role in voice production because it results in the generation of pulses containing a large quantity of high harmonics which enrich speech. Therefore we have to add impact conditions to Eqs. (8.29). These conditions may be obtained as follows. Because the plates are assumed to be absolutely rigid, then only the plate edges can collide. For definiteness, first we consider the collision of the lower plate edges. Let an impulse of force $\mathcal{F}\Delta t$, causing changes in the plate momentum $m\Delta\dot{h}_a$ and the plate angular momentum $J\Delta\dot{\varphi}$, arise as a result of the collision. Thus, $\mathcal{F}\Delta t = m\Delta\dot{h}_a$, $\mathcal{F}a\Delta t = J\Delta\dot{\varphi}$. It follows from this that

$$\Delta\dot{\varphi} = \frac{ma}{J}\Delta\dot{h}_a. \tag{8.31}$$

On the other hand, it is easily shown that

$$\Delta\dot{h}_a + a\Delta\dot{\varphi} = \Delta\dot{h}_0. \tag{8.32}$$

If the velocity restitution coefficient after impact is $R \leq 1$ then

$$\Delta\dot{h}_0 = \dot{h}_0^+ - \dot{h}_0^- = -(1+R)\dot{h}_0^-, \tag{8.33}$$

where $\dot{h}_0^+$ is the value of $\dot{h}_0$ after impact, and $\dot{h}_0^-$ is the value of $\dot{h}_0$ before impact. From (8.31)–(8.33) we find the following conditions for the collision of the lower plate edges:

$$\Delta\dot{h}_a = -\frac{(1+R)\dot{h}_0^-}{1+ma^2/J}, \quad \Delta\dot{\varphi} = -\frac{ma}{J}\frac{(1+R)\dot{h}_0^-}{1+ma^2/J}. \tag{8.34}$$

In a similar manner we obtain the conditions for the collision of the upper plate edges:

$$\Delta\dot{h}_a = -\frac{(1+R)\dot{h}_b^-}{1+m(b-a)^2/J}, \quad \Delta\dot{\varphi} = \frac{m(b-a)}{J}\frac{(1+R)\dot{h}_b^-}{1+m(b-a)^2/J}. \tag{8.35}$$

From (8.34) and (8.35) we can find the conditions for the collision of the plate edges in terms of S_0 and S_b:
for the lower edge

$$\begin{aligned} \Delta\dot{S}_0 &= \dot{S}_0^+ - \dot{S}_0^- = -(1+R)\dot{S}_0^- \\ \Delta\dot{S}_b &= \dot{S}_b^+ - \dot{S}_b^- = -(1+R)\dot{S}_b^- \frac{J - ma(b-a)}{J + ma^2}, \end{aligned} \tag{8.36}$$

for the upper edge

$$\Delta\dot{S}_b = -(1+R)\dot{S}_b^-, \quad \Delta\dot{S}_0 = -(1+R)\dot{S}_b^- \frac{J - ma(b-a)}{J + m(b-a)^2}. \tag{8.37}$$

We note that the impact may be of quasi-plastic character [246], i.e., the duration of the contact of the plate edges during the impact may be finite.

To calculate the aerodynamic force and its moment, we use, following Ishizaka and Flanagan, the Bernoulli law. In accordance with this law

$$p(y) + \frac{\rho u^2}{2S^2(y)} = p_0 + \frac{\rho u^2}{2S_0^2} = p_b + \frac{\rho u^2}{2S_b^2}, \tag{8.38}$$

where p_0 and p_b are $p(y)$ for $y = 0$ and $y = b$, respectively. We note that, in the approximation corresponding to the Bernoulli law, the volume velocity u is constant.

Equations (8.38) contain the unknown pressure at the slot input p_0 and the unknown pressure at the slot output p_b. However, they can be expressed in terms of the pressure drop ΔP, which is presumed to be given. Assuming the quasi-static approximation to be valid and using the formula for a dynamic pressure drop as flow is contracted gradually [128], we obtain as a zero approximation with respect to h/H:

$$p_0 = \Delta P - \zeta_1 \frac{\rho u^2}{2S_0^2}, \tag{8.39}$$

where $1 < \zeta_1 < 2$ is the coefficient depending on the profile of flow velocity at the slot input and, in general, on the Reynolds number. The second boundary condition can be obtained by using the formula for a dynamic pressure drop as flow diverges abruptly at the slot output [128]:

$$p_b = \zeta_2 \frac{\rho u^2}{2S_b^2}, \tag{8.40}$$

where ζ_2 is the coefficient depending on the velocity profile at the slot output and on the Reynolds number Re (for $\mathsf{Re} > 10^3$, if the velocity profile is uniform then $\zeta_2 = 0$, and if the velocity profile is a Poiseuille's profile then $\zeta_2 \approx 0.6$). It follows from (8.38), (8.39) and (8.40) that

$$\Delta P = \frac{\rho u^2}{2}\left(\frac{\zeta_1 - 1}{S_0^2} + \frac{\zeta_2 + 1}{S_b^2}\right) \tag{8.41}$$

and

$$p(y) = \frac{\rho u^2}{2}\left(\frac{1 + \zeta_2}{S_b^2} - \frac{b^2}{\left(S_0 b + (S_b - S_0)x\right)^2}\right). \tag{8.42}$$

Equation (8.41) allows us to express u in terms of ΔP, S_0 and S_b.

Substituting (8.42) into (8.28) and integrating over y we find F and M:

$$F = \frac{bl\rho u^2}{2S_b^2}\left(\zeta_2 + \frac{S_0 - S_b}{S_0}\right),$$
$$M = rF - \frac{b^2 l \rho u^2}{2(S_b - S_0)^2}\left(\ln\frac{S_b}{S_0} + \frac{S_0^2 - S_b^2}{2S_0 S_b}\right), \tag{8.43}$$

where $r = b/2 - a$.

Equations (8.29), (8.41) and (8.43) with the impact conditions (8.36), (8.37) make it possible to solve the problem of self-excitation of oscillations and to determine the shape of self-oscillations of the variables S_0, S_b and u.

A steady-state solution of Eqs. (8.29) is determined by the following algebraic equations:

$$\omega_1^2(S_0 - \tilde{S}_0) + \kappa_1(S_b - \tilde{S}_b) = 2l\left(\frac{F}{m} + \frac{aM}{J}\right),$$
$$\omega_2^2(S_b - \tilde{S}_b) + \kappa_2(S_0 - \tilde{S}_0) = 2l\left(\frac{F}{m} - \frac{(b-a)M}{J}\right). \tag{8.44}$$

(Here the subscripts 'st' are omitted for brevity.)

The parameters which are necessary for the simulation of self-oscillations of the human vocal folds can be estimated on the basis of information presented in the book [313]. We set the following values of the parameters: $\zeta_1 = 1.37$, $\zeta_2 = 0.2$, $\rho = 1.3\times10^{-3}\,\mathrm{gcm^{-3}}$, $m = 0.15\,\mathrm{g}$, $J = 0.004\,\mathrm{g\,cm^2}$, $k = 8\times10^4\,\mathrm{gs^{-2}}$, $K = 2400\,\mathrm{g\,cm^2s^{-2}}$, $\kappa = 10^3\mathrm{g\,cm/s^2}$, $\alpha = 38.159\,\mathrm{gs^{-1}}$, $\beta = 0.954\,\mathrm{gs^{-1}}$, $a = 0.15\,\mathrm{cm}$, $b = 0.5\,\mathrm{cm}$, $l = 1.6\,\mathrm{cm}$, $\tilde{h}_0 = 0.07\,\mathrm{cm}$, $\tilde{h}_b = 0.06\,\mathrm{cm}$. The parameter ΔP varies.

Examples of the dependencies on the pressure drop ΔP of the steady-state values of the cross-sectional area of the slot at its input and at its output, and of the flow volume velocity u, are given in Fig. 8.6.

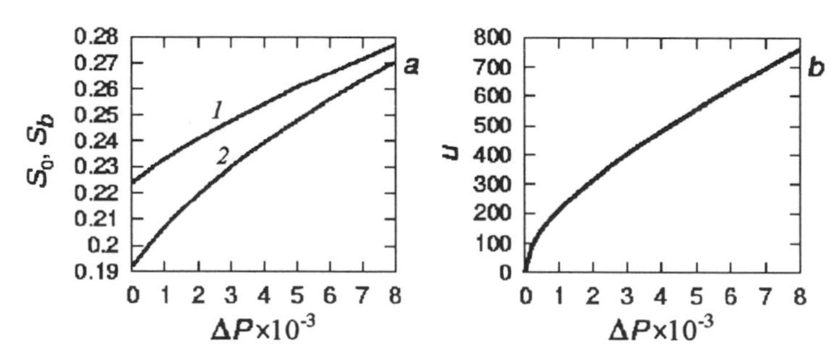

Fig. 8.6. (a) Plots of S_0 (curve 1) and of S_b (curve 2) versus ΔP, and (b) plot of u versus ΔP for the values of the parameters given in the text

Writing linearized equations for small deviations from the steady-state values found, we can find a condition for the self-excitation of the system and the oscillation frequency nearby the boundary of the self-excitation, much as it was done in the flutter problem. As a result we find that the self-excitation of oscillations occurs for $\Delta P > \Delta P_{\mathrm{cr}}$, where $\Delta P_{\mathrm{cr}} \approx 5000\,\mathrm{gcm^{-1}s^{-2}} \approx 5.1\,\mathrm{cm}$ of water. The critical value of u associated with this value of ΔP is $u_{\mathrm{cr}} \approx$

$550\,\mathrm{cm}^3\mathrm{s}^{-1}$. The frequency of self-oscillations is approximately equal to 120 Hz. The results obtained correspond to known experimental data.

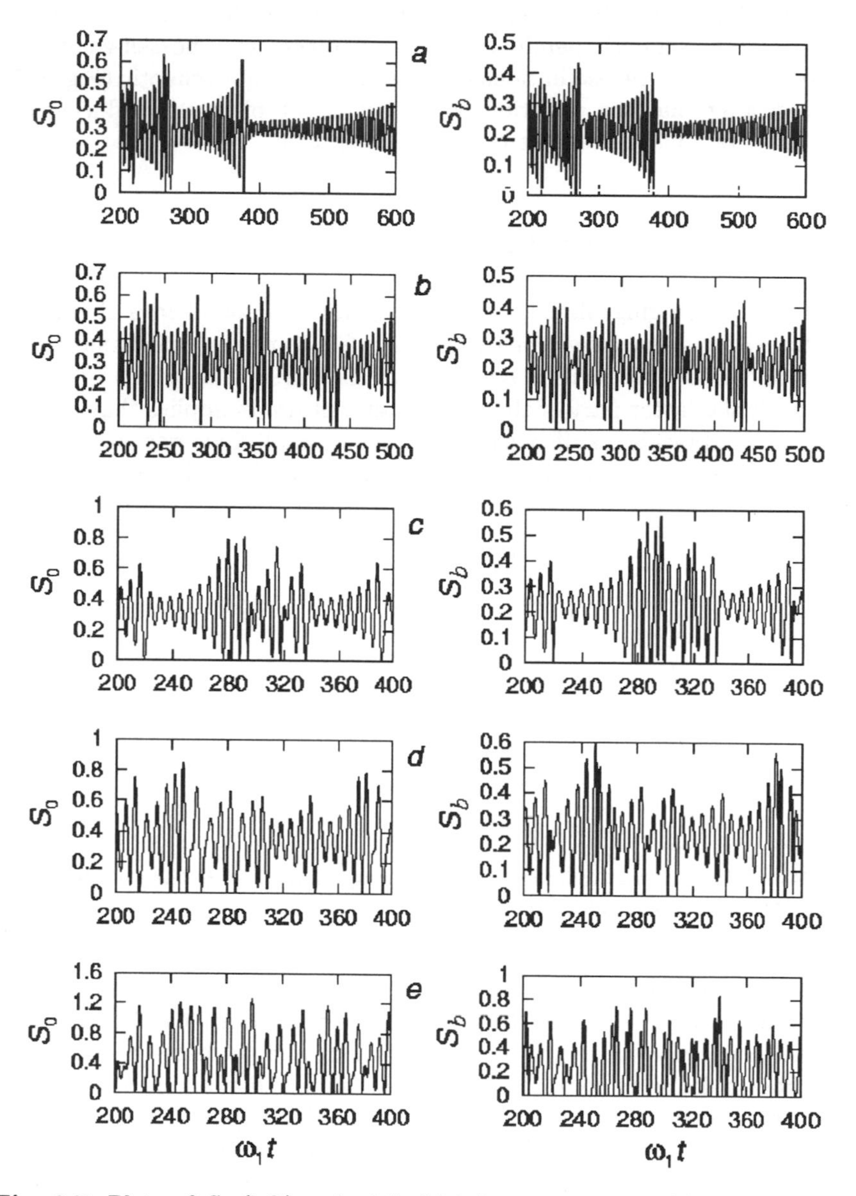

Fig. 8.7. Plots of S_0 (left) and of S_b (right) versus $\omega_1 t$ for (a) $\Delta P = 5500$, (b) $\Delta P = 6000$, (c) $\Delta P = 7000$, (d) $\Delta P = 8000$ and (e) $\Delta P = 12000$

Numerical simulation of Eqs. (8.29), (8.41) and (8.43) with the impact conditions (8.36), (8.37) shows that, even for very small excesses over the self-excitation threshold, the oscillations excited are chaotic (see Fig. 8.7 a). This is associated with the impacts. The mechanism of chaotization is similar to that for the Neimark pendulum. As the excess over the self-excitation threshold increases, the oscillations become more and more chaotic (Fig. 8.7). It is interesting that the duration of contact of the plate edges during the impact is finite, i.e., the impact is quasi-plastic [246].

8.4 The lumped model of a 'singing' flame

The effect of a 'singing' flame, like the excitation of sound in a Helmholtz resonator with non-uniformly heated walls, refers to thermo-mechanical phenomena and is described in Rayleigh's treatise [279]. This effect has been known since the latter part of the eighteenth century. It implies that if a gas burner is placed inside a sufficiently long tube of small diameter (see Fig. 8.8 a) then under certain conditions strong oscillations of the flame and the air

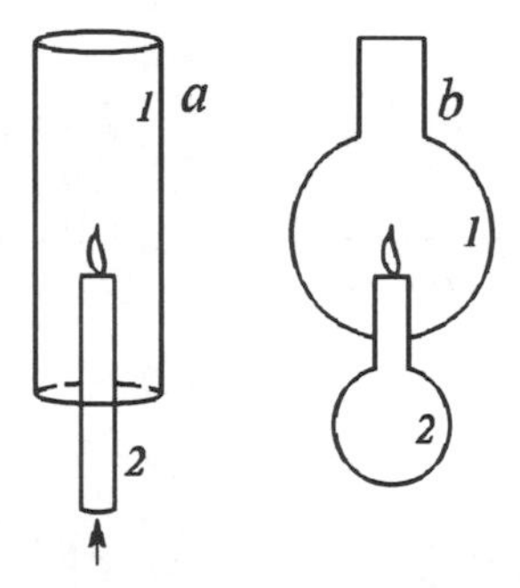

Fig. 8.8. (a) The schematic image of the 'singing' flame, and (b) the model of the 'singing' flame as two coupled Helmholtz resonators

in the tube arise. These oscillations result in the radiation of sound. On the basis of qualitative reasoning, Rayleigh showed that sound can be excited if the flame is placed close to the pressure antinode and oscillates so that more heat is liberated during compression than during rarefaction of the air in the tube. A quantitative solution of this problem on the basis of one-dimensional Euler's equations and of the relation between the air pressure in the tube and the heat liberated in combustion was first given by Neimark and Aronovich [249]. The simpler model based on substitution of two coupled Helmholtz resonators for the tube labeled 1 and the tube delivering the flame (see Fig. 8.8 b), was also suggested by Neimark [252]. Below we consider a modified version of this model.

Let the resonator labeled 1 be filled with air and the other resonator labeled 2 be filled with a fuel gas. The equations of oscillations of the air and the gas in the throat of the resonators can be written as

$$\rho_1 l_1 S_1 \ddot{x} = -\alpha_1 \dot{x} + S_1 \Delta p_1, \quad \rho_2 l_2 S_2 \ddot{y} = -\alpha_2 \dot{y} + S_2(\Delta p_2 - \Delta p_1), \quad (8.45)$$

where x and y are the displacements of the air and the gas in the throats of the first and second resonators respectively, l_1 and l_2 are the throat lengths, S_1 and S_2 are their cross-sectional areas, ρ_1 and ρ_2 are the densities of the air and the gas at atmospheric pressure, α_1 and α_2 are the coefficients of friction of the air and of the gas on the resonator walls, and Δp_1 and Δp_2 are the pressure deviations in the resonator cavities. Because the cavity of the first resonator is supplied with the heat Q liberated in gas combustion, we can write for Δp_1 a formula similar to (7.16):

$$\Delta p_1 = -\kappa_1 \frac{p_0}{V_1}(S_1 x - S_2 y) + \frac{\kappa_1 - 1}{V_1} Q, \quad (8.46)$$

where p_0 is atmospheric pressure, and V_1 is the volume of the first resonator. Processes occurring in the cavity of the second resonator are assumed to be adiabatic; therefore

$$\Delta p_2 = -\kappa_2 \frac{p_0}{V_2} S_2 y. \quad (8.47)$$

The balance equation for the heat Q can be written as

$$\frac{\mathrm{d}Q}{\mathrm{d}t} = \rho_2 c_{\mathrm{g}} S_2 T_{\mathrm{g}} \dot{y} - K\vartheta_1, \quad (8.48)$$

where T_{g} is the gas temperature in the cavity of the second resonator, c_{g} is the gas specific heat, ϑ_1 is the difference between the temperature in the cavity of the first resonator and the temperature of the environment, and K is the heat transfer factor. The first term on the right hand side of Eq. (8.48) describes the amount of heat that enters over a unit of time into the first resonator as a result of the gas combustion issuing from the throat of the second resonator, and the second term characterizes the emission of heat into the environment.

By using the first principle of thermodynamics we find the equation for ϑ_1:

$$\vartheta_1 = \frac{Q - p_0(S_1 x - S_2 y)}{m_1 c_v}, \quad (8.49)$$

where m_1 is the air mass in the cavity of the first resonator, and c_v is the air specific heat at constant volume. Substituting (8.49) into Eq. (8.48), we rewrite the latter as

$$\frac{\mathrm{d}Q}{\mathrm{d}t} + \gamma Q = b\dot{y} + b_1 x - b_2 y, \quad (8.50)$$

where $\gamma = K/(m_1 c_v)$, $b = \rho_2 c_{\mathrm{g}} S_2 T_{\mathrm{g}}$, $b_1 = \gamma p_0 S_1$, $b_2 = \gamma p_0 S_2$.

Further, substituting (8.46) and (8.47) into Eqs. (8.45), we transform these equations into the form

$$\ddot{x} + 2\delta_1\dot{x} + \omega_1^2 x = \sigma_1 y + k_1 Q, \quad \ddot{y} + 2\delta_2\dot{y} + \omega_2^2 y = \sigma_2 x - k_2 Q, \tag{8.51}$$

where

$$2\delta_j = \frac{\alpha_j}{\rho_j l_j S_j}, \quad \omega_1^2 = \frac{\kappa_1 p_0 S_1}{m_1 l_1}, \quad \omega_2^2 = \frac{\kappa_2 p_0 S_2}{m_2 l_2}\left(\frac{\kappa_1 V_2}{\kappa_2 V_1} + 1\right),$$
$$\sigma_j = \frac{\kappa_1 p_0 S_{j+1}}{m_j l_j}\frac{V_j}{V_1}, \quad k_j = \frac{\kappa_1 - 1}{m_j l_j}\frac{V_j}{V_1},$$

m_2 is the gas mass in the cavity of the second resonator, and the subscript j takes the values 1 and 2 with 1 in place of 3.

Equations (8.51) and (8.50) describe, in a linear approximation, a certain self-oscillatory system. According to the Routh–Hurwitz criterion, the condition of self-excitation of this system and the frequency of self-oscillations in the neighborhood of the boundary of self-excitation can be written as

$$(a_1 a_2 - a_3)(a_3 a_4 - a_2 a_5) - (a_1 a_4 - a5)^2 < 0, \tag{8.52}$$

$$\omega = \sqrt{\frac{a_1 a_4 - a_5}{a_1 a_2 - a_3}}, \tag{8.53}$$

where a_j ($j = 1, 2, \dots, 5$) are the coefficients of the characteristic equation written in the form

$$p^5 + a_1 p^4 + a_2 p^3 + a_3 p^2 + a_4 p + a_5 = 0.$$

For Eqs. (8.51), (8.50)

$$a_1 = \gamma + 2(\delta_1 + \delta_2), \quad a_2 = 2\gamma(\delta_1 + \delta_2) + \omega_1^2 + \omega_2^2 + 4\delta_1\delta_2 + bk_2,$$
$$a_3 = \gamma\left(\frac{\omega_1^2}{\kappa_1} + \omega_2^2\,\frac{V_2 + \kappa_2 V_1}{\kappa_1 V_2 + \kappa_2 V_1} + 4\delta_1\delta_2\right) + 2(\delta_1\omega_2^2 + \delta_2\omega_1^2 + \delta_1 b k_2),$$
$$a_4 = \frac{\kappa_1\kappa_2 p_0^2 S_1 S_2}{m_1 m_2 l_1 l_2} + 2\gamma\left(\delta_1\omega_2^2\,\frac{V_2 + \kappa_2 V_1}{\kappa_1 V_2 + \kappa_2 V_1} + \delta_2\,\frac{\omega_1^2}{\kappa_1}\right), \quad a_5 = \frac{\gamma\kappa_2 p_0^2 S_1 S_2}{m_1 m_2 l_1 l_2}.$$

It is taken into account here that

$$\omega_1^2 k_2 - \sigma_2 k_1 = 0, \quad \omega_2^2 k_1 - \sigma_1 k_2 = \frac{\kappa_2(\kappa_1 - 1)p_0 S_2}{m_1 m_2 l_1 l_2},$$

$$\omega_1^2\omega_2^2 - \sigma_1\sigma_2 = \frac{\kappa_1\kappa_2 p_0^2 S_1 S_2}{m_1 m_2 l_1 l_2}.$$

It is seen from the above that a_5 is positive, i.e., the instability of the system can only be of an oscillatory character when the condition (8.52) is fulfilled. This condition can be written as

$$\beta > \beta_{\mathrm{cr}}, \tag{8.54}$$

where $\beta = b(\kappa_1 - 1)\rho_1 l_1/(\kappa_1 p_0 S_1 \rho_2 l_2)$. An example of the dependence of β_{cr} on γ/ω_1 is given in Fig. 8.9 (curve 1). We see that for very small γ the excitation of oscillations is unfeasible because β_{cr} is very large. As γ increases, β_{cr} decreases sharply, attains its minimal value for $\gamma/\omega_1 \approx 1.4$

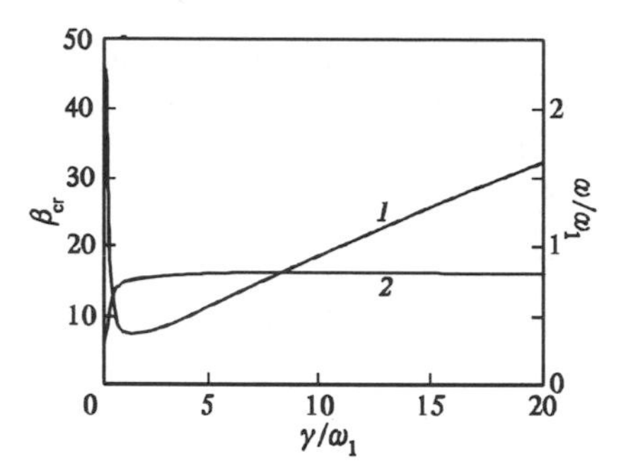

Fig. 8.9. The dependencies of β_{cr} (curve 1) and of ω/ω_1 (curve 2) in the vicinity of the boundary of the oscillation self-excitation on γ/ω_1 for $2\delta_1/\omega_1 = 0.1$, $\rho_1 l_1 = 4\rho_2 l_2$, $S_1 = 4S_2$, $\kappa_2 V_1 = 9\kappa_1 V_2$, $\delta_2/\delta_1 = 3$, $\kappa_1 = 1.4$

and then increases slowly. The dependence of the relative frequency in the neighborhood of the boundary of self-excitation on γ/ω_1 is also presented in Fig. 8.9 (curve 2). It is seen that the self-oscillation frequency is always less than ω_1.

8.5 A self-oscillatory system based on a ring Toda chain

A number of self-oscillatory systems based on ring homogeneous Toda chains were studied by Ebeling et al. [220, 78]. We consider a system that is described by the equations

$$\ddot{x}_j - f(z_j) + f(z_{j+1}) = \mu(a - \dot{x}_j^2)\dot{x}_j, \tag{8.55}$$

where $z_j = x_j - x_{j-1}$ is the strain of the jth spring, and $f(z) = -(\omega_0^2/4)(1 - e^{-z})$ is the nonlinear elastic force (see Ch. 5). Since the chain is closed in a ring, the condition

$$x_{j+N} = x_j, \tag{8.56}$$

where N is the number of the balls, must be fulfilled for any j.

Soliton-like solutions of Eqs. (8.55) for $\mu = 0$ were found by Toda [340] (see Ch. 5). It follows from these solutions that the velocity of the jth ball is

$$\dot{x}_j(\xi_j) = \frac{A}{m}\left[\mathrm{zn}\left(\frac{\mathbf{K}(k)}{\pi}\xi_j, k\right) - \mathrm{zn}\left(\frac{\mathbf{K}(k)}{\pi}(\xi_j - \beta), k\right)\right] + C, \tag{8.57}$$

where

$$A = \frac{m\omega\mathbf{K}(k)}{\pi},$$

$$\omega = \frac{\pi\omega_0}{2\mathbf{K}(k)}\left[1 - \left(1 - \frac{\mathbf{E}(k)}{\mathbf{K}(k)}\right)\mathrm{sn}^2\left(\frac{\mathbf{K}(k)}{\pi}\beta, k\right)\right]^{-1/2} \mathrm{sn}\left(\frac{\mathbf{K}(k)}{\pi}\beta, k\right), \tag{8.58}$$

$\omega_0 = 2\sqrt{\alpha/m}$, $\xi_j = \omega t - \beta j$, β is the phase shift between the oscillations of neighboring balls, and k and C are arbitrary constants. It is easily shown that $\dot{x}_j(t)$ is a periodic function of ξ_j of period 2π. Taking account of (8.58) we can rewrite (8.57) as

$$\dot{x}_j(\xi_j) = F(\xi_j, k) + C. \tag{8.59}$$

From the condition (8.56) we find possible values of β:

$$\beta = \beta_n = \frac{2\pi n}{N}, \quad n = 0, \ldots, N. \tag{8.60}$$

Thus, a chain from N elements possesses $N+1$ different modes of oscillations. It should be noted that the solution $\dot{x}_j = C$, which corresponds to values $n = 0$ and $n = N$, responds to uniform rotation of the chain as a whole. In the active chain described by Eq. (8.55) the modes found may each generate the corresponding attractor.

In the case of small dissipation ($\mu \ll 1$) the solution (8.57) can be considered as a generative one. This solution involves arbitrary constants k and C. k is the modulus of the elliptic functions, which determines the amplitude and the shape of self-oscillations for the corresponding mode, and C is the constant constituent of the ball velocity. To calculate k and C, we require that the energy and momentum conservation laws should be fulfilled on average for the oscillation period. To answer the first requirement, we can, for each j, multiply the jth equation from (8.55) by $\dot{x}_j$, then add all equations and average over time. As a result we obtain

$$\frac{\mathrm{d}}{\mathrm{d}t} \sum_{j=1}^{N} \int_0^{2\pi} \left(\frac{\dot{x}_j^2}{2} + \frac{\omega_0^2}{4} f(z_j) \right) \mathrm{d}\xi_j = \mu \sum_{j=1}^{N} \int_0^{2\pi} (a - \dot{x}_j^2) \dot{x}_j^2 \, \mathrm{d}\xi_j. \tag{8.61}$$

For calculating the integrals we should substitute (8.57) and

$$f(z_j) = \frac{A\omega}{\pi} \left[\mathbf{K}(k) \mathrm{dn}^2 \left(\frac{\mathbf{K}(k)}{\pi} \xi_j, k \right) - \mathbf{E}(k) \right]$$

into Eq. (8.61) and assume k and C to be constant. It is easily shown that all summands in Eq. (8.61) are identical. Equation (8.61) is one of the truncated equations for k and C.

Another truncated equation for k and C can be obtained from the averaged momentum conservation law. Adding Eqs. (8.55), taking into account that $\sum_{j=1}^{N} \left(f(z_{j+1}) - f(z_j) \right) = 0$ and $\sum_{j=1}^{N} \partial F(\xi_j, k/) \partial \xi_j = 0$ (see Ch. 5), and averaging over time we obtain

$$\sum_{j=1}^{N} \left(\frac{\partial \overline{F}}{\partial k} \frac{\mathrm{d}k}{\mathrm{d}t} + \frac{\mathrm{d}C}{\mathrm{d}t} \right) = \frac{\mu}{2\pi} \sum_{j=1}^{N} \int_0^{2\pi} (a - \dot{x}_j^2) \dot{x}_j \, \mathrm{d}\xi_j, \tag{8.62}$$

where

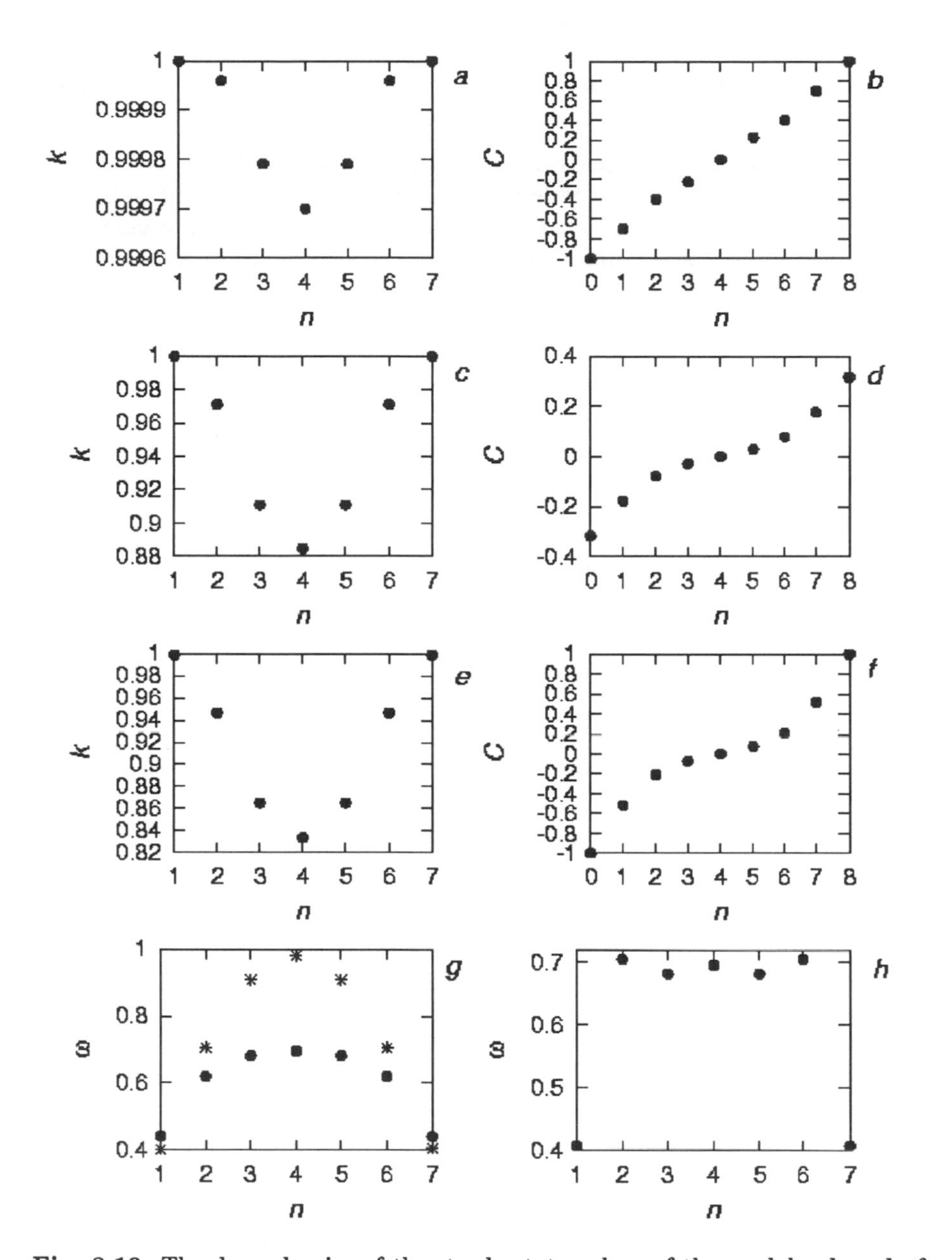

Fig. 8.10. The dependencies of the steady-state values of the modulus k and of the constant constituent of velocity C on n for $N = 8$: (a and b) $a = 1$, $\omega_0 = 1$, (c and d) $a = 0.1$, $\omega = 1$, and (e and f) $a = 1$, $\omega_0 = 4$; (g and h) the dependencies of the frequency ω on n for (g) $a = 1$, $\omega_0 = 1$ (circles) and $\omega_0 = 4$ (asterisks), and (h) $a = 0.1$, $\omega_0 = 1$

$$\overline{F} = \frac{1}{2\pi} \int_0^{2\pi} F(\xi_j, k, \omega)\, \mathrm{d}\xi_j. \tag{8.63}$$

As in Eq. (8.61), all summands in Eq. (8.62) are also identical.

Since the calculation of the left-hand sides of Eqs. (8.61) and (8.62) are rather complicated, we restrict ourselves to calculations of only steady-state values of k and C for the different modes of oscillations. Equations for these values can be found by equating the right-hand sides of Eqs. (8.61) and (8.62) to zero, i.e.,

$$\int_0^{2\pi} (a - \dot{x}_j^2)\dot{x}_j^2\, \mathrm{d}\xi_j = 0, \tag{8.64a}$$

$$\int_0^{2\pi} (a - \dot{x}_j^2)\dot{x}_j\, \mathrm{d}\xi_j = 0. \tag{8.64b}$$

Substituting (8.57) into (8.64b) we obtain a cubic equation for C which can be written as

$$C^3 + 3pC + 2q = 0, \tag{8.65}$$

where $p = r_2 - a/3$, $q = r_3/2$, and

$$r_n = \frac{A^n}{2\pi m^n} \int_0^{2\pi} \left[\mathrm{zn}\left(\frac{\mathbf{K}(k)}{\pi}\xi_j, k \right) - \mathrm{zn}\left(\frac{\mathbf{K}(k)}{\pi}(\xi_j - \beta), k \right) \right]^n \mathrm{d}\xi_j. \tag{8.66}$$

According to Cardano's formula a real root of Eq. (8.65) is

$$C = \left(\sqrt{q^2 + p^3} - q\right)^{1/3} - \left(\sqrt{q^2 + p^3} + q\right)^{1/3}. \tag{8.67}$$

Taking account of (8.66) and (8.65) we can rewrite Eq. (8.64a) as

$$ar_2 - r_4 - 3r_3C - 3r_2C^2 = 0. \tag{8.68}$$

By substituting (8.66) and (8.67) into (8.68) we obtain an equation for k which can be solved by means of graphical displays.

The results of the calculations for $N = 8$ are illustrated in Fig. 8.10, where the dependencies of the steady-state values of k, the constant constituent of velocity C, and the frequencies ω are shown for two values of a and ω_0. We see that the values of k and $|C|$ decrease monotonically as n increases from 1 to 4 (or decreases from 7 to 4). The values of C and k found nearly coincide with those calculated from the results of a direct computation of the initial equations (8.55) for $N = 8$.

The dependencies of $\tilde{x}_j \equiv x_j - Ct$ and $\dot{x}_j$ on $\tilde{\xi}_j \equiv (\xi_j - \beta/2)/(2\pi)$ for all possible oscillation modes are represented in Fig. 8.11 for $N = 8$, $a = 1$ and $\omega_0 = 1$. We see that the oscillations corresponding to the different modes are essentially nonharmonic. Oscillations of the ball velocities are closed in their

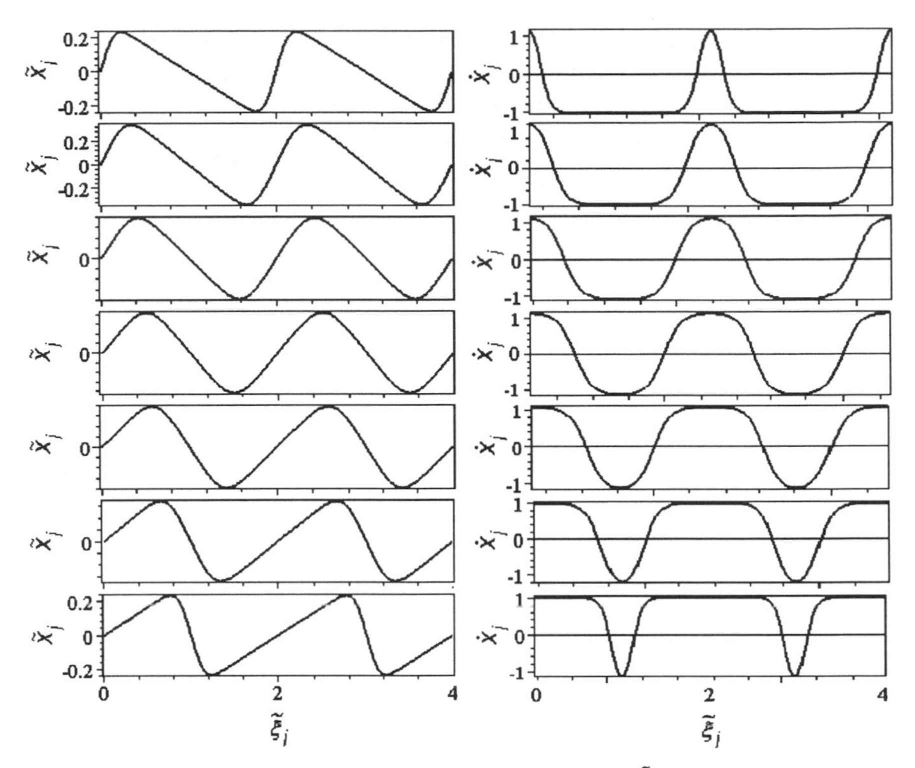

Fig. 8.11. The dependencies of $\tilde{x}_j \equiv x_j - Ct$ and $\dot{x}_j$ on $\tilde{\xi}_j \equiv (\xi_j - \beta/2)/(2\pi)$ for all possible oscillation modes in the case of $N = 8$, $a = 1$, $\omega_0 = 1$

shape either to a light autosoliton (for $n \leq 3$) or to a dark one (for $n \geq 5$). For $a = 1$ the oscillation frequency first increases, as n increases from 1 to 4, and then decreases as n increases from 4 to 7 (see Fig. 8.10 g), whereas for $a = 0.1$ the frequency has two minima: for $n = 3$ and $n = 5$ (see Fig. 8.10 h).

Table 8.1. An example of initial values of $\dot{x}_j$ such that different modes are excited

Mode number	$\dot{x}_1(0)$	$\dot{x}_2(0)$	$\dot{x}_3(0)$	$\dot{x}_4(0)$	$\dot{x}_5(0)$	$\dot{x}_6(0)$	$\dot{x}_7(0)$	$\dot{x}_8(0)$
0	−1	−1	−1	−1	−1	−1	−1	−1
1	1	1	−1	−1	−1	−1	−1	−1
2	1	1	1	−1	−1	−1	−1	−1
3	1	−1	1	−1	1	−1	−1	−1
4	1	−1	1	−1	1	−1	1	−1
5	−1	1	−1	1	−1	1	1	1
6	−1	−1	−1	1	1	1	1	1
7	−1	−1	1	1	1	1	1	1
8	1	1	1	1	1	1	1	1

Numerical simulation of Eqs. (8.55) for $N = 8$ shows that all of the modes indicated can be excited by varying the initial conditions. An example of initial values of $\dot{x}_j$ such that different modes are excited are given in Table 8.1, for $x_j(0) = 0$, $\omega_0 = 1$, $\mu = 0.1$, $a = 1$.

9. Synchronization and chaotization of self-oscillatory systems by an external harmonic force

The synchronization and chaotization of self-oscillatory systems by an external force can be considered as an important and widely occurring limiting case of the similar phenomena in coupled self-oscillatory systems (see the next chapter) when one of these systems is much more powerful than the other ones and its oscillations therefore depend only slightly on the oscillations of the other systems. Although equations corresponding to this case include time explicitly and hence are nonautonomous, we consider this case here because it is conceptually close to the problems considered in this part.

9.1 Synchronization of self-oscillations by an external periodic force in a system with one degree of freedom with soft excitation. Two mechanisms of synchronization

The phenomenon of synchronization of periodic self-oscillations by an external harmonic force was first studied by van der Pol [350] and Appleton [18] in the early 1920s. In 1930s thirties there appeared the publications of Mandelshtam and Papaleksi [227] in which it was shown that many nonlinear resonance phenomena can be observed in self-oscillatory systems with a periodic external force. These phenomena were called *resonances of the nth kind*. They consist in the following. If the frequency of the periodic force ω is close to a frequency which is n times larger or smaller than the frequency ω_0 of free self-oscillations, where n is an integer, then the synchronization of the self-oscillations can occur. This means that the frequency of the self-oscillations becomes exactly equal to ω/n or $n\omega$, respectively. The special case when $n = 1$ is called *the main resonance*. This is just the case which was considered by Appleton and by van der Pol.

For the sake of definiteness, in this section we consider different synchronization phenomena by using the example of a system described by the van der Pol–Duffing equation with a harmonic force

$$\ddot{x} - \mu(1 - \alpha x^2)\dot{x} + \omega_0^2(1 + \gamma x^2)x = \omega_0^2 B \cos \omega t. \tag{9.1}$$

9.1.1 The main resonance

Let ω be close to the natural frequency ω_0 of the system and the parameter μ be sufficiently small ($\mu \ll \omega_0$). In this case to calculate the synchronous regime we can use one of the approximate methods described in Ch. 2, e.g., the averaging method.

Substituting $x = A\cos(\omega t+\varphi)$, $\dot{x} = -A\omega\sin(\omega t+\varphi)$ into Eq. (9.1) we can obtain the exact equations for the amplitude A and the phase φ. If $\mu \ll \omega_0$ and $B \ll A_0$, where $A_0 = 2/\sqrt{\alpha}$ is the amplitude of free self-oscillations, then these equations can be averaged over time, see Ch. 2.

The time-averaged equations are

$$\dot{A} = \frac{\mu}{2}\left(1 - \frac{A^2}{A_0^2}\right)A - \frac{\omega_0 B}{2}\sin\varphi, \tag{9.2a}$$

$$\dot{\varphi} = \Delta + \frac{3}{8}\omega_0\gamma A^2 - \frac{\omega_0 B}{2A}\cos\varphi, \tag{9.2b}$$

where $\Delta \approx \omega_0 - \omega$ is the frequency mistuning. Equations (9.2a, 9.2b) describe both the synchronization regime, when the oscillations are periodic, and the beat regime, when the oscillations are quasi-periodic.

In the synchronization regime we can put $\dot{A} = 0$ and $\dot{\varphi} = 0$. In so doing we obtain a system of transcendental equations for the steady-state values of A and φ:

$$\begin{aligned}
&\frac{\mu}{2}\left(1 - \frac{A^2}{A_0^2}\right)A - \frac{\omega_0 B}{2}\sin\varphi = 0,\\
&\Delta + \frac{3}{8}\omega_0\gamma A^2 - \frac{\omega_0 B}{2A}\cos\varphi = 0.
\end{aligned} \tag{9.3}$$

Eliminating φ from (9.3) we find the following equation for A:

$$\left[\left(1 - \frac{A^2}{A_0^2}\right)^2 + \frac{4}{\mu^2}\left(\Delta + \frac{\Gamma A^2}{A_0^2}\right)^2\right]\frac{A^2}{A_0^2} = \frac{\omega_0^2 B^2}{\mu^2 A_0^2}, \tag{9.4}$$

where $\Gamma = (3/8)\omega_0\gamma A_0^2 = 3\omega_0\gamma/2\alpha$. The dependencies of A^2/A_0^2 on the relative mistuning Δ/μ are presented in Fig. 9.1 for a number of values of $\omega_0^2 B^2/(\mu^2 A_0^2)$ and Γ.

For small amplitudes of the external force and of small mistunings, when

$$\frac{\omega_0^2 B^2}{\mu^2 A_0^2} < 4/27$$

and

$$1 + \frac{36}{\mu^2}\left(\Delta + \frac{\Gamma A^2}{A_0^2}\right) - \left[1 - \frac{12}{\mu^2}\left(\Delta + \frac{\Gamma A^2}{A_0^2}\right)\right]^{3/2} \le \frac{27\omega_0^2 B^2}{2\mu^2 A_0^2}, \tag{9.5}$$

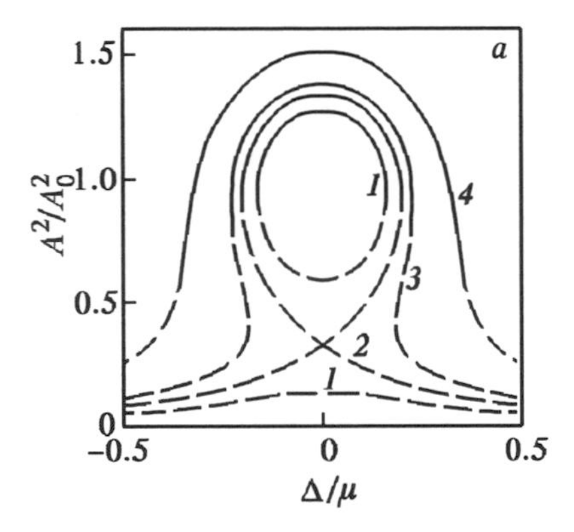

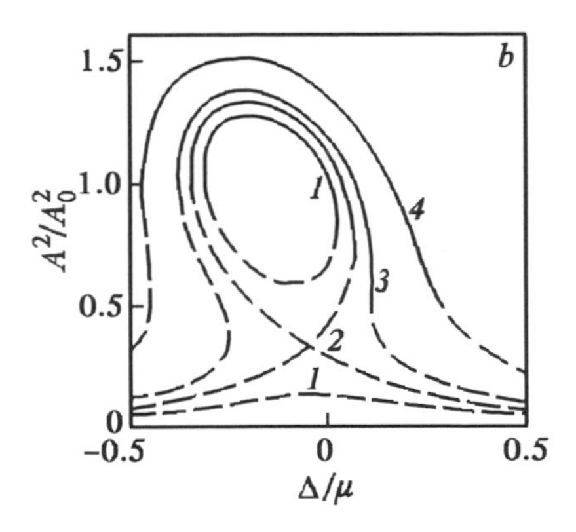

Fig. 9.1. The dependencies of A^2/A_0^2 on the relative mistuning Δ/μ for (a) $\Gamma = 0$ and (b) $\Gamma/\mu = 0.2$. Curves 1, 2, 3 and 4 correspond to $\omega_0^2 B^2/(\mu^2 A_0^2) = 0.1$, 4/27, 0.2 and 0.4, respectively. The stable parts of these dependencies are shown as solid lines, and the unstable parts are shown as dashed lines

the dependencies of A^2/A_0^2 on Δ/μ each break up into two curves: a closed curve of near-ellipsoidal shape and a curve of a resonance character for which $A^2 < A_0^2/2$ (see Fig. 9.1). For $\omega_0^2 B^2/(\mu^2 A_0^2) = 4/27$ both of the curves merge. However, for $\omega_0^2 B^2/(\mu^2 A_0^2) < 8/27$, in certain ranges of mistunings, for every value of Δ/μ there are three values of A^2/A_0^2. Only for $\omega_0^2 B^2/(\mu^2 A_0^2) \geq 8/27$ the dependencies of A^2/A_0^2 on Δ/μ become one-to-one.

Figure 9.2 gives the steady-state values of the phase φ as a function of the relative mistuning Δ/μ for $\Gamma = 0$ and the same values of b as in Fig. 9.1. These dependencies are found from the equation

$$\cos\varphi = \frac{2A}{\omega_0 B}\left(\Delta + \Gamma\frac{A^2}{A_0^2}\right). \tag{9.6}$$

We consider further the stability of the synchronization regime. For this we write equations for the deviations ξ and η from the steady-state values of A and φ ($\xi = A - A_{\mathrm{st}}$, $\eta = \varphi - \varphi_{\mathrm{st}}$). In the linear approximation these equations are[1]

$$\begin{aligned} \dot{\xi} &= \frac{\mu}{2}\left(1 - \frac{3A^2}{A_0^2}\right)\xi - \frac{\omega_0 B}{2}\cos\varphi\,\eta, \\ \dot{\eta} &= \frac{\omega_0 B}{2A^2}\cos\varphi\,\xi + \frac{\omega_0 B}{2A}\sin\varphi\,\eta. \end{aligned} \tag{9.7}$$

The characteristic equation corresponding to Eqs. (9.7), in view of Eqs. (9.3), is

[1] The subscript 'st' is omitted below.

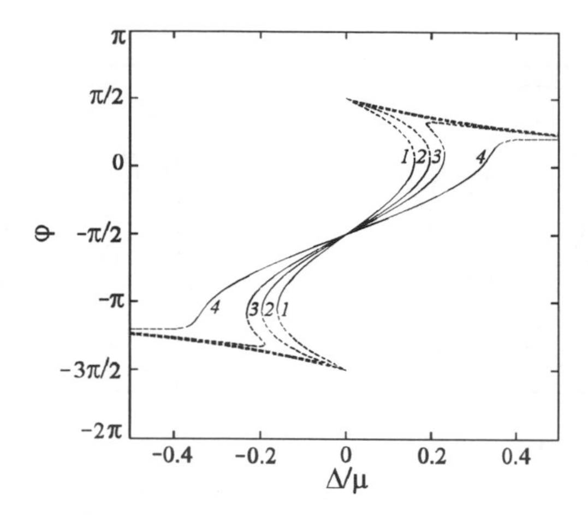

Fig. 9.2. The dependencies of φ on the relative mistuning Δ/μ for $\Gamma = 0$. Curves 1, 2, 3 and 4 correspond to $b = 0.1$, $4/27$ 0.2, and 0.4, respectively. The stable parts of these dependencies are shown as solid lines, and the unstable parts as dashed lines

$$p^2 - \mu\left(1 - \frac{2A^2}{A_0^2}\right)p + \frac{\mu^2}{4}\left(1 - \frac{A^2}{A_0^2}\right)\left(1 - \frac{3A^2}{A_0^2}\right) + \Delta^2 = 0. \tag{9.8}$$

From the requirement of positiveness for all the coefficients of Eq. (9.8) we find the stability conditions:

$$\frac{A^2}{A_0^2} \geq \frac{1}{2}, \tag{9.9a}$$

$$\left(1 - \frac{A^2}{A_0^2}\right)\left(1 - \frac{3A^2}{A_0^2}\right) + 4\frac{\Delta^2}{\mu^2} \geq 0. \tag{9.9b}$$

The stable parts of the dependencies of A^2/A_0^2 and φ on Δ/μ are shown in Figs. 9.1 and 9.2 as solid lines, whereas the unstable parts are shown as dashed lines.

From Fig. 9.1 it is seen that for small amplitudes of the external force, when $\omega_0^2 B^2/(\mu^2 A_0^2) < 8/27$, the stability of the synchronization regime is determined by the condition (9.9b), whereas for large amplitudes of the external force, when the dependencies of A^2/A_0^2 on Δ/μ become one-to-one, the condition (9.9b) is always fulfilled and the stability of the synchronization regime is determined by the condition (9.9a).

It is easy to verify that Eq. (9.9b) coincides with the equation of the geometric locus, where the tangents to the resonance curves are vertical. This means that, for any B, the stable parts of the resonance curves are one-to-one.

It should be noted that in Fig. 9.1 the boundaries of the stable parts determine the boundaries of the synchronization regions which we denote by

$\pm\Delta_s^{(\pm)}$. In the particular case when $\Gamma = 0$ we have $\Delta^{(+)} = \Delta_s^{(-)} = \Delta_s$. It is easily shown that in this case

$$\Delta_s = \begin{cases} \dfrac{\omega_0 B}{2A_0} & \text{for } \dfrac{\omega_0 B}{\mu A_0} \ll 1 \\ \dfrac{1}{4}\sqrt{\dfrac{8\omega_0^2 B^2}{A_0^2} - \mu^2} & \text{for } \dfrac{\omega_0^2 B^2}{\mu^2 A_0^2} \geq \dfrac{8}{27}. \end{cases} \tag{9.10}$$

Let us consider further the beat regime outside the synchronization region. We assume first that the external force is so small that $\omega_0 B/(\mu A_0) \ll 1$ [148]. In this case $|\Delta_s^{(\pm)}|/\mu \ll 1$ too, and Eqs. (9.2a), (9.2b) involve the fast variable A and the slow variable φ. Therefore we can put $\dot{A} = 0$ in Eq. (9.2a), i.e., we can seek a steady-state solution for the amplitude A. This solution is $A = A_0(1 + a)$, where $a = -\omega_0 B \sin\varphi/2(\mu A_0) \ll 1$. So, as a first approximation, the truncated equation for the phase φ is

$$\dot{\varphi} = \Delta + \Gamma - \Delta_0 \cos\varphi, \tag{9.11}$$

where $\Delta_0 = \omega_0 B/(2A_0)$. It is seen from Eq. (9.11) that $\Delta_s^{(\pm)} = \pm\,\Delta_0 - \Gamma$; for $\Gamma = 0$, $\Delta_s = \Delta_0$.

Equation (9.11) can be integrated. Its solution is

$$\varphi = 2\arctan\left[\frac{\sqrt{(\Delta+\Gamma)^2 - \Delta_0^2}}{\Delta + \Gamma + \Delta_0}\tan\left(\frac{\sqrt{(\Delta+\Gamma)^2 - \Delta_0^2}}{2}(t - t_0)\right)\right], \tag{9.12}$$

where t_0 is the instant at which $\varphi = 0$. Differentiating (9.12) with respect to time, we find

$$\dot{\varphi} = \frac{(\Delta+\Gamma)^2 - \Delta_0^2}{\Delta + \Gamma + \Delta_0 \cos\left(\sqrt{(\Delta+\Gamma)^2 - \Delta_0^2}\,(t - t_0)\right)}. \tag{9.13}$$

It is seen from (9.13) that $\dot{\varphi}$ is a periodic function of time with period $2\pi/\sqrt{(\Delta+\Gamma)^2 - \Delta_0^2}$. The absolute value of this function averaged over the period is called the *beat frequency*. It is

$$\omega_b = \overline{|\dot{\varphi}|} = \sqrt{(\Delta+\Gamma)^2 - \Delta_0^2}. \tag{9.14}$$

The oscillation amplitude outside the synchronization region is weakly modulated relative to its averaged value A_0. The shape of the modulation is rather complicated, but away from the synchronization region boundary it becomes near harmonic. The average frequency of the modulation is equal to ω_b. The transition from the beat regime to the synchronization regime, as Δ_0 is fixed and the mistuning Δ decreases (or, vice versa, Δ is fixed and Δ_0 increases), occurs at the cost of a gradual decrease of the beat frequency ω_b to zero (see Fig. 9.3 a and b). In this lies one of the mechanisms of synchronization which manifests itself for sufficiently small amplitudes of the external force [5, 334, 187, 48, 49].

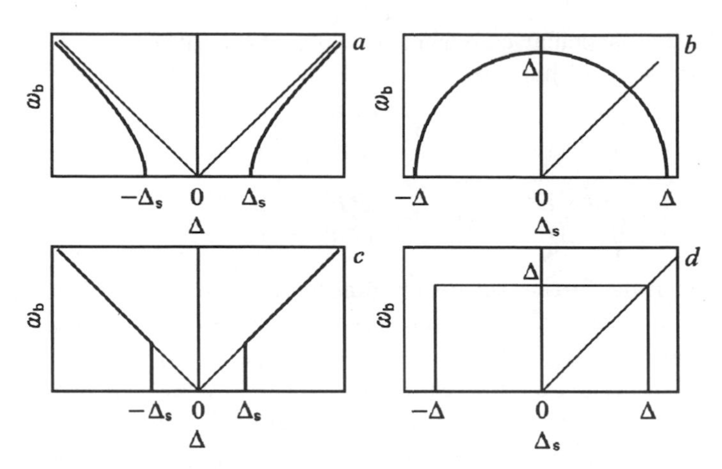

Fig. 9.3. The qualitative dependencies of the beat frequency ω_b on the mistuning Δ for $\Gamma = 0$ in the case when the synchronization region half-width Δ_s is fixed (a and c), and on Δ_s in the case when Δ is fixed (b and d), for small (a and b) and large (c and d) amplitudes of the external force

If the external force is sufficiently large, the dependencies of ω_b on Δ for fixed Δ_0 and on Δ_0 for fixed Δ, and the mechanism of synchronization, are of different character. Let us demonstrate this by a particular example in which $\Gamma = 0$, $\omega_0 B/(\mu A_0) \gg 1$. In this case, in the vicinity of the synchronization region, $|\omega - \omega_0| \gg \mu$. Substituting into Eq. (9.1)

$$x = F\cos\omega t + y, \tag{9.15}$$

where $F = \omega_0^2 B/(\omega_0^2 - \omega^2)$ is the amplitude of forced oscillations in the system described by Eq. (9.1) for $\mu = 0$, we obtain the following equation for y:

$$\ddot{y} + \omega_0^2 y = \mu\Big(1 - \alpha(F\cos\omega t + y)^2\Big)(\dot{y} - F\omega\sin\omega t). \tag{9.16}$$

A solution of this equation can be sought in the form

$$y = C\cos(\omega_0 t + \psi), \tag{9.17}$$

where C and ψ are slowly varying functions. By substituting (9.17) into (9.16) we find the following truncated equations for C and ψ:

$$\dot{C} = \frac{\mu}{2}\left(1 - \frac{C^2}{A_0^2} - \frac{2F^2}{A_0^2}\right)C, \quad \dot{\psi} = 0. \tag{9.18}$$

A steady-state solution of Eqs. (9.18) is $C = A_0\sqrt{1 - 2F^2/A_0^2}$. It follows from this that for $F = A_0/\sqrt{2}$, i.e., for

$$\Delta = \Delta_s = \omega_0\left(1 - \sqrt{1 - \sqrt{2}\,\frac{B}{A_0}}\right), \tag{9.19}$$

the amplitude of self-oscillations at the frequency ω_0 vanishes, which is to say that synchronization arises. The formula (9.19) for $B \ll A_0$ reduces to (9.10) for $\omega_0^2 B^2/(\mu^2 A_0^2) \gg 1$. Outside the synchronization region the beat frequency is

$$\omega_b = |\Delta|. \tag{9.20}$$

This mechanism of synchronization, which manifests itself for sufficiently large amplitudes of the external force, is known as *synchronization by quenching* [5, 334, 187, 48, 49]. Special attention must be given to the fact that in this case the onset of synchronization is accompanied by a jump the beat frequency from Δ_s to 0, see Fig. 9.3 c and d. In real systems, as a rule, hysteresis takes place, i.e., the transitions from the synchronization to the beat regime and from the beats to the synchronization regime occur for different values of the mistuning Δ (or for different values of Δ_s).

We note that the beat regime corresponds to a two-dimensional torus in the cylindrical phase space of Eq. (9.1). The transition from the beat to the synchronization regime, as the amplitude of the external force is small, is associated with the resonance on this torus, i.e., with the appearance on this torus of a closed trajectory [5]. For large amplitudes of the external force the limit cycle in the phase space corresponding to the synchronization regime does not lie on a torus. The transition from the synchronization to the beat regime is associated with the bifurcation of the loss of stability of this limit cycle and of the birth of the torus.

It should be borne in mind that in actual practice both mechanisms of synchronization take place simultaneously. Depending on the parameters one or the other mechanism is dominant.

9.1.2 Resonances of the nth kind

Subharmonic synchronization. Subharmonic synchronization occurs as a result of the interaction between forced oscillations at the frequency of the external force and harmonics of self-oscillations. The number of the harmonics, which have significant amplitudes, depends on the system nonlinearity. If the nonlinearity is of the order of a small parameter and can be described by a polynomial of the nth degree, then all harmonics of self-oscillations up to and including the nth have amplitudes not less than of first order with respect to the small parameter. Any one of these harmonics can be synchronized by the external force.

We consider a self-oscillatory system described by the equation

$$\ddot{x} - \mu(1 - \alpha x^2 + \beta x)\dot{x} + \omega_0^2 x = \omega_0^2 B \cos\omega t \tag{9.21}$$

and assume that the frequency ω is close to $n\omega_0$. Since the nonlinearity involves only a quadratic and a cubic term, synchronization is possible only for $n = 2$ and $n = 3$. We consider both of these cases.

Let us rewrite Eq. (9.21) in the form

$$\ddot{x} - \mu(1 - \alpha x^2 + \beta x)\dot{x} + \omega_0^2 x = \omega_0^2(n^2 - 1)F_n \cos\omega t, \tag{9.22}$$

where $F_n = B/(n^2 - 1)$ is the amplitude of the forced oscillations in the generative system ($\mu = 0$) for $\omega = n\omega_0$, $n = 2$, 3.

We make the change of variables similar to (9.15), i.e., we substitute into Eq. (9.22) $y = x - F\cos\omega t$ in place of x. The equation for y is conveniently rewritten as

$$\begin{aligned}\ddot{y} + \frac{\omega^2}{n^2}y = \mu\Big(1 - \alpha(y - F\cos\omega t)^2 + \beta(y - F\cos\omega t)\Big)& \\ \times(\dot{y} + \omega F\sin\omega t) - 2\omega_0\Delta(y - F\cos\omega t),&\end{aligned} \tag{9.23}$$

where $\Delta \approx \omega_0 - \omega/n$ is the frequency mistuning.

If μ, $|\Delta| \ll \omega_0$, we can use the averaging method to solve Eq. (9.23) approximately. Setting

$$y = A\cos\left(\frac{\omega}{n}t + \varphi\right), \quad \dot{y} = A\frac{\omega}{n}\sin\left(\frac{\omega}{n}t + \varphi\right),$$

we find the following averaged equations for A and φ:

- $n = 2$

$$\begin{aligned}\dot{A} &= \frac{\mu A}{2}\left(1 - \frac{A^2}{A_0^2} - \frac{2F^2}{A_0^2} - \frac{\beta F}{2}\cos 2\varphi\right),\\ \dot{\varphi} &= \frac{\mu\beta F}{4}\sin 2\varphi + \Delta,\end{aligned} \tag{9.24}$$

- $n = 3$

$$\begin{aligned}\dot{A} &= \frac{\mu A}{2}\left(1 - \frac{A^2}{A_0^2} - \frac{2F^2}{A_0^2} + \frac{FA}{A_0^2}\cos 3\varphi\right),\\ \dot{\varphi} &= -\frac{\mu FA}{2A_0^2}\sin 3\varphi + \Delta,\end{aligned} \tag{9.25}$$

where $A_0 = 2/\sqrt{\alpha}$.

The synchronization regimes are determined by steady-state solutions of Eqs. (9.24) and (9.25). Setting $\dot{A} = 0$ and $\dot{\varphi} = 0$ and eliminating φ, we find the equations for the steady-state values of amplitudes:

- $n = 2$

$$\left[\left(1 - \frac{A^2}{A_0^2} - \frac{2F^2}{A_0^2}\right)^2 - \frac{\beta^2 F^2}{4} + \frac{4\Delta^2}{\mu^2}\right]A^2 = 0, \tag{9.26}$$

- $n = 3$

$$\left[\left(1 - \frac{A^2}{A_0^2} - \frac{2F^2}{A_0^2}\right)^2 - \frac{F^2 A^2}{A_0^4} + \frac{4\Delta^2}{\mu^2}\right]A^2 = 0. \tag{9.27}$$

Equation (9.26) possesses the following three solutions:

$$A_1 = 0, \quad A_{2,3}^2 = A_0^2 \left(1 - \frac{2F^2}{A_0^2} \pm \sqrt{\frac{\beta^2 F^2}{4} - \frac{4\Delta^2}{\mu^2}}\right). \tag{9.28}$$

The first solution corresponds to the main resonance, when oscillations are executed at the frequency of the external force and self-oscillations are quenched. The second and the third solutions correspond to the subharmonic synchronization, when there exist self-oscillations at the frequency $\omega/2$, along with small forced oscillations at the frequency ω. The conditions for the existence of the second and the third solutions are

$$\frac{\Delta^2}{\mu^2} \leq \frac{\beta^2 F^2}{16}, \tag{9.29a}$$

$$1 - \frac{2F^2}{A_0^2} \pm \sqrt{\frac{\beta^2 F^2}{4} - \frac{4\Delta^2}{\mu^2}} > 0. \tag{9.29b}$$

Equation (9.27) possesses the following solutions:

$$A_1 = 0, \quad A_{2,3}^2 = A_0^2 \left[1 - \frac{3F^2}{2A_0^2} \pm \sqrt{\frac{F^2}{A_0^2}\left(1 - \frac{7F^2}{4A_0^2}\right) - \frac{4\Delta^2}{\mu^2}}\right]. \tag{9.30}$$

The conditions for the existence of the second and the third solutions are

$$\frac{\Delta^2}{\mu^2} \leq \frac{F^2}{4A_0^2}\left(1 - \frac{7F^2}{4A_0^2}\right), \tag{9.31a}$$

$$\frac{F^2}{A_0^2} \leq \frac{4}{7}. \tag{9.31b}$$

When these conditions are fulfilled, the corresponding subharmonic synchronization is possible.

Let us consider the stability of the solutions found. First of all, we consider the stability of the solution $A_1 = 0$. For this, transforming to Cartesian coordinates

$$u = \frac{A}{A_0}\cos\varphi, \quad v = \frac{A}{A_0}\sin\varphi$$

is convenient in Eqs. (9.24), (9.25). The linearized equations for u and v are:

- $n = 2$

$$\begin{aligned} \dot{u} &= \frac{\mu}{2}\left(1 - \frac{\beta F}{2} - \frac{2F^2}{A_0^2}\right)u - \Delta v, \\ \dot{v} &= \frac{\mu}{2}\left(1 + \frac{\beta F}{2} - \frac{2F^2}{A_0^2}\right)v + \Delta u, \end{aligned} \tag{9.32}$$

- $n = 3$

$$\dot{u} = \frac{\mu}{2}\left(1 - \frac{2F^2}{A_0^2}\right)u - \Delta v,$$
$$\dot{v} = \frac{\mu}{2}\left(1 - \frac{2F^2}{A_0^2}\right)v + \Delta u. \tag{9.33}$$

Writing the characteristic equations for Eqs. (9.32) and (9.33) we find the stability conditions:

- $n = 2$

$$\frac{2F^2}{A_0^2} - 1 \geq \sqrt{\frac{\beta^2 F^2}{4} - \frac{4\Delta^2}{\mu^2}}, \tag{9.34}$$

- $n = 3$

$$\frac{2F^2}{A_0^2} - 1 \geq 0. \tag{9.35}$$

It is easily seen that condition (9.34) is fulfilled if and only if the solutions $A_{2,3}$ described by (9.28) are imaginary, i.e., in this case only the main resonance is possible. But if condition (9.35) is fulfilled, one of the solutions $A_{2,3}$ described by (9.30 can be real. This means that in a certain region of the mistunings Δ and over a certain range of B both the main resonance and the subharmonic synchronization with $n = 3$ are possible.

Let us consider further the stability of the solutions $A_{2,3}$ corresponding to subharmonic synchronization. Substituting into Eqs. (9.24), (9.25) $\xi = A - A_{\text{st}}$, $\eta = \varphi - \varphi_{\text{st}}$ we obtain the following linearized equations for ξ and η:

- $n = 2$

$$\dot{\xi} = -\mu\frac{A^2}{A_0^2}\xi + 2A\Delta\eta,$$
$$\dot{\eta} = \mu\left(1 - \frac{2F^2}{A_0^2} - \frac{A^2}{A_0^2}\right)\eta, \tag{9.36}$$

- $n = 3$

$$\dot{\xi} = -\frac{\mu}{2}\left(1 - \frac{2F^2}{A_0^2} + \frac{A^2}{A_0^2}\right)\xi + 3A\Delta\eta,$$
$$\dot{\eta} = \frac{3\mu}{2}\left(1 - \frac{2F^2}{A_0^2} - \frac{A^2}{A_0^2}\right)\eta + \frac{\Delta}{A}\xi. \tag{9.37}$$

From here we find the stability conditions:

- $n = 2$

$$\frac{A^2}{A_0^2} \geq 1 - \frac{2F^2}{A_0^2}, \tag{9.38}$$

- $n = 3$

$$\frac{A^2}{A_0^2} \geq 1 - \frac{3F^2}{2A_0^2}. \tag{9.39}$$

The conditions (9.38) and (9.39) can be fulfilled for solutions (9.28) and (9.30) only with the plus sign at the radical.

The dependencies of A_2^2/A_0^2 on Δ/μ for $n = 2$ and $n = 3$ are given in Figs. 9.4 and 9.5, respectively. It is seen that, in contrast to the case of the

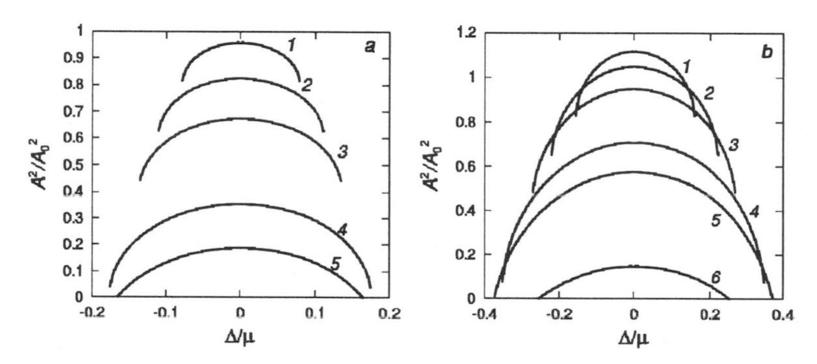

Fig. 9.4. The dependencies of A_2^2/A_0^2 on Δ/μ for $n = 2$ and (a) $\beta^2 A_0^2 = 1$, (b) $\beta^2 A_0^2 = 4$. Curves 1, 2, 3, 4, 5, and 6 correspond to $F^2/A_0^2 = 0.1$, 0.2, 0.3, 0.5, 0.6 and 0.9, respectively

main resonance, in the case of subharmonic synchronization the amplitude of the synchronized self-oscillations for $\Delta = 0$ decreases with increasing external force amplitude. This is explained by asynchronous suppression of the self-oscillations. The same effect takes place in the case of synchronization by harmonics of the external force (see below). The synchronization region width first increases as the external force amplitude increases and then decreases. The synchronization region width is maximal for

$$\frac{F^2}{A_0^2} = \begin{cases} \dfrac{1}{2} + \dfrac{\beta^2 A_0^2}{32} & \text{in the case of } n = 2, \\[2ex] \dfrac{2}{7} & \text{in the case of } n = 3. \end{cases} \tag{9.40}$$

This maximal value is

$$\Delta_{\max} = \mu \begin{cases} \dfrac{\beta A_0}{\sqrt{32}} \sqrt{1 - \dfrac{\beta^2 A_0^2}{64}} & \text{in the case of } n = 2, \\[2ex] \dfrac{1}{2\sqrt{7}} & \text{in the case of } n = 3. \end{cases} \tag{9.41}$$

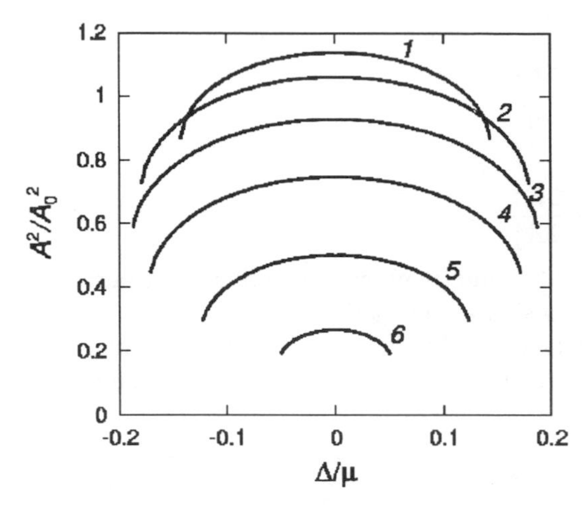

Fig. 9.5. The dependencies of A_2^2/A_0^2 on Δ/μ for $n = 3$. Curves 1, 2, 3, 4, 5, and 6 correspond to $F^2/A_0^2 = 0.1$, 0.2, 0.3, 0.4, 0.5 and 0.56, respectively

In the case of zeroth mistuning the dependencies of the oscillation amplitudes squared at the frequency ω (the curves labeled 1) and at the frequency ω/n (the curves labeled 2) on the external force amplitude squared are shown in Fig. 9.6 for $n = 2$ (a) and $n = 3$ (b). Unstable parts of these dependencies are shown as dashed lines. It is seen from these figures that the subharmon-

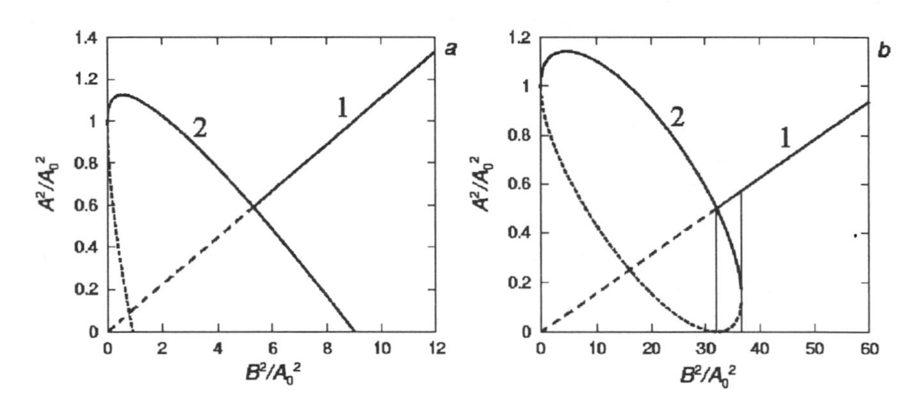

Fig. 9.6. The dependencies of A^2/A_0^2 at the frequency ω (curves 1) and at the frequency ω/n (curves 2) on B^2/A_0^2 for $\Delta = 0$; (a) $n = 2$, $\beta^2 A_0^2/4 = 1$ and (b) $n = 3$. The stable parts of the dependencies are shown as solid lines, and the unstable parts as dashed lines

ic synchronization is possible only for moderately small amplitudes of the external force, whereas for large amplitudes of the external force the main resonance takes place. For $n = 2$ the transition from the first regime to the second one with increasing external force amplitude occurs smoothly: the amplitude of the synchronous self-oscillations gradually reduces to zero and only forced oscillations at the frequency ω remain. For $n = 3$ the transition from one regime to another occurs by a jump. As the external force amplitude increases, the amplitude of the synchronous self-oscillations decreases to a certain value and then jumps to zero, whereupon only forced oscillations remain. As the external force amplitude decreases from large values, the main resonance regime is held to a smaller value of B than that corresponding to the onset of this regime, i.e., hysteresis takes place. Inside the hysteresis loop both the main resonance and the subharmonic synchronization are possible depending on the initial conditions.

It should be noted that in regimes of subharmonic synchronization n stable steady states with different values of the phase φ ($\varphi_1 = \varphi_0$, $\varphi_2 = \varphi_0 + 2\pi/n$, $\varphi_3 = \varphi_0 + 2\cdot 2\pi/n, \ldots, \varphi_n = \varphi_0 + (n-1)\cdot 2\pi/n$, where the value of φ_0 depends on the mistuning Δ) are possible. The excitation of oscillations with one or other of these phases depends on the initial conditions. Therefore such systems can be used as n-state triggers. In addition, subharmonic synchronization can be used for division of the oscillation frequency.

Synchronization by a harmonic of the external force. Let us consider the case when self-oscillations are synchronized by a certain harmonic of the external force. Such a phenomenon is often used for multiplication of the oscillation frequency. In this case Eq. (9.21) is conveniently rewritten as

$$\ddot{x} - \mu(1 - \alpha x^2 + \beta x)\dot{x} + \omega_0^2 x = \frac{n^2 - 1}{n^2}\omega_0^2 F \cos\omega t, \tag{9.42}$$

where

$$F = \frac{n^2}{n^2 - 1} B \tag{9.43}$$

is the amplitude of forced oscillations in the generative system ($\mu = 0$) for $\omega = \omega_0/n$, $n = 2, 3$. Substituting $x = y + F\cos\omega t$ into Eq. (9.42 we obtain the following equation for y:

$$\begin{aligned} \ddot{y} + n^2\omega^2 y = \mu\Big(1 - \alpha(y + F\cos\omega t)^2 + \beta(y + F\cos\omega t)\Big) \\ \times (\dot{y} - \omega F\sin\omega t) - 2\omega_0\Delta(y + F\cos\omega t), \end{aligned} \tag{9.44}$$

where $\Delta \approx \omega_0 - n\omega$ is the frequency mistuning.

As before, in the case of μ, $|\Delta| \ll \omega_0$, we can use the averaging method. Setting $y = A\cos(n\omega t + \varphi)$ in Eq. (9.44) we find the following averaged equations for A and φ:

$$\begin{aligned} \dot{A} &= \frac{\mu A}{2}\left(1 - \frac{A^2}{A_0^2} - \frac{2F^2}{A_0^2}\right) + \frac{\mu}{2}K_n\cos\varphi, \\ \dot{\varphi} &= -\frac{\mu K_n}{2A}\sin\varphi + \Delta, \end{aligned} \tag{9.45}$$

where $K_2 = \beta F^2/4$, $K_3 = -F^3/(3A_0^2)$.

Setting $\dot{A} = 0$ and $\dot{\varphi} = 0$ in Eqs. (9.45) and eliminating φ, we find the equations for the steady-state values of amplitudes determining the synchronization regime:

$$\left[\left(1 - \frac{A^2}{A_0^2} - \frac{2F^2}{A_0^2}\right)^2 + \frac{4\Delta^2}{\mu^2}\right] A^2 = K_n^2. \tag{9.46}$$

Equation (9.46) is similar to Eq. (9.4) for the main resonance. The stability conditions for solutions of Eq. (9.46) are also similar to those for the main resonance (9.9a and 9.9b). They are

$$\frac{A^2}{A_0^2} \geq \frac{1}{2}\left(1 - \frac{2F^2}{A_0^2}\right), \tag{9.47a}$$

$$\left(1 - \frac{2F^2}{A_0^2} - \frac{A^2}{A_0^2}\right)\left(1 - \frac{2F^2}{A_0^2} - \frac{3A^2}{A_0^2}\right) + 4\frac{\Delta^2}{\mu^2} \geq 0. \tag{9.47b}$$

It follows from (9.47a) that for $F^2 \geq A_0^2/2$ the solution is stable for any mistunings Δ.

The dependencies of A^2/A_0^2 on Δ/μ are illustrated in Figs. 9.7 ($n = 2$) and 9.8 ($n = 3$) for different values of the external force amplitudes.

It is seen from these figures that here, as in the case of subharmonic synchronization, the amplitude of the self-oscillations synchronized for $\Delta = 0$ decreases as the external force amplitude increases. The reason is

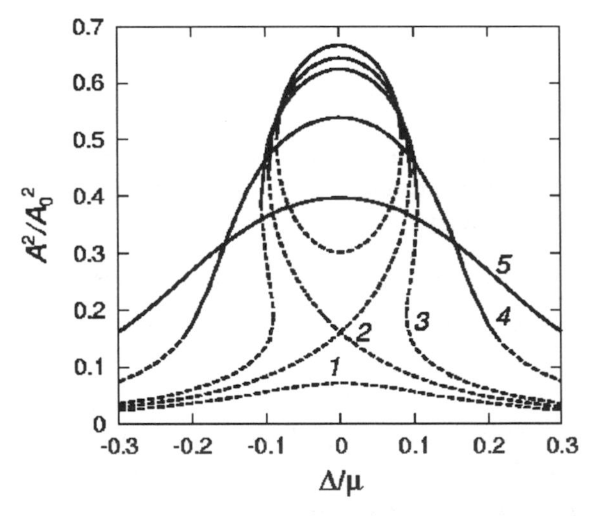

Fig. 9.7. The dependencies of A^2/A_0^2 on the relative mistuning Δ/μ in the case of the synchronization by the second harmonic of the external force for $\beta^2 A_0^2/4 = 1$. Curves 1, 2, 3, 4 and 5 correspond to $F^2/A_0^2 = 0.24$, 0.258463, 0.275, 0.35 and 0.5, respectively. The stable parts of these dependencies are shown as solid lines, and the unstable parts as dashed lines

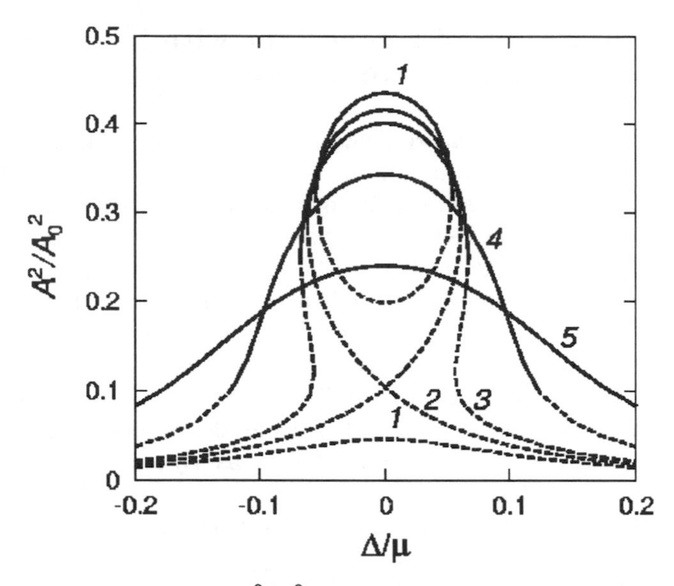

Fig. 9.8. The dependencies of A^2/A_0^2 on the relative mistuning Δ/μ in the case of the synchronization by the third harmonic of the external force. Curves 1, 2, 3, 4 and 5 correspond to $F^2/A_0^2 = 0.33$, 0.343812, 0.355, 0.4 and 0.5, respectively. The stable parts of these dependencies are shown as solid lines, and the unstable parts are shown as dashed lines

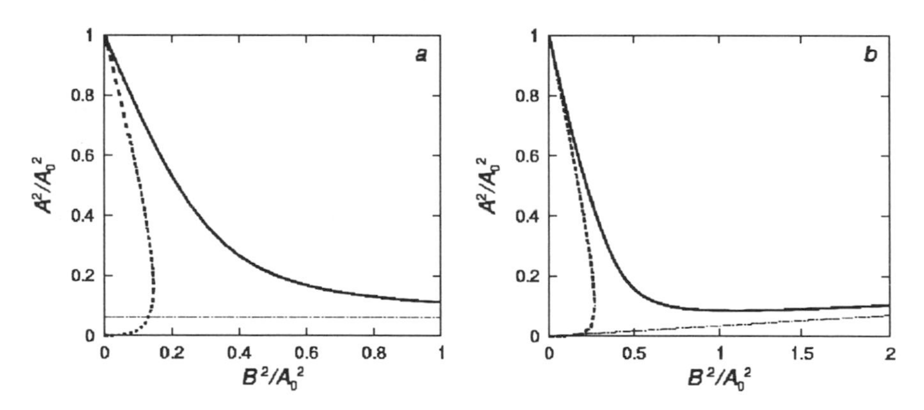

Fig. 9.9. The dependencies of A^2/A_0^2 at the frequency $n\omega$ on B^2/A_0^2 for $\Delta = 0$; (a) $n = 2$, $\beta^2 A_0^2/4 = 1$ and (b) $n = 3$. The stable parts of the dependencies are shown as solid lines, the unstable parts are shown as dashed lines, and the asymptotes are shown as dot-and-dash lines

asynchronous suppression of the self-oscillations. For $n = 2$ the amplitude of the self-oscillations tends to the limiting value $A_\infty = \beta A_0^2/8$ as B tends to infinity (see Fig. 9.9 a). For $n = 3$ the amplitude of the self-oscillations first decreases as B increases and then increases moderately tending to $F/6$ as $B \to \infty$ (see Fig. 9.9 b).

9.2 Synchronization of a generator with hard excitation. Asynchronous excitation of self-oscillations

In the preceding section synchronization of generators with soft excitation was considered. It was shown that in the case when the external force is non-resonant it suppresses self-oscillations and sets the generator to oscillate at its frequency. A different situation arises when a harmonic force acts on a generator with hard excitation [304]. If the external force frequency is close to the resonance one, from some threshold value of the external force amplitude onward, excitation of self-oscillations at the external force frequency occurs. The amplitude of these self-oscillations is much more than the amplitude of forced oscillations of the generative linear system. A non-resonance force can induce the excitation of self-oscillations at the natural frequency of the generative system. This phenomenon is known as *asynchronous excitation of self-oscillations* [228, 166].

The simplest equation of a generator with hard excitation is

$$\ddot{x} + \mu(1 - \alpha x^2 + \gamma x^4)\dot{x} + \omega_0^2 x = \omega_0^2 B \cos \omega t. \tag{9.48}$$

We consider first the case when the external force frequency is close to the natural frequency ω_0. In this case we seek a solution of Eq. (9.48) in the form $x = A\cos(\omega t + \varphi)$. By using the averaging method we obtain the following averaged equations for the amplitude A and the phase φ:

$$\begin{aligned} \dot{A} &= -\frac{\mu}{2} A \left(1 - \frac{\alpha A^2}{4} + \frac{\gamma A^4}{8}\right) - \frac{\omega_0 B}{2} \sin \varphi, \\ \dot{\varphi} &= \Delta - \frac{\omega_0 B}{2A} \cos \varphi, \end{aligned} \tag{9.49}$$

where $\Delta = \omega_0 - \omega$. These equations are conveniently rewritten in dimensionless variables

$$a = \sqrt{\frac{\gamma}{\alpha}} \frac{A}{2}, \quad b = \sqrt{\frac{\gamma}{\alpha}} \frac{B}{2}, \quad \eta = \frac{\gamma}{\alpha^2}$$

as

$$\begin{aligned} \dot{a} &= -\frac{\mu}{2\eta} a(\eta - a^2 + 2a^4) - \frac{\omega_0 b}{2} \sin \varphi, \\ \dot{\varphi} &= \Delta - \frac{\omega_0 b}{2a} \cos \varphi. \end{aligned} \tag{9.50}$$

A steady-state solution of Eqs. (9.51) is determined by the equation

$$\left((\eta - a^2 + 2a^4)^2 + \frac{4\eta^2 \Delta^2}{\mu^2}\right) = \frac{\omega_0^2 b^2}{\mu^2}. \tag{9.51}$$

There are three ranges of the parameter η, where the behavior of the resonance curves described by Eq. (9.51) differs qualitatively: $\eta < 1/8$, $1/8 < \eta <$

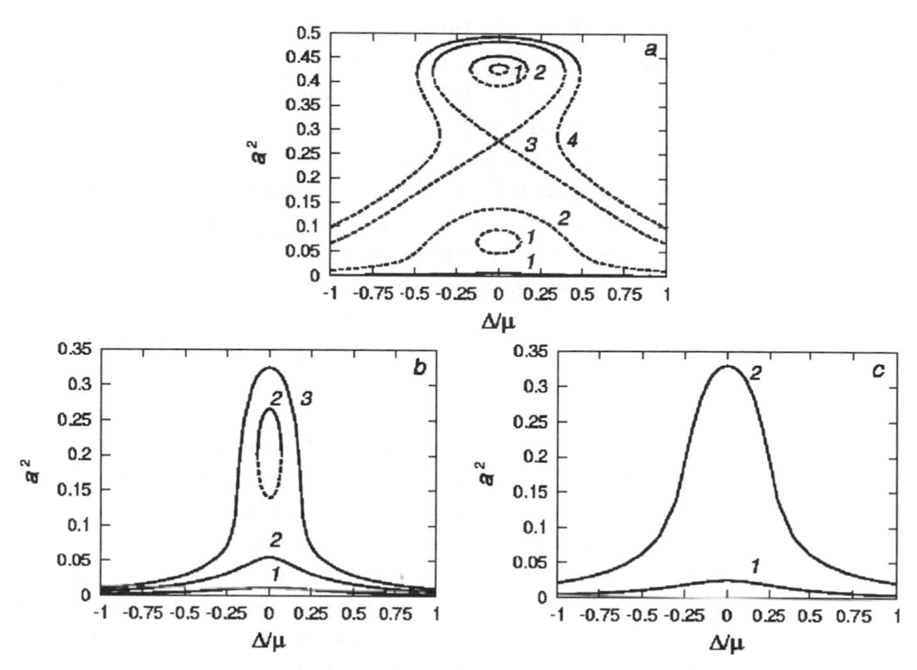

Fig. 9.10. The dependencies of a^2 on the relative mistuning Δ/μ for (a) $\eta = 1/16$, $\omega_0^2 b^2/\mu^2 = 0.005$, 0.05, 0.2642315 and 0.4 (curves 1, 2, 3 and 4, respectively); (b) $\eta = 3/16$, $\omega_0^2 b^2/\mu^2 = 0.01$, 0.03 and 0.05 (curves 1, 2 and 3, respectively); and (c) $\eta = 1/4$, $\omega_0^2 b^2/\mu^2 = 0.02$ and 0.1 (curves 1 and 2, respectively). The stable parts of these dependencies are shown as solid lines, and the unstable parts are shown as dashed lines

9/40, and $\eta > 9/40$. Examples of the resonance curves in these ranges are illustrated in Fig. 9.10 for different values of the external force amplitude.

Let us consider the stability of the steady-state solutions found. For this we write the linearized equations for small deviations from the steady-state values of the amplitude a and phase φ:

$$\dot{\xi} = -\frac{\mu}{2\eta}\xi(\eta - 3a^2 + 10a^4) - a\Delta\psi,$$

$$\dot{\psi} = \frac{\Delta}{a}\xi - \frac{\mu}{2\eta}\psi(\eta - a^2 + 2a^4), \tag{9.52}$$

where $\xi = a - a_{\text{st}}$, $\psi = \varphi - \varphi_{\text{st}}$. From (9.52) we find the following stability conditions:

$$a^4 - \frac{a^2}{3} + \frac{\eta}{6} \geq 0, \tag{9.53a}$$

$$(\eta - a^2 + 2a^4)(\eta - 3a^2 + 10a^4) + \frac{4\eta^2\Delta^2}{\mu^2} \geq 0. \tag{9.53b}$$

If $\eta < 1/6$ then condition (9.53a) is fulfilled for $a^2 \leq (1 - \sqrt{1-6\eta})/6$ and $a^2 \geq (1+\sqrt{1-6\eta})/6$. If $\eta > 1/6$ then condition (9.53a) is fulfilled everywhere.

For $\eta < 1/8$ condition (9.53b) is fulfilled everywhere, except in two regions on the plane Δ/μ, a^2 bounded by closed curves. One of these regions is located in the vicinity of the unstable cycle amplitude of the autonomous generator, and the other region is located in the vicinity of the stable cycle amplitude of the autonomous generator. For $1/8 \leq \eta \leq 9/40$ there is only one region bounded by a closed curve in which condition (9.53b) is not fulfilled. For $\eta > 9/40$ both conditions (9.53a) and (9.53b) are fulfilled everywhere. The analysis of conditions (9.53a) and (9.53b) shows that in the case when Eq. (9.51) possesses five roots, only the least and the greatest roots are stable; in the case when Eq. (9.51) possesses three roots, only the largest root is stable; and in the case when Eq. (9.51) possesses a single root, it is stable either in a limited region of the mistunings (for $\eta < 1/6$) or for any mistunings (for $\eta > 1/6$). In Fig. 9.10 the stable parts of the resonance curves are shown as solid lines and the unstable parts are shown as dashed lines.

The characteristic features of the behavior of the resonance curves in the ranges of the parameter η indicated above manifest themselves in the dependencies of the oscillation amplitude squared on the external force amplitude squared as well (see Fig. 9.11). For $\eta < 1/8$ (curves 1) and small external force amplitudes there exist two steady states: forced oscillations with small amplitude and synchronized self-oscillations with large amplitude; from a certain value of the external force amplitude onward, only synchronized self-oscillations are stable. For $1/8 < \eta < 9/40$ (curves 2) and small external force amplitudes only forced oscillations with moderately small amplitude are sta-

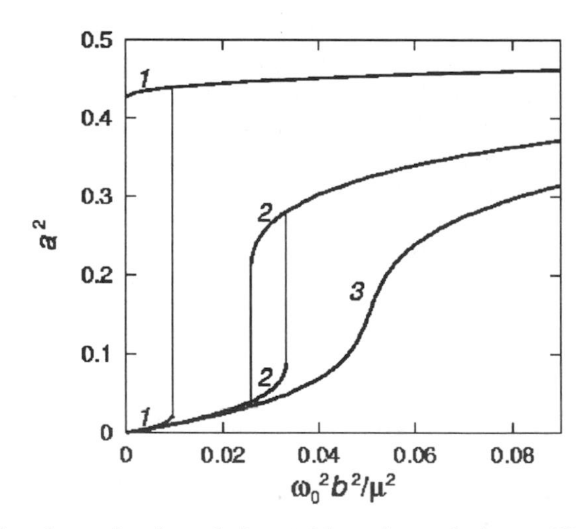

Fig. 9.11. The dependencies of the stable values of the oscillation amplitude squared a^2 at the external force frequency ω on $\omega_0^2 b^2/\mu^2$ for $\Delta = 0$, $\eta = 1/16$ (curves 1), $\eta = 3/16$ (curves 2) and $\eta = 1/4$ (curve 3)

ble; over a certain range of the external force amplitudes both the regime of forced oscillations and the regime of synchronized self-oscillations are stable; finally, from a certain value of the external force amplitude onward, only synchronized self-oscillations remain stable. As the external force amplitude decreases, hysteresis takes place. In the case of $\eta > 9/40$ the transition from forced oscillations to synchronized self-oscillations with increasing external force amplitude occurs smoothly (curve 3).

9.2.1 Asynchronous excitation of self-oscillations

As mentioned above, a non-resonance external force can induce the excitation of self-oscillations at the natural frequency of the generator. If the frequency of the external force is far from its resonance value then it is convenient to perform a change of variables of the form (9.15). In so doing we obtain the following equation for the variable y:

$$\ddot{y}+\omega_0^2 y = -\mu\left(1-\alpha(y+F\cos\omega t)^2+\gamma(y+F\cos\omega t)^4\right)(\dot{y}-F\omega\sin\omega t). \tag{9.54}$$

Setting $y = C\cos(\omega_0 t+\psi)$ we can write for C and ψ the truncated equations

$$\begin{aligned}\dot{C} &= -\frac{\mu}{2}\left(1-\frac{\alpha}{4}(C^2+2F^2)+\frac{\gamma}{8}(C^4+6C^2F^2+3F^4)\right)C,\\ \dot{\psi} &= 0.\end{aligned} \tag{9.55}$$

It is seen from this that the external force influences not only linear friction but nonlinear friction as well. Owing to this, both intensification and attenuation of self-oscillations can occur. If the influence on the nonlinear friction can be ignored then it follows from (9.55) that an external force with a small amplitude has to result in intensification of self-oscillations, whereas an external force with a large amplitude has to result in attenuation of self-oscillations.

The condition for asynchronous excitation of self-oscillations is

$$\frac{\alpha F^2}{2}-\frac{3\gamma F^4}{8}-1>0. \tag{9.56}$$

It follows from (9.56) that asynchronous excitation is possible only in a limited range of the external force amplitudes determined by the inequality

$$\frac{2\Delta^2}{3\mu^2}\left(1-\sqrt{1-6\eta}\right)<\frac{\omega_0^2b^2}{\mu^2}<\frac{2\Delta^2}{3\mu^2}\left(1+\sqrt{1-6\eta}\right). \tag{9.57}$$

This range is the greater the greater is the mistuning Δ. It follows from (9.57) that asynchronous excitation is possible only for $\eta < 1/6$, i.e., only for sufficiently large nonlinear negative friction.

If the force amplitude is fixed, then the inequality (9.56) determines the mistuning region in which asynchronous excitation occurs. It is

$$\frac{3}{2\left(1+\sqrt{1-6\eta}\right)}\,\frac{\omega_0^2 b^2}{\mu^2} < \frac{\Delta^2}{\mu^2} < \frac{3}{2\left(1-\sqrt{1-6\eta}\right)}\,\frac{\omega_0^2 b^2}{\mu^2}. \tag{9.58}$$

For mistunings smaller than the left-hand side of inequality (9.58), only forced oscillations with frequency ω can exist, whereas, for mistunings greater than the right hand side of inequality (9.58), depending on the initial conditions, both only forced oscillations and beats can exist.

9.3 Synchronization of the van der Pol generator with modulated natural frequency

Let us consider the action of a harmonic external force on a van der Pol generator whose natural frequency varies periodically according to the law $\tilde{\omega}_0 = \omega_0 + \Omega f(t)$, where $f(t)$ is a periodic function with period $\tau = 2\pi/\nu$ [165]. The equation describing oscillations of such a generator is

$$\ddot{x} - \mu(1-\alpha x^2)\dot{x} + \left(\omega_0 + \Omega f(t)\right)^2 x = \omega_0^2 B\cos\omega t. \tag{9.59}$$

In the absence of the external force the self-oscillations in the system described by Eq. (9.59) are quasi-periodic with two fundamental frequencies ω_0 and ν. With the external force, there exists a number of regions of mistunings between the external force frequency ω and the mean frequency of free oscillations ω_0 within which the mean frequency of self-oscillations is constant and equal to $\omega + n\nu$, where $n = 0, \pm 1, \pm 2, \ldots$. We called these regions of the mistunings the synchronization regions [165]. If the amplitude of the external force is sufficiently small and the frequency modulation is sufficiently large, then the synchronization regions are comparatively narrow and located in the vicinities of the mistuning values $\Delta = \Delta_n = n\nu$, where $\Delta = \omega_0 - \omega$. Within the synchronization regions the oscillations are quasi-periodic with two fundamental frequencies ω and ν.

Setting $x = A\cos(\omega t + \varphi)$ and assuming that $\mu \ll \omega$, $\Delta \ll \omega$, $\Omega \ll \omega$, $\nu \ll \omega$ we obtain from Eq. (9.59) the following truncated equations for the oscillation amplitude and phase:

$$\begin{aligned} \dot{A} &= \frac{\mu}{2}\left(1 - \frac{A^2}{A_0^2}\right)A - \frac{\omega_0 B}{2}\sin\varphi, \\ \dot{\varphi} &= \Delta - \frac{\omega_0 B}{2A}\cos\varphi + \Omega f(t), \end{aligned} \tag{9.60}$$

where $A_0 = 2/\sqrt{\alpha}$ is the amplitude of self-oscillations in the autonomous van der Pol generator with no frequency modulation.

If the amplitude B of the external force is sufficiently small and we are interested in sufficiently small values of the mistuning Δ, so that $\omega_0 B/(\mu A_0) \ll 1$ and $|\Delta|/\mu \ll 1$, then the oscillation amplitude A is close to A_0 and the equation for the phase φ can be written as

$$\dot{\varphi} = \Delta - \Delta_0 \cos\varphi + \Omega f(t), \tag{9.61}$$

where $\Delta_0 = \omega_0 B/2A_0$ is the half-width of the synchronization region in the absence of frequency modulation.

We note that several problems that are important in practice, such as the calculation of the voltage–current responses of a Josephson junction in a microwave field [19, 164] and of the frequency responses of a laser gyroscope placed on a vibrating base [160], lead to an equation of the form (9.61).

We consider first the simplest case when the function $f(t)$ is a sequence of positive and negative rectangular pulses of duration $\tau/2$, i.e.,

$$f(t) = \begin{cases} 1 & \text{for } m\tau \le t \le (m+1/2)\tau \\ -1 & \text{for } (m+1/2)\tau \le t \le (m+1)\tau \end{cases} \qquad (m = 0, \pm 1, \pm 2, \dots).$$

To find the synchronization regions we should calculate the average of the beat frequency $\dot{\varphi}$ over the modulation period τ as a function of the mistuning Δ. For this purpose it suffices to find the change of the phase φ in the time τ and to divide it by τ. In the synchronization regions the averaged beat frequency should be a constant equal to $n\nu$. Let the value of the phase φ at the origin of the $(m+1)$th modulation period be equal to φ_m. We integrate Eq. (9.61) on the time interval from $m\tau$ to $(m+1/2)\tau$ and then from $(m+1/2)\tau$ to $(m+1)\tau$. As a result we obtain the following equations:

$$\begin{aligned} &\arctan\left(a_1 \tan\frac{\varphi'_m}{2}\right) - \arctan\left(a_1 \tan\frac{\varphi_m}{2}\right) = b_1, \\ &\arctan\left(a_2 \tan\frac{\varphi'_m}{2}\right) - \arctan\left(a_2 \tan\frac{\varphi_{m+1}}{2}\right) = b_2, \end{aligned} \tag{9.62}$$

where

$$a_1 = \sqrt{\frac{\Omega + \Delta + \Delta_0}{\Omega + \Delta - \Delta_0}}, \quad a_2 = \sqrt{\frac{\Omega - \Delta - \Delta_0}{\Omega - \Delta + \Delta_0}},$$

$$b_1 = \frac{\tau}{4}\sqrt{(\Omega + \Delta)^2 - \Delta_0^2}\ \operatorname{sign}(\Omega + \Delta + \Delta_0),$$

$$b_2 = \frac{\tau}{4}\sqrt{(\Omega - \Delta)^2 - \Delta_0^2}\ \operatorname{sign}(\Omega - \Delta - \Delta_0),$$

and φ'_m is the value of the phase for $t = (m+1/2)\tau$.

Eliminating φ'_m from Eqs. (9.62) we express the phase change in the $(m+1)$th modulation period $\Delta\varphi_m = \varphi_{m+1} - \varphi_m$ in terms of the initial phase value φ_m:

$$\Delta\varphi_m = 2\arctan\frac{a_2(1+a_1^2)\tan b_1 - a_1(1+a_2^2)\tan b_2 + R\sin(\varphi_m+\chi)}{2a_1a_2 + (a_1^2+a_2^2)\tan b_1 \tan b_2 - R\cos(\varphi_m+\chi)}, \tag{9.63}$$

where

$$R = \sqrt{(a_1^2-a_2^2)^2(\tan b_1\tan b_2)^2 + \Big(a_2(1-a_1^2)\tan b_1 - a_1(1-a_2^2)\tan b_2\Big)^2},$$

$$\tan\chi = \frac{a_2(1-a_1^2)\tan b_1 - a_1(1-a_2^2)\tan b_2}{(a_1^2-a_2^2)\tan b_1\tan b_2}.$$

Since the parameters of Eq. (9.63) do not depend on m, the subscript m can be omitted. The solution of the finite-difference equation (9.63) allows us to find both the synchronization regions and the dependence of the beat frequency on the mistuning.

As mentioned above, in the nth synchronization region the phase change during one modulation period has to be $n\nu\tau = 2\pi n$. Hence in the synchronization regime the phase shift at the beginning of the $(m+1)$th modulation period is equal to $2\pi nm$. Let us substitute into Eq. (9.63) the new variable

$$\vartheta_{mn} = \varphi_m + \chi - 2\pi nm. \tag{9.64}$$

Omitting the subscript m we obtain for the following equation ϑ_n:

$$\Delta\vartheta_n = 2\arctan\frac{a_2(1+a_1^2)\tan b_1 - a_1(1+a_2^2)\tan b_2 + R\sin\vartheta_n}{2a_1a_2 + (a_1^2+a_2^2)\tan b_1\tan b_2 - R\cos\vartheta_n} - 2\pi n. \tag{9.65}$$

In the synchronization regime $\Delta\vartheta_n$ should be equal to zero, i.e.,

$$a_2(1+a_1^2)\tan b_1 - a_1(1+a_2^2)\tan b_2 + R\sin\vartheta_n = 0. \tag{9.66}$$

Equation (9.66) determines the phase ϑ_n in the nth synchronization region. The boundaries of this synchronization region are found from the equations

$$a_2(1+a_1^2)\tan b_1 - a_1(1+a_2^2)\tan b_2 = \pm R. \tag{9.67}$$

An approximate analytical solution of Eqs. (9.67) can be obtained only in a number of specific cases. First of all we consider the case when the amplitude Ω of the frequency modulation is much more than the half-width Δ_0 of the synchronization region in the absence of the frequency modulation. In this case, for mistunings Δ not too close to Ω and for not too large values of $\Delta_0\tau$, the conditions

$$\frac{\Delta_0}{\big|\Omega-|\Delta|\big|} \ll 1, \qquad \frac{\Delta_0^2\tau}{\big|\Omega-|\Delta|\big|} \ll 1 \tag{9.68}$$

are fulfilled. With these conditions Eqs. (9.67) approximately take the form:

$$\tan b_1 - \tan b_2 = \pm\frac{R}{2}, \tag{9.69}$$

where

$$R = \frac{4\Delta_0 \Omega}{\Omega^2 - \Delta^2} \sqrt{\tan^2 b_1 \tan^2 b_2 + \frac{\left((\Omega - \Delta)\tan b_1 + (\Omega + \Delta)\tan b_2\right)^2}{\Omega^2}}, \tag{9.70}$$

$$b_1 = \frac{(\Omega + \Delta)\tau}{4}, \quad b_2 = \frac{(\Omega - \Delta)\tau}{4}. \tag{9.71}$$

Transforming the difference of tangents in Eq. (9.69) we obtain

$$\frac{\Delta_{sn}^{(\pm)}\tau}{2} = \pi n \pm \arctan \frac{R}{2(1 + \tan b_1 \tan b_2)} \quad (n = 0, \pm 1, \pm 2, \ldots), \tag{9.72}$$

where $\Delta_{sn}^{(\pm)}$ are the boundaries of the nth synchronization region. Owing to the conditions (9.68) $\Delta_{sn}^{(\pm)}\tau$ differ only slightly from $2\pi n$; therefore we can put $\Delta_{sn}^{(\pm)} = n\nu \pm \delta_{sn}$, where $\delta_{sn} \ll \nu$. Neglecting δ_{sn} in expressions (9.71) we obtain

$$\tan b_1 \approx \tan b_2 \approx \begin{cases} \tan(\Omega\tau/4) & \text{for even } n, \\ -\cot(\Omega\tau/4) & \text{for odd } n. \end{cases} \tag{9.73}$$

Taking account of (9.73) and (9.70), from (9.72) we find

$$\delta_{sn} = \frac{4\Delta_0\Omega}{|\Omega^2 - n^2\nu^2|\tau} \begin{cases} \left|\sin\dfrac{\Omega\tau}{4}\right| & \text{for even } n, \\ \left|\cos\dfrac{\Omega\tau}{4}\right| & \text{for odd } n. \end{cases} \tag{9.74}$$

It follows from expression (9.74) that the width of the nth synchronization region, which is equal to $2\delta_{sn}$, depends on n. The widths of the even and odd regions can differ markedly. In the special case when $\Omega\tau/4 = (2k+1)\pi/2$, where k is an integer, the odd regions vanish and only the even regions remain. The widths of these regions for small n first increase with increasing n, and then, for $n > \Omega/\nu$, they decrease. In another special case when $\Omega\tau/4 = k\pi$ the even regions vanish and only the odd regions remain. The dependence of their width on n is the same as in the first case.

With conditions (9.68) it is possible to calculate the beat frequency in vicinities of the synchronization regions. Assuming the change of the phase ϑ_n in the time τ to be small, we can substitute into Eq. (9.65) $\mathrm{d}\vartheta_n/\mathrm{d}t$ in place of $\Delta\vartheta_n/\tau$. In so doing this equation can be written as

$$\frac{\mathrm{d}\vartheta_n}{\mathrm{d}t} = \delta_n + \delta_{sn}\sin\vartheta_n, \tag{9.75}$$

where $\delta_n = \Delta - n\nu$ is the frequency mistuning relative to the center of the nth synchronization region, and δ_{sn} is determined by (9.74). Equation (9.75) is similar to Eq. (9.11); therefore, in view of (9.64), we obtain for the averaged beat frequency ω_{b} in the vicinity of the nth synchronization region $\omega_{\mathrm{b}} = n\nu + \sqrt{\delta_n^2 - \delta_{sn}^2}$. An example of the dependence of the relative beat frequency ω_{b}/ν on the relative mistuning Δ/ν is given in Fig. 9.12 a.

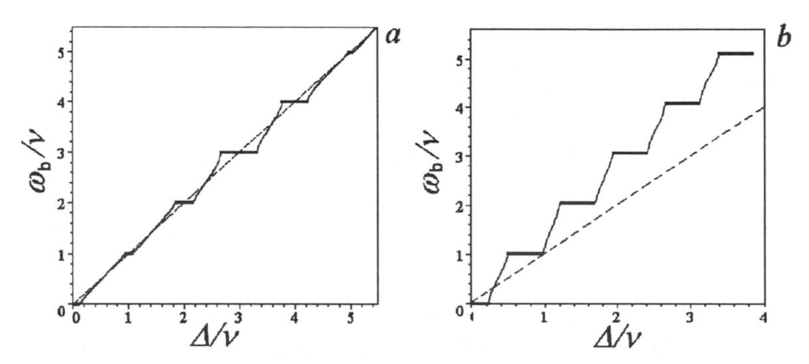

Fig. 9.12. The dependencies of the relative beat frequency ω_{b}/ν on the relative mistuning Δ/ν for (a) $\Delta_0/\Omega = 0.2$, $\nu/\Omega = 0.3$, and (b) $\Delta_0/\Omega = 1/\sqrt{2}$, $\nu/\Omega = \sqrt{2}/(2k+1)$, where k is any integer

Another specific case, when an approximate analytical solution of Eqs. (9.67) can be found, is the case of sufficiently small mistunings when the condition

$$|\Delta| \ll |\Omega - \Delta_0| \tag{9.76}$$

is fulfilled. For $n \neq 0$ condition (9.76) can be fulfilled only when $\Delta_{sn}^{(\pm)} \ll |\Omega - \Delta_0|$. Hence the solution obtained below is valid only for not very large n.

Under condition (9.76), in Eq. (9.67) we can put

$$a_1 = \sqrt{\frac{\Omega + \Delta_0}{\Omega - \Delta_0}}, \quad a_2 = \sqrt{\frac{\Omega - \Delta_0}{\Omega + \Delta_0}},$$
$$b_{1,2} = \left(\sqrt{\Omega^2 - \Delta_0^2} \pm \frac{\Omega\Delta}{\sqrt{\Omega^2 - \Delta_0^2}}\right)\frac{\tau}{4}. \tag{9.77}$$

Substituting (9.77) into Eqs. (9.67) we rewrite the latter in the form

$$\sin^2 \frac{\Omega\Delta\tau}{2\sqrt{\Omega^2 - \Delta_0^2}} - \frac{\Delta_0^2}{\Omega^2 - \Delta_0^2}\cos^2\frac{\Omega\Delta\tau}{2\sqrt{\Omega^2 - \Delta_0^2}} - \frac{\Delta_0^4}{\Omega^2(\Omega^2 - \Delta_0^2)}\cos^2\frac{\sqrt{\Omega^2 - \Delta_0^2}\,\tau}{2}$$
$$+ \frac{2\Delta_0^2}{\Omega^2 - \Delta_0^2}\cos\frac{\sqrt{\Omega^2 - \Delta_0^2}\,\tau}{2}\cos\frac{\Omega\Delta\tau}{2\sqrt{\Omega^2 - \Delta_0^2}} - \frac{\Delta_0^2}{\Omega^2} = 0. \tag{9.78}$$

Resolving Eq. (9.78) with respect to $\sin\left(\Omega\Delta\tau/2\sqrt{\Omega^2-\Delta_0^2}\right)$ we find

$$\sin\frac{\Omega\Delta\tau}{2\sqrt{\Omega^2-\Delta_0^2}} = \pm 2\frac{\Delta_0}{\Omega}\sqrt{1-\frac{\Delta_0^2}{\Omega^2}\sin^2\frac{\sqrt{\Omega^2-\Delta_0^2}\,\tau}{4}}\,\sin\frac{\sqrt{\Omega^2-\Delta_0^2}\,\tau}{4}, \tag{9.79a}$$

$$\sin\frac{\Omega\Delta\tau}{2\sqrt{\Omega^2-\Delta_0^2}} = \pm 2\frac{\Delta_0}{\Omega}\sqrt{1-\frac{\Delta_0^2}{\Omega^2}\cos^2\frac{\sqrt{\Omega^2-\Delta_0^2}\,\tau}{4}}\,\cos\frac{\sqrt{\Omega^2-\Delta_0^2}\,\tau}{4}. \tag{9.79b}$$

For $\Delta_0 \ll \Omega$ the solution (9.79a) can be reduced to (9.74) for even n, whereas the solution (9.79b) can be reduced to (9.74) for odd n. This suggests that for any relations between Δ_0 and Ω the solutions (9.79a) and (9.79b) are valid for even and odd n, respectively.

From (9.79a) and (9.79b) the following expressions for the boundaries of the nth synchronization region can be easily obtained:

$$\Delta_{sn}^{(\pm)} = \frac{2\sqrt{\Omega^2-\Delta_0^2}}{\Omega\tau}\left[\pi n \pm \arcsin\left(2\frac{\Delta_0}{\Omega}q\right)\right] \tag{9.80}$$

where

$$q = \begin{cases} \sqrt{1-\dfrac{\Delta_0^2}{\Omega^2}\sin^2\dfrac{\sqrt{\Omega^2-\Delta_0^2}\,\tau}{4}}\,\sin\dfrac{\sqrt{\Omega^2-\Delta_0^2}\,\tau}{4} & \text{for even } n, \\ \sqrt{1-\dfrac{\Delta_0^2}{\Omega^2}\cos^2\dfrac{\sqrt{\Omega^2-\Delta_0^2}\,\tau}{4}}\,\cos\dfrac{\sqrt{\Omega^2-\Delta_0^2}\,\tau}{4} & \text{for odd } n. \end{cases}$$

The width of the nth synchronization region is equal to $\Delta_{sn}^{(+)} - \Delta_{sn}^{(-)}$ and its center is located at $\Delta = \Delta_n = n\nu\sqrt{\Omega^2-\Delta_0^2}/\Omega$.

As an example, the synchronization regions calculated by (9.80) for $\Omega^2 = 2\Delta_0^2$, and $\sqrt{\Omega^2-\Delta_0^2}\tau = \Delta_0\tau = (2k+1)\pi$, where k is any integer, are illustrated in Fig. 9.12 b. For these parameters the widths of even and odd synchronization regions are the same and equal to $\sqrt{2}\nu/3$, and the centers of the synchronization regions are located at $\Delta = \Delta_n = n\nu/\sqrt{2}$. Thus, in this case the widths of the synchronization regions are comparable with the distances between them.

Let us consider further the case of harmonic frequency modulation. In this case Eq. (9.61) becomes

$$\dot{\varphi} = \Delta - \Delta_0\cos\varphi + \Omega\cos\nu t. \tag{9.81}$$

Generally, Eq. (9.81) cannot be solved analytically. Therefore we solve it by using successive approximations. For this we assume that $\epsilon \equiv \Delta_0/(\Omega - |\Delta|)$ is a small parameter. As a zero approximation with respect to ϵ, a solution of Eq. (9.81) is

$$\varphi = \Delta t + \frac{\Omega}{\nu} \sin \nu t + \varphi_m, \tag{9.82}$$

where φ_m is the phase at the beginning of the $(m+1)$th modulation period. Substituting (9.82) into the right-hand side of Eq. (9.81) we find the following equation of the first approximation:

$$\dot{\varphi} = \Delta + \Omega \cos \nu t - \Delta_0 \cos\left(\Delta t + \frac{\Omega}{\nu} \sin \nu t + \varphi_m\right). \tag{9.83}$$

By integrating this equation we find the change of phase in the modulation period $\tau = 2\pi/\nu$:

$$\begin{aligned}\Delta\varphi_m = \Bigg\{ \Delta - \frac{\Delta_0}{2} \Bigg[\Bigg(\boldsymbol{J}_{\Delta/\nu}\left(-\frac{\Omega}{\nu}\right) + \cos\left(\frac{\Delta}{\nu}\pi\right) \boldsymbol{J}_{\Delta/\nu}\left(\frac{\Omega}{\nu}\right) \\ - \sin\left(\frac{\Delta}{\nu}\pi\right) \boldsymbol{E}_{\Delta/\nu}\left(\frac{\Omega}{\nu}\right)\Bigg) \cos\varphi_m - \Bigg(\boldsymbol{E}_{\Delta/\nu}\left(-\frac{\Omega}{\nu}\right) \\ + \cos\left(\frac{\Delta}{\nu}\pi\right) \boldsymbol{E}_{\Delta/\nu}\left(\frac{\Omega}{\nu}\right) + \sin\left(\frac{\Delta}{\nu}\pi\right) \boldsymbol{J}_{\Delta/\nu}\left(\frac{\Omega}{\nu}\right)\Bigg) \sin\varphi_m \Bigg] \Bigg\}, \end{aligned} \tag{9.84}$$

where $\boldsymbol{J}_\mu(z)$ and $\boldsymbol{E}_\mu(z)$ are the Anger and the Weber functions, respectively.

We set in expression (9.84) $\varphi_m = \vartheta_{mn} + 2\pi nm$, $\Delta = \delta_n + 2\pi n/\tau$ and assume that $\delta_n \tau \ll 1$. In this case expression (9.84) is significantly simplified and we obtain

$$\frac{\Delta\vartheta_{mn}}{\tau} = \delta_n - \Delta_0 J_n\left(\frac{\Omega}{\nu}\right) \cos\vartheta_{mn}, \tag{9.85}$$

where $J_n(z)$ is the Bessel function. It follows from (9.85) that the half-width of the nth synchronization region is

$$\delta_{sn} = \Delta_0 \left| J_n\left(\frac{\Omega}{\nu}\right) \right|. \tag{9.86}$$

Hence, the widths of even and odd synchronization regions alternate as Bessel functions of even and odd order. If the condition $\delta_n \tau \ll 1$ is not fulfilled, the calculation of the synchronization regions based on expression (9.84) is more complicated.

We note that for finding an approximate solution of Eq. (9.81) the condition of smallness of the parameter ϵ is not necessary; the condition $\delta_n \tau \ll 1$ is sufficient. Under this condition Eq. (9.81) can be solved approximately by the averaging method [310]. Substituting $\varphi = n\nu t + \vartheta_n$ into Eq. (9.81) we obtain the following equation for ϑ_n:

$$\dot{\vartheta}_n = \delta_n - \Delta_0 \cos(n\nu t + \vartheta_n) + \Omega \cos \nu t. \tag{9.87}$$

According to the averaging method, in the second approximation we should substitute into the right-hand side of Eq. (9.87) $\vartheta_n + \vartheta_n^{(\mathrm{vibr})}$ in place of ϑ_n, where

$$\vartheta_n^{(\text{vibr})} = \begin{cases} \dfrac{\Omega}{\nu}\sin\nu t & \text{for } n = 0, \\ \dfrac{\Omega}{\nu}\sin\nu t - \dfrac{\Delta_0}{n\nu}\sin(n\nu t + \vartheta_n) & \text{for } n \neq 0. \end{cases}$$

After this the right-hand side should be averaged over time. In so doing we find for ϑ_n the following averaged equations:

$$\dot{\vartheta}_0 = \delta_0 - \Delta_0 \overline{\cos\left(\frac{\Omega}{\nu}\sin\nu t + \vartheta_0\right)} = \delta_0 - \Delta_0 J_0\left(\frac{\Omega}{\nu}\right)\cos\vartheta_0 \quad (n = 0), \tag{9.88a}$$

$$\dot{\vartheta}_n = \delta_n - \Delta_0 \overline{\cos\left(\frac{\Omega}{\nu}\sin\nu t + \vartheta_n + n\nu t - \frac{\Delta_0}{n\nu}\sin(n\nu t + \vartheta_n)\right)} \quad (n \neq 0). \tag{9.88b}$$

We see that for $n = 0$ the result coincides with that obtained above, but for $n \neq 0$ there is a difference. For example, in the case of $n = 1$ Eq. (9.88b) takes the form

$$\dot{\vartheta}_1 = \delta_1 - \frac{\kappa^2 - \Omega^2 + \Delta_0^2}{2\kappa} J_1\left(\frac{\kappa}{\nu}\right), \tag{9.89}$$

where $\kappa = \sqrt{\Omega^2 + \Delta_0^2 - 2\Omega\Delta_0\cos\vartheta_1}$. Synchronization regimes correspond to $\dot{\vartheta}_1 = 0$, i.e., to the mistunings δ_1 described by the equation

$$\delta_1 = \frac{\kappa^2 - \Omega^2 + \Delta_0^2}{2\kappa} J_1\left(\frac{\kappa}{\nu}\right). \tag{9.90}$$

The boundaries of the synchronization region are determined from the conditions for the minimum and maximum of the right-hand side of Eq. (9.38) with respect to ϑ_1. They are

$$\delta_1^{(+)} = \max\begin{cases} \Delta_0 J_1\left(\dfrac{\Omega + \Delta_0}{\nu}\right), \\ \Delta_0 \operatorname{sign}(\Delta_0 - \Omega) J_1\left(\dfrac{|\Omega - \Delta_0|}{\nu}\right), \\ \dfrac{\kappa_0^2 - \Omega^2 + \Delta_0^2}{2\kappa_0} J_1\left(\dfrac{\kappa_0}{\nu}\right), \end{cases} \tag{9.91a}$$

$$\delta_1^{(-)} = \min\begin{cases} \Delta_0 J_1\left(\dfrac{\Omega + \Delta_0}{\nu}\right), \\ \Delta_0 \operatorname{sign}(\Delta_0 - \Omega) J_1\left(\dfrac{|\Omega - \Delta_0|}{\nu}\right), \\ \dfrac{\kappa_0^2 - \Omega^2 + \Delta_0^2}{2\kappa_0} J_1\left(\dfrac{\kappa_0}{\nu}\right), \end{cases} \tag{9.91b}$$

where κ_0 are the roots of the equation

$$(\Omega^2 - \Delta_0^2) J_1\left(\frac{\kappa_0}{\nu}\right) = \frac{\kappa_0}{2\nu}(\Omega^2 - \kappa_0^2 - \Delta_0^2)\, J_0\left(\frac{\kappa_0}{\nu}\right)$$

satisfying the condition $|\Delta_0 - \Omega| \le \kappa_0 \le \Delta_0 + \Omega$. It follows from (9.91a) and (9.91b) that the synchronization region for $n = 1$ is generally located asymmetrically with respect to $\Delta_1 = \nu$.

In a similar way we can calculate synchronization regions for $n = 2, 3, \ldots$.

9.4 Synchronization of periodic oscillations in systems with inertial nonlinearity

Let us now consider synchronization of periodic self-oscillations in a system with inertial nonlinearity. Self-oscillations in the autonomous system are described by Eqs. (7.71), in the case of a thermistor, or by Eqs. (7.81), (7.82) with $f(x) = K\vartheta(x)x$, where $\vartheta(x)$ is Heaviside's step function, in the case of a linear detector. Let the external force $\omega_0^2 B \cos\omega t$ act upon this system. If μ, $\gamma \ll \omega_0$, $\alpha_1 T \ll \alpha_0$ then we obtain the following truncated equations for the oscillation amplitude A, the phase φ and the temperature T:

$$\begin{aligned} \dot{A} &= \frac{1}{2}(\mu - bT)A - \frac{\omega_0 B}{2}\sin\varphi, \quad \dot{\varphi} = \Delta - \frac{\omega_0 B}{2A}\cos\varphi, \\ \dot{T} &= -\gamma T + \frac{\alpha_0}{2}A^2, \end{aligned} \tag{9.92}$$

where $\Delta = \omega_0 - \omega$ is the mistuning between the natural frequency ω_0 and the external force frequency ω.

In the case of a linear detector the corresponding truncated equations are

$$\begin{aligned} \dot{A} &= \frac{1}{2}(\mu - bV)A - \frac{\omega_0 B}{2}\sin\varphi, \quad \dot{\varphi} = \Delta - \frac{\omega_0 B}{2A}\cos\varphi, \\ \dot{V} &= -\gamma\left(V - \frac{K}{\pi}A\right). \end{aligned} \tag{9.93}$$

Steady-state solutions of Eqs. (9.92) and (9.93) corresponding to the synchronization regimes are:

- in the case of a thermistor

$$\begin{aligned} &\cos\varphi = \frac{2\Delta A}{\omega_0 B}, \quad T = \frac{\alpha_0}{2\gamma}A^2, \\ &\left[\left(1 - \frac{A^2}{A_0^2}\right)^2 + 4\frac{\Delta^2}{\mu^2}\right]\frac{A^2}{A_0^2} = \frac{\omega_0^2 B^2}{\mu^2 A_0^2}, \end{aligned} \tag{9.94}$$

where $A_0 = \sqrt{2\mu\gamma/(\alpha_0 b)}$ is the amplitude of free self-oscillations;

- in the case of a linear detector

$$\cos\varphi = \frac{2\Delta A}{\omega_0 B}, \quad V = \frac{KA}{\pi},$$

$$\left[\left(1 - \frac{A}{A_0}\right)^2 + 4\frac{\Delta^2}{\mu^2}\right]\frac{A^2}{A_0^2} = \frac{\omega_0^2 B^2}{\mu^2 A_0^2}, \tag{9.95}$$

where $A_0 = \pi\mu/(Kb)$.

Equation (9.94) coincides with (9.4) for $\Gamma = 0$. Therefore the resonance curves described by this equation are the same as in Fig. 9.1 a. However, as indicated below, the stability regions for these curves are different.

The resonance curves constructed by solving Eq. (9.95) for a number of values of the external force amplitude are given in Fig. 9.13 a.

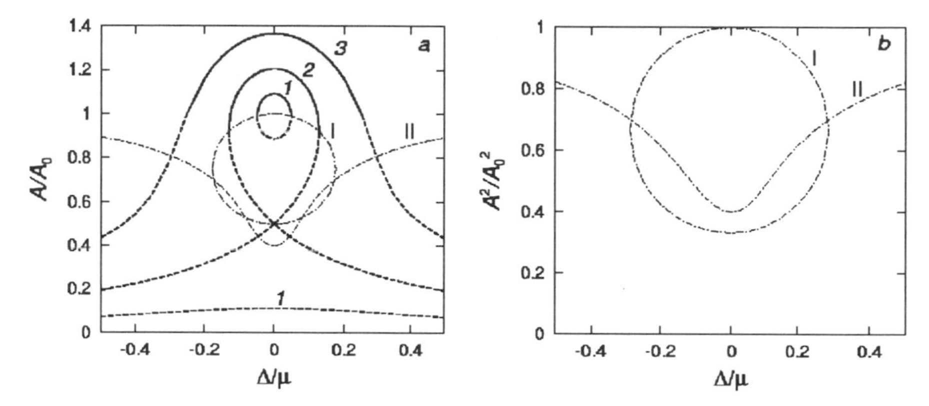

Fig. 9.13. (a) The dependencies of A/A_0 on Δ/μ constructed by solving Eq. (9.95) for $\omega_0 B/(\mu A_0) = 0.1$, 0.25, and 0.5 (curves 1, 2, 3, respectively); the stable parts of the dependencies are shown as solid lines, and the unstable parts are shown as dashed lines. The curves labeled I in a and b correspond to the boundary of the inequality (9.96), and the curves labeled II corresponds to the boundary of the inequality (9.97) for $\gamma/\mu = 0.3$

Investigation of the stability of the steady-state solution found leads to the characteristic equation

$$p^3 + (\gamma - 2\delta_2)p^2 + \left(\delta_2^2 - \gamma(\delta_1 + \delta_2) + \Delta^2\right)p + \gamma(\delta_1\delta_2 + \Delta^2) = 0,$$

where:

- in the case of a thermistor

$$\delta_1 = (\mu/2)(1 - 3A^2/A_0^2), \quad \delta_2 = (\mu/2)(1 - A^2/A_0^2)$$

- in the case of a linear detector

$$\delta_1 = (\mu/2)(1 - 2A/A_0), \quad \delta_2 = (\mu/2)(1 - A/A_0).$$

Using the Routh–Hurwitz criterion we find the following stability conditions:

$$\delta_1\delta_2 + \Delta^2 \geq 0, \tag{9.96}$$

$$\gamma\delta_2(\delta_1 + 3\delta_2) - \gamma^2(\delta_1 + \delta_2) - 2\delta_2(\delta_2^2 + \Delta^2) \geq 0. \tag{9.97}$$

In the case of the generator with a thermistor condition (9.96) coincides with (9.9b), whereas condition (9.97) depends on γ/μ and differs essentially from (9.9a). The boundaries of the inequalities (9.96) and (9.97) for $\gamma/\mu = 0.3$ are shown in Fig. 9.13 (a) in the case of a linear detector and in (b) in the case of a thermistor.

We consider further beat regimes outside the synchronization region by the example of the generator with a thermistor. For the consideration of such regimes it is convenient in Eqs. (9.92) to turn to the Cartesian coordinates $u = (A/A_0)\cos\varphi$ and $v = (A/A_0)\sin\varphi$. In these coordinates Eqs. (9.92) become

$$\dot{u} = \frac{1}{2}(\mu - bT)u - \Delta v, \quad \dot{v} = \frac{1}{2}(\mu - bT)v + \Delta u - \frac{\omega_0 B}{2A_0},$$
$$\dot{T} = -\gamma\left(T - \frac{\mu\gamma}{b}(u^2 + v^2)\right). \tag{9.98}$$

Substituting into Eqs. (9.98)

$$u = C\sin(\Delta t + \psi) + \frac{\omega_0 B}{2A_0\Delta}, \quad v = -C\cos(\Delta t + \psi), \tag{9.99}$$

we obtain the following equations for C, ψ and T:

$$\dot{C} = \frac{1}{2}(\mu - bT)\left(C + \frac{\omega_0 B}{2A_0\Delta}\sin(\Delta t + \psi)\right),$$
$$\dot{\psi} = \frac{\omega_0 B}{4A_0\Delta C}(\mu - bT)\cos(\Delta t + \psi), \tag{9.100}$$
$$\dot{T} = -\gamma\left[T - \frac{\mu}{b}\left(C^2 + \frac{\omega_0^2 B^2}{4A_0^2\Delta^2} + \frac{\omega_0 B}{A_0\Delta}C\sin(\Delta t + \psi)\right)\right].$$

Equations (9.100) should be solved differently, depending on the relations between γ, Δ and μ. If $\gamma \ll |\Delta|$, Eqs. (9.100) can be averaged over time. A steady-state solution of the averaged equations is

$$T = \frac{\mu}{b}, \quad C^2 = 1 - \frac{\omega_0^2 B^2}{4A_0^2\Delta^2}. \tag{9.101}$$

This solution exists and is stable for $|\Delta| \geq \omega_0 B/(2A_0)$. From this it follows that under the condition $\gamma \ll \omega_0 B/(2A_0)$ the half-width of the synchronization region Δ_s is equal to $\omega_0 B/(2A_0)$.

Near to a boundary of the synchronization region the value of C is small. As the first approximation with respect to C we find from the third equation of (9.100)

$$T = \frac{\mu}{b}\left(1 - \gamma\,\frac{\omega_0 B}{A_0 \Delta^2}\,C\cos(\Delta t + \psi)\right). \tag{9.102}$$

Substituting (9.102) in the second equation of (9.100) and averaging over time we find

$$\bar{\dot{\psi}} = \frac{\gamma\mu\omega_0^2 B^2}{8A_0^2\Delta^3} = \frac{\gamma\mu\Delta_s^2}{2\Delta^3}. \tag{9.103}$$

It follows from here that the beat frequency $\bar{\dot{\varphi}}$ is

$$\bar{\dot{\varphi}} = \Delta\left(1 + \frac{\gamma\mu\Delta_s^2}{2\Delta^4}\right). \tag{9.104}$$

For $\gamma \to 0$ the beat frequency tends to Δ. This means that outside the synchronization region the oscillations are a sum of forced oscillations at the external force frequency ω and self-oscillations at the frequency ω_0. It is easily shown that the self-oscillation amplitude is $A_0 C = \sqrt{A_0^2 - \omega_0^2 B^2/(4\Delta^2)}$. For $|\Delta| = \Delta_s$ synchronization by quenching occurs. The generator behaves just like a van der Pol generator under the action of an external force of great amplitude.

We note that the expressions (9.103) and (9.104) are valid for

$$\mu\gamma\Delta_s^2 \ll \Delta^4. \tag{9.105}$$

This condition is fulfilled the better the greater is the mistuning Δ. Nearby the synchronization region boundary (9.105) becomes

$$\mu\gamma \ll \Delta_s^2. \tag{9.106}$$

It cannot be fulfilled even for $\gamma \ll \Delta_s$ if $\Delta_s \ll \mu$. In this case the use of the averaging method for the solution of Eqs. (9.100) is not valid, and outside the synchronization region oscillations can be of a moderately complicated shape. This is corroborated by the experimental and numerical investigations performed by Kaptsov [144]. He has shown that a so-called *spike mode* is observed over a certain range of the inertiality parameter γ. An example of the oscillations in such a mode of operation, obtained by numerical simulation of the equations

$$\begin{aligned} &\ddot{I} - (\mu - bT)\dot{I} + \omega_0^2 I + bI\dot{T} = \omega_0^2 B\cos\omega t, \\ &\dot{T} + \gamma T = \alpha_0 I^2, \end{aligned} \tag{9.107}$$

is given in Fig. 9.14. A distinguishing feature of the spike mode is the presence of short spikes in the oscillation amplitude. Between these spikes the amplitude executes small oscillations with a frequency slightly above the mistuning Δ. The spike frequency ω_{sp} gradually increases with increasing mistuning, being always somewhat less than $\sqrt{\Delta^2 - \Delta_s^2}$ (Fig. 9.14 d). The spike magnitude decreases as the mistuning increases, and for a certain value of Δ the spikes

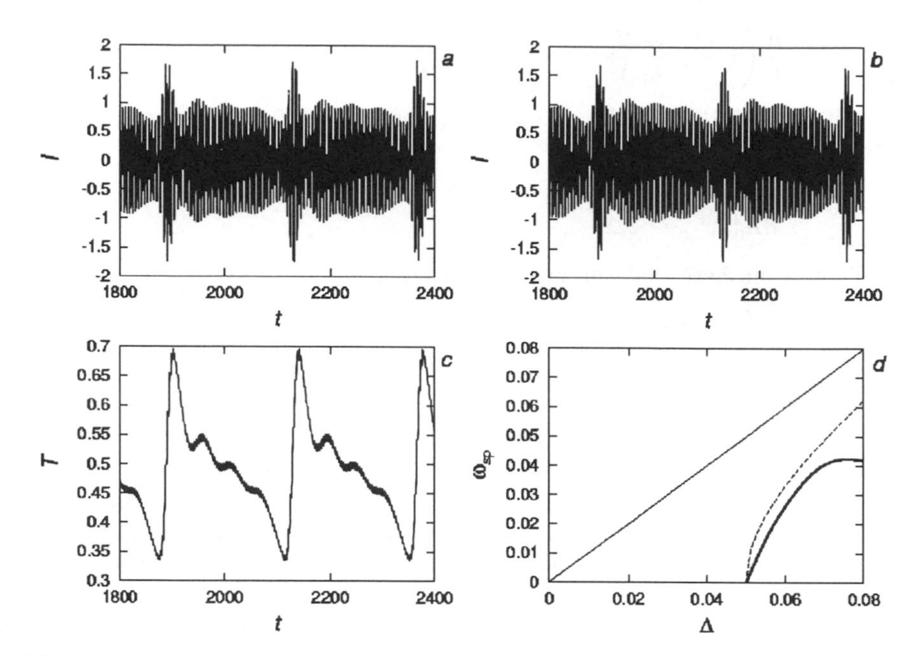

Fig. 9.14. (a, b, c) An example of oscillations of a non-autonomous generator with inertial nonlinearity in spike mode for $\omega_0 = 1$, $\mu = 0.5$, $\alpha_0 = 0.0265$, $\gamma = 0.025$, $B = 0.1$ ($\Delta_\mathrm{s} = 0.05$), $\omega = 1.06$, and (d) the dependencies of ω_sp (the solid line) and of $\sqrt{\Delta^2 - \Delta_\mathrm{s}^2}$ (the dashed line) on the mistuning Δ

vanish; only weak amplitude modulation remains. As γ decreases the mistuning range in which the spikes exist shrinks, the spike frequency and the width of the spikes, for a fixed mistuning, decrease, whereas their magnitude increases. Conversely, as γ increases the mistuning range in which the spikes exist expands, the spike frequency, for a fixed mistuning, increases and the spike magnitude decreases. For sufficiently large γ the spikes vanish; only the modulation of amplitude with frequency $\sqrt{\Delta^2 - \Delta_\mathrm{s}^2}$ remains. Hence, for large γ the generator behaves as an inertialess one.

We consider further the case of moderately large values of γ. If $\gamma \gg \mu \gg |\Delta|$, the variable T can be eliminated from Eqs. (9.92) and we obtain for the beat frequency the expression

$$\omega_\mathrm{b} = \overline{|\dot{\varphi}|} = \sqrt{\Delta^2 - \Delta_\mathrm{s}^2}\,, \tag{9.108}$$

where $\Delta_\mathrm{s} = \omega_0 B/(2A_0)$. This expression coincides with (9.14) for $\Gamma = 0$.

If $\gamma \gg |\Delta| \gg \mu$, the variable T in Eqs. (9.100) can be considered as a 'fast' one. In this case we can find a stationary solution for it assuming C and ψ to be time independent. For this we set

$$T = T_0 + T_1 \cos(\Delta t + \psi) + T_2 \sin(\Delta t + \psi). \tag{9.109}$$

Substituting (9.109) into the third equation of (9.100) we find

$$T_0 = \frac{\mu}{b}\left(C^2 + \frac{\omega_0^2 B^2}{4A_0^2 \Delta^2}\right), \quad T_1 = -\frac{\mu\omega_0 B}{bA_0\gamma} C, \quad T_2 = \frac{\mu\omega_0 B}{bA_0\Delta} C. \quad (9.110)$$

Substituting further (9.109), (9.110) into the first two equations of (9.100) and averaging over time we obtain the equations for C and ψ:

$$\dot{C} = \frac{\mu}{2}\left(1 - C^2 - \frac{\omega_0^2 B^2}{2A_0^2 \Delta^2}\right) C, \quad (9.111a)$$

$$\dot{\psi} = \frac{\mu\omega_0^2 B^2}{8A_0^2 \gamma \Delta}. \quad (9.111b)$$

A steady-state solution of Eq. (9.111a) is

$$C^2 = 1 - \frac{\omega_0^2 B^2}{2A_0^2 \Delta^2}. \quad (9.112)$$

This solution exists and is stable for $|\Delta| \geq \omega_0 B/(\sqrt{2}A_0)$. Hence, in the case of $\gamma \gg \omega_0 B/(\sqrt{2}A_0)$ the half-width of the synchronization region is $\Delta_s = \omega_0 B/(\sqrt{2}A_0)$. Equation (9.111b) describes a correction to the beat frequency Δ outside the synchronization region. Taking account of this correction we obtain the following expression for the beat frequency:

$$\omega_b = \overline{\dot{\varphi}} = \Delta\left(1 + \frac{\mu}{4\gamma}\frac{\Delta_s^2}{\Delta^2}\right). \quad (9.113)$$

We see that the beat frequency is greater than the mistuning Δ. Based on the limiting expressions (9.104) and (9.113) we obtain the interpolated expression for the beat frequency:

$$\omega_b = \Delta\sqrt{1 + \mu\gamma\frac{\Delta_s^2}{\Delta^2(\Delta^2 + 2\gamma^2)}}. \quad (9.114)$$

Expression (9.114) is valid for $\Delta_s \gg \min(\gamma, \mu)$, i.e., for sufficiently large external force amplitudes. In this case synchronization occurs by quenching. In another limiting case, when $\Delta_s \ll \gamma, \mu$, the beat frequency is described by (9.108) and synchronization occurs by way of a gradual decrease of the beat frequency to zero. Thus, in the case when a periodic external force acts upon a generator with inertial nonlinearity the mechanisms of synchronization considered above by the example of the van der Pol–Duffing generator manifest themselves as well.

9.5 Chaotization of periodic self-oscillations by an external force

It follows from (9.19) that in a nonautonomous van der Pol–Duffing generator synchronization at the main frequency can occur only for $B \leq A_0/\sqrt{2}$. Otherwise, either synchronization of another kind or chaotization should take place. It is very interesting that the phenomenon of chaotization for the van der Pol equation was detected as early as the 1940s by Cartwright and Littlewood [65] and Levinson [209], long before the appearance of the notion of dynamical chaos. By using qualitative methods it was shown that the van der Pol equation with an external harmonic force can have irregular stationary solutions for large values of the parameter μ. Later, such solutions were actually found by computation of this equation and the van der Pol–Duffing equation (9.1). It was numerically shown for Eq. (9.1) that in the case of $\mu \ll \omega_0$ chaotization can occur only for $\gamma \neq 0$ [343, 72]. [2] But if the oscillations of the autonomous generator are close to the relaxation oscillations, which exist for $\mu > \omega_0$, then chaotization is possible for $\gamma = 0$ as well [71, 261, 104]. This fact conforms to the results of Cartwright and Littlewood, and Levinson. For example, a chaotic solution of the van der Pol equation was found for $\mu/\omega_0 = 3$, $\omega/\omega_0 = 2.7$, $B/A_0 = 1.25$ [71]. We note that the possibility of the existence of chaotic solutions for the van der Pol equation with $\mu/\omega_0 \gg 1$ was proved theoretically both by qualitative methods [208] and by the technique of symbolic dynamics [111].

Chaotization of oscillations by a periodic external force is also observed in generators with inertial nonlinearity. The first results were obtained by Kaptsov [144]. He discovered that, in the transition from the spike mode operation to the weak amplitude modulation regime, as the parameter of inertiality γ decreases, the oscillation amplitude and phase oscillate chaotically. Later a similar effect was studied numerically and experimentally by Anischenko et al. [9, 11, 13, 15]. It was shown that, as μ increases, quasi-period-doubling bifurcations of the torus emerge, and a chaotic attractor appears. Similar results are also observed if the amplitude B of the external force increases for fixed μ. For fixed B the number of torus-doubling bifurcations until the torus is destroyed is finite, and depends on B: the smaller B, the greater the number of the bifurcations. This result correlates well with the general conclusions of Kaneko [139, 140]. He has shown by the example of the two-dimensional map

$$x_{n+1} = 1 - Ax_n^2 + \epsilon \sin 2\pi y_n, \quad y_{n+1} = y_n + C + \epsilon x_n$$

that the number of torus-doubling bifurcations, occurring before transition to chaos as the parameter A varies, is finite and determined by the coupling parameter ϵ.

It is evident that chaotization of oscillations in a generator with inertial nonlinearity can also occur under parametric action with a frequency of the order of the frequency of the relaxation process. This phenomenon was studied numerically and experimentally by Bezaeva et al., see [37, 38, 253].

[2] What actually happens is that chaotization is possible for $\gamma = 0$ as well, but for very large values of the external force amplitudes.

9.6 Synchronization of chaotic self-oscillations. The synchronization threshold and its relation to the quantitative characteristics of the attractor

In the case when an autonomous self-oscillatory system executes chaotic oscillations, the action on it of a harmonic external force can lead to a transition from chaotic to periodic oscillations with a period which is a multiple of the force period. Such a transition was called by Kuznetsov et al. *synchronization of chaotic oscillations* [162]. It is important to keep in mind that the transition occurs only beyond a certain critical value of the force amplitude. This critical value depends resonantly on the force frequency. Therefore, it makes sense to define the synchronization threshold as the minimal value of the amplitude of a harmonic force for which synchronization takes place at some frequency. This threshold is strictly zero for self-oscillatory systems executing periodic oscillations in the absence of the external force [47]. However, if an autonomous self-oscillatory system executes chaotic oscillations then the synchronizing action should overcome the tendency to chaos which exists in the system. This is possible only for a finite value of the force amplitude. Therefore a nonzero synchronization threshold can serve as a criterion of chaos in dynamical systems. By numerical simulation of several simple three-dimensional systems it was shown, see [162, 253], that the synchronization threshold B_{thr} is related to the maximal Lyapunov exponent λ by the universal power dependence $B_{\text{thr}} = C\lambda^{\chi}$, where $C = \text{const}$ and $\chi \approx 0.33$. Study of more complicated systems, in particular of the Mackey–Glass system [174], showed that the maximal Lyapunov exponent must be replaced by the Kolmogorov metric entropy.

It should be noted that the systems studied in the works cited have a single attractor. The case when an autonomous system has several attractors is far more complex. The results of numerical simulation of self-oscillations of two coupled generators with inertial self-excitation under a harmonic external force are given in [77]. It is shown that, firstly, the power dependence of the synchronization threshold on the Kolmogorov entropy is true only close to the boundary of the appearance of chaotic oscillations, and, secondly, the exponent χ is not universal.

Numerical determination of the synchronization regions is a relatively complicated problem because the external force frequencies associated with these regions are not known in advance. A simple algorithm suggested by Rosenblum [288] for the fast determination of one of the external force frequencies for which synchronization occurs can somewhat alleviate this problem. The algorithm is based on calculating the instantaneous frequency of a time series by means of a Hilbert transform and revealing the most probable value of this frequency.

We emphasize that in experimental studies of chaotic self-oscillations measuring the synchronization threshold is far simpler than for other quantitative characteristics. Such measurements were performed by Bezaeva et al. [37, 38] and Bumyalene et al. [63]. In the latter it was found that the value of the synchronization threshold is correlated to the attractor fractal dimension.

As an example we consider the synchronization of chaotic oscillations in a system with inertial excitation described by the equations

$$\ddot{x} + 2\delta\dot{x} + x = -ky - bx^3 + B\cos 2\pi\omega t, \quad \dot{y} + y = x - x^2 - x^3. \tag{9.115}$$

Equations (9.115) were simulated for $k = 20$, $b = 17.5$ and $2\delta = 0.75$ [171]. Different synchronization regions were found. In these regions the chaotic attractor is replaced by two-, three-, four- or six-revolution cycles depending on B and ω. It is interesting to note that two mechanisms of synchronization, which were discussed in the first section of this chapter, also manifest themselves in the synchronization of chaotic oscillations. Really, numerical calculations have shown that transitions from synchronization regions to regions of chaos, as the external force frequency ω varies, occur differently for various amplitudes of the external force. For moderately small B a limit cycle inside the synchronization regions is located on the corresponding two-dimensional torus in the system phase space. The transition from a synchronization region, as ω varies, occurs through disrupting the resonance on the torus and the appearance of beats, i.e., of quasi-periodic oscillations. As ω varies further, a finite number of doubling bifurcations of the torus occurs, whereupon chaos arises.

For large B, there are no tori inside the synchronization regions. The transition from the synchronization region to the region of chaos, as ω varies, occurs via a sequence of cycle period-doubling bifurcations.

9.7 Synchronization of vortex formation in the case of transverse flow around a vibrated cylinder

From time immemorial navigators have known that, when a long thin body is streamlined by air, excitation of sound occurs: they observed that, when a strong wind blows, sail masts, ropes and other extended objects begin to emit sounds. These sounds came to be known as *Aeolian tones*. In more recent times it was found that the excitation of such sounds is associated with the formation of a periodic vortex structure beyond the streamlined body. This structure was studied in detail by von Karman [146] and is referred to by his name. Vortices create nonhomogeneities of the fluid velocity field that result in the emission of sound [211, 212]. Von Karman's vortex wake is two rows of vortices rotating in opposite directions and spaced relative to each other in staggered order, see Fig. 9.15. Von Karman showed that the ratio of the distance λ between two neighboring vortices which are rotating in the same direction to the distance h between the vortex rows is constant, namely, $h/\lambda = 0.283$. It should be noted that this relation is a result of the constancy

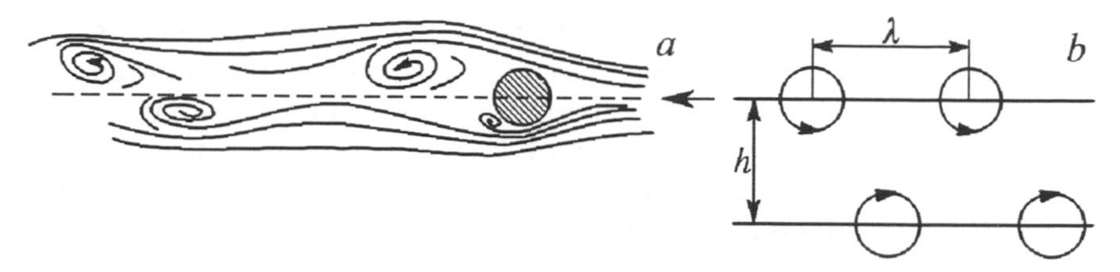

Fig. 9.15. (a) Von Karman's vortex wake, and (b) its schematic image

of Strouhal's number St which was introduced by von Strouhal [323] in experimental studies of sound generated by a rotating wire. Strouhal detected that a certain combination of the sound frequency f, the wire diameter D, and the wire motion velocity V, namely, $\mathrm{St} = fD/V$, remains a constant which is approximately equal to 0.22 as V varies over a wide range. A qualitative explanation of Strouhal's experiments was given by Rayleigh [277].

Before considering the synchronization phenomenon we will dwell on the transverse flow around an immovable cylinder. A large body of experimental research strongly testifies that, as in Strouhal's experiments, over a wide range of flow velocities the frequency of vortex formation is associated with the Strouhal number $\mathrm{St} \approx 0.22$. The formation of vortices results in pulsations of the flow velocity and pressure. An example of the pressure pulsations on the surface of cylinder in a section orthogonal to the flow is given in Fig. 9.16 [84]. If the cylinder is placed within some channel then the vortex formation

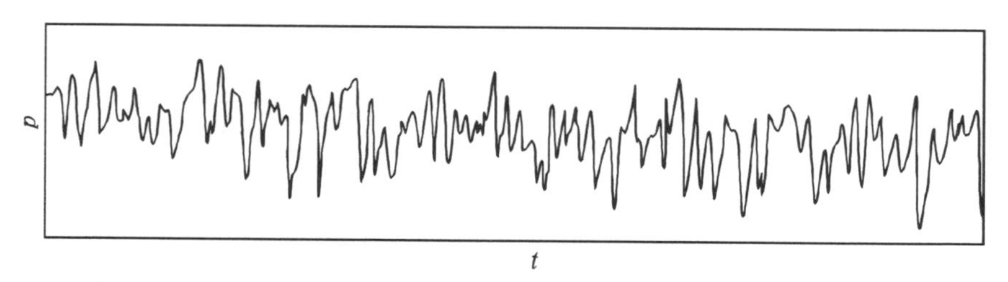

Fig. 9.16. Oscillogram of the pressure pulsations on the surface of immovable cylinder in a section orthogonal to the flow

and sound generation caused by this can be considerably intensified. Intensification occurs if the frequency of vortex formation is close to one of the natural frequencies of acoustic oscillations in the channel. This phenomenon is sometimes spoken of as *aero-acoustic resonance* [129, 130, 25].

In the case of flow around a cylinder vibrated transversely to the flow with given frequency and amplitude values, the frequency of vortex separation f_s essentially depends on the frequency f_0 of vibration of the cylinder. If f_0 essentially differs from the frequency of vortex separation f_{s0} in the case of immovable cylinder then f_s is close to f_{s0}. As f_0 approaches to f_{s0}

the frequency f_s changes approaching to f_0, i.e., synchronization occurs. The width of the synchronization region essentially depends on the vibration amplitude of the cylinder. The first experiments concerning synchronization of the vortex separation were performed by Pavlikhina and Smirnov [264] and Bishop and Hassan [43]. In addition to the synchronization at the main frequency, Bishop and Hassan found the synchronization at the second and third subharmonics of the vibration frequency of the cylinder.

Somewhat later synchronization of the vortex separation at the main frequency was studied in detail by Blumina and Fedyaevsky [51, 84]. The dependence of $\Delta\mathbf{St} = (f_s - f_0)D/U_0$ on $\Delta\mathbf{St}_0 = (f_{s0} - f_0)D/U_0$, where D is the diameter of the cylinder and U_0 is the flow velocity, is shown in Fig. 9.17. It is constructed on the basis of the experimental data presented in [84]. We see

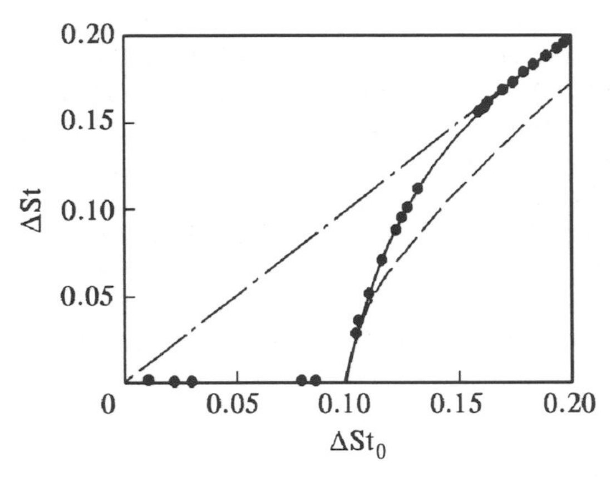

Fig. 9.17. The dependence of $\Delta\mathbf{St} = (f_s - f_0)D/U_0$ on $\Delta\mathbf{St}_0 = (f_{s0} - f_0)D/U_0$. The dashed line shows the dependence $\Delta\mathbf{St} = \sqrt{(\Delta\mathbf{St}_0)^2 - (\Delta\mathbf{St}_0)_s^2}$

that this dependence resembles the dependence of the beat frequency on the frequency mistuning in the case of synchronization of a van der Pol generator by an external periodic force (see above). For comparison, the dependence

$$\Delta\mathbf{St} = \sqrt{(\Delta\mathbf{St}_0)^2 - (\Delta\mathbf{St}_0)_s^2}\,, \tag{9.116}$$

where $(\Delta\mathbf{St}_0)_s$ is the half-width of the synchronization region, which should hold in the case of synchronization by a small periodic external force, is shown as a dashed line. We see that the experimental dependence is steeper than (9.116). This is caused by the fact that the amplitude of vibration of the cylinder was sufficiently large.

9.8 Synchronization of relaxation self-oscillations

Classical examples of relaxation generators are a thyratron generator [334] and an alternating water source known as Tantal's vessel [322, 260]. Schematic images of these devices are illustrated in Fig. 9.18 a and b. The shape of

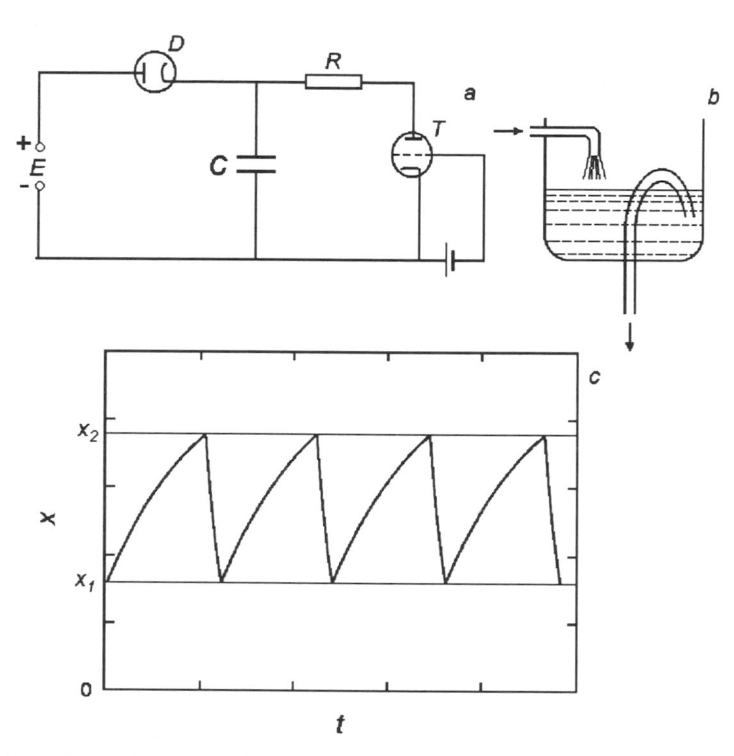

Fig. 9.18. Schematic images (a) of a thyratron generator and (b) of Tantal's vessel, and (c) the shape of oscillations of the amount of water in Tantal's vessel and of the voltage across the capacitor C

oscillations of the amount of water in Tantal's vessel and of the voltage across the capacitor C is shown in Fig. 9.18 c.

Synchronization of a relaxation generator by a periodic external force was first considered by Teodorchik [329, 330]. For definiteness, we consider synchronization of a thyratron generator by an external force of period T. It is convenient to introduce dimensionless time $\tau = t/T$. The growing and descending portions of the function $x(t)$ will be called the direct and reverse moves, respectively. These portions may be described by functions $x_{\mathrm{d}}(\tau_1, x_1)$ and $x_{\mathrm{r}}(\tau_2, x_2)$, where τ_1 and τ_2 are the times reckoned from the beginning of the direct and reverse moves, respectively, x_1 and x_2 are the minimal and maximal values of $x(\tau)$, which will be called the lower and higher limit levels, respectively. For the thyratron generator the firing potential of the thyratron is the higher limit level, and the quenching potential is the lower limit level.

If an external periodic voltage is applied to the thyratron grid, the limit levels will turn out to be modulated periodically, i.e., $x_1 = x_1(\tau) = x_{10}+e_1(\tau)$ and $x_2 = x_2(\tau) = x_{20} + e_2(\tau)$, where x_{10} and x_{20} are the lower and higher limit levels in the absence of the external voltage. Owing to this modulation the durations of the direct and reverse moves will change. For the calculation of the synchronization regime we may use the point map technique.

Let us denote the durations of the direct and reverse moves on the nth step by τ_{1n} and τ_{2n}, and the fractions of the external voltage period corresponding to the beginning of the nth direct and reverse moves by φ_n and φ'_n. Then the equations relating τ_{1n} and τ_{2n} to φ_n and φ'_n are

$$\tau_{1n} = \varphi'_n - \varphi_n + p_n, \quad \tau_{2n} = \varphi_{n+1} - \varphi'_n + r_n, \tag{9.117}$$

where p_n and r_n are integers. It can be easily shown that $x_{\mathrm{d}}\big(\tau_{1n}, x_1(\varphi_n)\big) = x_2(\varphi'_n)$ and $x_{\mathrm{r}}\big(\tau_{2n}, x_2(\varphi'_n)\big) = x_1(\varphi_{n+1})$. From this we obtain the following equations relating φ_{n+1} to φ_n:

$$\begin{aligned} f_1(\varphi_n, \varphi'_n) &\equiv x_{\mathrm{d}}\big(\varphi'_n - \varphi_n + p_n, x_{10} + e_1(\varphi_n)\big) - x_{20} - e_2(\varphi'_n) = 0, \\ f_2(\varphi_{n+1}, \varphi'_n) &\equiv x_{\mathrm{r}}\big(\varphi_{n+1} - \varphi'_n + r_n, x_{20} + e_2(\varphi'_n)\big) - x_{10} - e_1(\varphi_{n+1}) = 0. \end{aligned} \tag{9.118}$$

Equations (9.118) describe the required point map.

Searching synchronous regimes reduces to searching fixed points of the map (9.118) and investigation of their stability. The coordinates of a fixed point φ_{s}, φ'_{s} are found from the conditions

$$\varphi_n = \varphi_{n+1} = \varphi_{\mathrm{s}}, \quad \varphi'_n = \varphi'_{\mathrm{s}}$$

for any n. Linearizing Eqs. (9.118) with respect to small deviations $\xi_n = \varphi_n - \varphi_{\mathrm{s}}$, $\eta_n = \varphi'_n - \varphi'_{\mathrm{s}}$ from the fixed point φ_{s}, φ'_{s}, eliminating η_n and using the Koenigs theorem (see Ch. 3) we find the following stability condition for the fixed point φ_{s}, φ'_{s}:

$$\left| \frac{\dfrac{\partial f_1}{\partial \varphi_n} \dfrac{\partial f_2}{\partial \varphi'_n}}{\dfrac{\partial f_1}{\partial \varphi'_n} \dfrac{\partial f_2}{\partial \varphi_{n+1}}} \right|_{\varphi_n = \varphi_{n+1} = \varphi_{\mathrm{s}},\, \varphi'_n = \varphi'_{\mathrm{s}}} \le 1. \tag{9.119}$$

Following [308] we calculate the synchronization regime in the case when the oscillations of an autonomous thyratron generator are of triangular shape with period $T = T_1 + T_2$, where T_1 and T_2 are the durations of the direct and reverse moves. We assume that only the higher limit level is modulated, and the modulation is of triangular shape too.[3] The period of the external action can be set equal to unity. In this case Eqs. (9.118) become

[3] Synchronization of a relaxation generator by a harmonic external force is considered in [371].

$$f_1(\varphi_n, \varphi_n') \equiv x_{10} + \frac{x_{20} - x_{10}}{T_1} (\varphi_n' - \varphi_n + p_n) - x_{20} - e_2(\varphi_n') = 0,$$
$$f_2(\varphi_{n+1}, \varphi_n') \equiv x_{20} + e_2(\varphi_n') - \frac{x_{20} - x_{10}}{T_2} (\varphi_{n+1} - \varphi_n' + r_n) - x_{10} = 0, \tag{9.120}$$

where

$$e_2(\varphi_n') = \begin{cases} -\dfrac{B}{2} + \dfrac{B}{\theta_1} \varphi_n' & \text{for } 0 \le \varphi_n' \le \theta_1, \\ \dfrac{B}{2} - \dfrac{B}{\theta_2} (\varphi_n' - \theta_1) & \text{for } \theta_1 \le \varphi_n' \le 1, \end{cases} \tag{9.121}$$

B is the amplitude of the external voltage, and θ_1 and θ_2 are the durations of the direct and reverse moves of this voltage ($\theta_1 + \theta_2 = 1$).

Taking account of (9.121) we find the following equations for fixed points:

- for $0 \le \varphi_s' \le \theta_1$

$$x_{10} + \frac{x_{20} - x_{10}}{T_1} (\varphi_s' - \varphi_s + p_s) - x_{20} + \frac{B}{2} - \frac{B}{\theta_1} \varphi_s' = 0,$$
$$x_{20} - \frac{B}{2} + \frac{B}{\theta_1} \varphi_s' - \frac{x_{20} - x_{10}}{T_2} (\varphi_s - \varphi_s' + r_s) - x_{10} = 0, \tag{9.122}$$

- for $\theta_1 \le \varphi_s' \le 1$

$$x_{10} + \frac{x_{20} - x_{10}}{T_1} (\varphi_s' - \varphi_s + p_s) - x_{20} - \frac{B}{2} + \frac{B}{\theta_2} (\varphi_s' - \theta_1) = 0,$$
$$x_{20} + \frac{B}{2} - \frac{B}{\theta_2} (\varphi_s' - \theta_1) - \frac{x_{20} - x_{10}}{T_2} (\varphi_s - \varphi_s' + r_s) - x_{10} = 0, \tag{9.123}$$

where p_s and r_s are the values of p_n and r_n in the synchronization regime. It is easily shown that $p_s + r_s$ is equal to the number of external voltage periods in one period of the synchronous oscillations of the generator. It follows from Eqs. (9.122) and (9.123) that

$$\varphi_s' = \begin{cases} \theta_1 \left[\dfrac{1}{2} - \dfrac{x_{20} - x_{10}}{B} \left(1 - \dfrac{p_s + r_s}{T} \right) \right] & \text{for } 0 \le \varphi_s' \le \theta_1, \\ \theta_1 + \theta_2 \left[\dfrac{1}{2} + \dfrac{x_{20} - x_{10}}{B} \left(1 - \dfrac{p_s + r_s}{T} \right) \right] & \text{for } \theta_1 \le \varphi_s' \le 1, \end{cases} \tag{9.124a}$$

$$\varphi_s = \begin{cases} \varphi_s' + \dfrac{p_s T_2 - r_s T_1}{T} & \text{for } 0 \le \varphi_s' \le \theta_1, \\ \varphi_s' + \dfrac{p_s T_2 - r_s T_1}{T} & \text{for } \theta_1 \le \varphi_s' \le 1. \end{cases} \tag{9.124b}$$

The solutions (9.124a) and (9.124b) exist if

$$B \geq 2(x_{20} - x_{10}) \left| \frac{p_s + r_s}{T} - 1 \right|. \tag{9.125}$$

The stability condition (9.119) in the case under consideration becomes

$$\left. \begin{array}{ll} \left| \dfrac{(x_{20} - x_{10})\theta_1 + BT_2}{(x_{20} - x_{10})\theta_1 - BT_1} \right| & \text{for} \quad 0 \leq \varphi_s' \leq \theta_1 \\ \left| \dfrac{(x_{20} - x_{10})\theta_2 - BT_2}{(x_{20} - x_{10})\theta_2 + BT_1} \right| & \text{for} \quad \theta_1 \leq \varphi_s' \leq 1 \end{array} \right\} \leq 1. \tag{9.126}$$

For $0 \leq \varphi_s' \leq \theta_1$ condition (9.126) is equivalent to

$$B \geq \frac{2(x_{20} - x_{10})\theta_1}{T_1 - T_2}, \tag{9.127}$$

and for $\theta_1 \leq \varphi_s' \leq 1$, $T_1 \geq T_2$ it is fulfilled for any B.

In addition to the stability condition, the existence of the synchronization regime is limited by the so-called self-shielding effect [308]. This effect is observed for sufficiently large amplitudes of the external voltage, when the slope of the plot of $x_d(t)$ is less than the slope of the plot $e_2(t)$, i.e., for

$$B \geq B_s = \frac{(x_{20} - x_{10})\theta_1}{T_1}. \tag{9.128}$$

Thus, a synchronization regime can exist only for

$$B < B_s. \tag{9.129}$$

It follows from (9.127)–(9.129) that the synchronization regime with $0 \leq \varphi_s' \leq \theta_1$ is unstable when it exists.

The existence of the synchronization regime with $\theta_1 \leq \varphi_s' \leq 1$ is determined by the condition of the absence of self-shielding, namely

$$\theta_1 + \theta_2 \left[\frac{1}{2} + \frac{x_{20} - x_{10}}{B} \left(1 - \frac{p_s + r_s}{T} \right) \right] \geq \frac{BT_1}{BT_1 + (x_{20} - x_{10})\theta_2}. \tag{9.130}$$

Condition (9.130) gives the upper boundaries of the synchronization regions:

$$\frac{B}{x_{20} - x_{10}} \leq \frac{\theta_1}{T_1} + \frac{\theta_2}{2T_1} + 1 - \frac{p_s + r_s}{T} + \sqrt{\left(\frac{\theta_1}{T_1} + \frac{\theta_2}{2T_1} + 1 - \frac{p_s + r_s}{T} \right)^2 + \frac{2\theta_2}{T_1} \left(1 - \frac{p_s + r_s}{T} \right)}. \tag{9.131}$$

The synchronization regions for $p_s + r_s = 1$ (the main resonance), $p_s + r_s = 2$ (the subharmonic resonance of the second kind), and $p_s + r_s = 3$ (the subharmonic resonance of the third kind) are shown in Fig. 9.19 for $T_1 = 2T/3$, $T_2 = T/3$, $\theta_1 = \theta_2 = 1/2$. These regions are constructed from (9.125) and (9.131).

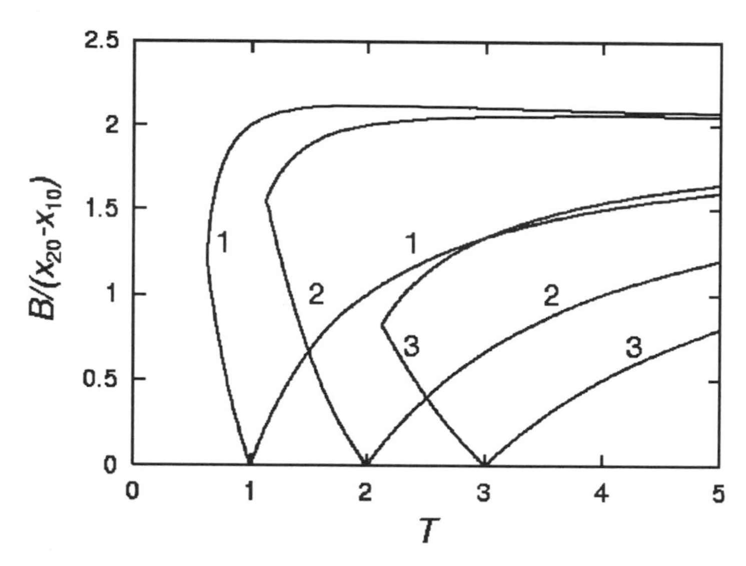

Fig. 9.19. The synchronization regions for $p_s + r_s = 1$ (curves 1), $p_s + r_s = 2$ (curves 2), and $p_s + r_s = 3$ (curves 3) for $T_1 = 2T/3$, $T_2 = T/3$, $\theta_1 = \theta_2 = 1/2$

10. Interaction of two self-oscillatory systems. Synchronization and chaotization of self-oscillations

In this and the next chapters we consider systems involving two or more self-oscillatory subsystems which are described by their 'own' phase coordinates and weakly interact with each other. Such subsystems we call *weakly coupled*. If we separate a system into subsystems which strongly interact with each other, then such a separation loses its significance and the initial system should be considered as a unified one.

Synchronization of periodic self-oscillations in coupled systems has been known for a long time. Huygens was probably the first scientist who observed this phenomenon as early as the seventeenth century [126]. He discovered that a pair of pendulum clocks suspended from a light beam which rests on two supports are synchronized, i.e., their pendulums vibrate with equal frequencies and with opposite phases. Huygens realistically understood that the synchronization is caused by the imperceptible motion of the beam.

In the middle of the nineteenth century, in his famous treatise '*The Theory of Sound*', Rayleigh [279] described the interesting phenomenon of mutual synchronization of organ pipes. He discovered that two organ pipes with closely located mouths sound in unison.

In the early twentieth century, after studies by van der Pol [350] and Appleton [18], a boom began in this field. Synchronization phenomena were discovered in many electrical, electro-mechanical, and radio engineering systems. A rich variety of papers appeared on this subject. In the 1960–1970s the appearance of laser gyroscopes caused a rebirth of interest in synchronization phenomena (see, e.g., the monograph [372]). The operation of these devices is based on measuring the difference between frequencies of waves traveling in opposite directions in a ring laser. The synchronization of these waves plays a harmful role because it significantly restricts the sensitivity of the gyroscopes.

10.1 Mutual synchronization of periodic self-oscillations with close frequencies

As an example we consider the mutual synchronization of two coupled generators described by the equations

$$\ddot{x}_1 - \mu_1(1 - 4\alpha_1 x_1^2 + \beta_1 x_1)\dot{x}_1 + \omega_1^2 x_1 = c_{11}\ddot{x}_2 + c_{12}\dot{x}_2 + c_{13}x_2,$$
$$\ddot{x}_2 - \mu_2(1 - 4\alpha_2 x_2^2 + \beta_2 x_2)\dot{x}_2 + \omega_2^2 x_2 = c_{21}\ddot{x}_1 + c_{22}\dot{x}_1 + c_{23}x_1. \tag{10.1}$$

If $\mu_{1,2}$ and c_{ij} are sufficiently small and the frequencies ω_1 and ω_2 are close ($|\omega_1 - \omega_2| \ll \omega_{1,2}$), then a solution of Eqs. (10.1) can be sought as $x_{1,2} = A_{1,2}\cos(\omega t + \varphi_{1,2})$, where $\omega = (\omega_1 + \omega_2)/2$, and $A_{1,2}$ and $\varphi_{1,2}$ are slowly varying functions. The truncated equations for the amplitudes $A_{1,2}$ and the phases $\varphi_{1,2}$ are

$$\dot{A}_{1,2} = \frac{\mu_{1,2}}{2}(1 - \alpha_{1,2}A_{1,2}^2)A_{1,2} \mp c_{1,2}A_{2,1}\sin(\Phi \pm \vartheta_{1,2}), \tag{10.2}$$

$$\dot{\varphi}_{1,2} = \pm\frac{\Delta}{2} - \frac{c_{1,2}A_{2,1}}{A_{1,2}}\cos(\Phi \pm \vartheta_{1,2}), \tag{10.3}$$

where $\Phi = \varphi_1 - \varphi_2$ is the phase difference, $\Delta = \omega_1 - \omega_2$ is the frequency mistuning,

$$c_1 = \frac{\sqrt{(c_{11}\omega^2 - c_{13})^2 + c_{12}^2\omega^2}}{2\omega}, \quad c_2 = \frac{\sqrt{(c_{21}\omega^2 - c_{23})^2 + c_{22}^2\omega^2}}{2\omega}$$

are the moduli of the linear coupling factors, and

$$\vartheta_1 = \arctan\frac{c_{12}\omega}{c_{11}\omega^2 - c_{13}}, \quad \vartheta_2 = \arctan\frac{c_{22}\omega}{c_{21}\omega^2 - c_{23}}$$

are the phases of the linear coupling factors.

The following equation for the phase difference Φ can be obtained from Eqs. (10.3):

$$\dot{\Phi} = \Delta - \frac{c_1 A_2}{A_1}\cos(\Phi + \vartheta_1) + \frac{c_2 A_1}{A_2}\cos(\Phi - \vartheta_2). \tag{10.4}$$

This equation combined with Eqs. (10.2) gives the closed system of equations for $A_{1,2}$ and Φ.

Equations (10.1) are written for the case of linear coupling between the generators. In the general case nonlinear coupling is also possible. For example, coupling between different modes in lasers which is caused by their interaction in the active medium is nonlinear [151, 152, 372]. For this reason we consider thereafter the equations for the amplitudes $A_{1,2}$ of more general form than Eqs. (10.2), namely,

$$\dot{A}_{1,2} = \frac{\mu_{1,2}}{2}\left(1 - \alpha_{1,2}A_{1,2}^2 - b_{1,2}A_{2,1}^2\right)A_{1,2} \mp c_{1,2}A_{2,1}\sin(\Phi \pm \vartheta_{1,2}), \tag{10.5}$$

where the coefficients $b_{1,2}$ describe nonlinear coupling between the generators.

10.1.1 The case of weak linear coupling

If the linear coupling factors are sufficiently small, so that

$$c_{1,2} \ll \mu_{1,2} \frac{|\alpha_{1,2} - b_{1,2}|}{\alpha_{1,2}}, \tag{10.6}$$

then the values of the amplitudes of synchronous oscillations are slightly affected by the coupling factors. Therefore let us first find steady-state solutions of Eqs. (10.5) in the absence of the linear coupling. In this case Eqs. (10.5) possess four steady-state solutions:

$$A_{10} = A_{20} = 0, \tag{10.7a}$$

$$A_{10} = 0, \quad A_{20} = \frac{1}{\sqrt{\alpha_2}}, \tag{10.7b}$$

$$A_{10} = \frac{1}{\sqrt{\alpha_1}}, \quad A_{20} = 0, \tag{10.7c}$$

$$A_{10} = \sqrt{\frac{\alpha_2 - b_1}{\alpha_1\alpha_2 - b_1 b_2}}, \quad A_{20} = \sqrt{\frac{\alpha_1 - b_2}{\alpha_1\alpha_2 - b_1 b_2}}. \tag{10.7d}$$

If the self-excitation conditions for both of the generators are fulfilled then the solution (10.7a) is unstable. The conditions of stability for the solutions (10.7b) and (10.7c) are, respectively,

$$\alpha_2 < b_1, \quad \alpha_1 < b_2. \tag{10.8}$$

In the range of parameters where the conditions (10.8) are both fulfilled, depending on the initial conditions, one of the generators is excited, while oscillations of the other generator are entirely quenched. Such quenching of the oscillations of one of the generators, caused by the sufficiently large nonlinear coupling between them, is known as *competitive quenching*. If neither of the conditions (10.8) are fulfilled then the solution (10.7d) is stable. Let us consider just this case.

Substituting (10.7d) into Eq. (10.4) and setting $\dot{\Phi} = 0$ we obtain an equation which can be written as

$$\Delta - \Delta_s \cos(\Phi + \theta) = 0, \tag{10.9}$$

where

$$\Delta_s = \sqrt{c_1^2 \frac{\alpha_1 - b_2}{\alpha_2 - b_1} + c_2^2 \frac{\alpha_2 - b_1}{\alpha_1 - b_2} - 2c_1 c_2 \cos(\vartheta_1 + \vartheta_2)} \tag{10.10}$$

is the half-width of the synchronization region in the first approximation with respect to the linear coupling factors, and

$$\tan\theta = \frac{c_1(\alpha_1 - b_2)\sin\vartheta_1 + c_2(\alpha_2 - b_1)\sin\vartheta_2}{c_1(\alpha_1 - b_2)\cos\vartheta_1 - c_2(\alpha_2 - b_1)\cos\vartheta_2}.$$

Equation (10.9) for a fixed value of Δ has two solutions:

$$\Phi + \theta = \pm\arccos\frac{\Delta}{\Delta_s}. \tag{10.11}$$

To find the stability condition for these solutions we should use Eqs. (10.5) and (10.4). As a result, in the first approximation with respect to the linear coupling factors, we obtain the following stability condition:

$$c_1 \sqrt{\frac{\alpha_1 - b_2}{\alpha_2 - b_1}} \sin(\Phi + \vartheta_1) - c_2 \sqrt{\frac{\alpha_2 - b_1}{\alpha_1 - b_2}} \sin(\Phi - \vartheta_2) \leq 0. \tag{10.12}$$

From this it follows that only the solution (10.11) with the '−' sign is stable. Let us clarify the meaning of this result by the example of when $\vartheta_1 = \vartheta_2 = \vartheta = 0$. If the generators are not identical then we can always separate them into the driving generator and the driven one. So, if $c_1 A_2/A_1 > c_2 A_1/A_2$ then the second generator is the driving one. The stability condition found implies that the phase of the oscillations of the driven generator should be behind the phase of the oscillations of the driving generator. In addition, it can be shown that the correction to the amplitude caused by the linear coupling is positive for the driven generator and negative for the driving generator.

In the second approximation with respect to the linear coupling factors the stability condition is rather complicated. We write it only for the case of identical generators, when

$$\mu_1 = \mu_2 = \mu, \quad \alpha_1 = \alpha_2 = \alpha, \quad b_1 = b_2 = b. \tag{10.13}$$

In this case the stability condition is

$$\frac{\mu(\alpha - b)}{\alpha + b}\Big(c_1 \sin(\Phi + \vartheta_1) - c_2 \sin(\Phi - \vartheta_2)\Big) + \frac{\alpha - b}{\alpha + b}\Big(c_1 \sin(\Phi + \vartheta_1) + c_2 \sin(\Phi - \vartheta_2)\Big)^2 - c_1^2 + c_2^2 + 2c_1c_2 \cos(\vartheta_1 + \vartheta_2) \leq 0. \tag{10.14}$$

It follows from (10.10) that $\Delta_s = 0$ for $\alpha_1 = \alpha_2 = \alpha$, $b_1 = b_2 = b$, $c_1 = c_2 = c$, $\vartheta_1 + \vartheta_2 = 0$. In this case we must use a second approximation for calculating the width of the synchronization region. For this we should find the small corrections to (10.7d) caused by the linear coupling factors satisfying to condition (10.6). Setting in Eqs. (10.5) $A_1 = A_{10} + a_1$, $A_2 = A_{20} + a_2$, where $a_1 \ll A_{10}$, $a_2 \ll A_{20}$, we find

$$a_1 = -\frac{\mu_2 c_1 \alpha_2 A_{20}^2 \sin(\Phi + \vartheta_1) + \mu_1 c_2 b_1 A_{10}^2 \sin(\Phi - \vartheta_2)}{\mu_1 \mu_2 (\alpha_2 - b_1) A_{20}},$$

$$a_2 = \frac{\mu_1 c_2 \alpha_1 A_{10}^2 \sin(\Phi - \vartheta_2) + \mu_2 c_1 b_2 A_{20}^2 \sin(\Phi + \vartheta_1)}{\mu_1 \mu_2 (\alpha_1 - b_2) A_{10}}. \tag{10.15}$$

In the specific case when $\alpha_1 = \alpha_2 = \alpha$, $b_1 = b_2 = b$, $c_1 = c_2 = c$, $\vartheta_1 = -\vartheta_2 = \vartheta$ we find from (10.15)

$$a_1 = -a_2 = -\frac{(\mu_1 + \mu_2) c \alpha A_0 \sin(\Phi + \vartheta)}{\mu_1 \mu_2 (\alpha - b)}, \tag{10.16}$$

where $A_0 = 1/\sqrt{\alpha + b}$. Substituting (10.16) into Eq. (10.4) and setting $\dot{\Phi} = 0$ we obtain

$$\Delta - \frac{(\mu_1+\mu_2)(\alpha+b)c^2}{\mu_1\mu_2(\alpha-b)} \sin 2(\Phi+\vartheta) = 0. \tag{10.17}$$

We find from here

$$\Delta_{\mathrm{s}} = \frac{(\mu_1+\mu_2)(\alpha+b)c^2}{\mu_1\mu_2(\alpha-b)}. \tag{10.18}$$

The stability condition for this case is fulfilled for both solutions (10.11). For example, as follows from (10.14), for $\mu_1 = \mu_2$ this condition is

$$1 + \frac{\alpha-b}{\alpha+b}\sin^2(\Phi+\vartheta) \geq 0. \tag{10.19}$$

It should be noted that inside the synchronization region the oscillation amplitudes and frequencies change, even though the difference between the frequencies remains constant. To trace these changes, we will assume that conditions (10.13) are fulfilled and consider three specific cases:

1. $\vartheta_1 + \vartheta_2 = \pm\pi$. As follows from (10.15), (10.9) and (10.10), in this case

$$a_{1,2} = -\frac{\alpha c_{1,2} - bc_{2,1}}{\mu(\alpha-b)} A_0 \sin(\Phi+\vartheta_1), \quad \Delta_{\mathrm{s}} = c_1 + c_2.$$

The stability condition (10.14) becomes

$$\mu\frac{\alpha-b}{\alpha+b}(c_1+c_2)\sin(\Phi+\vartheta_1) - (c-1-c_2)^2 \leq 0. \tag{10.20}$$

It follows from here that for $c_1 \neq c_2$ in the neighborhood of synchronization region boundaries, where $|\sin(\Phi+\vartheta_1)| = \sqrt{1-\Delta^2/\Delta_{\mathrm{s}}^2}$ is small, both solutions (10.11) are stable. But if $c_1 = c_2$, everywhere over the synchronization region only the solution (10.11) with the '$-$' sign is stable.

For $c_1 = c_2 = c$ we have $a_1 = a_2 = (c/\mu)A_0\sqrt{1-\Delta^2/\Delta_{\mathrm{s}}^2}$, i.e., inside the synchronization region the amplitudes of both of the generators are identical and they peak for $\Delta = 0$. But if, for example, $c_1 > c_2$, i.e., the first generator is driven, whereas the second generator is driving, then the amplitude of the first generator is more than that of the second one. What is more, if $bc_1 > \alpha c_2$, the amplitude of the second generator is smaller than the amplitude of free self-oscillations.

The frequency of synchronous oscillations can be approximately calculated from Eqs. (10.3) setting $A_1 \approx A_2 \approx A_0$. As a result we find

$$\omega_{\mathrm{s}} = \omega_1 - c_1\cos(\Phi+\vartheta) = \frac{c_1\omega_2 + c_2\omega_1}{c_1+c_2}. \tag{10.21}$$

It is seen that the frequency ω_{s} lies between the frequencies of the autonomous generators, closer to the frequency of the driving generator.

2. $\vartheta_1 + \vartheta_2 = 0$. In this case

$$a_{1,2} = \mp\frac{\alpha c_{1,2} + bc_{2,1}}{\mu(\alpha-b)} A_0 \sin(\Phi+\vartheta_1), \quad \Delta_{\mathrm{s}} = |c_1 - c_2|. \tag{10.22}$$

We note that the expression for Δ_s given above is valid for $|c_1 - c_2| \gg (\alpha + b)c_{1,2}^2/(\mu(\alpha - b))$.

The stability condition (10.14) becomes

$$\mu \frac{\alpha - b}{\alpha + b} (c_1 - c_2) \sin(\Phi + \vartheta_1) - (c - 1 - c_2)^2 \le 0.$$

It follows from here that in the neighborhood of synchronization region boundaries both solutions (10.11) are stable, whereas in the vicinity of the center of the synchronization region only the solution for which

$$\sin(\Phi + \vartheta_1) = \sqrt{1 - \frac{\Delta^2}{\Delta_s^2}} \operatorname{sign}(c_2 - c_1)$$

is stable.

It is seen from (10.22) that, as $|\Delta|$ decreases, the amplitude of the driven generator increases and the amplitude of the driving generator decreases. The frequency of synchronous oscillations is

$$\omega_s = \frac{c_1\omega_2 - c_2\omega_1}{c_1 - c_2}.$$

In contrast to the first case, here ω_s does not lie between the frequencies of the autonomous generators. However, as in the first case, it is closer to the frequency of the driving generator. For example, if $\omega_2 > \omega_1$, then $\omega_s > \omega_2 > \omega_1$ for $c_1 > c_2$ and $\omega_s < \omega_1 < \omega_2$ for $c_1 < c_2$.

In the case of $c_1 = c_2 = c$, when the synchronization region width is described by (10.18), the behavior of the generators is more complicated [334]. Inside the synchronization region jumps in the amplitude and frequency are possible. Such jumps were observed experimentally by Esafov [334].

3. $\vartheta_1 = \vartheta_2 = \vartheta$. To trace the most interesting features of the system behavior, we set $c_1 = c_2 = c$ and $b = 0$. In this case the stability condition (10.14) becomes

$$\mu \sin\vartheta \cos\Phi \le 2c \cos^2\vartheta \cos^2\Phi. \tag{10.23}$$

If $|\sin\vartheta| \gg c/\mu$, then only one of solutions (10.11) is stable. But if $|\sin\vartheta|$ is less than or of the order of c/μ, then from (10.4) and (10.15) we obtain the following equation for Φ in the synchronization regime:

$$\Delta + 2c \sin\vartheta \sin\Phi - \frac{2c^2}{\mu} \sin 2\Phi = 0. \tag{10.24}$$

It follows from here that in the vicinity of the center of the synchronization region three regimes are possible: (i) almost in-phase regime when Φ is small, (ii) almost anti-phase regime when Φ is close to π, and (iii) the regime in which $\cos\Phi \approx (\mu/(2c)) \sin\vartheta$. The stability of these regimes is determined by the condition (10.23). We see that regime (iii) is always unstable, whereas, for sufficiently small $\sin\vartheta$, both in-phase and anti-phase regimes are stable. As the mistuning increases $|\cos\Phi|$ decreases and the stability condition (10.23) for one of these regimes ceases to be fulfilled. Since different signs of $\cos\Phi$ correspond to these regimes, the transitions from one regime to another are accompanied by jumps in the oscillation amplitudes and frequencies.

Let us further consider the beat regimes. In the case of weak linear coupling, when $A_1 \approx A_{10}$ and $A_2 \approx A_{20}$, Eq. (10.4) allows us to calculate the beat frequency ω_b nearby the boundaries of the synchronization region. Like the van der Pol generator driven by a harmonic external force, the beat frequency is

$$\omega_b = \sqrt{\Delta^2 - \Delta_s^2}. \tag{10.25}$$

The expression (10.25) is valid for $\Delta \ll \mu_{1,2}$. For $\Delta \sim \mu_{1,2}$ the amplitudes $A_{1,2}$ cannot be eliminated from Eq. (10.4) and this equation should be solved in combination with (10.5). However, the calculations in this region of the mistunings is alleviated by the facts that for $|\Delta| \gg \Delta_s$ the beat frequency $\dot\Phi$ differs only slightly from Δ and the amplitudes A_1 and A_2 execute small oscillations about A_{10} and A_{20}. Therefore we can set

$$\dot\Phi = \Delta - \delta, \quad A_1 = A_{10} + a1, \quad A_2 = A_{20} + a_2, \tag{10.26}$$

where $\delta = \Delta_s \cos(\Delta t + \theta) \ll |\Delta|$, $a_1 \ll A_{10}$, and $a_2 \ll A_{20}$. For $a_{1,2}$ we obtain from (10.5) the following equations:

$$\begin{aligned} \dot a_1 &= -\mu_1 A_{10}\Big(\alpha_1 A_{10} a1 + b_1 A_{20} a_2\Big) - c_1 A_{20}\sin(\Delta t + \vartheta_1), \\ \dot a_2 &= -\mu_2 A_{20}\Big(b_2 A_{10} a1 + \alpha_2 A_{20} a_2\Big) + c_2 A_{10}\sin(\Delta t - \vartheta_2). \end{aligned} \tag{10.27}$$

A stationary solution of these equations can be written as

$$\begin{aligned} a_1 &= C_1\cos(\Delta t + \theta) + C_2\sin(\Delta t + \theta), \\ a_2 &= D_1\cos(\delta t + \theta) + D_2\sin(\Delta t + \theta), \end{aligned} \tag{10.28}$$

where C_1, C_2, D_1 and D_2 are found from the system of linear algebraic equations

$$\begin{aligned} \Delta C_1 - \mu_1\alpha_1 A_{10}^2 C_2 - \mu_1 b_1 A_{10}A_{20}D_2 &= c_1 A_{20}\cos(\vartheta_1 - \theta), \\ \mu_1\alpha_1 A_{10}^2 C_1 + \Delta C_2 + \mu_1 b_1 A_{10}A_{20}D_1 &= -c_1 A_{20}\sin(\vartheta_1 - \theta), \\ -\mu_2 b_2 A_{10}A_{20}C_2 + \Delta D_1 - \mu_2\alpha_2 A_{20}^2 D_2 &= -c_2 A_{10}\cos(\vartheta_2 + \theta), \\ \mu_2 b_2 A_{10}A_{20}C_1 + \mu_2\alpha_2 A_{20}^2 D_1 + \Delta D_2 &= -c_2 A_{10}\sin(\vartheta_2 + \theta). \end{aligned} \tag{10.29}$$

We substitute (10.26) into the right-hand side of Eq. (10.4) and expand it in a series correct to second order. Taking account of (10.28) and averaging Eq. (10.4) over time we find

$$\omega_{\mathrm{b}} = \bar{\dot{\Phi}} = \Delta - \frac{\Delta_{\mathrm{s}}^2}{2\Delta} + \frac{A_{10}D_1 - A_{20}C_1}{2A_{10}A_{20}\Delta_{\mathrm{s}}}\left(c_1^2\frac{A_{20}^2}{A_{10}^2} - c_2^2\frac{A_{10}^2}{A_{20}^2}\right) + \frac{A_{10}D_2 - A_{20}C_2}{A_{10}A_{20}\Delta_{\mathrm{s}}}c_1c_2\sin(\vartheta_1+\vartheta_2), \tag{10.30}$$

where Δ_{s} is defined by (10.10). Formula (10.30) describes the dependence of the beat frequency on the mistuning away from the synchronization region.

Generally, a solution of Eqs. (10.29) is moderately cumbersome. In this connection we consider only particular cases. In the limiting case of large mistunings, when

$$\Delta \gg \mu_{1,2}, \tag{10.31}$$

we find from Eqs. (10.29)

$$\frac{A_{10}D_1 - A_{20}C_1}{A_{10}A_{20}} = \frac{1}{\Delta\Delta_{\mathrm{s}}}\left(c_1^2\frac{A_{20}^2}{A_{10}^2} - c_2^2\frac{A_{10}^2}{A_{20}^2}\right),$$
$$\frac{A_{10}D_2 - A_{20}C_2}{A_{10}A_{20}} = \frac{2}{\Delta\Delta_{\mathrm{s}}}c_1c_2\sin(\vartheta_1+\vartheta_2). \tag{10.32}$$

Substituting (10.32) into (10.30) we obtain

$$\omega_{\mathrm{b}} = \Delta + \frac{2c_1c_2}{\Delta}\cos(\vartheta_1+\vartheta_2). \tag{10.33}$$

It is seen that under condition (10.31) the dependence of the beat frequency on the mistuning depends only on the linear coupling factors. If $\vartheta_1+\vartheta_2 = 0$ then $\omega_{\mathrm{b}} = \Delta + 2c_1c_2/\Delta > \Delta$, and if $\vartheta_1+\vartheta_2 = \pm\pi$ then $\omega_{\mathrm{b}} = \Delta - 2c_1c_2/\Delta < \Delta$.

If condition (10.31) is not fulfilled, a rather simple expression for the beat frequency is obtained only in the specific case of identical generators, when conditions (10.13) are realized. In this case we find from Eqs. (10.29)

$$\frac{A_{10}D_1 - A_{20}C_1}{A_{10}A_{20}} = \frac{\Delta(\alpha+b)^2(c_1^2-c_2^2) + 2\mu(\alpha^2-b^2)c_1c_2\sin(\vartheta_1+\vartheta_2)}{\Delta_{\mathrm{s}}\left(\Delta^2(\alpha+b)^2+\mu^2(\alpha-b)^2\right)},$$
$$\frac{A_{10}D_2 - A_{20}C_2}{A_{10}A_{20}} = \frac{2\Delta(\alpha+b)^2c_1c_2\sin(\vartheta_1+\vartheta_2) - \mu(\alpha^2-b^2)(c_1^2-c_2^2)}{\Delta_{\mathrm{s}}\left(\Delta^2(\alpha+b)^2+\mu^2(\alpha-b)^2\right)}. \tag{10.34}$$

Substituting (10.34) into (10.30) we find

$$\omega_{\mathrm{b}} = \Delta - \frac{\Delta_{\mathrm{s}}^2}{2\Delta} + \frac{\Delta(\alpha+b)^2\left(c_1^2+c_2^2+2c_1c_2\cos(\vartheta_1+\vartheta_2)\right)}{2\left(\Delta^2(\alpha+b)^2+\mu^2(\alpha-b)^2\right)}. \tag{10.35}$$

It follows from (10.35) that in the case of $\vartheta_1+\vartheta_2 = 0$, when $\Delta_{\mathrm{s}} = |c_1-c_2|$, the plot $\omega_{\mathrm{b}}(\Delta)$ intersects the asymptote $\omega_{\mathrm{b}} = \Delta$ at the point

$$\Delta^* = \frac{\mu(\alpha - b)\Delta_s}{2\sqrt{c_1 c_2}(\alpha + b)}. \tag{10.36}$$

For $\Delta < \Delta^*$ the plot $\omega_b(\Delta)$ lies under the asymptote, whereas for $\Delta > \Delta^*$ it lies above the asymptote. In the case of $\vartheta_1 + \vartheta_2 = \pm\pi$, when $\Delta_s = c_1 + c_2$, the plot $\omega_b(\Delta)$ lies everywhere under the asymptote, monotonically approaching it as Δ increases. The qualitative dependencies of ω_b on Δ in the two cases indicated are illustrated in Fig. 10.1 a.

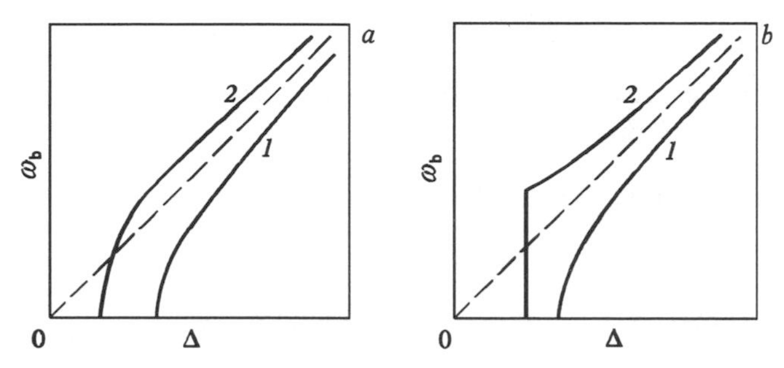

Fig. 10.1. The qualitative dependencies of ω_b on Δ in the cases of (a) weak and (b) strong linear coupling for $\vartheta_1 + \vartheta_2 = \pm\pi$ (curves 1) and for $\vartheta_1 + \vartheta_2 = 0$ (curves 2)

10.1.2 The case of strong linear coupling

Let us further consider the case of rather strong linear coupling between the generators and assume that the generators are identical, i.e., $\mu_1 = \mu_2 = \mu$, $\alpha_1 = \alpha_2 = \alpha$, $b_1 = b_2 = b$, $c_1 = c_2 = c$. The coupling is strong when $c \gg \mu$. We assume that the nonlinear coupling is sufficiently small, so that $b < \alpha$, and restrict our consideration to the specific cases when $\vartheta_1 + \vartheta_2 = \pm\pi$ and $\vartheta_1 + \vartheta_2 = 0$.

1. $\vartheta_1 + \vartheta_2 = \pm\pi$. In this case a steady-state solution of Eqs. (10.4), (10.5) describing the synchronization regime is

$$A_1^2 = A_2^2 = \frac{1}{\alpha + \beta} + \frac{2c}{\mu}\sqrt{1 - \frac{\Delta^2}{4c^2}}, \quad \cos(\Phi + \vartheta_1) = \frac{\Delta}{2c}. \tag{10.37}$$

The solution (10.37) exists and is stable for $\Delta \le 2c$. Hence, the half-width of the synchronization region is $\Delta_s = 2c$. We note that in the case in question the half-width of the synchronization region is determined by the same expression, as for weak coupling.

To find the solution outside the synchronization region it is convenient to introduce the new variables

$$x = A_1^2 + A_2^2, \quad y = A_1^2 - A_2^2,$$
$$u = 4A_1A_2\sin(\Phi + \vartheta_1), \quad v = 4A_1A_2\sin(\Phi + \vartheta_1).$$

It is easily seen that these variables are related by the algebraic equation

$$4(x^2 - y^2) = u^2 + v^2. \tag{10.38}$$

With x, y, u and v in place of $A_{1,2}$ and $\varphi_{1,2}$ Eqs. (10.3), (10.5) become

$$\begin{gathered} \dot{x} = \mu\left(1 - \frac{\alpha+\beta}{2}x\right)x - \mu\frac{\alpha-\beta}{2}y^2 + cu, \quad \dot{y} = \mu(1-\alpha x)y, \\ \dot{u} = \mu\left(1 - \frac{\alpha+\beta}{2}x\right)u + \Delta v + 4cx, \quad \dot{v} = \mu\left(1 - \frac{\alpha+\beta}{2}x\right)v - \Delta u. \end{gathered} \tag{10.39}$$

In the zero approximation with respect to the small parameter μ/c the solution of Eqs. (10.39) is

$$\begin{gathered} x = x_0 + \frac{c}{\Omega}B\sin(\Omega t + \psi), \quad y = y_0, \\ u = B\cos(\Omega t + \psi), \quad v = v_0 - \frac{\Delta}{\Omega}B\sin(\Omega t + \psi), \end{gathered} \tag{10.40}$$

where $\Omega = \sqrt{\Delta^2 - 4c^2}$, and x_0, y_0, v_0, B and ψ are arbitrary constants. It follows from (10.38) that x_0, y_0, v_0 and B are related by the following equations:

$$4cx_0 + \Delta v_0 = 0, \quad B^2 = 4\left(\frac{\Omega^2 x_0^2}{\Delta^2} - y_0^2\right). \tag{10.41}$$

Let us consider the solution (10.40) as the formulas for the substitution into Eqs. (10.39) of the slow variables x_0, y_0, v_0, B, and ψ in place of the fast variables x, y, u, and v. Taking account of Eqs. (10.41) we can use only x_0, y_0 and ψ as independent variables. The averaged equations for these variables are

$$\dot{x}_0 = \mu x_0\left(1 - \frac{\alpha+\beta}{2}x_0\right), \quad \dot{y}_0 = \mu y_0(1 - \alpha x_0), \quad \dot{\psi} = 0. \tag{10.42}$$

A steady-state solution of Eqs. (10.42) is

$$x_0 = \frac{2}{\alpha+\beta}, \quad y_0 = 0. \tag{10.43}$$

The fact that the difference between the intensities of oscillations is equal to zero signifies that the intensities are modulated in phase.

The beat frequency ω_b can be found from the relation

$$\dot{\Phi} = \frac{\dot{u}v - \dot{v}u}{4(x^2 - y^2)}. \tag{10.44}$$

Substituting (10.40), (10.41), (10.43) into (10.44) and averaging over time, we find that $\omega_b = \Omega = \sqrt{\Delta^2 - 4c^2}$. Hence the character of the dependence of the beat frequency on the mistuning Δ in the case under consideration is the same as for a small coupling factor.

2. In the case of $\vartheta_1 + \vartheta_2 = 0$ a steady-state solution of Eqs. (10.4), (10.5) can be found only approximately. As a zero approximation with respect to μ/c we obtain

$$\begin{aligned}
A_1^2 &= \frac{\Delta^2 + 4c^2}{2}\left(\alpha(\Delta^2 + 2c^2) + 2\beta c^2\right)\left(1 \pm \frac{\Delta}{\sqrt{\Delta^2 + 4c^2}}\right),\\
A_2^2 &= \frac{\Delta^2 + 4c^2}{2}\left(\alpha(\Delta^2 + 2c^2) + 2\beta c^2\right)\left(1 \mp \frac{\Delta}{\sqrt{\Delta^2 + 4c^2}}\right),\\
&\cos(\Phi + \vartheta_1) = \mp 1.
\end{aligned} \tag{10.45}$$

It can be shown that the solution (10.45) is stable for $\Delta^2 - 2c^2 \le 0$. It follows from this that the half-width of the synchronization region is $\Delta_s = \sqrt{2}\,c$.

To calculate the beat frequency outside the synchronization region we perform the same substitution as in the preceding case. In so doing we obtain the following equations for x, y, u and v:

$$\begin{aligned}
\dot{x} &= \mu\left(1 - \frac{\alpha+\beta}{2}x\right)x - \mu\frac{\alpha-\beta}{2}y^2, \quad \dot{y} = \mu(1 - \alpha x)y + cu,\\
\dot{u} &= \mu\left(1 - \frac{\alpha+\beta}{2}x\right)u + \Delta v - 4cy, \quad \dot{v} = \mu\left(1 - \frac{\alpha+\beta}{2}x\right)v - \Delta u.
\end{aligned} \tag{10.46}$$

In the zero approximation with respect to μ/c the solution of Eqs. (10.46) is

$$\begin{aligned}
x &= x_0, \quad y = y_0 + \frac{c}{\Omega}B\sin(\Omega t + \psi),\\
u &= B\cos(\Omega t + \psi), \quad v = v_0 - \frac{\Delta}{\Omega}B\sin(\Omega t + \psi),
\end{aligned} \tag{10.47}$$

where $\Omega = \sqrt{\Delta^2 + 4c^2}$, and x_0, y_0, v_0, B and ψ are arbitrary constants. It follows from (10.38) that x_0, y_0, v_0 and B are related by the following equations:

$$v_0 = \frac{4c}{\Delta}y_0, \quad y_0^2 = \frac{\Delta^2}{\Omega^2}\left(x_0^2 - \frac{B^2}{4}\right). \tag{10.48}$$

Considering, as before, the solution (10.47) as the formulas for the substitution into Eqs. (10.46) of the slow variables x_0, y_0, v_0, B and ψ in place of the fast variables x, y, u and v, and taking x_0, B and ψ as independent variables, we obtain for these variables the following averaged equations:

$$\begin{aligned}
\dot{x}_0 &= \mu\left[\left(1 - \frac{\alpha+\beta}{2}x_0\right)x_0 - \frac{\alpha-\beta}{2\Omega^2}\left(\Delta^2 x_0^2 - (\Delta^2 - 2c^2)B^2\right)\right],\\
\dot{B} &= \mu\left(1 - \left((3\alpha+\beta)\Omega^2 - (\alpha-\beta)\Delta^2\right)\frac{x_0}{4\Omega^2}\right)B,\\
\dot{\psi} &= 0.
\end{aligned} \tag{10.49}$$

Equations (10.49) have two steady-state solutions:

$$x_0 = \frac{4\Omega^2}{(3\alpha+\beta)\Omega^2 - (\alpha-\beta)\Delta^2}\,,$$
$$B = \frac{8\Omega^2}{(3\alpha+\beta)\Omega^2 - (\alpha-\beta)\Delta^2}\,, \tag{10.50a}$$

$$x_0 = \frac{2\Omega^2}{\alpha(\Omega^2+\Delta^2)+4\beta c^2}\,, \quad B=0. \tag{10.50b}$$

It follows from (10.48) that $y_0 = v_0 = 0$ for the solution (10.50a) and $y_0 = \pm x_0\Delta/\Omega$, $v_0 = \pm 4x_0 c/\Omega$ for the solution (10.50b). Taking account of (10.47), we find from (10.50a)

$$A_1^2 = \frac{x_0}{2}\left(1+\frac{2c}{\Omega}\sin(\Omega t+\psi)\right), \quad A_2^2 = \frac{x_0}{2}\left(1-\frac{2c}{\Omega}\sin(\Omega t+\psi)\right),$$

$$\tan(\Phi+\vartheta_1) = -\frac{\Omega}{\Delta}\cot(\Omega t+\psi).$$

From this it is seen that the solution (10.50a) describes the beat regime. In this regime the oscillation intensities are modulated with frequency Ω and with opposite phases. The amplitude of the modulation of the intensities is equal to $2c/\Omega$. The solution (10.50b) describes the synchronization regime for which the expressions for the intensities A_1^2, A_2^2 and for the phase difference Φ coincide with (10.45).

It can be shown that the beat frequency outside the synchronization region, as in the preceding case, is equal to Ω. For $\Delta = \Delta_s$, $\Omega = \sqrt{6}\,c$, i.e., in the transition from the synchronization region to the beat regime, the beat frequency changes by a jump. The dependencies of the beat frequency on the mistuning Δ for both of the cases considered are presented in Fig. 10.1 b.

We see from Fig. 10.1 that two mechanisms of synchronization considered in the preceding section manifest themselves in the case of mutual synchronization of two generators only for $\vartheta_1 + \vartheta_2 = 0$.

10.2 Mutual synchronization of self-oscillations with multiple frequencies

It is known that two self-oscillatory systems with multiple frequencies can be synchronized [293, 373, 346, 347, 291, 149]. If these systems are described by Eqs. (10.1) then synchronization is possible when the frequencies ω_1 and ω_2 differ from one another by a factor of two or three. This phenomenon is similar to the resonances of the nth kind considered in the preceding chapter.

As an example we consider two generators described by the equations

$$
\begin{aligned}
&\ddot{x}_1 - \mu_1 F_1(x_1)\dot{x}_1 + \omega_1^2 x_1 = c_1 \ddot{x}_2, \\
&\ddot{x}_2 - \mu_2 F_2(x_2)\dot{x}_2 + \omega_2^2 x_2 = c_2 \ddot{x}_1,
\end{aligned}
\tag{10.51}
$$

where

$$
F_{1,2}(x_{1,2}) = 1 - 4\alpha_{1,2} x_{1,2}^2 + \beta_{1,2} x_{1,2}. \tag{10.52}
$$

If the coupling is sufficiently small and the frequencies ω_1 and ω_2 differ essentially from one another, then the motion of each of the generators can be represented as a sum of self-oscillations at the frequency close to the corresponding natural frequency and small forced oscillations at the frequency of another generator. Since the forced oscillations are small, their amplitudes can be found from the corresponding linear equations. Let us denote the terms describing the self-oscillations by $y_{1,2}(t)$. Then the forced oscillations can be approximately described by the terms $k_{1,2} y_{2,1}(t)$, where

$$
k_{1,2} = \frac{c_{1,2}\omega_{2,1}^2}{\omega_{2,1}^2 - \omega_{1,2}^2}. \tag{10.53}
$$

Thus, it makes sense to substitute into Eqs. (10.51) the new variables y_1 and y_2 defined as

$$
x_1 = y_1 + k_1 y_2, \quad x_2 = y_2 + k_2 y_1, \tag{10.54}
$$

where $k_{1,2}$ are determined by (10.53). After the substitution, as a first approximation with respect to $k_{1,2}$, we obtain the following equations for $y_{1,2}$:

$$
\begin{aligned}
&\ddot{y}_1 + \omega_1^2 y_1 = \mu_1 F_1(y_1 + k_1 y_2)(\dot{y}_1 + k_1 \dot{y}_2) - \mu_2 k_1 \frac{\omega_1^2}{\omega_2^2} F_2(y_2)\dot{y}_2, \\
&\ddot{y}_2 + \omega_2^2 y_2 = \mu_2 F_2(y_2 + k_2 y_1)(\dot{y}_2 + k_2 \dot{y}_1) - \mu_1 k_2 \frac{\omega_2^2}{\omega_1^2} F_1(y_1)\dot{y}_1.
\end{aligned}
\tag{10.55}
$$

Setting $y_{1,2} = A_{1,2}\cos(\omega_{1,2} t + \varphi_{1,2})$ and using the averaging method we find from Eqs. (10.55) the truncated equations for the amplitudes and phases:

$$
\begin{aligned}
\dot{A}_1 &= \overline{\mu_1 F_1(A_1 \cos\psi_1 + k_1 A_2 \cos\psi_2)\left(A_1 \sin\psi_1 + \frac{k_1\omega_2}{\omega_1} A_2 \sin\psi_2\right)\sin\psi_1} \\
&\quad - \mu_2 k_1 \frac{\omega_1}{\omega_2} A_2 \overline{F_2(A_2\cos\psi_2)\sin\psi_1 \sin\psi_2}, \\
\dot{A}_2 &= \overline{\mu_2 F_2(A_2 \cos\psi_2 + k_2 A_1 \cos\psi_1)\left(A_2 \sin\psi_2 + \frac{k_2\omega_1}{\omega_2} A_1 \sin\psi_1\right)\sin\psi_2} \\
&\quad - \mu_1 k_2 \frac{\omega_2}{\omega_1} A_1 \overline{F_1(A_1\cos\psi_1)\sin\psi_1 \sin\psi_2},
\end{aligned}
\tag{10.56}
$$

$$\begin{aligned}
\dot{\varphi}_1 &= \mu_1 \overline{F_1(A_1\cos\psi_1 + k_1 A_2\cos\psi_2)\left(\sin\psi_1 + \frac{k_1\omega_2 A_2}{\omega_1 A_1}\sin\psi_2\right)\cos\psi_1} \\
&\quad - \mu_2 k_1 \frac{\omega_1 A_2}{\omega_2 A_1}\overline{F_2(A_2\cos\psi_2)\sin\psi_2\cos\psi_1}, \\
\dot{\varphi}_2 &= \mu_2 \overline{F_2(A_2\cos\psi_2 + k_2 A_1\cos\psi_1)\left(\sin\psi_2 + \frac{k_2\omega_1 A_1}{\omega_2 A_2}\sin\psi_1\right)\cos\psi_2} \\
&\quad - \mu_1 k_2 \frac{\omega_2 A_1}{\omega_1 A_2}\overline{F_1(A_1\cos\psi_1)\sin\psi_1\cos\psi_2},
\end{aligned} \tag{10.57}$$

where $\psi_{1,2} = \omega_{1,2}t + \varphi_{1,2}$.

- Let $\omega_2 = 2\omega_1 - \Delta$, where $|\Delta| \ll \omega_1$ is the mistuning. In this case we obtain from (10.56) and (10.57)

$$\begin{aligned}
\dot{A}_1 &= \frac{\mu_1}{2}\left(1 - \frac{A_1^2}{A_{10}^2}\right)A_1 + m_1\frac{A_1 A_2}{A_{10}}\cos\Phi, \\
\dot{A}_2 &= \frac{\mu_2}{2}\left(1 - \frac{A_2^2}{A_{20}^2}\right)A_2 + m_2\frac{A_1^2}{2A_{10}}\cos\Phi,
\end{aligned} \tag{10.58}$$

$$\dot{\Phi} = \Delta - \left(2m_1\frac{A_2}{A_{10}} + m_2\frac{A_1^2}{2A_2A_{10}}\right)\sin\Phi, \tag{10.59}$$

$$\dot{\varphi}_1 = -\frac{m_1 A_2}{A_{10}}\sin\Phi, \tag{10.60}$$

where $A_{10} = 1/\sqrt{\alpha_1}$ and $A_{20} = 1/\sqrt{\alpha_2}$ are the amplitudes of free self-oscillations, $\Phi = \Delta t + 2\varphi_1 - \varphi_2$ is the generalized phase shift, and $m_{1,2} = \mu_1\beta_1 A_{10}c_{1,2}/3$. We emphasize that Eqs. (10.58) and (10.59) combine into a closed system.

By a synchronization regime is meant that the oscillation frequency of the first generator, which is equal to $\omega_1 + \dot{\varphi}_1$, is exactly half the oscillation frequency of the second generator, which is equal to $\omega_2 + \dot{\varphi}_2$. Hence, the condition for the synchronization regime is $2(\omega_1 t + \varphi_1) - (\omega_2 t + \varphi_2) = \text{const}$, or $\Phi = \text{const}$.

In the case of weak coupling, when $m_1 A_{20} \ll \mu_1 A_{10}$, $m_2 A_{10} \ll \mu_2 A_{20}$, the oscillation amplitudes A_1 and A_2 are close to A_{10} and A_{20}, i.e., $A_1 = A_{10} + a_1$, $A_2 = A_{20} + a_2$, where $a_1 \ll A_{10}$, $a_2 \ll A_{20}$. For $|\Delta| \ll \mu_1$

$$a_1 = \frac{m_1 A_{20}}{\mu_1}\cos\Phi, \quad a_2 = \frac{m_2 A_{10}}{2\mu_2}\cos\Phi. \tag{10.61}$$

Substituting into Eq. (10.59) $A_1 \approx A_{10}$ and $A_2 \approx A_{20}$ we find the equation for Φ which is conveniently rewritten as

$$\dot{\Phi} = \Delta - \Delta_s\sin\Phi, \tag{10.62}$$

where

$$\Delta_s = 2m_1 \frac{A_{20}}{A_{10}} + m_2 \frac{A_{10}}{2A_{20}} = \frac{\mu_1 \beta_1 A_{10}}{3} \left(2c_1 \frac{A_{20}}{A_{10}} + c_2 \frac{A_{10}}{2A_{20}} \right). \quad (10.63)$$

It is seen from (10.63) that the width of the synchronization region is proportional to the coefficient of the quadratic term for the generator with smaller frequency. This is associated with the fact that synchronization appears as a result of the interaction between the main harmonic of the second generator and the second harmonic of the first generator; the latter is proportional to the coefficient β_1.

- In the case when $\omega_2 = 3\omega_1 - \Delta$, where $|\Delta| \ll \omega_1$, we obtain from (10.56) and (10.57) the following equations:

$$\dot{A}_1 = \frac{\mu_1}{2} \left(1 - \frac{A_1^2}{A_{10}^2} \right) A_1 - m_1 \frac{A_1^2 A_2}{A_{10}^2} \cos \Phi,$$
$$\dot{A}_2 = \frac{\mu_2}{2} \left(1 - \frac{A_2^2}{A_{20}^2} \right) A_2 - m_2 \frac{A_1^3}{3A_{10}^2} \cos \Phi, \quad (10.64)$$

$$\dot{\Phi} = \Delta + \left(3m_1 \frac{A_1 A_2}{A_{10}^2} + m_2 \frac{A_1^3}{3A_2 A_{10}^2} \right) \sin \Phi, \quad (10.65)$$

$$\dot{\varphi}_1 = \frac{m_1 A_1 A_2}{A_{10}^2} \sin \Phi, \quad (10.66)$$

where $\Phi = \Delta t + 3\varphi_1 - \varphi_2$, $m_{1,2} = 9\mu_1 c_{1,2}/16$. In the case of weak coupling, when $m_1 A_{20} \ll \mu_1 A_{10}$, $m_2 A_{10} \ll \mu_2 A_{20}$, we obtain for Φ an equation of the form of (10.62), where

$$\Delta_s = 3m_1 \frac{A_{20}}{A_{10}} + m_2 \frac{A_{10}}{3A_{20}}. \quad (10.67)$$

10.3 Parametric synchronization of two generators with different frequencies

We consider two generators described by Eqs. (10.52) and assume that the friction factor is modulated by a signal of the frequency Ω which is close to the frequency difference $\omega_1 - \omega_2$. With this assumption we can rewrite Eqs. (10.52) as

$$\ddot{x}_1 - \mu_1 F_1(x_1)\dot{x}_1 + \omega_1^2 x_1 = c_1 \ddot{x}_2,$$
$$\ddot{x}_2 - \mu_2 F_2(x_2)\dot{x}_2 + \omega_2^2 x_2 = c_2 \ddot{x}_1, \quad (10.68)$$

where

$$F_{1,2}(x_{1,2}) = 1 + \kappa_{1,2} \sin(\Omega t - \vartheta_{1,2}) - 4\alpha_{1,2} x_{1,2}^2 + \beta_{1,2} x_{1,2}. \quad (10.69)$$

We seek a solution of Eqs. (10.68) in the form (10.54), where $k_{1,2}$ is determined by (10.53) and $y_{1,2} = A_{1,2}\cos(\omega_{1,2}t + \varphi_{1,2})$. Setting $\omega_1 - \omega_2 = \Omega + \Delta$, where $|\Delta| \ll \Omega$, we find the following truncated equations for $A_{1,2}$ and $\varphi_{1,2}$:

$$\begin{aligned}
\dot{A}_1 &= \frac{\mu_1}{2}\left(1 - \frac{A_1^2}{A_{10}^2}\right) A_1 - mk_1 A_2 \sin(\Phi + \chi), \\
\dot{A}_2 &= \frac{\mu_2}{2}\left(1 - \frac{A_2^2}{A_{20}^2}\right) A_2 + mk_2 A_1 \sin(\Phi + \chi), \\
\dot{\varphi}_1 &= -mk_1 \frac{A_2}{A_1} \cos(\Phi + \chi), \\
\dot{\varphi}_2 &= -mk_2 \frac{A_1}{A_2} \cos(\Phi + \chi),
\end{aligned} \tag{10.70}$$

where $A_{10} = 1/\sqrt{\alpha_1}$, $A_{20} = 1/\sqrt{\alpha_2}$ are the amplitudes of free self-oscillations, $\Phi = \Delta t + \varphi_1 - \varphi_2$ is the generalized phase shift,

$$\begin{aligned}
m &= \frac{1}{4}\sqrt{\mu_1^2\kappa_1^2 \frac{\omega_2^2}{\omega_1^2} + \mu_2^2\kappa_2^2 \frac{\omega_1^2}{\omega_2^2} - 2\mu_1\mu_2\kappa_1\kappa_2 \cos(\vartheta_1 - \vartheta_2)}\,, \\
\tan\chi &= \frac{\mu_1\kappa_1\omega_2^2 \sin\vartheta_1 - \mu_2\kappa_2\omega_1^2 \sin\vartheta_2}{\mu_1\kappa_1\omega_2^2 \cos\vartheta_1 - \mu_2\kappa_2\omega_1^2 \cos\vartheta_2}\,.
\end{aligned} \tag{10.71}$$

In the case of weak coupling, when $A_1 \approx A_{10}$, $A_2 \approx A_{20}$, we obtain the closed equation for Φ:

$$\dot{\Phi} = \Delta - Km\cos(\Phi + \chi), \tag{10.72}$$

where

$$K = k_1 \frac{A_{20}}{A_{10}} - k_2 \frac{A_{10}}{A_{20}}\,.$$

By synchronization regime is meant the regime where the difference of the self-oscillation frequencies is equal to the frequency Ω. In this regime $\dot{\Phi} = 0$, and from (10.72) we find the half-width of the synchronization region: $\delta_s = Km$. In the specific case, when $\mu_1\kappa_1\omega_2/\omega_1 = \mu_2\kappa_2\omega_1/\omega_2 \equiv \mu\kappa$, we obtain

$$\Delta_s = \frac{\mu\kappa K}{2}\left|\sin\frac{\vartheta_1 - \vartheta_2}{2}\right|. \tag{10.73}$$

It follows from (10.73) that for $\vartheta_1 = \vartheta_2$ the synchronization region in the first approximation vanishes.

10.4 Chaotization of self-oscillations in two coupled generators

If the coupling between generators is sufficiently strong then chaotization can occur instead of synchronization. The chaotization phenomenon for two coupled generators with inertial nonlinearity was found in numerical and experimental studies by Anischenko et al. [10] and by Kalyanov and Lebedev [137]. A similar effect is described by Sbitnev [298, 299] for the case of the interaction between two neuron populations, each involving neurons of two kinds, *activators* (x_i) and *inhibitors* (y_i).

Chaotization of self-oscillations also occurs in a system of two coupled electro-mechanical vibrators, see Fig. 10.2 [187]. Such vibrators in the free state will be considered in Ch. 12. The equations of this system are

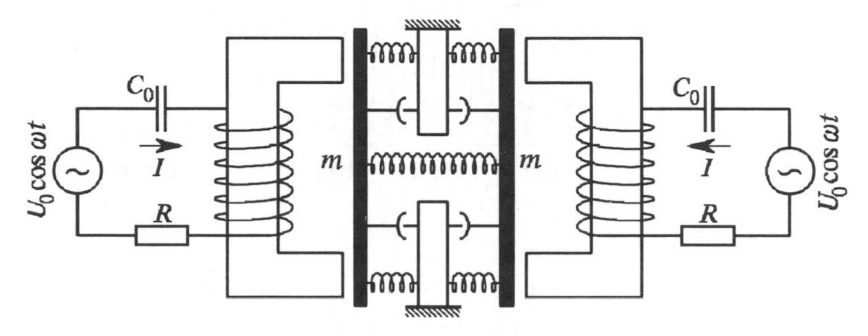

Fig. 10.2. Two coupled electro-mechanical vibrators

$$\frac{\mathrm{d}^2}{\mathrm{d}t^2}\Big(L(x_{1,2})I_{1,2}\Big) + R\,\frac{\mathrm{d}I_{1,2}}{\mathrm{d}t} + \frac{I_{1,2}}{C_0} = U_0\omega \sin \omega t,$$

$$m\,\frac{\mathrm{d}^2 x_{1,2}}{\mathrm{d}t^2} + \alpha\,\frac{\mathrm{d}x_{1,2}}{\mathrm{d}t} + k_{1,2}x_{1,2} - cx_{2,1} = F(x_{1,2}, I_{1,2}), \tag{10.74}$$

where $L(x) = L_0(1 + a_1x + a_2x^2 + a_3x^3)$, $F(x, I) = (I^2/2)\mathrm{d}L(x)/\mathrm{d}x$.

Equations (10.74) were numerically simulated by Landa and Rosenblum [187] for the following values of the parameters: $\omega = 1$, $1/\sqrt{L_0C_0} = 0.9$, $R/2L_0 = 0.1$, $\alpha/2m = 0.005$, $\sqrt{k_1/m} = 0.05$, $\sqrt{k_2/m} = 0.049$, $a_1 = 1$, $a_2 = -0.1$, $a_3 = 0.1$, $U_0 = 0.03$. The coupling factor c was varied. The results of the simulation are presented in Fig. 10.3. As the coupling is absent ($c = 0$) oscillations of the vibrators are periodic with slightly different periods (a). Beyond a certain value of the coupling factor the synchronization of oscillations occurs (b). As the coupling factor increases the synchronization regime is retained but the period of each of the vibrators is doubled (c). As the coupling factor increases further the synchronous oscillations again become periodic with the main period (d). Finally, if the coupling is sufficiently strong ($c > 0.0028$) then oscillations become chaotic, see Fig. 10.3 e.

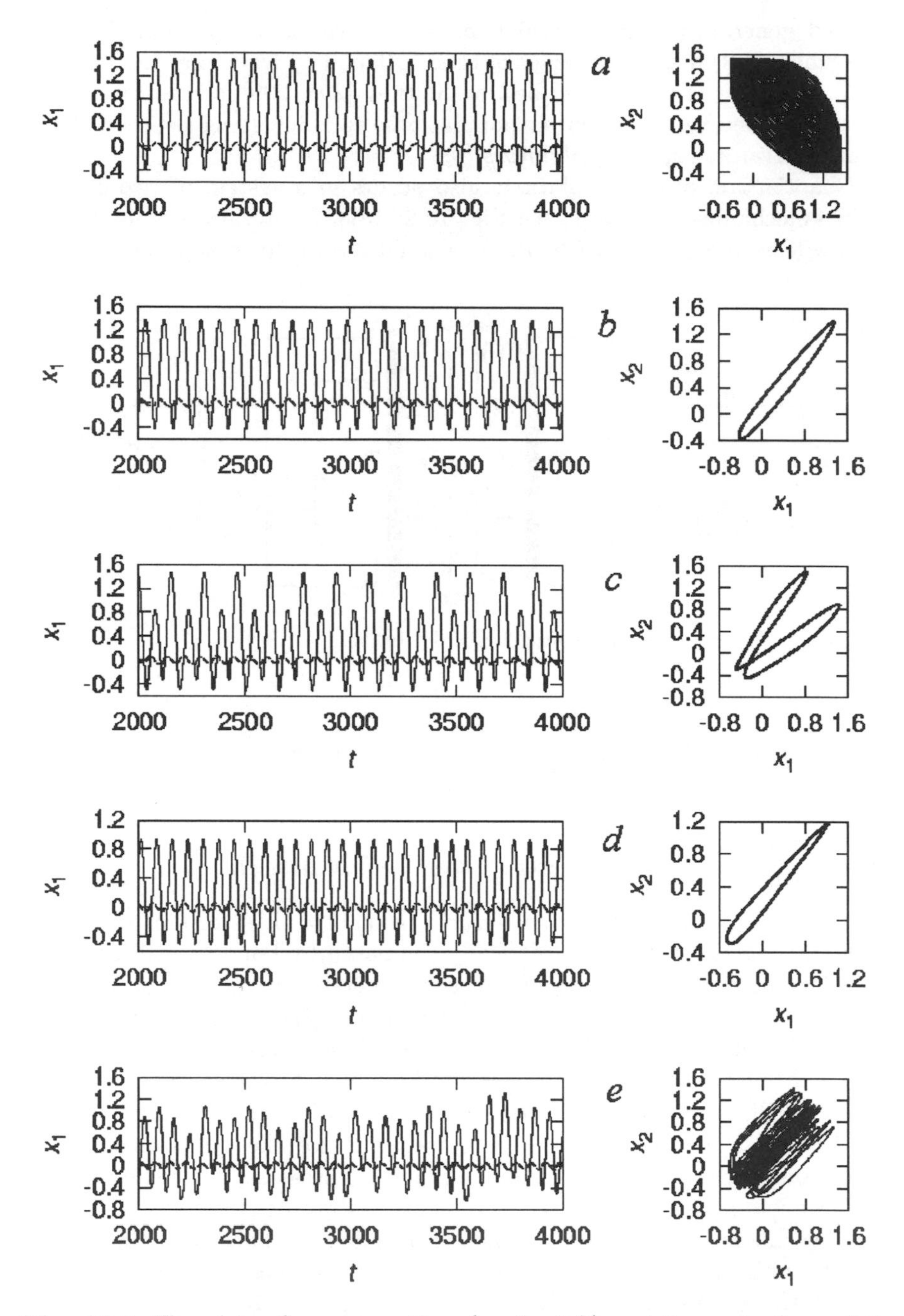

Fig. 10.3. The plots of x_1 versus time (on the left) and the projections of the phase trajectories on the x_1, x_2 plane (on the right) for (a) $c = 0$, (b) $c = 0.0005$, (c) $c = 0.0015$, (d) $c = 0.0025$ and (e) $c = 0.003$. Dashed lines show the plots of $I_1(t)$

10.5 Interaction of generators of periodic and chaotic oscillations

When generators of periodic and chaotic oscillations interact, they can be synchronized with the result that oscillations of both of the generators become either periodic or chaotic. An example of such an interaction is considered in [171]. The equations of two such generators were numerically simulated in this work. One of the generators is described by the van der Pol equation and the other, being a system with inertial excitation, is described by Eqs. (7.1). The equations of the interacting generators are

$$\ddot{x} + 0.75\dot{x} + x = -20y - 17.5x^3 + c_1 u, \quad \dot{y} + y = x - x^2 - x^3,$$
$$\ddot{u} - 0.1(1 - u^2)\dot{u} + \omega^2 u = c_2 x. \tag{10.75}$$

The case $c_2 = 0$ is equivalent to the action of a periodic external force with a certain amplitude B on the generator with inertial excitation. Such an action was considered in the preceding chapter. The value of B is determined by the coupling factor c_1. In particular, $c_1 = 1$ corresponds to $B = 2$. Equations (10.75) were simulated first for $c_1 = 1$, $\omega = \pi$, and varying c_2. It has been found that for $c_2 < c_{\text{thr}} \approx 1.2$ the solution of Eqs. (10.75) is periodic with period $T \approx 4$, twice as large as that of oscillations of the autonomous van der Pol generator. This means that the chaotic oscillations of the first generator are synchronized (captured) by the periodic oscillations of the second one. For $c_2 = c_{\text{thr}}$ the synchronization is broken, and beats arise. They exist over a narrow range of the values of c_2. As c_2 increases further oscillations of the first generator become chaotic, and, in its turn, force oscillations of the van der Pol generator to become chaotic too. It should be noted that the regions of chaotic and periodic oscillations alternate as c_2 increases. In the regions of chaotic oscillations the power spectra of both of the generators are similar in width and in the location of the peaks, especially in low-frequency regions. This is demonstrated in Fig. 10.4 b and c. For comparison, the power spectrum of the autonomous generator with inertial excitation is shown in Fig. 10.4 a.

It is interesting that for $c_2 = 3$ the cross-correlation function $R_{xu}(\tau)$ is asymmetric with respect to $\tau = 0$ and has a well-marked peak for $\tau = -0.23$ (Fig. 10.5); that is to say, there is a 'phase shift' between $x(t)$ and $u(t)$.

In the case when the frequency ω of the autonomous van der Pol generator is close to the frequency of the autonomous generator with inertial excitation corresponding to the maximum of its spectrum, the van der Pol generator becomes chaotic even for moderately small c_2. As c_2 increases further, the width of the van der Pol generator power spectrum increases, whereas the width of the power spectrum of the generator with inertial excitation decreases. As a result, the widths of the power spectra become approximately the same, see Fig. 10.4 d.

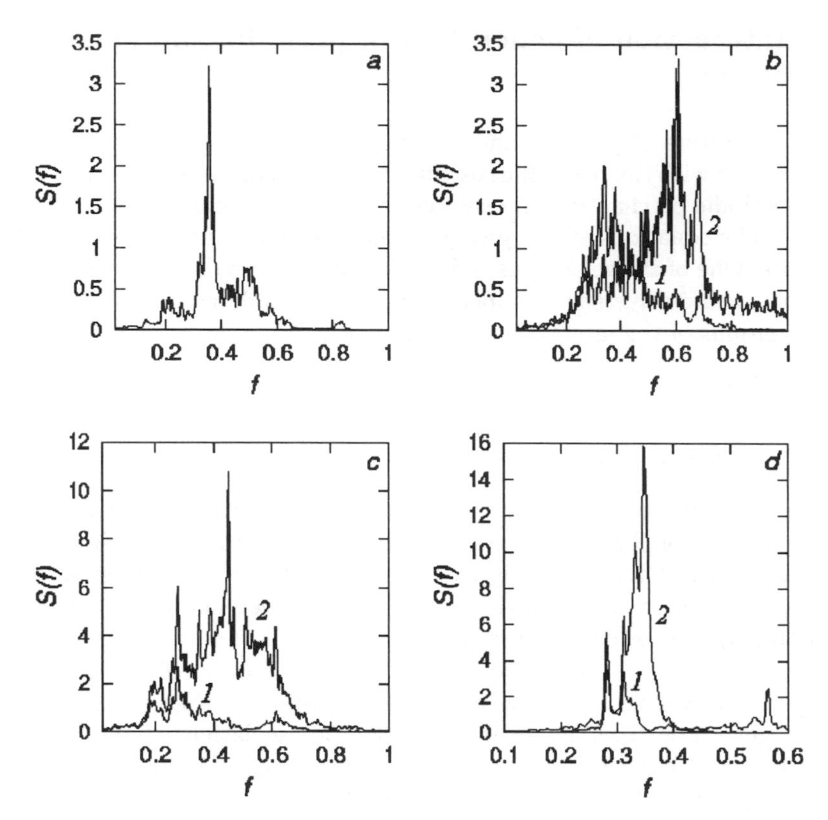

Fig. 10.4. (a) The power spectrum of the autonomous generator with inertial excitation; (b, c, d) the power spectra of $x(t)$ (curves 1) and $u(t)$ (curves 2) for (b) $\omega = \pi$, $c_2 = 3$, (c) $\omega = \pi$, $c_2 = 10$ and (d) $\omega = \pi/\sqrt{2}$, $c_2 = 1.5$

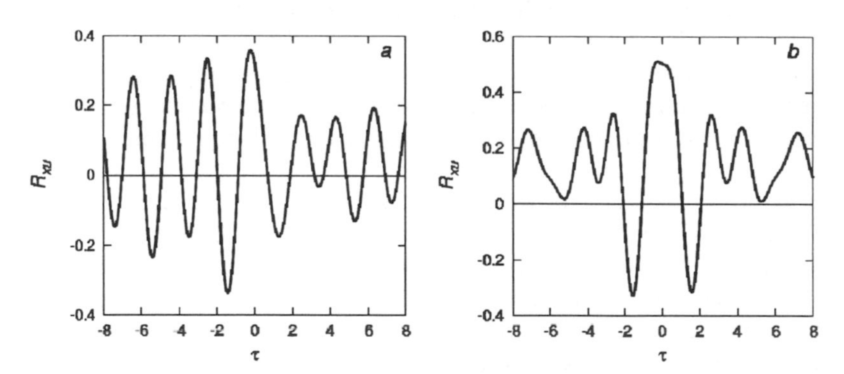

Fig. 10.5. The plots of the cross-correlation function $R_{xu}(\tau)$ for $\omega = \pi$, (a) $c_2 = 3$ and (b) $c_2 = 10$

10.6 Interaction of generators of chaotic oscillations

Interesting phenomena are observed when both of the interacting generators, which can be both identical and different, execute chaotic oscillations. Examples of such an interaction are given in [171]. First of all, let us consider two identical generators with inertial excitation described by the equations

$$\begin{aligned} &\ddot{x}_{1,2} + 0.75\dot{x}_{1,2} + x_{1,2} = -20y_{1,2} - 17.5x_{1,2}^3 + cx_{2,1}, \\ &\dot{y}_{1,2} + y_{1,2} = x_{1,2} - x_{1,2}^2 - x_{1,2}^3. \end{aligned} \tag{10.76}$$

If coupling between the generators is absent ($c = 0$) and the initial conditions are different, then their oscillations are chaotic and really uncorrelated (see Fig. 10.6 a). As the coupling factor c increases the oscillations become more and more correlated. An important point is that the regions of chaotic oscillations alternate with the regions of periodic ones. For example, over the ranges $0.9 \leq c \leq 1.3$ and $1.7 \leq c \leq 2.3$ the coupled oscillations are periodic. The evolution of the power spectra and of the cross-correlation functions with increasing coupling factor in the regions of chaotic oscillations is represented in Fig. 10.6. We see from Fig. 10.6 that the main peak of the power spectra is shifted successively into the low-frequency region as the coupling factor increases. It should be noted that the cross-correlation coefficient peaks at $\tau = 0$, which is to say that oscillations are executed 'in phase'.

In a body of works devoted to synchronization of chaotic systems (see, e.g., [265]) it is asserted that synchronization occurs when oscillations in the interacting systems become identical. In the case of the system under consideration the oscillations do not become identical up to $c = 3$.[1] Therefore in this case it is necessary to use other quantitative characteristics of the extent to which oscillations are synchronized. In [186, 187] it was shown that the correlation dimension of the attractor of a coupled system d_c can be regarded as such a characteristic. What is more, it was proposed to define the synchronization of chaotic oscillations as the coincidence of the attractor dimension of a coupled system with the partial dimension of interacting subsystems.[2] This definition is valid for self-oscillatory systems executing both periodic and chaotic oscillations. By a partial dimension we mean the dimension which is calculated in the phase subspace associated with the corresponding partial system. For example, the system described by Eqs. (10.76) involves two coupled three-dimensional subsystems with phase coordinates $x_{1,2}$, $\dot{x}_{1,2}$, $y_{1,2}$. By calculating the correlation dimension in this six-dimensional phase space we find the dimension for the coupled system. But, by calculating the correlation dimensions in the three-dimensional subspaces x_i, $\dot{x}_i$, y_i, where $i = 1, 2$, we

[1] For $c > 3$ solutions of Eqs. (10.76) go to infinity.

[2] It should be noted that the partial dimensions of interacting subsystems, even if they are essentially different, tend to be equalized with increasing coupling factor.

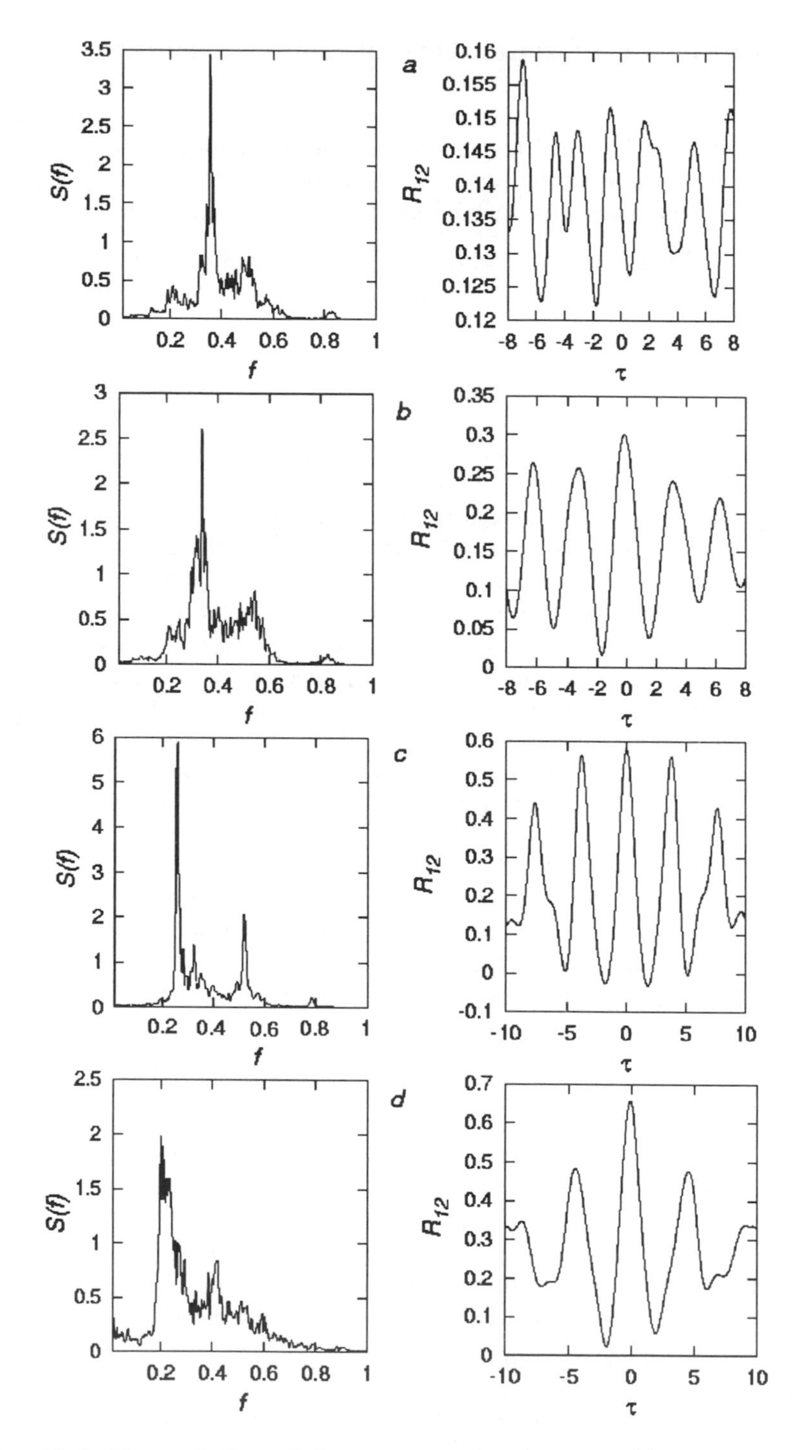

Fig. 10.6. The evolution of the power spectra (on the left) and of the cross-correlation coefficient $R_{12}(\tau)$ (on the right) in the case of two coupled identical generators with inertial excitation: (a) $c = 0$, (b) $c = 0.5$, (c) $c = 1.5$, (d) $c = 2.5$

find the partial dimensions. It is evident that in the case when the partial subsystems are identical their partial dimensions are the same.

As an example of the interaction of two identical and near-identical generators of chaotic self-oscillations, we consider two coupled Rössler oscillators which are described by the equations

$$\begin{aligned} \dot{x}_{1,2} &= -y_{1,2} - b_{1,2}z_{1,2} + cx_{2,1}, \\ \dot{y}_{1,2} &= x_{1,2} + ay_{1,2}, \\ z_{1,2} &= d - ez_{1,2} + x_{1,2}z_{1,2}, \end{aligned} \tag{10.77}$$

where $a = 0.15$, $d = 0.2$, $e = 10$, and the parameters $b_{1,2}$ and c are varied. For $b_1 = b_2 = 1$ the partial systems are identical. In this case the dependencies of the correlation dimension of the coupled system d_{c} and the partial dimension d_{p} on the coupling factor c are given in Fig. 10.7 a. We see that

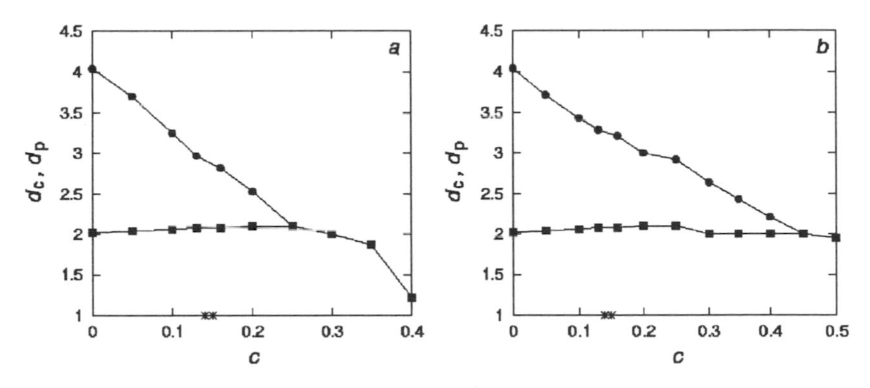

Fig. 10.7. The dependencies of the correlation dimension of the coupled system d_{c} (circles) and the partial dimensions d_{p} (squares) on the coupling factor c for the coupled (a) identical Rössler oscillators and (b) different ones

the dimension of the coupled system becomes equal to the dimension of the partial systems d_{p} from $c \approx 0.25$ onwards. Hence, the mutual synchronization of self-oscillatory systems executing chaotic oscillations, like the synchronization of such systems by an external periodic force, is of a threshold character. We note that for $c \geq 0.25$ the oscillations of both Rössler oscillators become identical. This agrees with the definition of synchronization given in [265].

Another example is the system described by Eqs. (10.76) which was discussed above. The dependencies of the correlation dimension of the coupled system and of the partial dimension on the coupling factor c for this system are shown in Fig. 10.8 a. We see that synchronization in the sense indicated above is not achieved for this system.

As an example of the interaction of essentially different systems executing chaotic oscillations, we consider the interaction of two different generators with inertial excitation described by the equations [171]

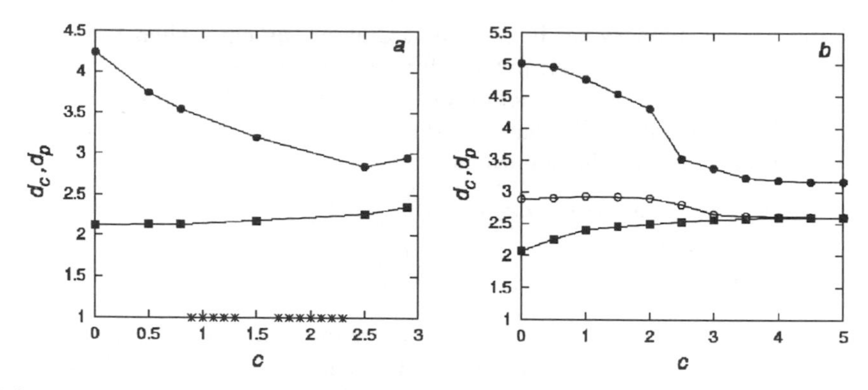

Fig. 10.8. The dependencies of the correlation dimension of the coupled system (black circles) and of the partial dimensions on the coupling factor c (a) for the system described by Eqs. (10.76) and (b) for the system described by Eqs. (10.78) (the partial dimensions for the first and the second subsystems are shown by light circles and squares, respectively)

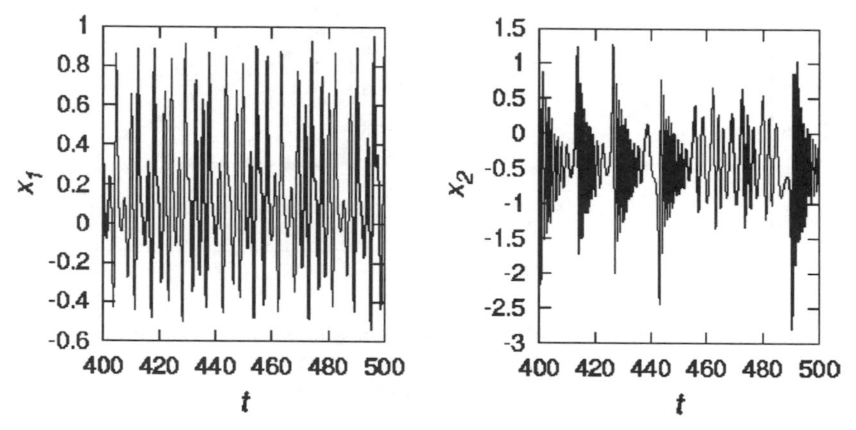

Fig. 10.9. The plots of $x_1(t)$ and $x_2(t)$ for the system described by Eqs. (10.78)

$$\begin{aligned} \ddot{x}_1 + 0.75\dot{x}_1 + x_1 &= -20y_1 - 17.5x_1^3 + cx_2, \quad \dot{y}_1 + y_1 = x_1 - x_1^2 - x_1^3, \\ \ddot{x}_2 + 0.5\dot{x}_2 + x_2 &= -10y_2 - 20x_2y_2 + cx_1, \quad \dot{y}_2 + 0.1y_2 = x_2 + x_2^2. \end{aligned} \tag{10.78}$$

In the absence of the coupling the oscillations of these generators are chaotic, uncorrelated and different in their form, see Fig. 10.9. Their power spectra are different in width and in the location of their peaks (Fig. 10.10 a). As the coupling factor c increases the difference between the widths of the spectra decreases and their peaks become closer in frequency. This is seen from Fig. 10.10, where the evolution of the power spectra with increasing coupling factor is shown. Furthermore, the correlation between oscillations

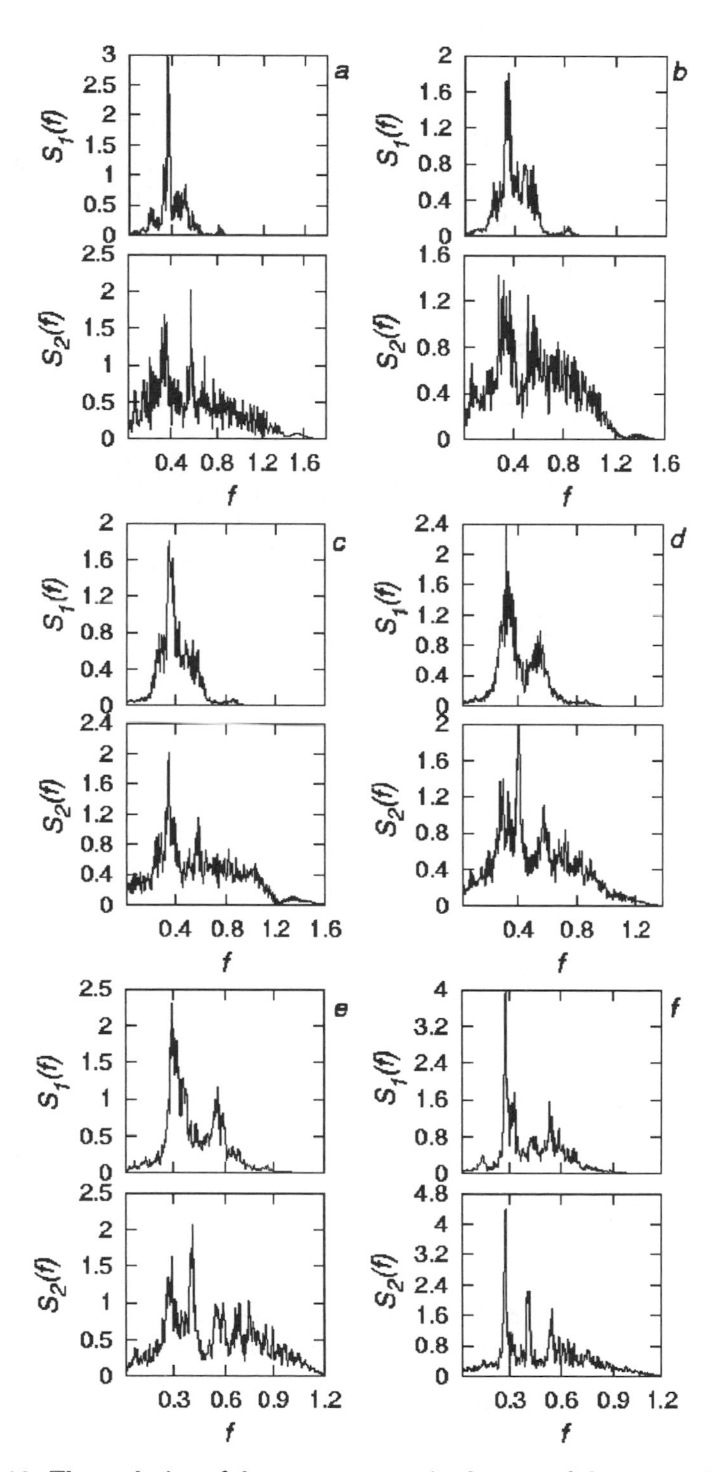

Fig. 10.10. The evolution of the power spectra in the case of the system described by Eqs. (10.79): (a) $c = 0$, (b) $c = 0.5$, (c) $c = 1$, (d) $c = 1.5$, (e) $c = 2$, (f) $c = 2.5$

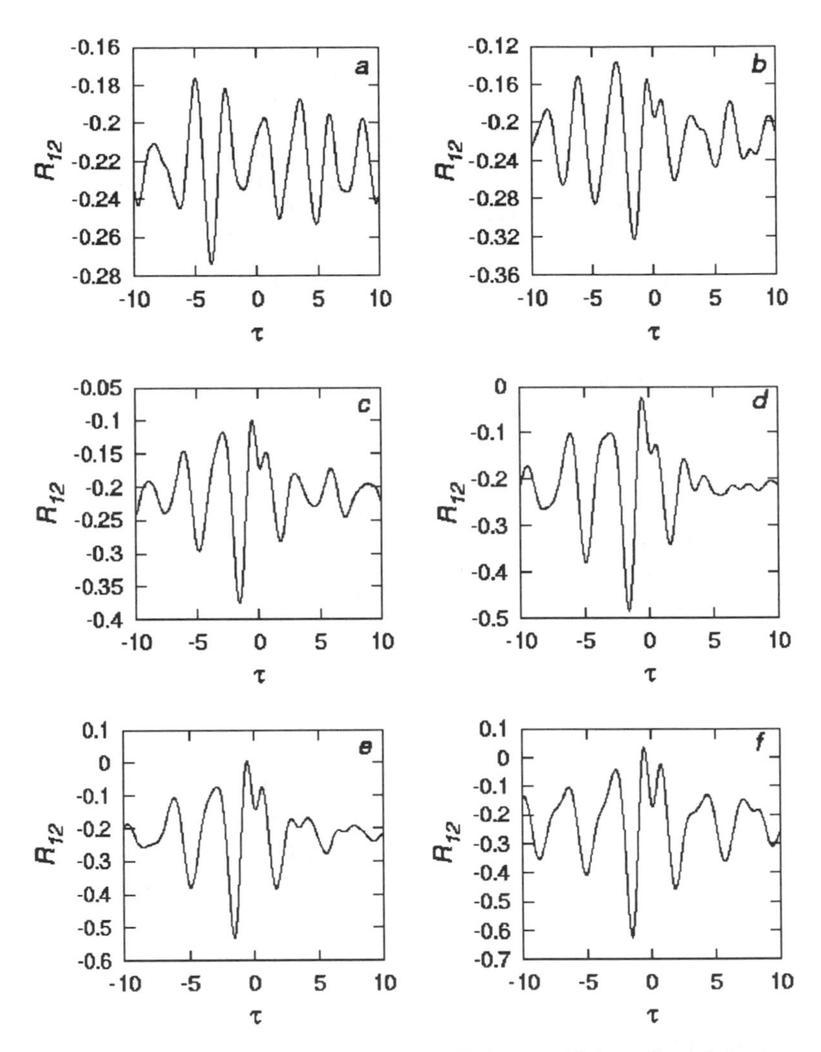

Fig. 10.11. The evolution of the cross-correlation coefficient $R_{12}(\tau)$ in the case of the system described by Eqs. (10.78): (a) $c = 0$, (b) $c = 0.5$, (c) $c = 1$, (d) $c = 1.5$, (e) $c = 2$, (f) $c = 2.5$

of the subsystems increases as the coupling factor increases. The evolution of the cross-correlation coefficient is illustrated in Fig. 10.11.

For the system described by Eqs. (10.78), in the absence of coupling the partial dimensions d_{p1} and d_{p2} differ essentially from one another. As the coupling factor increases, the correlation dimension of the coupled system d_c decreases first slowly and then, from a certain value of the coupling factor, the rate of decrease of d_c rises steeply. As a result, d_c approaches the value of the partial dimension (Fig. 10.8 b).

The smooth decrease of the correlation dimension d_c in the case of coupled identical or slightly different subsystems and the steep decrease of this dimension in the case of coupled essentially different subsystems are associated with the two mechanisms of synchronization discussed above. We see that both of the mechanisms manifest themselves for systems with chaotic oscillations as well with periodic oscillations.

10.7 Mutual synchronization of two relaxation generators

We consider below the mutual synchronization of two coupled relaxation generators of two different kinds. Depending on the kind of generator different methods are more appropriate for the calculation of synchronous regimes.

10.7.1 Mutual synchronization of two coupled relaxation generators of triangular oscillations

The problem of mutual synchronization of two coupled relaxation generators of triangular oscillations was first solved by Sidorova [307] using the point map technique. We denote the output oscillations of the first and second generators by $x(t)$ and $y(t)$, respectively. Let the durations of the direct and reverse moves for the first generator in the autonomous regime be T_1 and T_2, and, respectively, for the second one be θ_1 and θ_2. The oscillation period of the first generator is $T = T_1 + T_2$ and of the second generator is $\theta = \theta_1 + \theta_2$. Oscillations of both generators in the autonomous regime are shown in Fig. 10.12 a. The ratio of the periods $\xi = \theta/T$ plays the role of the frequency mistuning. The amplitudes of the autonomous oscillations of the generators are $A = x_2 - x_1$ and $B = y_2 - y_1$. The voltage x attenuated

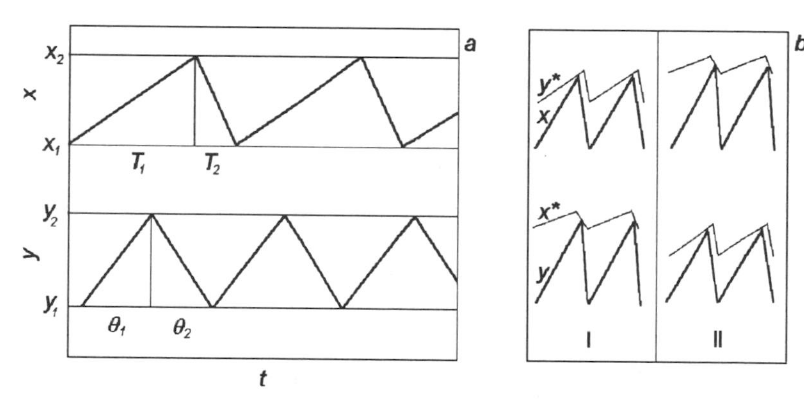

Fig. 10.12. (a) An example of oscillations of both generators in the autonomous regime, and (b) two stable synchronization regimes

by the factor k modulates the higher limit level of the generator y, and the voltage y attenuated by the factor k modulates the higher limit level of the generator x.

It is shown in [307] that four synchronization regimes are possible, but two of them are always unstable. Two stable regimes are illustrated in Fig. 10.12 b. The second regime differs from the first one in that the generators x and y change places. Therefore it is sufficient to consider only the first regime. We assume that the synchronizing voltages $x^*(t)$ and $y^*(t)$ are

$$x^*(t) = y_2 + k\left(x(t) - \frac{A}{2}\right), \quad y^*(t) = x_2 + k\left(y(t) - \frac{B}{2}\right). \tag{10.79}$$

Let us denote the durations of the direct and reverse moves of the first and second generators on the nth step by $\tau_{1n}^{(1,2)}$ and $\tau_{2n}^{(1,2)}$, and the time difference between the tops of the triangles on the same nth step by Δt_n. It is easily seen that

$$\Delta t_{n+1} - \Delta t_n = \tau_{1,n+1}^{(2)} - \tau_{1,n+1}^{(1)} + \tau_{2n}^{(2)} - \tau_{2n}^{(1)}. \tag{10.80}$$

Besides, $\tau_{1n}^{(1)}$, $\tau_{2n}^{(1)}$, $\tau_{1n}^{(2)}$ and $\tau_{2n}^{(2)}$ are related by the following evident relations:

$$\frac{\tau_{1n}^{(1)}}{T_1} = \frac{\tau_{2n}^{(1)}}{T_2}, \quad \frac{\tau_{1n}^{(2)}}{\theta_1} = \frac{\tau_{2n}^{(2)}}{\theta_2}. \tag{10.81}$$

Equating further the voltages $x(t)$ and $y(t)$ to $y^*(t)$ and $x^*(t)$ at the instants when they coincide, we obtain the equations

$$\begin{gathered} A\left(\frac{\tau_{1n}^{(1)}}{T_1} - 1\right) + \frac{kB}{2} = \frac{kB}{\theta_1}\left(\tau_{1n}^{(2)} - \Delta t_n\right), \\ B\left(\frac{\tau_{1n}^{(2)}}{\theta_1} - 1\right) + \frac{kA}{2} = kA\left(\frac{\tau_{1n}^{(1)}}{T_1} - \frac{\Delta t_n}{T_2}\right). \end{gathered} \tag{10.82}$$

The set of equations (10.80), (10.81) and (10.82) describes the point map $\tau_{1,n+1}^{(1)} = f_{11}(\tau_{1n}^{(1)}, \tau_{2n}^{(1)}, \tau_{1n}^{(2)}, \tau_{2n}^{(2)}, \Delta t_n)$, $\tau_{2,n+1}^{(1)} = f_{12}(\tau_{1n}^{(1)}, \tau_{2n}^{(1)}, \tau_{1n}^{(2)}, \tau_{2n}^{(2)}, \Delta t_n)$, $\tau_{1,n+1}^{(2)} = f_{21}(\tau_{1n}^{(1)}, \tau_{2n}^{(1)}, \tau_{1n}^{(2)}, \tau_{2n}^{(2)}, \Delta t_n)$, $\tau_{2,n+1}^{(2)} = f_{22}(\tau_{1n}^{(1)}, \tau_{2n}^{(1)}, \tau_{1n}^{(2)}, \tau_{2n}^{(2)}, \Delta t_n)$, $\Delta_{n+1}t = F(\tau_{1n}^{(1)}, \tau_{2n}^{(1)}, \tau_{1n}^{(2)}, \tau_{2n}^{(2)}, \Delta t_n)$. A fixed point of this point map corresponds to a synchronization regime. The coordinates of a fixed point are

$$\begin{gathered} \tau_{1s}^{(1)} = \frac{T_1\xi}{2}\,\frac{(2A - kB)A\theta_1 - (2B - kA)BT_2}{(\xi A - kB)A\theta_1 - (B - k\xi A)BT_2}, \\ \tau_{2s}^{(1)} = \frac{T_2}{T_1}\,\tau_{1s}^{(1)}, \quad \tau_{1s}^{(2)} = \frac{\theta_1}{\xi T_1}\,\tau_{1s}^{(1)}, \quad \tau_{2s}^{(2)} = \frac{\theta_2}{\xi T_1}\,\tau_{1s}^{(1)}, \\ \Delta t_s = \frac{\theta_1 T_2}{2k}\,\frac{(2 - k^2)AB(\xi - 1) + k(A^2\xi - B^2)}{(\xi A - kB)A\theta_1 - (B - k\xi A)BT_2}. \end{gathered} \tag{10.83}$$

The boundaries of the synchronization region can be found from the conditions for the existence and stability of the fixed point (10.83) [166]. For reasons of awkwardness, we will not do this.

10.7.2 Mutual synchronization of two Rayleigh relaxation generators

We consider two coupled Rayleigh relaxation generators described by the equations [284]

$$\begin{aligned}
\dot{x}_1 &= y_1 + c_1(x_2 - x_1),\\
\dot{y}_1 &= -x_1 + \epsilon(1 - y_1^2)y_1 + c_2(y_2 - y_1),\\
\dot{x}_2 &= \xi y_2 + c_1(x_1 - x_2),\\
\dot{y}_2 &= -x_2 + \epsilon(1 - y_2^2)y_2 + c_2(y_1 - y_2),
\end{aligned} \tag{10.84}$$

where ξ characterizes the frequency mistuning between the generators. For $c_1 = c_2 = 0$ Eqs. (10.84) reduce to two Rayleigh equations (6.28). For $\epsilon \gg 1$ we can put $\dot{y}_1$ and $\dot{y}_2$ to be equal to zero. Substituting $\tau = t/\epsilon$ and $z_i = x_i/\epsilon$ into Eqs. (10.84) we obtain

$$\begin{aligned}
\frac{\mathrm{d}z_1}{\mathrm{d}\tau} &= y_1 + \epsilon c_1(z_2 - z_1),\\
z_1 &= y_1 - y_1^3 + \frac{c_2}{\epsilon}(y_2 - y_1),\\
\frac{\mathrm{d}z_2}{\mathrm{d}\tau} &= \xi y_2 + \epsilon c_1(z_1 - z_2),\\
z_2 &= y_2 - y_2^3 + \frac{c_2}{\epsilon}(y_1 - y_2).
\end{aligned} \tag{10.85}$$

We consider the case of the weakly coupled generators when $c_1 \ll 1/\epsilon$ and $c_2 \ll \epsilon$. Substituting z_1 and z_2 into the first and third equations of (10.85), respectively, and neglecting terms of second order with respect to the coupling we find the following equations for y_1 and y_2:

$$\begin{aligned}
(1 - 3y_1^2)\frac{\mathrm{d}y_1}{\mathrm{d}\tau} &= y_1 + \epsilon c_1(y_2 - y_2^3 - y_1 + y_1^3) + \frac{c_2}{\epsilon}\left(\frac{\mathrm{d}y_1}{\mathrm{d}\tau} - \frac{\mathrm{d}y_2}{\mathrm{d}\tau}\right),\\
(1 - 3y2^2)\frac{\mathrm{d}y_2}{\mathrm{d}\tau} &= \xi y_2 + \epsilon c_1(y_1 - y_1^3 - y_2 + y_2^3) + \frac{c_2}{\epsilon}\left(\frac{\mathrm{d}y_2}{\mathrm{d}\tau} - \frac{\mathrm{d}y_1}{\mathrm{d}\tau}\right).
\end{aligned} \tag{10.86}$$

As the generators are uncoupled the shapes of oscillations of the variables z and y are as shown in Fig. 6.1. We denote them by $Z(\tau)$ and $Y(\tau)$, respectively. In the case of weak coupling under consideration the shape of oscillations of the coupled generators is slightly different from the shape of oscillations of the free generators. Therefore we can seek a solution of Eqs. (10.86) in the form

$$y_{1,2} = Y(\tau + \varphi_{1,2}), \tag{10.87}$$

where $\varphi_{1,2}$ are slowly time-varying functions. Considering (10.87) as the formulas of the change of variables we obtain from (10.86) the following equations for the functions $\varphi_1(\tau)$ and $\varphi_2(\tau)$ [166]:

$$\begin{aligned}
\frac{\mathrm{d}\varphi_1}{\mathrm{d}\tau} &= \epsilon c_1 \frac{Z(\tau+\varphi_2) - Z(\tau+\varphi_1)}{Y(\tau+\varphi_1)} \\
&\quad + \frac{c_2}{\epsilon Y(\tau+\varphi_1)} \left(\frac{Y(\tau+\varphi_1)}{1-3Y^2(\tau+\varphi_1)} - \frac{Y(\tau+\varphi_2)}{1-3Y^2(\tau+\varphi_2)} \right), \\
\frac{\mathrm{d}\varphi_2}{\mathrm{d}\tau} &= \xi - 1 + \epsilon c_1 \frac{Z(\tau+\varphi_1) - Z(\tau+\varphi_2)}{Y(\tau+\varphi_2)} \\
&\quad + \frac{c_2}{\epsilon Y(\tau+\varphi_2})\left(\frac{Y(\tau+\varphi_2)}{1-3Y^2(\tau+\varphi_2)} - \frac{Y(\tau+\varphi_1)}{1-3Y^2(\tau+\varphi_1)} \right).
\end{aligned} \tag{10.88}$$

Here it is taken into account that $Z = Y - Y^3$.

Averaging Eqs. (10.88) over the oscillation period we find the truncated equations

$$\frac{\mathrm{d}\varphi_1}{\mathrm{d}\tau} = \epsilon c_1 \Phi(\varphi_2 - \varphi_1) + \frac{c_2}{\epsilon} F(\varphi_2 - \varphi_1), \tag{10.89a}$$

$$\frac{\mathrm{d}\varphi_2}{\mathrm{d}\tau} = \Delta + \epsilon c_1 \Phi(\varphi_1 - \varphi_2) + \frac{c_2}{\epsilon} F(\varphi_1 - \varphi_2), \tag{10.89b}$$

where $\Delta = \xi - 1$ and

$$\begin{aligned}
\Phi(\varphi) &= \frac{1}{T} \int_0^T \frac{Z(\tau+\varphi) - Z(\tau)}{Y(\tau)} \, \mathrm{d}\tau, \\
F(\varphi) &= \frac{1}{T} \int_0^T \frac{1}{Y(\tau)} \left(\frac{Y(\tau)}{1-3Y^2(\tau)} - \frac{Y(\tau+\varphi)}{1-3Y^2(\tau+\varphi)} \right) \mathrm{d}\tau.
\end{aligned} \tag{10.90}$$

The plots of the functions $\Phi(\varphi)$ and $F(\varphi)$, as well as their even ($\Phi_\mathrm{e}(\varphi)$, $F_\mathrm{e}(\varphi)$) and odd constituents ($\Phi_\mathrm{o}(\varphi)$, $F_\mathrm{o}(\varphi)$) which will be needed for further calculations, are presented in Fig. 10.13 for $\epsilon = 10$. It can be seen that the dependencies of Φ and F on φ/T may be approximated by sinusoidal and sawtooth curves, respectively.

Subtracting Eq. (10.89a) from (10.89b) and denoting the phase difference $\varphi_2 - \varphi_1$ by φ we obtain the following equation for φ:

$$\frac{\mathrm{d}\varphi}{\mathrm{d}\tau} = \Delta + 2\epsilon c_1 \Phi_\mathrm{o}(\varphi) + \frac{2c_2}{\epsilon} F_\mathrm{o}(\varphi). \tag{10.91}$$

A synchronization regime corresponds to a stationary solution of Eq. (10.91), i.e., $\mathrm{d}\varphi/\mathrm{d}\tau = 0$. It can be seen from Fig. 10.13 that this equation possesses

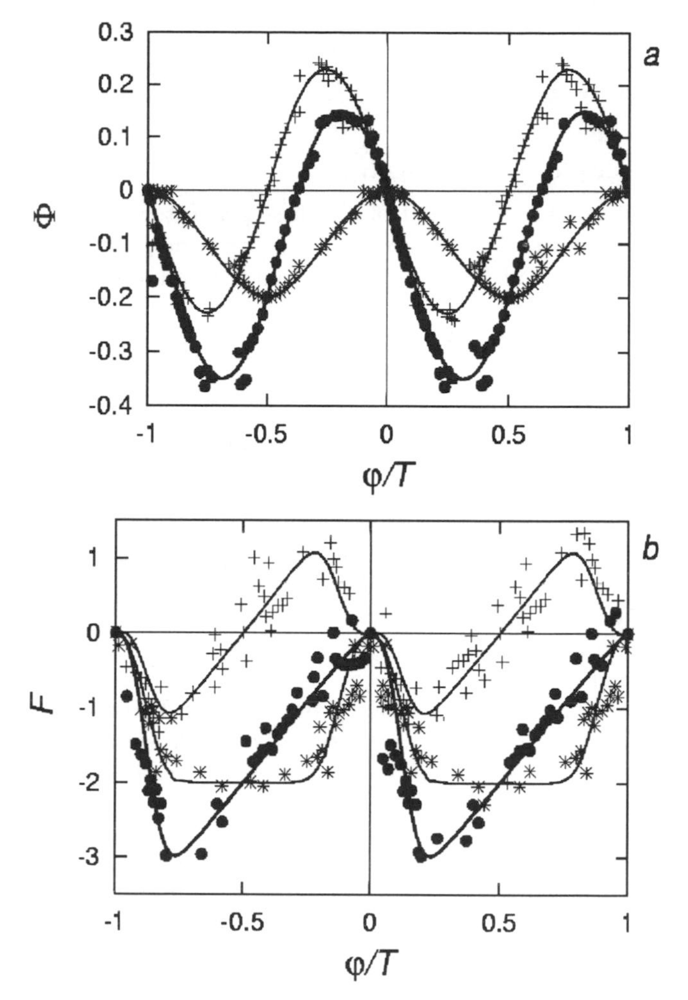

Fig. 10.13. Plots of (a) Φ (circles), Φ_e (asterisks) and Φ_o (crosses) and (b) F (circles), F_e (asterisks) and F_o (crosses) versus φ/T. The dependencies of Φ and F on φ/T may be approximated by $-0.25\sin(2\pi(x-0.065)) - .1$ and a sawtooth curve, respectively (see heavy lines)

four stationary solutions. However, only two of these solutions are stable. It is evident that the stability condition for a stationary solution $\varphi = \varphi_0$ is

$$\left(\epsilon c_1 \frac{\mathrm{d}\Phi_o}{\mathrm{d}\varphi} + \frac{c_2}{\epsilon}\frac{\mathrm{d}F_o}{\mathrm{d}\varphi}\right)\Bigg|_{\varphi=\varphi_0} \leq 0. \tag{10.92}$$

One of the stable solutions corresponds to near in-phase oscillations of the generators, whereas the other solution corresponds to near anti-phase oscillations. The synchronization region is determined from the condition

$$|\Delta| \leq \max \left| 2\epsilon c_1 \Phi_o(\varphi) + \frac{2c_2}{\epsilon} F_o(\varphi) \right| . \tag{10.93}$$

It can be easily shown that the period of the synchronous oscillations is

$$T_s = \frac{T}{1 + d\varphi_1/d\tau} = \frac{T}{1 + d\varphi_2/d\tau}, \tag{10.94}$$

where T is the oscillation period of the first generator in the autonomous regime (in the dimensionless time τ) and $d\varphi_1/d\tau = d\varphi_2/d\tau$ can be found by adding Eqs. (10.89a) and (10.89b). As a result we find

$$\frac{d\varphi_1}{d\tau} = \frac{\Delta}{2} + \epsilon c_1 \Phi_e(\varphi_0) + \frac{c_2}{\epsilon} F_e(\varphi_0). \tag{10.95}$$

Let us consider two particular cases:

- $c_1 \neq 0$, $c_2 = 0$. In this case the synchronization region is

$$|\Delta| \leq \max |2\epsilon c_1 \Phi_o(\varphi)| \approx \frac{\epsilon c_1}{2}. \tag{10.96}$$

- $c_1 = 0$, $c_2 \neq 0$. In this case the synchronization region is

$$|\Delta| \leq \max \left| \frac{2c_2}{\epsilon} F_o(\varphi) \right| \approx \frac{2c_2}{\epsilon}. \tag{10.97}$$

We see that in the first case the width of the synchronization region for relaxation generators ($\epsilon \gg 1$) is much more than for Thomsonian ones ($\epsilon \ll 1$),[3] whereas in the second case, conversely, the width of the synchronization region for generators operating in the relaxation regime is much less than for ones operating in the quasi-harmonic regime.

[3] In the case of Thomsonian generators the width of the synchronization region is approximately equal to $c_1 + c_2$.

11. Interaction of three or more self-oscillatory systems

Studies of the interaction of three or more self-oscillatory systems began many years ago [333, 263, 232, 221, 283]. In particular, interest in this phenomenon was caused by the elaborations of frequency standards based on many generators coupled in a unified synchronized system (see, e.g., [28]). The accuracy of such frequency standards was studied in [357, 150, 222, 158].

In systems of coupled generators different synchronization regimes are possible. Among these are:

- the regime when all generators execute synchronous oscillations with the same frequencies,
- the regime when different generators execute synchronous oscillations with multiple frequencies,
- the synchronization regime when the differences of the frequencies of neighboring generators are the same.

11.1 Mutual synchronization of three generators

We illustrate the above by the example of three generators described by the equations

$$\begin{aligned}
&\ddot{x}_1 - \mu_1(1 - 4\alpha_1 x_1^2 + \beta_1 x_1)\dot{x}_1 + \omega_1^2 x_1 \\
&\quad = c_{12}^{(1)} x_2 + c_{12}^{(2)} \dot{x}_2 + c_{13}^{(1)} x_3 + c_{13}^{(2)} \dot{x}_3, \\
&\ddot{x}_2 - \mu_2(1 - 4\alpha_2 x_2^2 + \beta_2 x_2)\dot{x}_2 + \omega_2^2 x_2 \\
&\quad = c_{21}^{(1)} x_1 + c_{21}^{(2)} \dot{x}_1 + c_{23}^{(1)} x_3 + c_{23}^{(2)} \dot{x}_3, \\
&\ddot{x}_3 - \mu_3(1 - 4\alpha_3 x_3^2 + \beta_3 x_3)\dot{x}_3 + \omega_3^2 x_3 \\
&\quad = c_{31}^{(1)} x_1 + c_{31}^{(2)} \dot{x}_1 + c_{32}^{(1)} x_2 + c_{32}^{(2)} \dot{x}_2.
\end{aligned} \tag{11.1}$$

11.1.1 The case of close frequencies

If $\mu_{1,2,3}$ and $c_{ij}^{(1,2)}$ are sufficiently small and the frequencies ω_1, ω_2, and ω_3 are close, then a solution of Eqs. (11.1) can be sought as $x_{1,2,3} = A_{1,2,3}\cos(\omega t + \varphi_{1,2,3})$, where $\omega = (\omega_1 + \omega_2 + \omega_3)/3$, and $A_{1,2,3}$ and $\varphi_{1,2,3}$ are slowly varying functions. The truncated equations for the amplitudes $A_{1,2,3}$ and the phases $\varphi_{1,2,3}$ are

$$
\begin{aligned}
\dot{A}_1 &= \frac{\mu_1}{2}(1-\alpha_1 A_1^2)A_1 + c_{12}A_2\sin(\Phi_{12}+\vartheta_{12}) + c_{13}A_3\sin(\Phi_{13}+\vartheta_{13}),\\
\dot{A}_2 &= \frac{\mu_2}{2}(1-\alpha_2 A_2^2)A_2 + c_{21}A_1\sin(\Phi_{21}+\vartheta_{21}) + c_{23}A_3\sin(\Phi_{23}+\vartheta_{23}),\\
\dot{A}_3 &= \frac{\mu_3}{2}(1-\alpha_3 A_3^2)A_3 + c_{31}A_1\sin(\Phi_{31}+\vartheta_{31}) + c_{32}A_2\sin(\Phi_{32}+\vartheta_{32}),
\end{aligned}
\tag{11.2}
$$

$$
\begin{aligned}
\dot{\varphi}_1 &= \Delta_1 + \frac{c_{12}A_2}{A_1}\cos(\Phi_{12}+\vartheta_{12}) + \frac{c_{13}A_3}{A_1}\cos(\Phi_{13}+\vartheta_{13}),\\
\dot{\varphi}_2 &= \Delta_2 + \frac{c_{21}A_1}{A_2}\cos(\Phi_{12}-\vartheta_{21}) + \frac{c_{23}A_3}{A_2}\cos(\Phi_{23}+\vartheta_{23}),\\
\dot{\varphi}_3 &= \Delta_3 + \frac{c_{31}A_1}{A_3}\cos(\Phi_{13}-\vartheta_{31}) + \frac{c_{32}A_2}{A_3}\cos(\Phi_{23}-\vartheta_{32}),
\end{aligned}
\tag{11.3}
$$

where $\Phi_{ij} = \varphi_i - \varphi_j$, $\Delta_1 = \dfrac{2\omega_1-(\omega_2+\omega_3)}{3}$, $\Delta_2 = \dfrac{2\omega_2-(\omega_1+\omega_3)}{3}$, $\Delta_3 = \dfrac{2\omega_3-(\omega_1+\omega_2)}{3}$, $c_{ij} = \dfrac{1}{2}\sqrt{\left(c_{ij}^{(1)}\right)^2/\omega^2 + \left(c_{ij}^{(2)}\right)^2}$, $\tan\vartheta_{ij} = -\dfrac{\omega c_{ij}^{(2)}}{c_{ij}^{(1)}}$, $i, j = 1, 2, 3$. Taking account of the relation $\Phi_{13} = \Phi_{12} + \Phi_{23}$, it is possible to eliminate one equation from Eqs. (11.3).

In the case of weak coupling between the generators, when $c_{ij} \ll \mu_{ij}$, we can put in Eqs. (11.3) $A_i = A_{i0} = 1/\sqrt{\alpha_i}$. As a result we obtain the following equations for Φ_{12} and Φ_{23} in the synchronization regime:

$$
\begin{aligned}
&\Delta_{12} + \frac{c_{12}A_{20}}{A_{10}}\cos(\Phi_{12}+\vartheta_{12}) + \frac{c_{13}A_{30}}{A_{10}}\cos(\Phi_{12}+\Phi_{23}+\vartheta_{13})\\
&\quad - \frac{c_{21}A_{10}}{A_{20}}\cos(\Phi_{12}-\vartheta_{21}) - \frac{c_{23}A_{30}}{A_{20}}\cos(\Phi_{23}+\vartheta_{23}) = 0,\\
&\Delta_{23} + \frac{c_{21}A_{10}}{A_{20}}\cos(\Phi_{12}-\vartheta_{21}) + \frac{c_{23}A_{30}}{A_{20}}\cos(\Phi_{23}+\vartheta_{23})\\
&\quad - \frac{c_{31}A_{10}}{A_{30}}\cos(\Phi_{12}+\Phi_{23}-\vartheta_{31}) - \frac{c_{32}A_{20}}{A_{30}}\cos(\Phi_{23}-\vartheta_{32}) = 0,
\end{aligned}
\tag{11.4}
$$

where $\Delta_{ij} = \Delta_i - \Delta_j = \omega_i - \omega_j$.

Analytical solutions of Eqs. (11.4) can be found only in specific cases. We consider two such cases: (i) the generators are identical and $\Delta_1 = \Delta_2 = \Delta_3 = 0$, (ii) the generators are coupled in a chain, i.e., each of the generators has an effect only on the neighboring one.

(i) Setting in Eqs. (11.4) $A_{10} = A_{20} = A_{30}$, $c_{12} = c_{21} = c_{13} = c_{31} = c_{23} = c_{32} = c$, $\vartheta_{12} = \vartheta_{21} = \vartheta_{13} = \vartheta_{31} = \vartheta_{23} = \vartheta_{32} = \vartheta$, and $\omega_1 = \omega_2 = \omega_3$, we obtain

$$\sin\frac{\Phi_{12}}{2}\left[2\sin\vartheta\cos\frac{\Phi_{12}}{2}+\sin\left(\Phi_{23}+\frac{\Phi_{12}}{2}+\vartheta\right)\right]=0,$$
$$\sin\frac{\Phi_{23}}{2}\left[2\sin\vartheta\cos\frac{\Phi_{23}}{2}-\sin\left(\Phi_{12}+\frac{\Phi_{23}}{2}-\vartheta\right)\right]=0. \tag{11.5}$$

Equations (11.5) possess the following solutions:

$$\Phi_{12}=\Phi_{23}=\Phi_{13}=0, \tag{11.6a}$$
$$\Phi_{12}=\Phi_{23}=-\Phi_{13}=\pm 2\pi/3, \tag{11.6b}$$
$$\Phi_{12}=0, \Phi_{13}=\Phi_{23}=\pm 2\arctan(3\tan\vartheta). \tag{11.6c}$$

For the investigation of stability of the solutions found we should find the steady-state oscillation amplitudes in the first approximation with respect to c/μ and substitute them into Eqs. (11.3). As a result we find that solution (11.6c) is stable only for very small values of $|\vartheta|$, when $-c/\mu \leq \sin\vartheta \leq 3c/\mu$, whereas solutions (11.6a) and (11.6b) are stable for $\sin\vartheta \geq -3c/\mu$ and $\sin\vartheta \leq 3c/(2\mu)$, respectively. Thus, for sufficiently small $|\vartheta|$ all the solutions (11.6a), (11.6b) and (11.6c) are stable. The first solution corresponds to in-phase oscillations, whereas the second solution corresponds to the regime when there is the same phase shift between all the generators which is equal to $2\pi/3$.

It should be noted that even in the case of zero mistuning, which we consider, the frequency of synchronous oscillations does not coincide with the frequency of the autonomous generators. It follows from Eqs. (11.3) that for the in-phase regime the correction to the frequency is equal to $2c\cos\vartheta$, and for the regime with the phase shift the correction to the frequency is equal to $-c\cos\vartheta$.

(ii) In this case the coupling factors c_{13} and c_{31} are equal to zero and Eqs. (11.4) take the form

$$\Delta_{12}+\frac{c_{12}A_{20}}{A_{10}}\cos(\Phi_{12}+\vartheta_{12})-\frac{c_{21}A_{10}}{A_{20}}\cos(\Phi_{12}-\vartheta_{21}) -\frac{c_{23}A_{30}}{A_{20}}\cos(\Phi_{23}+\vartheta_{23})=0,$$
$$\Delta_{23}+\frac{c_{21}A_{10}}{A_{20}}\cos(\Phi_{12}-\vartheta_{21})+\frac{c_{23}A_{30}}{A_{20}}\cos(\Phi_{23}+\vartheta_{23}) -\frac{c_{32}A_{20}}{A_{30}}\cos(\Phi_{23}-\vartheta_{32})=0. \tag{11.7}$$

We find a solution of these equations only for $\vartheta_{12}=\vartheta_{21}=\vartheta_{23}=\vartheta_{32}=0$ and for $\vartheta_{12}=\vartheta_{21}=\vartheta_{23}=\vartheta_{32}=-\pi/2$.

In the case of $\vartheta_{12}=\vartheta_{21}=\vartheta_{23}=\vartheta_{32}=0$ Eqs. (11.7) become

$$\begin{aligned} &\Delta_{12} + C_{12}\cos\Phi_{12} - \frac{c_{23}A_{30}}{A_{20}}\cos\Phi_{23} = 0, \\ &\Delta_{23} + \frac{c_{21}A_{10}}{A_{20}}\cos\Phi_{12} + C_{23}\cos\Phi_{23} = 0, \end{aligned} \tag{11.8}$$

where

$$C_{12} = \frac{c_{12}A_{20}}{A_{10}} - \frac{c_{21}A_{10}}{A_{20}}, \quad C_{23} = \frac{c_{23}A_{30}}{A_{20}} - \frac{c_{32}A_{20}}{A_{30}}.$$

A solution of Eqs. (11.8) is

$$\begin{aligned} \cos\Phi_{12} &= -\frac{(\Delta_{12}+\Delta_{23})c_{23}A_{30}^2 - \Delta_{12}c_{32}A_{20}^2}{c_{12}c_{23}A_{30}^2 - c_{12}c_{32}A_{20}^2 + c_{21}c_{32}A_{10}^2}\,\frac{A_{10}}{A_{20}}, \\ \cos\Phi_{23} &= -\frac{\Delta_{23}c_{12}A_{20}^2 - (\Delta_{12}+\Delta_{23})c_{21}A_{10}^2}{c_{12}c_{23}A_{30}^2 - c_{12}c_{32}A_{20}^2 + c_{21}c_{32}A_{10}^2}\,\frac{A_{30}}{A_{20}}. \end{aligned} \tag{11.9}$$

Let us first find the synchronization region boundaries for a fixed value of Δ_{12}. If

$$\left|\frac{c_{12}A_{20}^2 - c_{21}A_{10}^2}{c_{23}A_{10}A_{30}}\right| \le 1,$$

then the synchronization region boundaries $\Delta_{23}^{(\pm)}$ for a fixed value of Δ_{12} should be found from the condition $|\cos\Phi_{12}| = 1$. As a result we obtain

$$\Delta_{23}^{(\pm)} = -\left(1 - \frac{c_{32}A_{20}^2}{c_{23}A_{30}^2}\right)\Delta_{12} \pm \frac{A_{20}}{A_{10}}\left|c_{12}\left(1 - \frac{c_{32}A_{20}^2}{c_{23}A_{30}^2}\right) + \frac{c_{21}c_{32}A_{10}^2}{c_{23}A_{30}^2}\right|. \tag{11.10}$$

But if

$$\left|\frac{c_{12}A_{20}^2 - c_{21}A_{10}^2}{c_{23}A_{10}A_{30}}\right| \ge 1,$$

then the synchronization region boundaries $\Delta_{23}^{(\pm)}$ for a fixed value of Δ_{12} should be found from the condition $|\cos\Phi_{23}| = 1$. As a result we obtain

$$\Delta_{23}^{(\pm)} = -\frac{c_{21}A_{10}^2}{c_{12}A_{20}^2 - c_{21}A_{10}^2}\Delta_{12} \pm \frac{A_{20}}{A_{30}}\left|\frac{c_{12}(c_{23}A_{30}^2 - c_{32}A_{20}^2) + c_{21}c_{32}A_{10}^2}{c_{12}A_{20}^2 - c_{21}A_{10}^2}\right|. \tag{11.11}$$

In both of these cases the width of the synchronization region is independent of Δ_{12}. For identical generators we find from (11.10) that $\Delta_{23}^{(\pm)} = \pm c$.

In a similar manner, we can calculate the synchronization region boundaries for a fixed value of Δ_{23}. If

$$\left|\frac{c_{23}A_{30}^2 - c_{32}A_{20}^2}{c_{21}A_{10}A_{30}}\right| \le 1,$$

then the synchronization region boundaries $\Delta_{12}^{(\pm)}$ for a fixed value of Δ_{23} should be found from the condition $|\cos\Phi_{23}| = 1$. As a result we obtain

$$\Delta_{12}^{(\pm)} = -\left(1 - \frac{c_{12}A_{20}^2}{c_{21}A_{10}^2}\right)\Delta_{23} \pm \frac{A_{20}}{A_{30}}\left|c_{32}\left(1 - \frac{c_{12}A_{20}^2}{c_{21}A_{10}^2}\right) + \frac{c_{12}c_{23}A_{30}^2}{c_{21}A_{10}^2}\right|. \tag{11.12}$$

Otherwise the synchronization region boundaries $\Delta_{12}^{(\pm)}$ for a fixed value of Δ_{23} should be found from the condition $|\cos\Phi_{12}| = 1$. As a result we obtain

$$\Delta_{12}^{(\pm)} = -\frac{c_{23}A_{30}^2}{c_{23}A_{30}^2 - c_{32}A_{20}^2}\Delta_{23} \pm \frac{A_{20}}{A_{10}}\left|\frac{c_{12}(c_{23}A_{30}^2 - c_{32}A_{20}^2) + c_{21}c_{32}A_{10}^2}{c_{23}A_{30}^2 - c_{32}A_{20}^2}\right|. \tag{11.13}$$

It follows from the results obtained that the synchronization region on the plane Δ_{12}, Δ_{23} is enclosed by the parallelogram with sides described by the equations (11.10) (or (11.11)) and (11.12) (or (11.13)), see Fig. 11.1 a. Only in the specific case of identical generators the parallelogram degenerates into

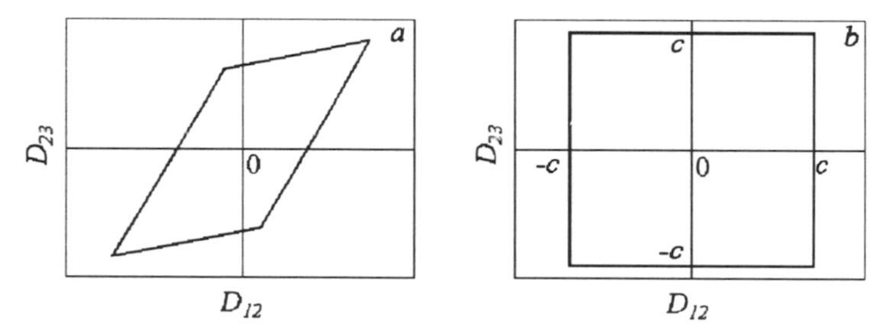

Fig. 11.1. The synchronization regions for $\vartheta_{12} = \vartheta_{21} = \vartheta_{23} = \vartheta_{32} = 0$: (a) the general case, (b) the case of identical generators

a square, see Fig. 11.1 b.

In the case of $\vartheta_{12} = \vartheta_{21} = \vartheta_{23} = \vartheta_{32} = -\pi/2$ Eqs. (11.7) become

$$\begin{aligned} &\Delta_{12} + D_{12}\sin\Phi_{12} - \frac{c_{23}A_{30}}{A_{20}}\sin\Phi_{23} = 0, \\ &\Delta_{23} - \frac{c_{21}A_{10}}{A_{20}}\sin\Phi_{12} + D_{23}\sin\Phi_{23} = 0, \end{aligned} \tag{11.14}$$

where

$$D_{12} = \frac{c_{12}A_{20}}{A_{10}} + \frac{c_{21}A_{10}}{A_{20}}, \quad D_{23} = \frac{c_{23}A_{30}}{A_{20}} + \frac{c_{32}A_{20}}{A_{30}}.$$

A solution of Eqs. (11.14) is

$$\sin\Phi_{12} = -\frac{(\Delta_{12}+\Delta_{23})c_{23}A_{30}^2 + \Delta_{12}c_{32}A_{20}^2}{c_{12}c_{23}A_{30}^2 + c_{12}c_{32}A_{20}^2 + c_{21}c_{32}A_{10}^2}\frac{A_{10}}{A_{20}},$$

$$\sin\Phi_{23} = -\frac{\Delta_{23}c_{12}A_{20}^2 + (\Delta_{12}+\Delta_{23})c_{21}A_{10}^2}{c_{12}c_{23}A_{30}^2 + c_{12}c_{32}A_{20}^2 + c_{21}c_{32}A_{10}^2}\frac{A_{30}}{A_{20}}. \tag{11.15}$$

The synchronization region boundaries for a fixed value of Δ_{12} and for a fixed value of Δ_{23} can be found in perfect analogy to the preceding case. If

$$\left|\frac{c_{23}A_{30}^2 + c_{32}A_{20}^2}{c_{21}A_{10}A_{30}}\right| \le 1,$$

then the synchronization region boundaries $\Delta_{12}^{(\pm)}$ for a fixed value of Δ_{23} should be found from the condition $|\sin\Phi_{23}| = 1$. As a result we obtain

$$\Delta_{12}^{(\pm)} = -\left(1+\frac{c_{12}A_{20}^2}{c_{21}A_{10}^2}\right)\Delta_{23} \pm \frac{A_{20}}{A_{30}}\left|c_{32}\left(1+\frac{c_{12}A_{20}^2}{c_{21}A_{10}^2}\right)+\frac{c_{12}c_{23}A_{30}^2}{c_{21}A_{10}^2}\right|. \tag{11.16}$$

But if

$$\left|\frac{c_{23}A_{30}^2 + c_{32}A_{20}^2}{c_{21}A_{10}A_{30}}\right| \ge 1,$$

then the synchronization region boundaries $\Delta_{12}^{(\pm)}$ for a fixed value of Δ_{23} should be found from the condition $|\sin\Phi_{12}| = 1$. As a result we obtain

$$\Delta_{12}^{(\pm)} = -\frac{c_{23}A_{30}^2}{c_{23}A_{30}^2 + c_{32}A_{20}^2}\Delta_{23} \pm \frac{A_{20}}{A_{10}}\left|\frac{c_{12}(c_{23}A_{30}^2 + c_{32}A_{20}^2) + c_{21}c_{32}A_{10}^2}{c_{23}A_{30}^2 + c_{32}A_{20}^2}\right|. \tag{11.17}$$

Similarly, if

$$\left|\frac{c_{12}A_{20}^2 + c_{21}A_{10}^2}{c_{23}A_{10}A_{30}}\right| \le 1,$$

then the synchronization region boundaries $\Delta_{23}^{(\pm)}$ for a fixed value of Δ_{12} should be found from the condition $|\sin\Phi_{12}| = 1$. As a result we obtain

$$\Delta_{23}^{(\pm)} = -\left(1+\frac{c_{32}A_{20}^2}{c_{23}A_{30}^2}\right)\Delta_{12} \pm \frac{A_{20}}{A_{10}}\left|c_{12}\left(1+\frac{c_{32}A_{20}^2}{c_{23}A_{30}^2}\right)+\frac{c_{32}c_{21}A_{10}^2}{c_{23}A_{30}^2}\right|. \tag{11.18}$$

Otherwise the synchronization region boundaries $\Delta_{23}^{(\pm)}$ for a fixed value of Δ_{12} should be found from the condition $|\sin\Phi_{23}| = 1$. As a result we find

$$\Delta_{23}^{(\pm)} = -\frac{c_{21}A_{10}^2}{c_{12}A_{20}^2 + c_{21}A_{10}^2}\Delta_{12} \pm \frac{A_{20}}{A_{30}}\left|\frac{c_{12}(c_{23}A_{30}^2 + c_{32}A_{20}^2) + c_{21}c_{32}A_{10}^2}{c_{12}A_{20}^2 + c_{21}A_{10}^2}\right|. \tag{11.19}$$

For identical generators we find from (11.17) and (11.19) that $\Delta_{12}^{(\pm)} = -\Delta_{23}/2 \pm 3c/2$, $\Delta_{23}^{(\pm)} = -\Delta_{12}/2 \pm 3c/2$.

Like the preceding case, the synchronization region on the plane Δ_{12}, Δ_{23} is enclosed by the parallelogram with sides described by the equations (11.16) (or (11.17)) and (11.18) (or (11.19)), see Fig. 11.2 a. But in the specific case

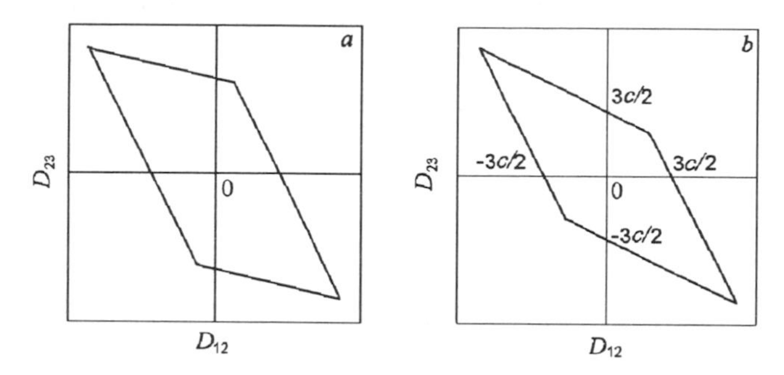

Fig. 11.2. The synchronization regions for $\vartheta_{12} = \vartheta_{21} = \vartheta_{23} = \vartheta_{32} = -\pi/2$: (a) the general case, (b) the case of identical generators

of identical generators the parallelogram degenerates into a rhombus, see Fig. 11.2 b, but not into a square.

11.1.2 The case of close differences of the frequencies of neighboring generators

We consider here three generators described by Eqs. (11.1) with $c_{ij}^{(2)} = 0$. Let the frequencies of these generators satisfy the relation

$$\omega_2 - \omega_1 = \omega_3 - \omega_2 + \Delta \equiv \Omega + \Delta, \tag{11.20}$$

where $\Omega = \omega_3 - \omega_2 \gg \mu_{1,2,3}$ and $\Delta \ll \Omega$. In this case, as shown below, there exists a synchronization regime when the difference between the frequencies of the first generator and the second one is equal to the difference between the frequencies of the second generator and the third one.

Owing to the condition $\Omega \gg \mu_{1,2,3}$ we can suppose that oscillations of each of the generators are a sum of self-oscillations at the frequency close to the natural one and of small forced oscillations at the frequencies of two other generators. If the self-oscillations in the ith generator are set as $A_i \cos(\omega_i t + \varphi_i)$, then the forced oscillations in this generator at the frequency of the jth generator are approximately equal to $k_{ij} A_j \cos(\omega_j t + \varphi_j)$, where $k_{ij} = c_{ij}^{(1)}/(\omega_i^2 - \omega_j^2)$. We assume that $|k_{ij}| \ll 1$ for all i and j. Thus, a solution of Eqs. (11.1) can be sought as

$$
\begin{aligned}
x_1 &= A_1\cos(\omega_1 t+\varphi_1)+k_{12}A_2\cos(\omega_2 t+\varphi_2)+k_{13}A_3\cos(\omega_3 t+\varphi_3),\\
x_2 &= A_2\cos(\omega_2 t+\varphi_2)+k_{21}A_1\cos(\omega_1 t+\varphi_1)+k_{23}A_3\cos(\omega_3 t+\varphi_3),\\
x_3 &= A_3\cos(\omega_3 t+\varphi_3)+k_{31}A_1\cos(\omega_1 t+\varphi_1)+k_{32}A_2\cos(\omega_2 t+\varphi_2).
\end{aligned}
\tag{11.21}
$$

Substituting (11.21) into Eqs. (11.1) and using the Krylov–Bogolyubov method we can find the truncated equations for the amplitudes and phases. Correct to second order with respect to the coupling factor, these equations are

$$
\begin{aligned}
\dot{A}_1 &= \frac{\mu_1}{2}\left(1-\frac{A_1^2}{A_{10}^2}\right)A_1, \quad \dot{A}_3 = \frac{\mu_3}{2}\left(1-\frac{A_3^2}{A_{30}^2}\right)A_3,\\
\dot{A}_2 &= \frac{\mu_2}{2}\left(1-\frac{A_2^2}{A_{20}^2}\right)A_2 - \frac{A_1A_2A_3}{A_{20}^2}\,\rho\cos\Phi,
\end{aligned}
\tag{11.22}
$$

$$
\dot{\varphi}_1 = 0, \quad \dot{\varphi}_3 = 0, \quad \dot{\varphi}_2 = \frac{A_1A_3}{A_{20}^2}\,\rho\sin\Phi,
\tag{11.23}
$$

where $\rho = \mu_2 k_{21} k_{23}$, $\Phi = \Delta t + 2\varphi_2 - \varphi_1 - \varphi_3$. The equation for Φ follows from (11.23):

$$
\dot{\Phi} = \Delta + 2\,\frac{A_1A_3}{A_{20}^2}\,\rho\sin\Phi.
\tag{11.24}
$$

By substituting into (11.24) a steady-state solution of Eqs. (11.22) we obtain

$$
\dot{\Phi} = \Delta + \Delta_{\mathrm{s}}\sin\Phi,
\tag{11.25}
$$

where

$$
\Delta_{\mathrm{s}} = 2\,\frac{A_{10}A_{30}}{A_{20}^2}\,\rho
$$

is the half-width of the synchronization region. We note that Δ_{s} is of second order with respect to the coupling factors.

11.2 Synchronization of N coupled generators with close frequencies

11.2.1 Synchronization of N coupled van der Pol generators

We first consider synchronization of N coupled van der Pol generators described by the equations

$$
\ddot{x}_i - \mu_i(1-\alpha_i x_i^2)\dot{x}_i + \omega_i^2 x_i = \sum_{j\neq i}\left(c_{ij}^{(1)}x_j + c_{ij}^{(2)}\dot{x}_j + c_{ij}^{(3)}\ddot{x}_j\right)
\tag{11.26}
$$

$$
(i = 1, 2, \ldots, N).
$$

We assume that the frequencies ω_i are close to each other, i.e., $|\omega_i - \omega_j| \ll \omega_k$ for all i, j and k. Then a solution of Eqs. (11.26) can be sought in the form

$$x_i = A_i \cos(\omega t + \varphi_i), \quad \dot{x}_i = -A_i \omega \sin(\omega t + \varphi_i), \tag{11.27}$$

where $\omega = (1/N) \sum_{i=1}^{N} \omega_i$. By using the averaging method we obtain the following truncated equations for A_i and φ_i:

$$\dot{A}_i = \frac{\mu_i}{2} \left(1 - \frac{A_i^2}{A_{i0}^2}\right) + \sum_{j \neq i} c_{ij} A_j \sin(\Phi_{ij} + \vartheta_{ij}), \tag{11.28}$$

$$\dot{\varphi}_i = \Delta_i + \sum_{j \neq i} c_{ij} \frac{A_j}{A_i} \cos(\Phi_{ij} + \vartheta_{ij}), \tag{11.29}$$

where $\Phi_{ij} = \varphi_i - \varphi_j$, $\Delta_i = \omega_i - \omega$,

$$c_{ij} = \frac{1}{2\omega} \sqrt{\left(c_{ij}^{(1)} - \omega^2 c_{ij}^{(3)}\right)^2 + \left(c_{ij}^{(1)}\right)^2},$$

$$\sin \vartheta_{ij} = \frac{c_{ij}^{(2)}}{2c_{ij}}, \quad \cos \vartheta_{ij} = -\frac{c_{ij}^{(1)} - \omega^2 c_{ij}^{(3)}}{2c_{ij}}.$$

Assuming the coupling to be weak and substituting into Eqs. (11.29) A_{i0} in place of A_i we find the following equations for Φ_{ij} in the synchronization regime:

$$\begin{aligned} &\Delta_{i,i+1} + \sum_{j \neq i} c_{ij} \frac{A_{j0}}{A_{i0}} \cos(\Phi_{ij} + \vartheta_{ij}) \\ &\qquad - \sum_{j \neq i+1} c_{i+1,j} \frac{A_{j0}}{A_{i+1,0}} \cos(\Phi_{i+1,j} + \vartheta_{i+1,j}) = 0, \end{aligned} \tag{11.30}$$

where $i = 1, 2, \ldots, N-1$, $\Delta_{i,i+1} = \omega_i - \omega_{i+1} = \Delta_i - \Delta_{i+1}$.

An analytical solution of Eqs. (11.30) can be found only in the specific case of the identical generators, when $A_{i0} = A_0$, $c_{ij} = c$, $\vartheta_{ij} = \vartheta$ and $\omega_i = \omega$ for all i and j. In this case Eqs. (11.30) become

$$\sin \frac{\Phi_{i,i+1}}{2} \left(\sum_{j \neq i,i+1} \sin\left(\frac{\Phi_{ij} + \Phi_{i+1,j}}{2} + \vartheta\right) + 2 \sin \vartheta \cos \frac{\Phi_{i,i+1}}{2} \right) = 0$$
$$(i = 1, 2, \ldots, N-1). \tag{11.31}$$

Partial solutions of Eqs. (11.30) are

$$\Phi_{ij} = 0, \tag{11.32a}$$

$$\Phi_{ij} = \pm (j - i) \frac{2\pi}{N}. \tag{11.32b}$$

By analogy with the case of three generators, one might expect that solutions (11.32a) and (11.32b) are stable for $\sin \vartheta \geq -Nc/\mu$ and $\sin \vartheta \leq Nc/(2\mu)$, respectively. The frequency of the synchronous oscillations in the in-phase regime is equal to $\omega + (N-1)c \cos \vartheta$, whereas in the second regime it is equal to $\omega - c \cos \vartheta$.

11.2.2 Synchronization of pendulum clocks suspended from a common beam

The problem of synchronization of pendulum clocks suspended from a common beam (Fig. 11.3), going back to studies by Huygens [126], differs from

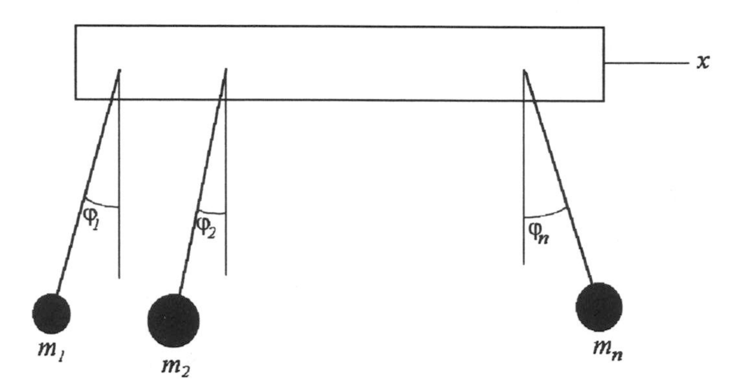

Fig. 11.3. The schematic image of pendulum clocks suspended from a common beam

that considered above in the characteristic of coupling only: the clocks are coupled via the beam but not directly. Nevertheless, this difference is not crucial and both problems are conceptually equivalent.

For the sake of simplicity assume that the oscillations of clock pendulums are described by the van der Pol equations and the beam moves as a whole along the x-axis. With these suppositions the equations of small oscillations of the pendulums and the beam are

$$\ddot{\varphi}_k + \omega_k^2 \varphi_k = \frac{m_k l_k}{I_k} \ddot{x} + \mu_k (1 - 4\alpha_k \varphi_k^2) \dot{\varphi}_k \quad (k = 1, 2, \ldots, n), \tag{11.33a}$$

$$\ddot{x} + 2\delta \dot{x} + \Omega^2 x = \sum_{k=1}^{n} \frac{m_k l_k}{M} \ddot{\varphi}_k, \tag{11.33b}$$

where φ_k is the angle between the kth pendulum and the vertical, ω_k, m_k, l_k and I_k are the natural frequency, mass, length and moment of inertia of the kth pendulum, respectively, M is the mass of the beam, Ω and δ are the natural frequency and damping factor of the beam oscillations.

If the natural frequencies of the pendulums are close to each other, then in the synchronization regime a solution of Eqs. (11.33a) can be sought as $\varphi_k = A_k \cos(\omega t + \vartheta_k)$, where $\omega = \sum_{k=1}^{n} \omega_k / n$. In this case a stationary solution of Eq. (11.33b) is

$$x = A \cos(\omega t + \vartheta), \tag{11.34}$$

where

$$A = \left\{ \frac{\omega^4}{(\Omega^2 - \omega^2)^2 + 4\delta^2\omega^2} \left[\left(\sum_{k=1}^{n} \frac{m_k l_k}{M} A_k \cos(\vartheta_k + \chi) \right)^2 + \left(\sum_{k=1}^{n} \frac{m_k l_k}{M} A_k \sin(\vartheta_k + \chi) \right)^2 \right] \right\}^{1/2}, \tag{11.35}$$

$$\tan\vartheta = \frac{\sum_{k=1}^{n} m_k l_k A_k \sin(\vartheta_k + \chi)}{\sum_{k=1}^{n} m_k l_k A_k \cos(\vartheta_k + \chi)}, \quad \sin\chi = -\frac{2\delta\omega}{\sqrt{(\Omega^2 - \omega^2)^2 + 4\delta^2\omega^2}}.$$

In view of (11.34) and (11.35), we obtain the following truncated equations for A_k and ϑ_k:

$$\begin{aligned} \dot{A}_k &= \frac{\mu_k}{2}(1 - \alpha_k A_k^2)A_k + \sum_{j=1}^{n} c_{kj} A_j \sin(\Phi_{kj} + \chi), \\ \dot{\vartheta}_k &= \Delta_k + \sum_{j=1}^{n} c_{kj} \frac{A_j}{A_k} \cos(\Phi_{kj} + \chi), \end{aligned} \tag{11.36}$$

where

$$c_{kj} = \frac{m_k m_j l_k l_j \omega^3}{2 I_k M \sqrt{(\Omega^2 - \omega^2)^2 + 4\delta^2\omega^2}}$$

is the coupling factor between the kth and jth clocks, $\Delta_k = \omega_k - \omega$ is the frequency mistuning for the kth pendulum, and $\Phi_{kj} = \vartheta_k - \vartheta_j$ is the corresponding phase shift.

Equations (11.36) are conveniently rewritten as

$$\dot{A}_k = \frac{\mu_k'}{2}\left(1 - \frac{A_k^2}{A_{k0}^2}\right) A_k + \sum_{j \neq k} c_{kj} A_j \sin(\Phi_{kj} + \chi), \tag{11.37}$$

$$\dot{\vartheta}_k = \Delta_k' + \sum_{j \neq k} c_{kj} \frac{A_j}{A_k} \cos(\Phi_{kj} + \chi), \tag{11.38}$$

where $\mu_k' = \mu_k + 2c_{kk}\sin\chi$, $A_{k0}^2 = \mu_k'/(\alpha_k\mu_k)$, $\Delta_k' = \Delta_k + c_{kk}\cos\chi$. These equations coincide in form with Eqs. (11.28), (11.29). Therefore all the results obtained in the preceding section are also valid in the case under consideration. In particular, if the clocks are identical and the frequency Ω is not much different from ω, then oscillations with the phase shift $\pm 2\pi/N$ have to be stable, because $\sin\vartheta < 0$. This correlates with Huygens' observations [126] concerning with the behavior of two pendulum clocks. However, away from resonance, when $|\sin\vartheta|$ is sufficiently small, both in-phase and anti-phase oscillations of two pendulum clocks are stable. Such oscillations were observed by Blekhman [46].

11.3 Synchronization and chaotization of self-oscillations in chains of coupled generators

11.3.1 Synchronization of N van der Pol generators coupled in a chain

Assuming that the generators each have an effect only on the succeeding ones and that the parameters of all generators except their natural frequencies are identical, we write the equations for the chain as

$$\ddot{x}_j - \mu(1 - 4\alpha x_j^2)\dot{x}_j + \omega_j^2 x_j = c_1 x_{j-1} + c_2 \dot{x}_{j-1} \quad (j = 1, 2, \ldots, N). \tag{11.39}$$

Using this problem as an example, we demonstrate the Krylov–Bogolyubov method in a complex form.

If the coupling factors c_1 and c_2 are sufficiently small and $\mu \ll \omega_j$ then a solution of Eqs. (11.39) can be sought in the form

$$x_j = \frac{1}{2}\left(R_j \mathrm{e}^{i\omega t} + \text{ c.c.}\right), \tag{11.40}$$

where $\omega = (1/N)\sum_{j=1}^{N}\omega_j$, and R_j is the complex oscillation amplitude of the jth generator. In the first approximation with respect to the small parameters, the equation for R_j is

$$\dot{R}_j = \frac{\mu}{2}\left(1 + 2\mathrm{i}\,\frac{\Delta_j}{\mu} - \frac{\alpha}{4}\,|R_j|^2\right)R_j + \frac{1}{2}\left(c_2 - \frac{\mathrm{i}c_1}{\omega}\right)R_{j-1}, \tag{11.41}$$

where $\Delta_j = \omega_j - \omega$ is the mistuning between the natural frequency of the jth generator and the mean frequency ω. In order that Eq. (11.41) is valid for all j from 1 to N, we must impose the boundary condition $R_0 = 0$.

Equation (ref11.41) is equivalent to the two following equations for the real amplitude A_j and the phase φ_j:

$$\dot{A}_j = \frac{\mu}{2}\left(1 - \alpha A_j^2\right)A_j - \frac{1}{2}\left(\frac{c_1}{\omega}\,\sin\Phi_j - c_2\cos\Phi_j\right)A_{j-1}, \tag{11.42}$$

$$\dot{\varphi}_j = \Delta_j - \frac{1}{2}\left(\frac{c_1}{\omega}\,\cos\Phi_j + c_2\sin\Phi_j\right)\frac{A_{j-1}}{A_j}, \tag{11.43}$$

where $\Phi_j = \varphi_j - \varphi_{j-1}$ is the phase shift between the jth and the $(j-1)$th generators. Since the coupling is assumed to be weak, in Eq. (11.43) we can put $A_j \approx A_0 = 1/\sqrt{\alpha}$. In so doing we obtain the following equations for steady-state values of Φ_j which are related to the synchronization regime:

$$\Delta_{2,1} - \frac{1}{2}\left(\frac{c_1}{\omega}\cos\Phi_2 + c_2\sin\Phi_2\right) = 0,$$
$$\Delta_{j,j-1} - \frac{1}{2}\left(\frac{c_1}{\omega}(\cos\Phi_j - \cos\Phi_{j-1}) + c_2(\sin\Phi_j - \sin\Phi_{j-1})\right) = 0 \quad (j \geq 3), \tag{11.44}$$

where $\Delta_{j,j-1} = \Delta_j - \Delta_{j-1} = \omega_j - \omega_{j-1}$. Equations (11.44) allow us to find both the values of Φ_j and the width of the synchronization region.

11.3.2 Synchronization and chaotization of self-oscillations in a chain of N coupled van der Pol–Duffing generators

If the chain consists of van der Pol–Duffing generators, then results of a radically different kind from those discussed above are obtained. Such a chain is described by the equations

$$\ddot{x}_j - \mu(1 - 4\alpha x_j^2)\dot{x}_j + \omega_j^2(1 + \gamma x_j^2)x_j = c_1 x_{j-1} + c_2\dot{x}_{j-1}$$
$$(j = 1, 2, \ldots, N). \tag{11.45}$$

Proceeding as for the van der Pol generators, we find the following equation for the complex amplitude R_j:

$$\dot{R}_j = \frac{\mu}{2}\left(1 + 2\mathrm{i}\frac{\Delta_j}{\mu} - \left(\alpha - \frac{3\mathrm{i}\omega\gamma}{4\mu}\right)|R_j|^2\right)R_j + \frac{1}{2}\left(c_2 - \frac{\mathrm{i}c_1}{\omega}\right)R_{j-1}. \tag{11.46}$$

Owing to the unidirectional coupling between the generators their amplitudes increase as the number j increases. Because the frequency of free oscillations of the jth generator depends on the amplitude, this frequency also has to increase with increasing j. [1] Therefore beyond a certain number j the generators cease to be synchronized, even with small Δ_j. As a result a beat regime arises initially and it can then be turned into a chaotic regime. Such a transition was observed by Gaponov-Grekhov et al. in numerical studies of Eqs. (11.46) [99]. The transition from one regime to the other with increasing number j was determined from changes in the Lyapunov dimension [2] D_L of the attractor in the phase space of the j leading generators. An example of the dependence of D_L on j taken from [99] is shown in Fig. 11.4. It can be seen that only first three generators are synchronized. From the fourth generator to the ninth one there is the beat regime. And from the tenth generator onwards the chaotic regime occurs. It is very important that the dimension does not increase monotonically but saturates. This fact is apt to be indicative of synchronization of the chaotic oscillations (see below).

[1] For the of definiteness, we assume $\gamma > 0$.
[2] The definition of the Lyapunov dimension is given in Ch. 3.

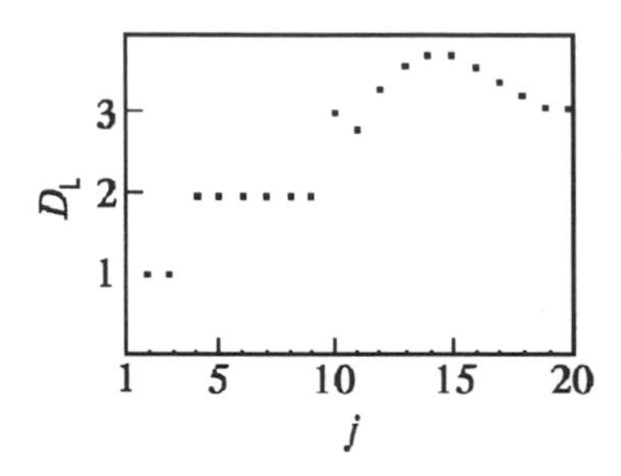

Fig. 11.4. The dependence of the Lyapunov dimension D_{L} on the number of generators j for $\mu = 1$, $\alpha = 32$, $3\omega\gamma = 40$, $\Delta_j = c_1/(2\omega) = 0.855$, and $c_2 = 1$

11.3.3 Synchronization of chaotic oscillations in a chain of generators with inertial nonlinearity

Synchronization of chaotic self-oscillations in a chain of generators with inertial nonlinearity was studied experimentally and numerically by Anischenko et al. The results of these studies are generalized in the book [14]. The following equations were simulated numerically:

$$\begin{aligned} &\ddot{x}_j - (1.16 - V_j)\dot{x}_j + x_j = c x_{j-1}, \\ &\dot{V}_j + 0.3V_j = \gamma\vartheta(x_j)x_j^2 \quad (j = 1, 2, \ldots, 10), \end{aligned} \tag{11.47}$$

where $\vartheta(x_j)$ is the Heaviside step function. The coupling factor c was varied. For these values of the parameters the generators each execute chaotic self-oscillations, being autonomous. The Lyapunov dimension of the attractor corresponding to these oscillations is $D_{\mathrm{L}}^{(0)} = 2.187$. If the generators were independent, the dimension $D_{\mathrm{L}}(j)$ of the system of j generators would be $jD_{\mathrm{L}}^{(0)}$. With the coupling, because of the synchronization effect the dimension $D_{\mathrm{L}}(j)$ of the chain of j generators has to be less than $jD_{\mathrm{L}}^{(0)}$. Moreover, the dimension has to be gradually saturated as j increases. The dependence of D_{L} on j for different values of c taken from [14] is shown in Fig. 11.5. We see that the saturation of the dimension occurs for the smaller j the larger is the coupling factor. The limiting value of the dimension is also the smaller the larger is the coupling factor. It would be expected that with further increase of the coupling factor the dimension D_{L} for any j becomes equal to the partial dimension $D_{\mathrm{L}}^{(0)}$.

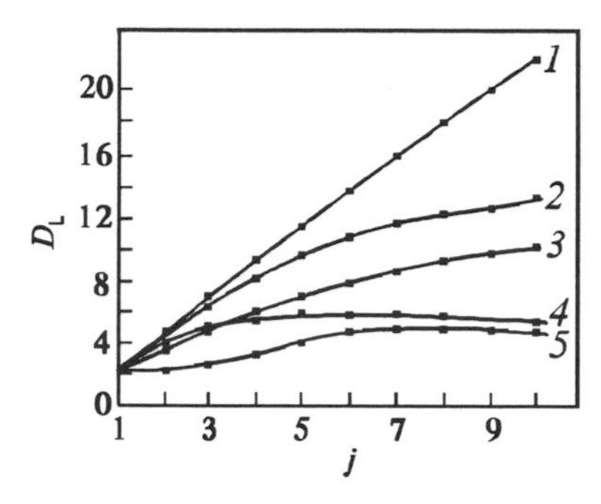

Fig. 11.5. The dependence of the Lyapunov dimension D_L on the number of generators j for: $c \leq 10^{-3}$ (curve 1), $c = 0.05$ (curve 2), $c = 0.10$ (curve 3), $c = 0.20$ (curve 4), and $c = 1.00$ (curve 5)

Part II

OSCILLATIONS IN NONAUTONOMOUS SYSTEMS

12. Oscillations of nonlinear systems excited by external periodic forces

12.1 A periodically driven nonlinear oscillator

Let us consider a nonlinear oscillator driven by a periodic external force with period $T = 2\pi/\omega$. Let the equation of this oscillator be

$$\ddot{x} + 2\delta\dot{x} + \omega_0^2 g(x) = f(t), \tag{12.1}$$

where $f(t)$ is a periodic function that can be represented in the form of a Fourier series:

$$f(t) = \sum_{n=1}^{\infty} \Big(B_n \cos n\omega t + C_n \sin n\omega t \Big). \tag{12.2}$$

Without loss of generality we can set $C_1 = 0$.

Owing to the superposition principle in the linear case, when $g(x) = x$, we can independently consider the response of the system to each term of the series (12.2). Therefore in this case it will suffice to find solutions of the equation

$$\ddot{x} + 2\delta\dot{x} + \omega_0^2 x = B\mathrm{e}^{\mathrm{i}\omega t}. \tag{12.3}$$

A partial solution of Eq. (12.3) describing forced oscillations is

$$x = A\mathrm{e}^{\mathrm{i}(\omega t + \varphi)}, \tag{12.4}$$

where

$$A = \frac{B}{\sqrt{(\omega^2 - \omega_0^2)^2 + 4\delta^2\omega^2}}, \qquad \tan\varphi = \frac{2\delta\omega}{\omega^2 - \omega_0^2}. \tag{12.5}$$

Let us analyze the solution found. For small frequencies of the external force the derivatives $\ddot{x}$ and $\dot{x}$ can be neglected in Eq. (12.3); hence $x \approx (B/\omega_0^2)\mathrm{e}^{\mathrm{i}\omega t}$. Thus, for small ω, oscillations of the variable x are approximately in phase with the external force. For sufficiently large ω, we have $x \approx -(B/\omega^2)\mathrm{e}^{\mathrm{i}\omega t}$, i.e., oscillations of the variable x are approximately in anti-phase with the external force. For $\omega = \omega_0$ the phase shift φ between x and the external force is equal to $-\pi/2$. This value of the external force frequency is commonly called *the resonance frequency*. If the damping factor δ is sufficiently small ($\delta \ll \omega_0$), the amplitude A is maximal for $\omega = \omega_0$ and equal to $A_0 = B/(2\delta\omega_0)$. The dependencies on ω of the phase shift φ and of the amplitudes of the displacement $|x| = A$, of the velocity $|\dot{x}| = A\omega$, and of the acceleration $|\ddot{x}| = A\omega^2$ are given in Fig. 12.1 for two values of the quantity $Q = \omega_0/2\delta$, known as the Q-factor.

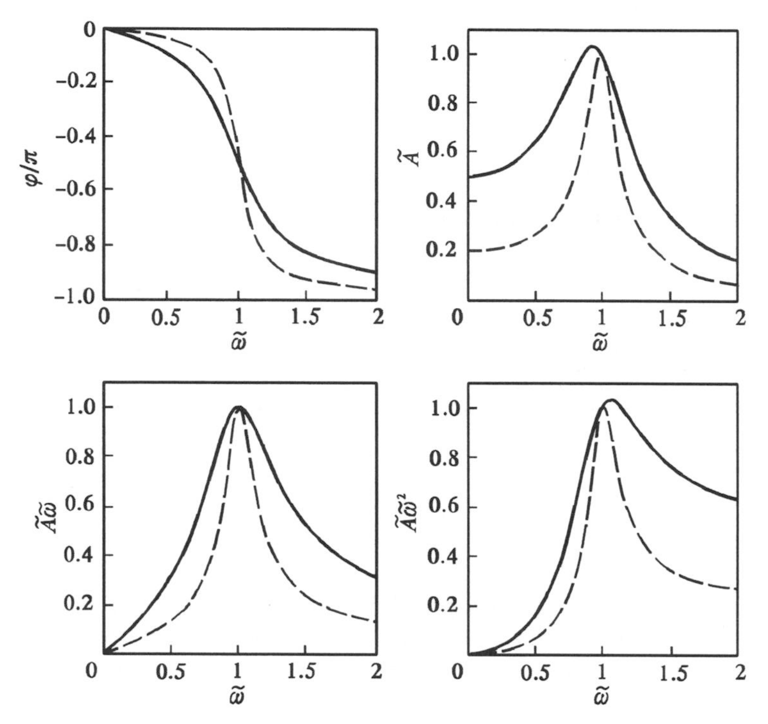

Fig. 12.1. The dependencies on the relative external force frequency $\tilde{\omega} = \omega/\omega_0$ of the phase shift φ and of the relative amplitudes of the displacement $\tilde{A} = |x|/A_0$, of the velocity $\tilde{A}\tilde{\omega} = |\dot{x}|/(A_0\omega_0)$, and of the acceleration $\tilde{A}\tilde{\omega}^2 = |\ddot{x}|/(A_0\omega_0^2)$ for the linear oscillator. The solid lines correspond to $Q = 2$, the dashed lines correspond to $Q = 5$

How will the results change if a slight nonlinearity of the oscillator is accounted for? First and foremost the superposition principle will cease to be valid. Therefore in the general case we cannot restrict ourselves to consideration of only a harmonic force acting upon the oscillator. For the sake of definiteness we set the function $g(x)$ in Eq. (12.1) in the polynomial form

$$g(x) = (1 + \beta x + \gamma x^2)x. \tag{12.6}$$

Assuming that the oscillator Q-factor is moderately large and that the external force amplitude is sufficiently small, we rewrite Eq. (12.1) in the following form:

$$\ddot{x} + \omega_0^2 x = \epsilon\Big(-2\delta\dot{x} - \omega_0^2\beta x^2 - \omega_0^2\gamma x^3 + f(t)\Big), \tag{12.7}$$

where ϵ is a small parameter which should be put equal to unity in the resultant expressions.

Let us find approximate solutions of Eq. (12.7) in the specific cases corresponding to the main resonance ($\omega \sim \omega_0$), subharmonic resonances ($\omega \sim n\omega_0$, $n = 2, 3, \ldots$), and superharmonic resonances ($\omega \sim \omega_0/n$, $n = 2, 3, \ldots$).

12.1.1 The main resonance

In this case, according to the Krylov–Bogolyubov asymptotic method, a solution of Eq. (12.7) can be sought in the form

$$x = A\cos\psi + \epsilon u_1(A,\psi) + \ldots\,, \tag{12.8}$$

where $\psi = \omega t + \varphi$, $u_1(A,\psi), \ldots$ are unknown functions free from resonance terms; A and φ are the oscillation amplitude and phase obeying the equations

$$\dot{A} = \epsilon f_1(A,\varphi) + \ldots\,, \quad \dot{\varphi} = -\Delta + \epsilon F_1(A,\varphi) + \ldots\,, \tag{12.9}$$

$\Delta = \omega - \omega_0$ is the frequency mistuning, and $f_1(A,\varphi), \ldots$, $F_1(A,\varphi), \ldots$ are unknown functions to be determined from the conditions for the absence of resonance terms in the functions $u_1(A,\psi), \ldots$. Substituting (12.8), in view of Eqs. (12.9), into Eq. (12.7) and equating the coefficients of ϵ, we obtain the equation for the function $u_1(A,\psi)$:

$$\begin{aligned} \omega^2 \frac{\partial^2 u_1}{\partial \psi^2} + \omega_0^2 u_1 = & \left(2\omega_0 f_1 - A\frac{\partial F_1}{\partial \varphi}\Delta\right)\sin\psi \\ & + \left(2\omega_0 A F_1 - \frac{\partial f_1}{\partial \varphi}\Delta\right)\cos\psi + 2\delta A\omega_0 \sin\psi \\ & - A^2\omega_0^2\left(\beta + \gamma A\cos\psi\right)\cos\psi + f(t). \end{aligned} \tag{12.10}$$

From the condition for the absence of resonance terms in the function $u_1(A,\psi)$ we find the following equations for f_1 and F_1:

$$\begin{aligned} 2f_1 - A\frac{\Delta}{\omega_0}\frac{\partial F_1}{\partial \varphi} &= -2\delta A - \frac{B_1}{\omega_0}\sin\varphi, \\ 2AF_1 + \frac{\Delta}{\omega_0}\frac{\partial f_1}{\partial \varphi} &= \frac{3}{4}\omega_0\gamma A^3 - \frac{B_1}{\omega_0}\cos\varphi, \end{aligned} \tag{12.11}$$

where B_1 is the amplitude of the fundamental harmonic of the external force (see (12.2)).

A partial solution of Eq. (12.11) is

$$f_1 = -\,\delta A - \frac{B_1\sin\varphi}{\omega + \omega_0}, \quad F_1 = \frac{3}{8}\omega_0\gamma A^2 - \frac{B_1\cos\varphi}{(\omega+\omega_0)A}.$$

It follows from this and from (12.9) that the equations for A and φ in the first approximation with respect to ϵ are

$$\dot{A} = -\delta A - \frac{B_1 \sin\varphi}{\omega_0 + \omega}, \quad \dot{\varphi} = -\Delta + \frac{3}{8}\omega_0 \gamma A^2 - \frac{B_1 \cos\varphi}{(\omega_0 + \omega)A}. \tag{12.12}$$

Taking account of (12.11) we find from Eq. (12.10) the function $u_1(A, \psi)$:

$$u_1(A,\psi) = -\frac{\beta A^2}{2} + \frac{\omega_0^2}{2(4\omega^2 - \omega_0^2)}\beta A^2 \cos 2\psi + \frac{\omega_0^2}{4(9\omega^2 - \omega_0^2)}\gamma A^3 \cos 3\psi$$
$$- \sum_{n=2}^{\infty} \frac{1}{n^2\omega^2 - \omega_0^2}\Big(B_n \cos n(\psi - \varphi) + C_n \sin n(\psi - \varphi)\Big). \tag{12.13}$$

It is seen from (12.13) that in the first approximation $x(t)$ involves, in addition to the fundamental frequency, a constant constituent and higher harmonics.

A steady-state solution of Eqs. (12.12), determining the amplitude and phase of forced oscillations at the fundamental frequency, can be found from the following equations:

$$\left[\delta^2 + \left(\Delta - \frac{3}{8}\omega_0\gamma A^2\right)^2\right] A^2(\omega_0 + \omega)^2 = B_1^2, \tag{12.14}$$

$$\tan\varphi = \delta\left(\Delta - \frac{3}{8}\omega_0\gamma A^2\right)^{-1}. \tag{12.15}$$

It is seen from this that the resonance value of the frequency ω, defined from the condition $\varphi = -\pi/2$, depends on the amplitude A, namely, $\omega_r = \omega_0\big(1+(3/8)\gamma A^2\big)$. The plots of $\tilde{A} = \omega_0^2 A/B_1$ and φ versus ω/ω_0 are illustrated in Fig. 12.2 for $Q = \omega_0/2\delta = 5$.

The stability of the steady-state solutions found can be analyzed by putting in Eqs. (12.12) $A = A_{\rm st} + a$, $\varphi = \varphi_{\rm st} + \phi$, where a and ϕ are small deviations from the steady-state values. The linearized equations for a and ϕ are

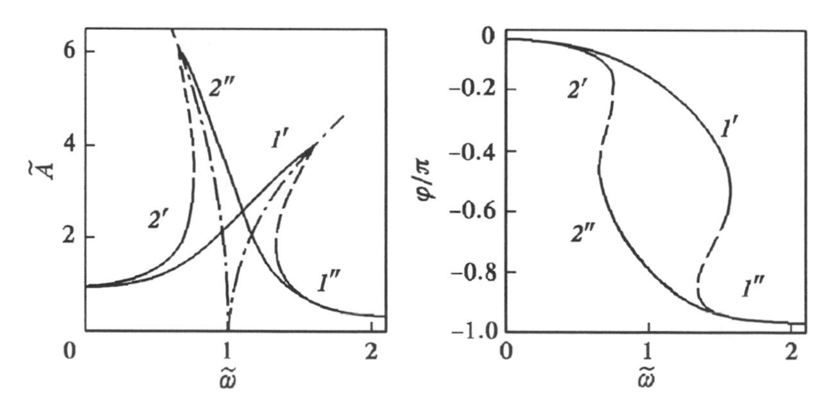

Fig. 12.2. Plots of $\tilde{A} = \omega_0^2 A/B_1$ and φ versus $\tilde{\omega} = \omega/\omega_0$ for $Q = \omega_0/2\delta = 5$. The curves labeled 1′ 1″ correspond to $\gamma B_1^2/\omega_0^4 = 0.1$, and those labeled 2′ 2″ correspond to $\gamma B_1^2/\omega_0^4 = -0.025$. Unstable parts are shown by dashed lines. Dot-and-dash lines show the skeleton curves determined by the equation $\omega/\omega_0 - 1 = (3/8)\gamma A^2$

$$\dot{a} = \frac{\partial f}{\partial A} a + \frac{\partial f}{\partial \varphi} \phi, \quad \dot{\phi} = \frac{\partial F}{\partial A} a + \frac{\partial F}{\partial \varphi} \phi, \tag{12.16}$$

where

$$f = -\delta A - \frac{B_1 \sin\varphi}{\omega_0 + \omega}, \quad F = -\Delta + \frac{3}{8}\omega_0 \gamma A^2 - \frac{B_1 \cos\varphi}{(\omega_0 + \omega) A}, \tag{12.17}$$

and all derivatives are calculated for $A = A_{\mathrm{st}}$, $\varphi = \varphi_{\mathrm{st}}$.[1] The following characteristic equation is found from (12.16):

$$p^2 - \left(\frac{\partial f}{\partial A} + \frac{\partial F}{\partial \varphi}\right) p + \frac{\partial f}{\partial A}\frac{\partial F}{\partial \varphi} - \frac{\partial f}{\partial \varphi}\frac{\partial F}{\partial A} = 0. \tag{12.18}$$

Because $\partial f/\partial A = -\delta$ and $\partial F/\partial \varphi = B_1 \sin\varphi/(\omega_0 + \omega)A = -\delta$, the coefficient of p is always positive. Therefore we obtain a single condition for instability:

$$\frac{\partial f}{\partial A}\frac{\partial F}{\partial \varphi} - \frac{\partial f}{\partial \varphi}\frac{\partial F}{\partial A} < 0. \tag{12.19}$$

To write this condition in a more convenient form we differentiate the equations $f = 0$ and $F = 0$ with respect to ω:

$$\frac{\partial f}{\partial \omega} + \frac{\partial f}{\partial A}\frac{\mathrm{d}A}{\mathrm{d}\omega} + \frac{\partial f}{\partial \varphi}\frac{\mathrm{d}\varphi}{\mathrm{d}\omega} = 0, \quad \frac{\partial F}{\partial \omega} + \frac{\partial F}{\partial A}\frac{\mathrm{d}A}{\mathrm{d}\omega} + \frac{\partial F}{\partial \varphi}\frac{\mathrm{d}\varphi}{\mathrm{d}\omega} = 0.$$

Eliminating $\mathrm{d}\varphi/\mathrm{d}\omega$ from these equations, we find

$$\left(\frac{\partial f}{\partial A}\frac{\partial F}{\partial \varphi} - \frac{\partial f}{\partial \varphi}\frac{\partial F}{\partial A}\right)\frac{\mathrm{d}A}{\mathrm{d}\omega} = \frac{\partial F}{\partial \omega}\frac{\partial f}{\partial \varphi} - \frac{\partial f}{\partial \omega}\frac{\partial F}{\partial \varphi}. \tag{12.20}$$

It follows from (12.17) that

$$\frac{\partial f}{\partial \omega} = \frac{B_1 \sin\varphi}{(\omega_0 + \omega)^2}, \quad \frac{\partial F}{\partial \omega} = -1 + \frac{B_1 \cos\varphi}{(\omega_0 + \omega)^2 A},$$

$$\frac{\partial f}{\partial \varphi} = -\frac{B_1 \cos\varphi}{\omega_0 + \omega}, \quad \frac{\partial F}{\partial \varphi} = \frac{B_1 \sin\varphi}{(\omega_0 + \omega) A}.$$

Substituting these expressions into (12.20) and taking into account the equation $F = 0$, we obtain

$$\left(\frac{\partial f}{\partial A}\frac{\partial F}{\partial \varphi} - \frac{\partial}{\partial \varphi}\frac{\partial F}{\partial A}\right)\frac{\mathrm{d}A}{\mathrm{d}\omega} = \left(-\Delta + \frac{3}{8}\omega_0 \gamma A^2 - \frac{B_1^2}{(\omega_0 + \omega)^3 A^2}\right) A. \tag{12.21}$$

It follows from (12.14) that

$$\frac{B_1^2}{(\omega_0 + \omega)^3 A^2} \sim \frac{1}{\omega}\left(\Delta - \frac{3}{8}\omega_0 \gamma A^2\right)^2;$$

hence for small Δ this term can be dropped. In so doing expression (12.21) becomes

$$\left(\frac{\partial f}{\partial A}\frac{\partial F}{\partial \varphi} - \frac{\partial f}{\partial \varphi}\frac{\partial F}{\partial A}\right)\frac{\mathrm{d}A}{\mathrm{d}\omega} = \left(-\Delta + \frac{3}{8}\omega_0 \gamma A^2\right) A.$$

[1] The subscript ‘st’ is omitted below.

It follows from this and from (12.19) that the steady-state solution is unstable if $\mathrm{d}A/\mathrm{d}\omega > 0$ for $\omega > \omega_0(A)$ and $\mathrm{d}A/\mathrm{d}\omega < 0$ for $\omega < \omega_0(A)$, where $\omega_0(A) = \omega_0 + (3/8)\omega_0\gamma A^2$. The dependencies of $\omega = \omega_0(A)$ on A, called *skeleton curves*, are shown in Fig. 12.2 by dot-and-dash lines.

It should be noted that in the linear case ($\gamma = 0$) the approximate expressions (12.14), (12.15) are distinguished from the exact ones (12.5) by terms of order ϵ.

12.1.2 Subharmonic resonances

Let us seek a periodic solution of Eq. (12.7) with fundamental frequency ν that is n times smaller than the external force frequency ω, where $n = 2, 3, \ldots$. As will be seen from the following, the subharmonic resonance for Eq. (12.7) can be calculated in the second and higher approximations with respect to the small parameter ϵ only. In doing so it is convenient to assume that the damping factor δ is also of second order with respect to ϵ. As before, we seek a solution of Eq. (12.7) in the form

$$x = A\cos\psi + \epsilon u_1(A,\psi) + \epsilon^2 u_2(A,\psi) + \ldots\,, \tag{12.22}$$

where $\psi = \nu t + \varphi$, $\nu = \omega/n$, and A and φ obey the equations

$$\begin{aligned} \dot{A} &= \epsilon f_1(A,\varphi) + \epsilon^2 f_2(A,\varphi) + \ldots\,, \\ \dot{\varphi} &= -\Delta + \epsilon F_1(A,\varphi) + \epsilon^2 F_2(A,\varphi) + \ldots, \end{aligned} \tag{12.23}$$

and $\Delta = \nu - \omega_0$ is the frequency mistuning. Substituting (12.22) into Eq. (12.7) and taking account of (12.23), we obtain in the first approximation:

$$\begin{aligned} \nu^2\frac{\partial^2 u_1}{\partial\psi^2} + \omega_0^2 u_1 &= \left(2\omega_0 f_1 - A\frac{\partial F_1}{\partial\varphi}\Delta\right)\sin\psi \\ &+ \left(2\omega_0 A F_1 + \frac{\partial f_1}{\partial\varphi}\Delta\right)\cos\psi \\ &- \beta A^2\omega_0^2(\cos\psi)^2 - \gamma A^3\omega_0^2(\cos\psi)^3 + f(t). \end{aligned} \tag{12.24}$$

From the condition for the absence of resonance terms on the right-hand side of this equation we find the following equations for f_1 and F_1:

$$f_1 = 0, \quad F_1 = \frac{3}{8}\omega_0\gamma A^2. \tag{12.25}$$

A solution of Eq. (12.24), in view of (12.25), is

$$\begin{aligned} u_1(A,\psi) = &-\frac{\beta A^2}{2} + \frac{\omega_0^2\beta A^2}{2(4\nu^2-\omega_0^2)}\cos 2\psi + \frac{\omega_0^2\gamma A^3}{4(9\nu^2-\omega_0^2)}\cos 3\psi \\ &-\sum_{k=1}^{\infty}\frac{1}{k^2n^2\nu^2-\omega_0^2}\Big(B_k\cos kn(\psi-\varphi) + C_k\sin kn(\psi-\varphi)\Big). \end{aligned} \tag{12.26}$$

Taking account of (12.25) and (12.26) we can obtain the equation for $u_2(A, \psi)$. From the condition for the absence of resonance terms in the right-hand side of this equation we find the following equations for f_2 and F_2:

$$f_2 = -\delta A - \delta_{n2} \frac{\omega_0^2 \beta A B_1}{2\nu(4\nu^2 - \omega_0^2)} \sin 2\varphi - \delta_{n3} \frac{3\omega_0^2 \gamma A^2 B_1}{4(9\nu^2 - \omega_0^2)(3\nu - \omega_0)} \sin 3\varphi, \tag{12.27}$$

$$F_2 = -\frac{8\nu^2 - 3\omega_0^2}{4(4\nu^2 - \omega_0^2)} \omega_0 \beta^2 A^2 - \frac{3(27\nu^2 - 7\omega_0^2)}{128(9\nu^2 - \omega_0^2)} \omega_0 \gamma^2 A^4$$
$$- \delta_{n2} \frac{\omega_0^2 \beta B_1}{2\nu(4\nu^2 - \omega_0^2)} \cos 2\varphi - \delta_{n3} \frac{3\omega_0^2 \gamma A B_1}{4(9\nu^2 - \omega_0^2)(3\nu - \omega_0)} \cos 3\varphi,$$

where δ_{jk} is the Kronecker delta.

It is seen from (12.27) that in the second approximation we can calculate the second and the third subharmonic resonances only (for the type of nonlinearity under consideration). Substituting (12.25) and (12.27) into Eqs. (12.23) and ignoring unessential terms, we obtain

$$\dot{A} = -\delta A - \delta_{n2} \frac{\omega_0^2 \beta A B_1}{2\nu(4\nu^2 - \omega_0^2)} \sin 2\varphi - \delta_{n3} \frac{3\omega_0^2 \gamma A^2 B_1}{4(9\nu^2 - \omega_0^2)(3\nu - \omega_0)} \sin 3\varphi, \tag{12.28}$$

$$\dot{\varphi} = -\Delta + \frac{3}{8} \omega_0 \gamma A^2 - \delta_{n2} \frac{\omega_0^2 \beta B_1}{2\nu(4\nu^2 - \omega_0^2)} \cos 2\varphi$$
$$- \delta_{n3} \frac{3\omega_0^2 \gamma A B_1}{4(9\nu^2 - \omega_0^2)(3\nu - \omega_0)} \cos 3\varphi.$$

Let us consider steady-state solutions of Eqs. (12.28) and their stability in two particular cases: (i) $n = 2$ and (ii) $n = 3$.

(i) $n = 2$. In this case the steady-state solution of Eqs. (12.28) is determined by the equations

$$\frac{\delta^2}{\omega^2} + \left(\frac{\Delta}{\omega_0} - \frac{3}{8} \gamma A^2\right)^2 = \frac{\omega_0^2 \beta^2 B_1^2}{4\nu^2(4\nu^2 - \omega_0^2)^2}, \tag{12.29}$$

$$\tan 2\varphi = \delta \left(\Delta - \frac{3}{8} \omega_0 \gamma A^2\right)^{-1}.$$

We investigate the stability of the steady-state solution by using Eqs. (12.28) in much the same manner as was done in the case of the main resonance. The equations for small deviations $a = A - A_{\mathrm{st}}$, $\phi = \varphi - \varphi_{\mathrm{st}}$ can be written as

$$\dot{a} = \frac{\partial f}{\partial A} a + \frac{\partial f}{\partial \varphi} \phi, \quad \dot{\phi} = \frac{\partial F}{\partial A} a + \frac{\partial F}{\partial \varphi} \phi,$$

where

$$f = -\left(\delta A + \frac{\omega_0^2 \beta A B_1}{2\nu(4\nu^2 - \omega_0^2)} \sin 2\varphi\right),$$

$$F = -\Delta + \frac{3}{8}\omega_0 \gamma A^2 - \frac{\omega_0^2 \beta B_1}{2\nu(4\nu^2 - \omega_0^2)} \cos 2\varphi. \tag{12.30}$$

It follows from (12.30) and (12.18) that, as before, the condition for instability of the steady-state solution can be written in the form (12.19). Using the formula (12.20) and taking into account that

$$\frac{\partial f}{\partial \omega} = \frac{1}{2}\frac{\partial f}{\partial \nu} = \frac{\omega_0^2 \beta B_1 (12\nu^2 - \omega_0^2)}{4\nu^2(4\nu^2 - \omega_0^2)^2} A \sin 2\varphi = -\frac{\delta A}{2}\frac{12\nu^2 - \omega_0^2}{\nu(4\nu^2 - \omega_0^2)},$$

$$\frac{\partial F}{\partial \omega} = -\frac{1}{2} + \frac{\omega_0^2 \beta B_1 (12\nu^2 - \omega_0^2)}{4\nu^2(4\nu^2 - \omega_0^2)^2} \cos 2\varphi,$$

$$\frac{\partial f}{\partial \varphi} = -\frac{\omega_0^2 \beta A B_1}{\nu(4\nu^2 - \omega_0^2)} \cos 2\varphi, \quad \frac{\partial F}{\partial \varphi} = \frac{\omega_0^2 \beta B_1}{\nu(4\nu^2 - \omega_0^2)} \sin 2\varphi,$$

we find

$$\left(\frac{\partial f}{\partial A}\frac{\partial F}{\partial \varphi} - \frac{\partial f}{\partial \varphi}\frac{\partial F}{\partial A}\right)\frac{\mathrm{d}A}{\mathrm{d}\omega} = \left(-\Delta + \frac{3}{8}\omega_0 \gamma A^2\right) A.$$

Hence, as for the main resonance, we conclude that the steady-state solution is unstable if $\mathrm{d}A/\mathrm{d}\omega > 0$ for $\nu > \omega_0(A)$ and $\mathrm{d}A/\mathrm{d}\omega < 0$ for $\nu < \omega_0(A)$.

(ii) $n = 3$. In this case the steady-state solution of Eqs. (12.28) is determined by the equations

$$\frac{\delta^2}{\omega^2} + \left(\frac{\Delta}{\omega_0} - \frac{3}{8}\gamma A^2\right)^2 = \frac{9\omega_0^2 \gamma^2 A^2 B_1^2}{16(3\nu - \omega_0)^4 (3\nu + \omega_0)^2},$$

$$\tan 3\varphi = \delta\left(\Delta - \frac{3}{8}\omega_0 \gamma A^2\right)^{-1}. \tag{12.31}$$

The plots of $\tilde{A} = \sqrt{(3/8)|\gamma|}\, A$ and of φ versus $\omega/\omega_0 = n\nu/\omega_0$, determined by Eqs. (12.29) (for $n = 2$) and (12.31) (for $n = 3$), are given in Fig. 12.3 for both positive and negative values of γ. As before, dashed lines show unstable parts and dot-and-dash lines show skeleton curves described by the equation $\omega = n\omega_0(A)$.

We note that the calculation of subharmonic resonances of higher orders for the Duffing equation was performed by Schmidt and Seisl [303].

12.1.3 Superharmonic resonances

If the external force involves higher harmonics with frequencies $n\omega$ then the calculation of superharmonic resonance can be performed in the first approximation with respect to the small parameter ϵ. In this case the analysis is

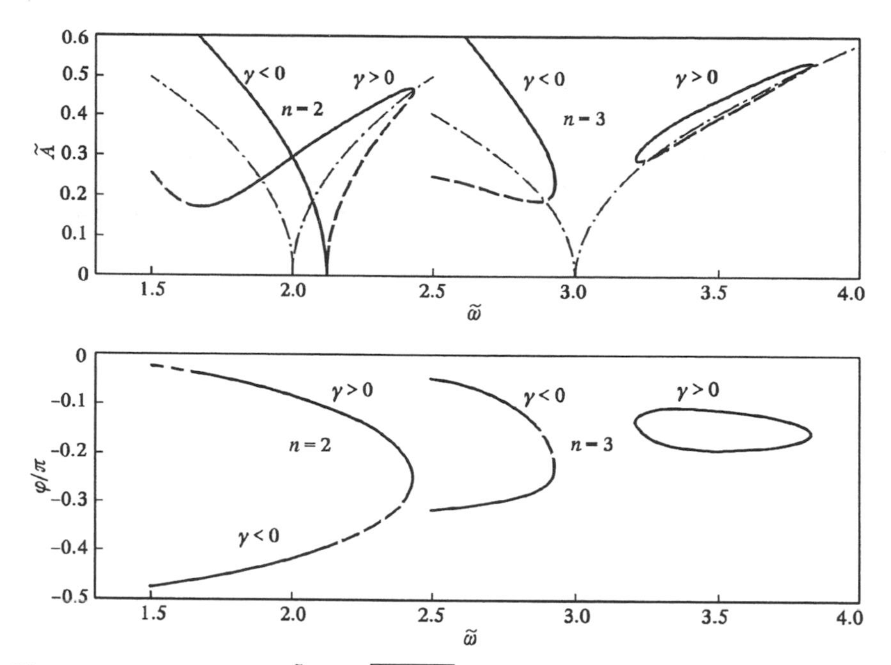

Fig. 12.3. The plots of $\tilde{A} = \sqrt{(3/8)|\gamma|}\,A$ and of φ versus $\tilde{\omega} = \omega/\omega_0 = n\nu/\omega_0$ for $Q = \omega_0/2\delta = 5$, $\beta^2 B_1^2/\omega_0^4 = 0.3$ (for $n = 2$), and $2|\gamma|B_1^2/\omega_0^4 = 9$ (for $n = 3$). Unstable parts are shown by dashed lines; dot-and-dash lines show skeleton curves

no different from the case of main resonance. But if the external force is free from higher harmonics then calculations of superharmonic resonance can be performed in the third and higher approximation only. For reasons of space we will not undertake these calculations.

12.2 Oscillations excited by an external force with a slowly time-varying frequency

The behavior of quasi-linear systems with slowly time-varying parameters was detailed in the monograph by Mitropol'sky [238]. Here we consider only one, but very important, problem of the passage of such a system through resonance.

First let us consider a linear oscillator subject to the action of an external force with a slowly time-varying frequency. Let the equation describing this oscillator be

$$\ddot{x} + 2\delta\dot{x} + \omega_0^2 x = B\cos\psi(t), \tag{12.32}$$

where

$$\psi(t) = \int_0^t \omega(\epsilon t)\, \mathrm{d}t + \chi,$$

and ϵ is a small parameter.

If the duration of the passage through the resonance region $t^* = \delta/\dot{\omega}$ is much more than the relaxation time $1/\delta$ then the process of the passage through the resonance can be considered as quasi-stationary; in this case the resonance curves are identical to those for an external force with constant frequency. Otherwise, distinctions may be very significant.

A particular solution of Eq. (12.32) corresponding to forced oscillations can be written in terms of the Green function $G(t,\tau)$ as

$$x(t) = \mathrm{Re}\left(B \int_{-\infty}^{t} G(t,\tau) \mathrm{e}^{\mathrm{i}\psi(\tau)}\, \mathrm{d}\tau\right), \tag{12.33}$$

where

$$G(t,\tau) = \frac{1}{\omega_1} \mathrm{e}^{-\delta(t-\tau)} \sin \omega_1 (t-\tau), \quad \omega_1 = \sqrt{\omega_0^2 - \delta^2}\,. \tag{12.34}$$

In the special case when $\psi(t) = \epsilon t^2 + \chi$, i.e., when the external force frequency depends on time only linearly, expression (12.33), in view of (12.34), can be written in terms of the complex error integrals [134] as

$$\begin{aligned} x(t) = \frac{B\sqrt{\pi}}{4\omega_1\sqrt{\epsilon}} \mathrm{Re}\Big\{ &\exp\left(-\delta t - \frac{\mathrm{i}}{4\epsilon}(\omega_1^2 - \delta^2) + \mathrm{i}\chi - \mathrm{i}\frac{\pi}{4}\right) \\ \times &\left[\exp\left(\frac{\omega_1 \delta}{2\epsilon} + \mathrm{i}\omega_1 t\right)\left(1 + \Phi\left(\sqrt{\frac{\epsilon}{\mathrm{i}}}\left(t - \frac{\omega_1 + \mathrm{i}\delta}{2\epsilon}\right)\right)\right)\right. \\ &\left.- \exp\left(-\frac{\omega_1 \delta}{2\epsilon} - \mathrm{i}\omega_1 t\right)\left(1 + \Phi\left(\sqrt{\frac{\epsilon}{\mathrm{i}}}\left(t + \frac{\omega_1 - \mathrm{i}\delta}{2\epsilon}\right)\right)\right)\right]\Big\}, \end{aligned} \tag{12.35}$$

where

$$\Phi(z) = \frac{2}{\sqrt{\pi}} \int_0^z \exp(-t^2)\, \mathrm{d}t.$$

It follows from the expression for $\psi(t)$ that the force frequency is $\omega(t) = \dot{\psi} = 2\epsilon t$. Hence the passage through the resonance region occurs at $t \sim t_0 = \omega_1/2\epsilon$. For $\delta t_0 \gg 1$ the second term in the square brackets of (12.35) can be ignored. In this case (12.35) can be represented as

$$x(t) = A(t) \cos\big(\psi(t) + \varphi(t)\big), \tag{12.36}$$

where $A(t)$ and $\varphi(t)$ are the slowly time-varying amplitude and phase, which is determined, as follows from (12.35), by

$$A(t) = \frac{B\sqrt{\pi}}{4\omega_1\sqrt{\epsilon}}\,\mathrm{e}^{-\delta(t-t_0)}\left|1 + \Phi\left(\sqrt{\frac{\epsilon}{\mathrm{i}}}\left(t - t_0 - \frac{\mathrm{i}\delta}{2\epsilon}\right)\right)\right|, \tag{12.37}$$

$$\varphi(t) = \frac{\delta^2}{4\epsilon} - \epsilon(t-t_0) - \frac{\pi}{4} + \arg\left[1 + \Phi\left(\sqrt{\frac{\epsilon}{\mathrm{i}}}\left(t - t_0 - \frac{\mathrm{i}\delta}{2\epsilon}\right)\right)\right]. \tag{12.38}$$

It is easily shown that for $\delta \gg \sqrt{\epsilon}$ expressions (12.37) and (12.38) coincide with (12.5). A distinction appears for $\delta \le \sqrt{\epsilon}$. In the opposite limiting case when $\delta \ll \sqrt{\epsilon}$ expression (12.37) can be simplified. With this in mind we note that, for $|t - t_0| \gg \delta/\epsilon$, the error integral can be represented by Fresnel integrals [134], namely, $\Phi(\sqrt{z/\mathrm{i}}) = \sqrt{2/\mathrm{i}}\big(C(z) + \mathrm{i}S(z)\big)$, where

$$C(z) = \sqrt{\frac{2}{\pi}}\int_0^{\sqrt{2z/\pi}} \cos t^2\,\mathrm{d}t, \quad S(z) = \sqrt{\frac{2}{\pi}}\int_0^{\sqrt{2z/\pi}} \sin t^2\,\mathrm{d}t.$$

As result we obtain

$$A(t) = \frac{B\sqrt{\pi}}{4\omega_1\sqrt{\epsilon}}\,\mathrm{e}^{-\delta(t-t_0)}\Bigg[\Big(C\big(\epsilon(t-t_0)^2\big) - S\big(\epsilon(t-t_0)^2\big)\Big)^2$$
$$+\Big(1 + C\big(\epsilon(t-t_0)^2\big) + S\big(\epsilon(t-t_0)^2\big)\mathrm{sign}(t-t_0)\Big)^2\Bigg]^{1/2}. \tag{12.39}$$

Examples of the time dependencies of the oscillation amplitude $A(t)$, in relative units, constructed from the formula (12.39) are given in Fig. 12.4 a for two values of the parameter $q = \omega_0^2/\epsilon$ (in order for the condition $\delta \ll \sqrt{\epsilon}$

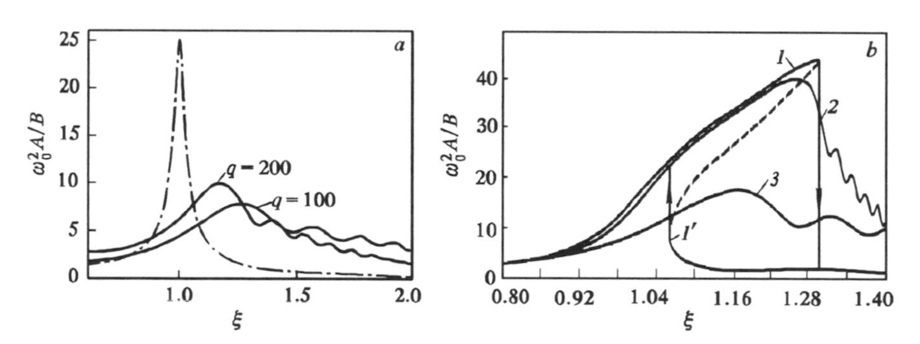

Fig. 12.4. Examples of the time dependencies of the oscillation amplitude $A(t)$ in the relative units $\omega_0^2 A/B$, $\xi = 2\epsilon t/\omega_1$ in the case of the linear time variation of the external force frequency: (a) for $\gamma = 0$, $\delta/\omega_0 = 0.02$, $q = 100$ and $q = 200$; and (b) for $B = 0.02\omega_0^2/\sqrt{\gamma}$, $\delta/\omega_0 = 0.01$, $q = \omega_0^2/\epsilon = \infty$ (curves 1, 1′), $q = 2000$ (curve 2) and $q = 400$ (curve 3). Dot-and-dash line in a corresponds to $q = \infty$

to be valid, the parameter q must be much less than ω_0^2/δ^2). For comparison the corresponding plot for $\delta \gg \sqrt{\epsilon}$ ($q \to \infty$) is presented as the dot-and-dash line in the same figure. It is seen from this figure that for $\delta \ll \sqrt{\epsilon}$ the smaller is q, i.e., the faster the frequency of the external force varies, the greater the resonance value of the frequency increases; and in the process the resonance becomes less pronounced. In addition, on the right slope of the resonance curve oscillations appear which are associated with oscillations of the Fresnel integrals. The 'period' of these oscillations is the smaller the greater is q. We note that similar dependencies were obtained by Lewis as early as 1932 [210].

We now turn our attention to slightly nonlinear oscillators with slowly time-varying parameters. As an example we consider the following equation:

$$\ddot{x} + 2\delta\dot{x} + \omega_0^2(1 + \gamma x^2)x = B\cos\psi(t). \tag{12.40}$$

For small damping ($\delta \ll \omega_0$) and sufficiently small amplitude of the external force Eq. (12.40) can be solved by the asymptotic method. In so doing a solution of Eq. (12.40) in the first approximation is $x(t) = A(t)\cos\big(\psi(t) + \varphi(t)\big)$, where A and φ must obey to the equations

$$\begin{aligned} \dot{A} &= -\,\delta A - \frac{B}{\omega_0 + \omega}\sin\varphi, \\ \dot{\varphi} &= \omega_0 - \omega + \frac{3}{8}\gamma\omega_0 A^2 - \frac{B}{A(\omega_0 + \omega)}\cos\varphi. \end{aligned} \tag{12.41}$$

The results of the numerical integration of Eqs. (12.41) presented in [238] show that the form of the resonance curves essentially depends on the direction in which the frequency ω varies. A number of examples of such curves calculated by us as the frequency ω increases are given in Fig. 12.4 b. They coincide with those presented in [238]. However, we did not find any curves similar those presented in [238] for the case when the frequency ω decreases (except for curve $1'$ corresponding to invariable ω). It is seen from Fig. 12.4 b that for $\delta \ll \sqrt{\epsilon}$, as in the linear case, on the far slope of the resonance curve (on the right slope as the action frequency increases and on the left one as it decreases) there are oscillations with a 'period' which is the smaller the greater is the parameter q.

In the monograph [238] several more complicated cases of the passage through resonance of systems with slowly time-varying parameters are also considered. But it is impossible to give here a complete description of these cases, for reasons of space.

12.3 Chaotic regimes in periodically driven nonlinear oscillators

As has been shown in many works, a periodic action upon a nonlinear oscillator need not entail its oscillations with a period which is a multiple of the action period. Chaotic oscillations are found to be possible. In experimental and numerical studies of these oscillations their spectrum seems to be continuous. Therefore an analogy arises between chaotic oscillations of periodically excited nonlinear oscillators and self-oscillatory systems with chaotic attractors. However, for the most part the observed continuous spectrum is dictated by the presence of weak noise, and in the absence of noise the oscillations would be periodic with a period divisible by the external force period.

12.3.1 Chaotic regimes in the Duffing oscillator

Many works are devoted to the study of chaotic oscillations described by the Duffing equation with a harmonic external force:

$$\ddot{x} + \alpha\dot{x} + \omega_0^2(1 + \gamma x^2)x = B\cos\omega t. \tag{12.42}$$

An overview of some these works is given in the book [253]. Of the recent works on this subject we point out [375, 335, 303, 344]. Here we dwell only on the main qualitative results obtained in studies of Eq. (12.42). It is evident that chaotic regimes can exist only for a sufficiently large amplitude of the external force, when the nonlinearity plays the essential role. As a rule, such regimes appear, as a parameter changes, via an infinite period-doubling bifurcation sequence in accordance with the Feigenbaum scenario. It was found that for negative γ, when there are three singular points on the phase plane of the corresponding autonomous oscillator, such regimes appear for significantly smaller amplitudes of the external force than for positive γ. An important point is that the region of existence of the chaotic regimes, as the external force frequency ω varies, is bounded; it is the wider the larger is the force amplitude (or the larger is the nonlinear parameter γ). It should be noted that within the region of chaotic oscillations there are narrow windows where oscillations are periodic with a period that is three, five and seven times larger than the period of the external force. Similar behavior is typical of systems with transitions to chaos in accordance with the Feigenbaum scenario.

12.3.2 Chaotic oscillations of a gas bubble in liquid under the action of a sound field

As an example of chaotic oscillations of a more complicated periodically driven nonlinear oscillator we consider a gas bubble in liquid under the action of a sound field. This problem is closely allied to the well-known cavitation problem. It has long been known from experimental results that, when cavitation arises, subharmonics in the power spectrum of the sound field appear and its strong noisiness is observed [82, 147]. For a long time these phenomena were ascribed to a very large number of bubbles each radiating a sound wave with a random phase. However, specially designed experiments by Lauterborn and co-workers [205], in which they were able to initiate a single bubble only in the liquid, showed that the noisiness occurs in this case as well.

We consider the following equation for the bubble oscillations [205, 262, 183]:

$$\left(1-\frac{\dot{R}}{a}\right)R\ddot{R}+\frac{3}{2}\dot{R}^2\left(1-\frac{\dot{R}}{3a}\right)-\frac{1}{\rho}\left(1+\frac{\dot{R}}{a}\right)P(R,\dot{R})-\frac{R}{\rho a}\frac{\mathrm{d}P}{\mathrm{d}t}=0, \tag{12.43}$$

where a is the sound velocity, ρ is the liquid density,

$$P(R,\dot{R})=\left(p_0-p_v+\frac{2\sigma}{R_0}\right)\left(\frac{R_0}{R}\right)^{3\kappa} \\ -\frac{2\sigma}{R}-4\eta\frac{\dot{R}}{R}-p_0+p_v-P_0\sin\omega t, \tag{12.44}$$

η is the dynamical viscosity of the liquid, σ is the surface tension factor, p_0 is the static pressure, p_v is the vapor pressure, R_0 is the equilibrium radius of the bubble, and κ is the polytropic exponent.

Equation (12.43) and expression (12.44) differ from (4.27) and (4.29) in that the vapor pressure in the bubble, the compressibility and viscosity of the liquid, as well as the presence of the external sound field described by the term $P_0 \sin\omega t$, are taken into account.

Equations (12.43), (12.44) were simulated by us [183] for the following set of parameters: $R_0 = 10$ μm, $P_0 = 90$ kPa, $p_0 = 100$ kPa, $p_v = 2.33$ kPa, $\sigma = 0.0725\ \mathrm{N\,m^{-1}}$, $\rho = 998\ \mathrm{kg\,m^{-3}}$, $\kappa = 4/3$, $\eta = 0.001\ \mathrm{N\,s\,m^{-3}}$, $a = 1500$ $\mathrm{m\,s^{-1}}$. The frequency of the sound field was varied from 60 to 72 kHz. It was found that over this range of ω the oscillations of the bubble radius are chaotic. Such oscillations and the projection of the corresponding phase trajectory on the plane $\tilde{R} = R/R_0$, $\mathrm{d}\tilde{R}/\mathrm{d}t$ are shown in Fig. 12.5.

12.3.3 Chaotic oscillations in the Vallis model

The Vallis model for the nonlinear interaction between ocean and atmosphere was considered in Ch. 7. It was shown that for a certain set of parameters chaotic self-oscillations are possible and the excitation of these self-oscillations is hard. Here we consider the case when a chaotic attractor is absent and the excitation of oscillations is caused by periodic variation of the tradewind velocity. The model is described by the equations

$$\begin{aligned}\dot{u} &= \frac{B}{2l}(T_\mathrm{e}-T_\mathrm{w})-C(u+u^*),\\ \dot{T}_\mathrm{w} &= \frac{u}{2l}(T_0-T_\mathrm{e})-A(T_\mathrm{w}-T^*),\\ \dot{T}_\mathrm{w} &= \frac{u}{2l}(T_\mathrm{w}-T_0)-A(T_\mathrm{e}-T^*),\end{aligned} \tag{12.45}$$

where $u^* = u_0 + u_1 \sin\omega t$ is a function describing the effect of the tradewinds.

The results of computation of Eqs. (12.45) for $A = 1\,\mathrm{year}^{-1}$, $l = 7500\,\mathrm{km}$, $C = 0.25\,\mathrm{month}^{-1}$, $B = 0.8562\,\mathrm{m^2\,s^{-1}\,K^{-1}}$, $T_0 = 0\,\mathrm{K}$, $T^* = 12\,\mathrm{K}$, $u_0 = u_1 =$

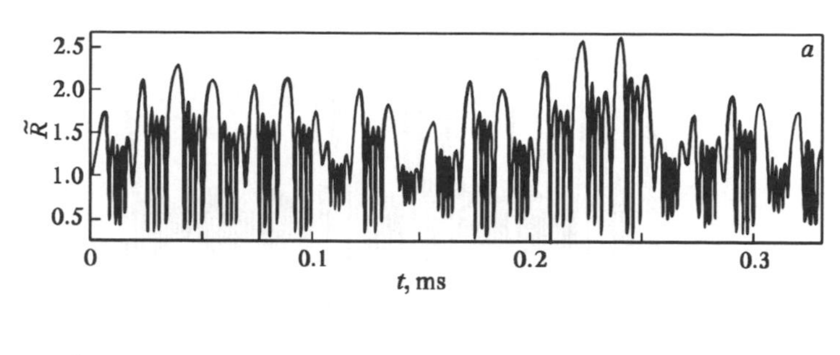

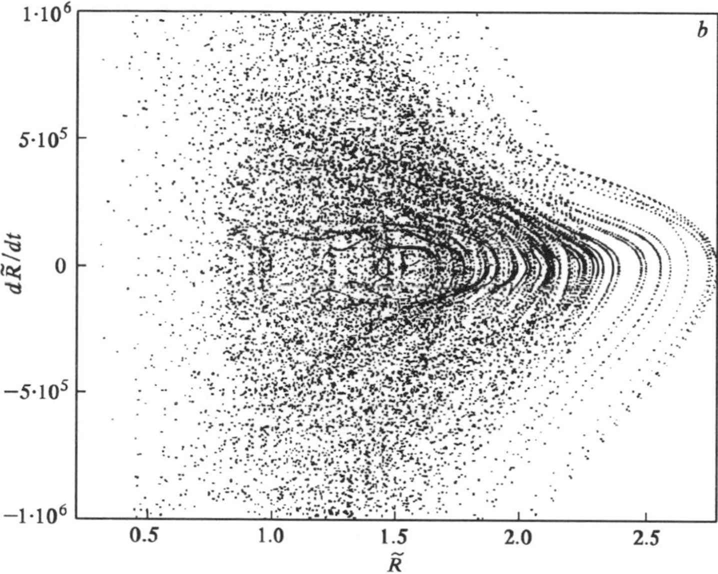

Fig. 12.5. An example of bubble chaotic oscillations for $\omega/2\pi = 65$ kHz: (a) the shape of the oscillations, (b) the projection of the corresponding phase trajectory on the plane $\tilde{R} = R/R_0$, $\mathrm{d}\tilde{R}/\mathrm{d}t$

$0.45\,\mathrm{m\,s^{-1}}$, $\omega = 2\pi\,\mathrm{year}^{-1}$ are given in Fig. 12.6. We see that these oscillations are similar in shape to the self-oscillations shown in Fig. 7.8.

12.4 Two coupled harmonically driven nonlinear oscillators

Let us consider two coupled quasilinear oscillators described by the equations

$$\begin{aligned} &\ddot{x}_1 + 2\delta_1\dot{x}_1 + \nu_1^2(1 + \beta_1 x_1 + \gamma_1 x_1^2)x_1 + c_1 x_2 = B_1 \cos\omega t, \\ &\ddot{x}_2 + 2\delta_2\dot{x}_2 + \nu_2^2(1 + \beta_2 x_2 + \gamma_2 x_2^2)x_2 + c_2 x_1 = 0. \end{aligned} \qquad (12.46)$$

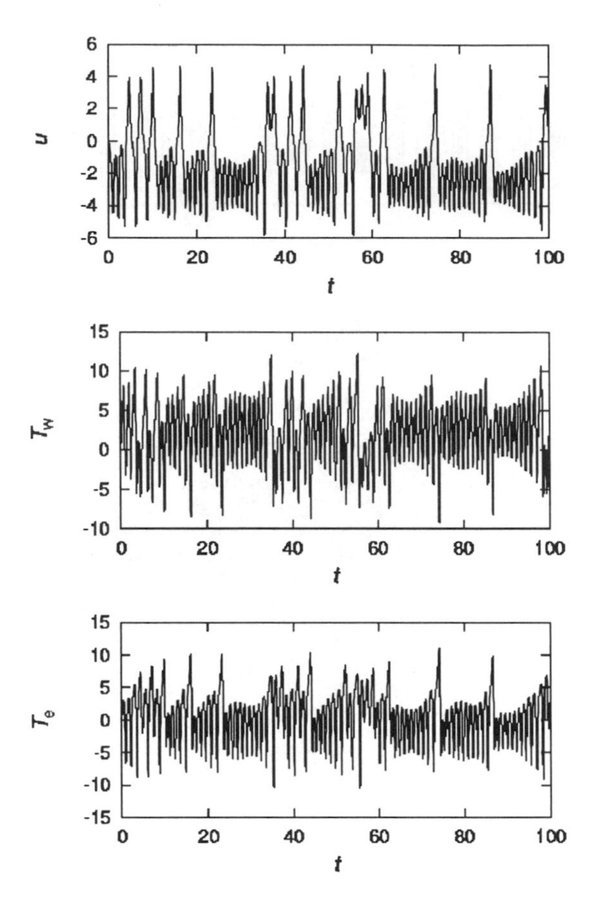

Fig. 12.6. Chaotic self-oscillations in the Vallis model caused by periodic variation of the tradewind velocity

12.4.1 The main resonance

As before, we use the Krylov–Bogolyubov asymptotic method for the approximate solution of Eqs. (12.46). Presuming that the coupling factors c_1 and c_2, as well as the difference between the natural frequencies $\xi = \nu_1 - \nu_2$, are sufficiently small, for $\omega \sim \nu_1 \sim \nu_2$ we can ignore the terms $\beta_{1,2}x_{1,2}$ and rewrite Eqs. (12.46) as

$$\ddot{x}_1 + \omega^2 x_1 = \epsilon\Big(-2\delta_1\dot{x}_1 - \nu_1^2\gamma_1 x_1^3 + (\omega^2 - \nu_1^2)x_1 - c_1 x_2 + B_1 \cos\omega t\Big), \tag{12.47}$$

$$\ddot{x}_2 + \omega^2 x_2 = \epsilon\Big(-2\delta_2\dot{x}_2 - \nu_2^2\gamma_2 x_2^3 + (\omega^2 - \nu_2^2)x_2 - c_2 x_1\Big),$$

where ϵ is a small parameter.

We seek a solution of Eqs. (12.47) in the form

$$x_1 = A_1 \cos\psi_1 + \epsilon u_1 + \dots, \quad x_2 = A_2 \cos\psi_2 + \epsilon u_2 + \dots, \tag{12.48}$$

where $\psi_{1,2} = \omega t + \varphi_{1,2}$, and $A_{1,2}$ and $\varphi_{1,2}$ obey the equations

$$\begin{aligned} \dot{A}_{1,2} &= \epsilon f_{1,2}(A_1, A_2, \varphi_1, \varphi_2) + \dots, \\ \dot{\varphi}_{1,2} &= \epsilon F_{1,2}(A_1, A_2, \varphi_1, \varphi_2) + \dots, \end{aligned} \tag{12.49}$$

$f_{1,2}, F_{1,2}, \dots, u_1, u_2, \dots$ are unknown functions. By substituting (12.48) into Eqs. (12.47) and equating coefficients of ϵ we obtain the following equations for u_1 and u_2:

$$\begin{aligned} \ddot{u}_1 + \omega^2 u_1 &= 2\delta_1 \omega A_1 \sin\psi_1 - \nu_1^2 \gamma_1 A_1^3 (\cos\psi_1)^3 + (\omega^2 - \nu_1^2) A_1 \cos\psi_1 \\ &\quad - c_1 A_2 \cos\psi_2 + 2\omega f_1 \sin\psi_1 + 2A_1 \omega F_1 \cos\psi_1 + B_1 \cos\omega t, \\ \ddot{u}_2 + \omega^2 u_2 &= 2\delta_2 \omega A_2 \sin\psi_2 - \nu_2^2 \gamma_2 A_2^3 (\cos\psi_2)^3 + (\omega^2 - \nu_2^2) A_2 \cos\psi_2 \\ &\quad - c_2 A_1 \cos\psi_1 + 2\omega f_2 \sin\psi_2 + 2A_2 \omega F_2 \cos\psi_2. \end{aligned}$$

From the conditions for the absence of resonance terms in the functions u_1 and u_2 we find

$$\begin{aligned} f_1 &= -\delta_1 A_1 + \frac{c_1 A_2}{2\omega} \sin\Phi - \frac{B_1}{2\omega} \sin\varphi_1, \\ F_1 &= -\Delta_1 + \frac{m_1 A_2}{2\omega A_1} \cos\Phi + \frac{3\omega\gamma_1 A_1^2}{8} - \frac{B_1}{2\omega A_1} \cos\varphi_1, \\ f_2 &= -\delta_2 A_2 - \frac{m_2 A_1}{2\omega} \sin\Phi, \quad F_2 = -\Delta_2 + \frac{m_2 A_1}{2\omega A_2} \cos\Phi + \frac{3\omega\gamma_2 A_2^2}{8}, \end{aligned} \tag{12.50}$$

where $\Phi = \varphi_1 - \varphi_2$, $\Delta_1 = \omega - \nu_1$, $\Delta_2 = \omega - \nu_2 = \Delta_1 + \nu_1 - \nu_2$.

We further substitute (12.50) into Eqs. (12.49) and put $\dot{A}_{1,2} = \dot{\varphi}_{1,2} = 0$. As a result we obtain four equations for steady-state values of A_1, A_2, φ_1 and Φ. These equations are conveniently written as

$$A_1^2 = \frac{4\omega^2}{m_2^2} A_2^2 \left[\delta_2^2 + \left(\Delta + \frac{\xi}{2} - \frac{3}{8} \omega\gamma_2 A_2^2 \right)^2 \right], \tag{12.51a}$$

$$\begin{aligned} &8\omega^2 \frac{c_1 A_2^2}{c_2} \left[\delta_1 \delta_2 - \left(\Delta - \frac{\xi}{2} - \frac{3}{8} \omega\gamma_1 A_1^2 \right) \left(\Delta + \frac{\xi}{2} - \frac{3}{8} \omega\gamma_2 A_2^2 \right) \right] \\ &\quad + c_1^2 A_2^2 + 4\omega^2 A_1^2 \left[\delta_1^2 + \left(\Delta - \frac{\xi}{2} - \frac{3}{8} \omega\gamma_1 A_1^2 \right)^2 \right] = B_1^2, \end{aligned} \tag{12.51b}$$

$$\begin{aligned} \tan\varphi_1 &= \left(\delta_1 + \frac{m_1 A_2^2}{m_2 A_1^2} \delta_2 \right) \left[\Delta - \frac{\xi}{2} - \frac{3}{8} \omega\gamma_1 A_1^2 \right. \\ &\quad \left. - \frac{c_1 A_2^2}{c_2 A_1^2} \left(\Delta + \frac{\xi}{2} - \frac{3}{8} \omega\gamma_2 A_2^2 \right) \right]^{-1}, \\ \tan\Phi_1 &= -\delta_2 \left(\Delta + \frac{\xi}{2} - \frac{3}{8} \omega\gamma_2 A_2^2 \right)^{-1}, \end{aligned} \tag{12.52}$$

where $\Delta = (\Delta_1 + \Delta_2)/2 = \omega - (\nu_1 + \nu_2)/2$ is the frequency mistuning.

We see that the amplitude A_1 is easily eliminated from Eqs. (12.51a), (12.51b) and a closed equation for A_2 can then be obtained. In the linear case this equation is

$$\left[c_1^2 + \frac{16\omega^4}{m_2^2}\left(\delta_1^2 + \left(\Delta - \frac{\xi}{2}\right)^2\right)\left(\delta_2^2 + \left(\Delta + \frac{\xi}{2}\right)^2\right) + 8\omega^2 \frac{c_1}{c_2}\left(\delta_1\delta_2 - \Delta^2 + \frac{\xi^2}{4}\right)\right] A_2^2 = B_1^2. \tag{12.53}$$

It follows from (12.53) that the amplitude A_2 has extrema at values of Δ that are the roots of the equation

$$4\Delta^3 + 2\Delta\left(\delta_1^2 + \delta_2^2 - \frac{\xi^2}{2} - \frac{c_1c_2}{2\omega^2}\right) + (\delta_1^2 - \delta_2^2)\xi = 0. \tag{12.54}$$

Equation (12.54) possesses either one or three real roots. If damping of natural oscillations is sufficiently large, and ξ as well as the coupling factors c_1, c_2 are sufficiently small, then Eq. (12.54) possesses only one real root. In this case the resonance curve has only a single maximum. But if Eq. (12.54) possesses three real roots then the resonance curve has two maxima and one minimum.

Let us demonstrate this by an example in which the natural frequencies are equal, i.e., $\xi = 0$. In this case Eq. (12.54) has one real root if $\delta_1^2 + \delta_2^2 \geq c_1c_2/2\omega^2$, i.e., if

$$\frac{1}{Q_1^2} + \frac{1}{Q_2^2} \geq 2\,\frac{c_1c_2}{\omega^4}\,, \tag{12.55}$$

where $Q_{1,2} = \omega/2\delta_{1,2}$ are the Q-factors of the oscillators. The resonance curves for this case are shown in Fig. 12.7 a and b. It is seen from this figure that the resonance curve $A_1(\omega)$ is wider then the curve $A_2(\omega)$. Furthermore, even if condition (12.55) is fulfilled, the curve $A_1(\omega)$, as opposed to $A_2(\omega)$, may have two peaks (Fig. 12.7 b). The resonance curves for the case when condition (12.55) is not fulfilled are given in Fig. 12.7 c. We see that the peaks on the curve $A_1(\omega)$ are much more pronounced and wider apart than on the curve $A_2(\omega)$.

Let us further consider the nonlinear case, and for the sake of simplicity suppose that $\gamma_1 = 0$, $\gamma_2 \neq 0$. Eliminating A_1 from Eqs. (12.51a) and (12.51b), we obtain the following equation for A_2:

$$\left[c_1^2 + \frac{16\omega^4}{c_2^2}\left(\delta_1^2 + \left(\Delta - \frac{\xi}{2}\right)^2\right)\left(\delta_2^2 + \left(\Delta + \frac{\xi}{2} - \frac{3}{8}\omega\gamma_2A_2^2\right)^2\right) + 8\omega^2\frac{c_1}{c_2}\left(\delta_1\delta_2 - \left(\Delta - \frac{\xi}{2}\right)\left(\Delta + \frac{\xi}{2} - \frac{3}{8}\omega\gamma_2A_2^2\right)\right)\right]A_2^2 = B_1^2.$$

This equation in combination with Eq. (12.51a) allows us to construct the resonance curves $A_1(\omega)$ and $A_2(\omega)$. These curves are presented in Fig. 12.8

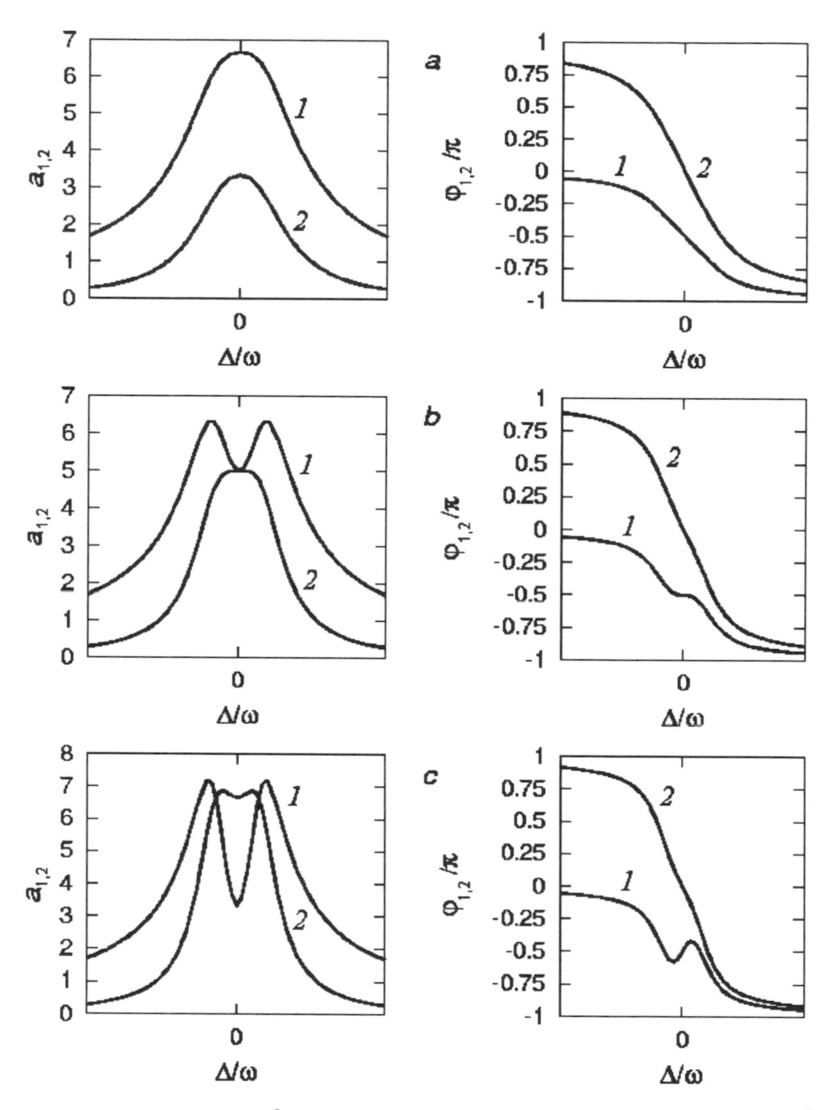

Fig. 12.7. Plots of $a_1 = \omega^2 A_1/B$ (curves 1 on the left), $a_2 = (\omega^2 A_2/B_1)\sqrt{c_1/c_2}$ (curves 2 on the left), φ_1 (curves 1 on the right) and φ_2 (curves 2 on the right) versus Δ/ω for $c_1 c_2/\omega^4 = 0.01$, $Q_1 = 10$ and (a) $Q_2 = 5$, (b) $Q_2 = 10$ and (c) $Q_2 = 20$

for $\gamma_2 B_1^2/c_1^2 = 40/3$ and the values of the remaining parameters equal to those for the linear case considered above. It is seen from this figure that with the nonlinearity the resonance curves are asymmetric relative to the sign of the frequency mistuning Δ. Furthermore, the nonlinearity significantly depresses the influence of the second oscillator on the first: in contrast to the linear case

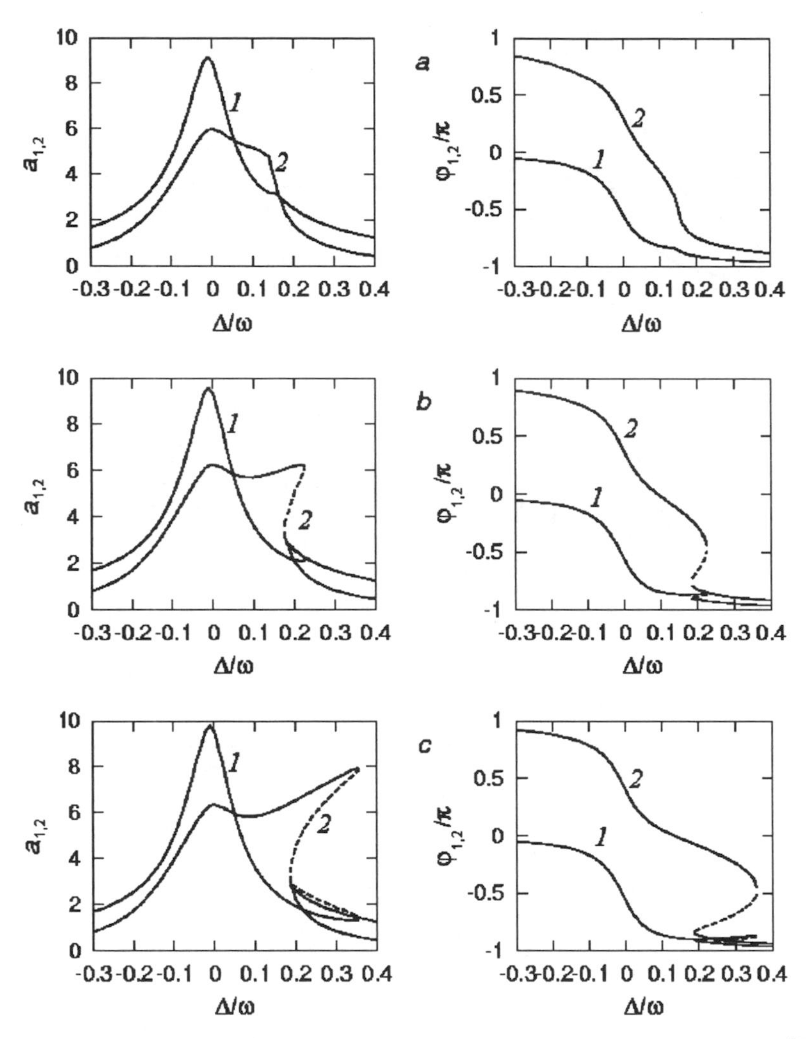

Fig. 12.8. Plots of $a_1 = \omega^2 A_1/B_1$ (curves 1 on the left), $a_2 = (3\omega^2 A_2/B_1)\sqrt{c_1/c_2}$ (curves 2 on the left), φ_1 (curves 1 on the right) and φ_2 (curves 2 on the right) versus Δ/ω for $\gamma_2 B_1^2/c_1^2 = 40/3$, $c_1 c_2/\omega^4 = 0.01$, $Q_1 = 10$ and (a) $Q_2 = 5$, (b) $Q_2 = 10$, (c) $Q_2 = 20$

the resonance curves $A_1(\omega)$ are almost the same as for uncoupled oscillators. Finally, the nonlinearity in the second oscillator significantly decreases its oscillation amplitude, and, as would be expected, causes hysteresis phenomena. However, in contrast to a system having one degree of freedom, in this case the hysteresis takes place only for a sufficiently large Q-factor of the nonlinear oscillator.

In conclusion we revert to the linear case and consider what happens to the results obtained if it is assumed that an external force with amplitude B_2 acts not upon the first oscillator but upon the second. It can easily be shown that in this case

$$A_2^2 = \frac{4\omega^2}{m_1^2}\left(\delta_1^2 + \Delta_1^2\right) A_1^2,$$

$$A_1^2\left(c_2^2 + \frac{16\omega^4}{c_1^2}(\delta_1^2 + \Delta_1^2)(\delta_2^2 + \Delta_2^2) + 8\omega^2\frac{c_1}{c_2}(\delta_1\delta_2 - \Delta_1\Delta_2)\right) = B_2^2.$$

Comparing these expressions with (12.51a), (12.53), we see that the oscillation amplitude in the first oscillator A_1, when an external force with amplitude B_2 acts upon the second oscillator, is equal to the oscillation amplitude in the second oscillator A_2, when an external force with amplitude $B_1 = c_1 B_2/c_2$ acts upon the first oscillator. This statement can be treated as a generalized formulation of the known reciprocity theorem which is usually formulated for the particular case of $c_1 = c_2$.

12.4.2 The combination resonance

The results of studies of so-called combination resonances in slightly nonlinear systems with two degrees of freedom in the case of an arbitrary nonlinearity are presented, for example, in [223, 368, 325]. We consider only a particular case when a system is described by the following equations:

$$\begin{aligned}
&\ddot{x}_1 + 2\delta_1\dot{x}_1 + \nu_1^2(1 + \beta_1 x_1 + \gamma_1 x_1^2)x_1 + c_1 x_2 = B\cos\omega t,\\
&\ddot{x}_2 + 2\delta_2\dot{x}_2 + \nu_2^2(1 + \beta_2 x_2 + \gamma_2 x_2^2)x_2 + c_2 x_1 = 0.
\end{aligned} \tag{12.56}$$

By a *combination resonance* in systems with two degrees of freedom is meant excitation of oscillations with fundamental frequencies ω_1, ω_2 and ω obeying the relation

$$n\omega = n_1\omega_1 + n_2\omega_2, \tag{12.57}$$

where n, n_1, n_2 are equal to $\pm 1, \pm 2, \pm 3, \ldots$.

We assume that the nonlinearity and damping factors in the system (12.56) are sufficiently small. In this case Eqs. (12.56) can be rewritten as

$$\begin{aligned}
&\ddot{x}_1 + \nu_1^2 x_1 + c_1 x_2 - B\cos\omega t = \epsilon\Big(-2\delta_1\dot{x}_1 - \nu_1^2(\beta_1 + \gamma_1 x_1)x_1^2\Big),\\
&\ddot{x}_2 + \nu_2^2 x_2 + c_2 x_1 = \epsilon\Big(-2\delta_2\dot{x}_2 - \nu_2^2(\beta_2 + \gamma_2 x_2)x_2^2\Big),
\end{aligned} \tag{12.58}$$

where ϵ is a small parameter. An approximate solution of Eqs. (12.58) is conveniently found by substituting new variables y_1 and y_2 which describe normal oscillations of the generative system (for $\epsilon = 0$, $B = 0$). For this purpose we set

$$x_1 = y_1 + y_2 + C_1 \cos\omega t,$$
$$x_2 = \frac{\omega_{01}^2 - \nu_1^2}{c_1} y_1 + \frac{\omega_{02}^2 - \nu_1^2}{c_1} y_2 + C_2 \cos\omega t, \tag{12.59}$$

where

$$C_1 = \frac{\nu_2^2 - \omega^2}{(\nu_1^2 - \omega^2)(\nu_2^2 - \omega^2) - c_1 c_2} B,$$
$$C_2 = -\frac{c_2}{(\nu_1^2 - \omega^2)(\nu_2^2 - \omega^2) - c_1 c_2} B, \tag{12.60}$$

and $\omega_{01,2} = \sqrt{(\nu_1^2 + \nu_2^2)/2 \pm \sqrt{(\nu_1^2 - \nu_2^2)^2/4 + c_1 c_2}}$ are the normal frequencies. By substituting (12.59) into Eqs. (12.58) we find the equations for y_1 and y_2 which are conveniently written in the form

$$\ddot{y}_1 + \omega_{01}^2 y_1 = \epsilon\left(\frac{2}{\omega_{01}^2 - \omega_{02}^2}\left[\delta_1(\omega_{02}^2 - \nu_1^2)(\dot{y}_1 + \dot{y}_2 - C_1\omega\sin\omega t)\right.\right.$$
$$\left.\left. - \delta_2\Big((\omega_{01}^2 - \nu_1^2)\dot{y}_1 + (\omega_{02}^2 - \nu_1^2)\dot{y}_2 - c_1 C_2\omega\sin\omega t\Big)\right] + \mathcal{F}_1\right),$$
$$\ddot{y}_2 + \omega_{02}^2 y_2 = \epsilon\left(\frac{2}{\omega_{01}^2 - \omega_{02}^2}\left[-\delta_1(\omega_{01}^2 - \nu_1^2)(\dot{y}_1 + \dot{y}_2 - C_1\omega\sin\omega t)\right.\right. \tag{12.61}$$
$$\left.\left. + \delta_2\Big((\omega_{01}^2 - \nu_1^2)\dot{y}_1 + (\omega_{02}^2 - \nu_1^2)\dot{y}_2 - m_1 C_2\omega\sin\omega t\Big)\right] + \mathcal{F}_2\right),$$

where $\mathcal{F}_{1,2}$ are nonlinear terms, which are not written out because of their awkwardness.

A solution of Eqs. (12.61) is sought in the form

$$y_1 = A_1\cos\psi_1 + \epsilon u_1 + \dots, \quad y_2 = A_2\cos\psi_2 + \epsilon v_1 + \dots, \tag{12.62}$$

where $\psi_{1,2} = \omega_{1,2}t + \varphi_{1,2}$, ω_1 and ω_2 are frequencies obeying the relation (12.57). In the first approximation we obtain the following equations for $A_{1,2}$ and $\varphi_{1,2}$:

$$\dot{A}_{1,2} = \frac{\delta_{1,2}(\omega_{02}^2 - \nu_1^2) - \delta_{2,1}(\omega_{01}^2 - \nu_1^2)}{\omega_{01}^2 - \omega_{02}^2} A_{1,2} + f_{1,2},$$
$$\dot{\varphi}_{1,2} = -\Delta_{1,2} + F_{1,2}, \tag{12.63}$$

where $\Delta_1 = \omega_1 - \omega_{01}$, $\Delta_2 = \omega_2 - \omega_{02} = (n\omega - n_1\omega_1)/n_2 - \omega_{02} = (\Delta - n_1\Delta_1)/n_2$, $\Delta = n\omega - n_1\omega_{01} - n_2\omega_{02}$ is the frequency mistuning, and $f_{1,2}$ and $F_{1,2}$ are functions which are solutions of the equations

$$\omega_{01,2} f_{1,2} - \frac{\Delta_{1,2} A_{1,2}}{2} \frac{\partial F_{1,2}}{\partial \varphi_{1,2}} = - \int_0^{2\pi} \mathcal{F}_{1,2}\Big|_{y_{1,2}=y_{10,20}} \sin \psi_{1,2} \, d\psi_{1,2},$$

$$\omega_{01,2} A_{1,2} F_{1,2} - \frac{\Delta_{1,2}}{2} \frac{\partial f_{1,2}}{\partial \varphi_{1,2}} = - \int_0^{2\pi} \mathcal{F}_{1,2}\Big|_{y_{1,2}=y_{10,20}} \cos \psi_{1,2} \, d\psi_{1,2}, \tag{12.64}$$

where $y_{10} = A_1 \cos \psi_1$, $y_{20} = A_2 \cos \psi_2$.

Since $\omega t = (n_1\psi_1 + n_2\psi_2)/n - \Phi$, where $\Phi = (n_1\varphi_1 + n_2\varphi_2)/n$, the functions $f_{1,2}$ and $F_{1,2}$ depend only on Φ but not on φ_1 and φ_2. Hence, putting $\dot{A}_{1,2} = \dot{\varphi}_{1,2} = 0$ in Eqs. (12.63), we can find steady-state values of A_1, A_2, Φ and one of the unknown frequencies ω_1 or ω_2 (the other frequency is determined by the relation (12.57)).

Let us consider two specific cases:

(i) $n = n_1 = 1$, $n_2 = \pm 1$, i.e., $\omega = \omega_1 \pm \omega_2$. In this case

$$\int_0^{2\pi} \mathcal{F}_1\Big|_{y_{1,2}=y_{10,20}} \sin \psi_1 \, d\psi_1 = \frac{\omega_{02}^2 - \nu_1^2}{2(\omega_{01}^2 - \omega_{02}^2)} (\nu_1^2 \beta_1 C_1 - \nu_2^2 \beta_2 C_2) A_2 \sin \Phi,$$

$$\int_0^{2\pi} \mathcal{F}_1\Big|_{y_{1,2}=y_{10,20}} \cos \psi_1 \, d\psi_1 = \frac{1}{2(\omega_{01}^2 - \omega_{02}^2)} \Bigg\{ \frac{3}{2} \Bigg[\nu_1^2 (\omega_{02}^2 - \nu_1^2) \gamma_1 \left(\frac{A_1^2}{2} + A_2^2 + C_1^2 \right) - \nu_2^2 (\omega_{01}^2 - \nu_1^2) \gamma_2 \left(\frac{(\omega_{01}^2 - \nu_1^2)^2}{2c_1^2} A_1^2 + \frac{(\omega_{02}^2 - \nu_1^2)^2}{c_1^2} A_2^2 + C_2^2 \right) \Bigg] A_1 + (\omega_{02}^2 - \nu_1^2)(\nu_1^2 \beta_1 C_1 - \nu_2^2 \beta_2 C_2) A_2 \cos \Phi \Bigg\}, \tag{12.65}$$

$$\int_0^{2\pi} \mathcal{F}_2\Big|_{y_{1,2}=y_{10,20}} \sin \psi_2 \, d\psi_2 = \frac{\omega_{01}^2 - \nu_1^2}{2(\omega_{01}^2 - \omega_{02}^2)} (\nu_1^2 \beta_1 C_1 - \nu_2^2 \beta_2 C_2) A_1 \sin \Phi,$$

$$\int_0^{2\pi} \mathcal{F}_2\Big|_{y_{1,2}=y_{10,20}} \cos \psi_2 \, d\psi_2 = \frac{1}{2(\omega_{01}^2 - \omega_{02}^2)} \Bigg\{ \frac{3}{2} \Bigg[\nu_1^2 (\omega_{01}^2 - \nu_1^2) \gamma_1 \left(A_1^2 + \frac{A_2^2}{2} + C_1^2 \right) - \nu_2^2 (\omega_{02}^2 - \nu_1^2) \gamma_2 \left(\frac{(\omega_{01}^2 - \nu_1^2)^2}{m_1^2} A_1^2 + \frac{(\omega_{02}^2 - \nu_1^2)^2}{2m_1^2} A_2^2 + C_2^2 \right) \Bigg] A_1 + (\omega_{01}^2 - \nu_1^2)(\nu_1^2 \beta_1 C_1 - \nu_2^2 \beta_2 C_2) A_1 \cos \Phi \Bigg\}.$$

By substituting (12.65) into (12.63) and setting $\dot{A}_{1,2} = \dot{\varphi}_{1,2} = 0$ we obtain the following equations for steady-state values of A_1, A_2, Φ and ω_1:

$$\left(\delta_1(\omega_{02}^2-\nu_1^2)-\delta_2(\omega_{01}^2-\nu_1^2)\right)A_1 \\ -\frac{\omega_{02}^2-\nu_1^2}{\omega_1+\omega_{01}}(\nu_1^2\beta_1C_1-\nu_2^2\beta_2C_2)A_2\sin\Phi=0,$$

$$\left(\delta_1(\omega_{01}^2-\nu_1^2)-\delta_2(\omega_{02}^2-\nu_1^2)\right)A_2 \\ \mp\frac{\omega_{01}^2-\nu_1^2}{\omega_2+\omega_{02}}(\nu_1^2\beta_1C_1-\nu_2^2\beta_2C_2)A_1\sin\Phi=0, \tag{12.66}$$

$$\Delta_1(\omega_{01}^2-\omega_{02}^2)+\frac{3}{4}\left[\frac{\nu_1^2(\omega_{02}^2-\nu_1^2)}{\omega_{01}}\gamma_1\left(\frac{A_1^2}{2}+A_2^2+C_1^2\right)\right. \\ \left.-\frac{\nu_2^2(\omega_{01}^2-\nu_1^2)}{c_1^2\omega_{01}}\gamma_2\left(\frac{(\omega_{01}^2-\nu_1^2)^2}{2}A_1^2+(\omega_{02}^2-\nu_1^2)^2A_2^2+c_1^2C_2^2\right)\right] \\ +\frac{\omega_{02}^2-\nu_1^2}{\omega_1+\omega_{01}}(\nu_1^2\beta_1C_1-\nu_2^2\beta_2C_2)\frac{A_2}{A_1}\cos\Phi=0,$$

$$\Delta_2(\omega_{01}^2-\omega_{02}^2)-\frac{3}{4}\left[\frac{\nu_1^2(\omega_{01}^2-\nu_1^2)}{\omega_{02}}\gamma_1\left(A_1^2+\frac{A_2^2}{2}+C_1^2\right)\right. \\ \left.-\frac{\nu_2^2(\omega_{02}^2-\nu_1^2)}{c_1^2\omega_{02}}\gamma_2\left((\omega_{01}^2-\nu_1^2)^2A_1^2+\frac{(\omega_{02}^2-\nu_1^2)^2}{2}A_2^2+c_1^2C_2^2\right)\right] \\ -\frac{\omega_{01}^2-\nu_1^2}{\omega_2+\omega_{02}}(\nu_1^2\beta_1C_1-\nu_2^2\beta_2C_2)\frac{A_1}{A_2}\cos\Phi=0, \tag{12.67}$$

where $\Delta_2=\pm(\Delta-\Delta_1)$, $\Delta=\omega-(\omega_{01}\pm\omega_{02})$.

Analysis of Eqs. (12.66), (12.67) shows that they possess nontrivial solutions only for $n_2=+1$, i.e., for $\omega=\omega_1+\omega_2$. Another necessary condition for the existence of nontrivial solutions is that the external force must exceed some threshold level.

As an example we find a nontrivial solution of Eqs. (12.66), (12.67) in the simplest case when $\nu_1=\nu_2=\nu$, $\delta_1=\delta_2=\delta$, $\beta_1=\beta_2=\beta$, $\gamma_1=\gamma_2=\gamma$, $c_1=c_2=c$. In this case from Eqs. (12.66) we find

$$\frac{A_1^2}{A_2^2}=\frac{\omega_2+\omega_{02}}{\omega_1+\omega_{01}},\qquad \sin\Phi=\frac{\delta\sqrt{(\omega_1+\omega_{01})(\omega_2+\omega_{02})}}{\nu^2\beta B}\left(\omega_{02}^2-\omega^2\right), \tag{12.68}$$

where $\omega_{01}=\sqrt{\nu^2+c}$, $\omega_{02}=\sqrt{\nu^2-c}$. Adding and subtracting Eqs. (12.67), in view of (12.60) and (12.68), we obtain the following equations for determining A_1^2 and ω_1:

$$\gamma A_1^2\left[\frac{1}{\omega_{01}}+\frac{2}{\omega_{02}}+\frac{\omega_1+\omega_{01}}{\omega_2+\omega_{02}}\left(\frac{2}{\omega_{01}}+\frac{1}{\omega_{02}}\right)\right] \\ =\frac{4}{3}\frac{\Delta-2\delta\cot\Phi}{\nu^2}-\left(\frac{1}{\omega_{01}}+\frac{1}{\omega_{02}}\right)\frac{(\nu^2-\omega^2)^2+c^2}{\left((\nu^2-\omega^2)^2-c^2\right)^2}\gamma B^2, \tag{12.69a}$$

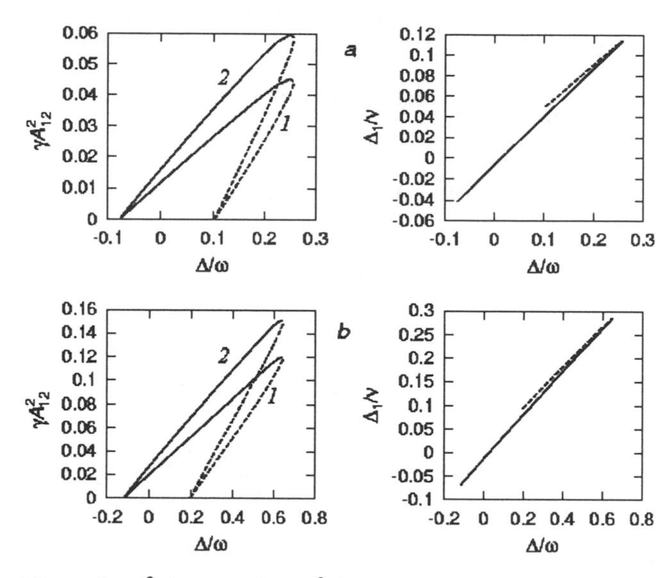

Fig. 12.9. Plots of γA_1^2 (curves 1), γA_2^2 (curves 2), and of Δ_1/ν versus Δ/ν in the case of the simplest combination resonance for $\gamma = \beta^2$, $c/\nu^2 = 0.3$, $2\delta/\nu = 0.1$, (a) $\gamma B^2/\nu^4 = 0.2$ and (b) $\gamma B^2/\nu^4 = 0.5$. Solid lines correspond to stable parts of the dependencies, dashed lines correspond to unstable ones

$$\gamma A_1^2 \left[\frac{1}{\omega_{01}} - \frac{2}{\omega_{02}} + \frac{\omega_1 + \omega_{01}}{\omega_2 + \omega_{02}} \left(\frac{2}{\omega_{01}} - \frac{1}{\omega_{02}}\right)\right]$$
$$= \frac{4}{3}\frac{2\Delta_1 - \Delta}{\nu^2} - \left(\frac{1}{\omega_{01}} - \frac{1}{\omega_{02}}\right) \frac{(\nu^2 - \omega^2)^2 + c^2}{\left((\nu^2 - \omega^2)^2 - c^2\right)^2} \gamma B^2, \tag{12.69b}$$

where

$$\cot \Phi = \pm \sqrt{\frac{\nu^4 \beta^2 B^2}{\delta^2 (\omega_1 + \omega_{01})(\omega_2 + \omega_{02})(\omega_{02}^2 - \omega^2)^2} - 1}\,.$$

The dependencies of $\gamma A_{1,2}^2$ and Δ_1/ν on Δ/ν found from (12.69a) and (12.69b) for two values of $\gamma B^2/\nu^4$ are given in Fig. 12.9. It is seen from this figure that there is a region of the frequency mistuning $\Delta^{(1)} \le \Delta \le \Delta^{(2)}$ in which the combination oscillations under consideration are excited for as small initial perturbations as wished. The values of $\Delta^{(1,2)}$ can be found from Eqs. (12.69a) and (12.69b) by putting $\gamma A_1^2 = 0$ there. The condition $\Delta^{(1)} = \Delta^{(2)}$ determines the threshold value of the external force amplitude B_{th}. It should also be noted that the amplitude of the low-frequency component (A_2) is more than the amplitude of the high-frequency one (A_1).

(ii) $n = 1$, $n_1 = 2$, $n_2 = \pm 1$, i.e., $\omega = 2\omega_1 \pm \omega_2$. In this case

$$\int_0^{2\pi} \mathcal{F}_1\Big|_{y_{1,2}=y_{10,20}} \sin\psi_1 \, d\psi_1 = \frac{3(\omega_{02}^2 - \nu_1^2)}{4(\omega_{01}^2 - \omega_{02}^2)} \left(\nu_1^2 \gamma_1 C_1 - \frac{(\omega_{01}^2 - \nu_1^2)\gamma_2 \nu_2^2 C_2}{c_1} \right) A_1 A_2 \sin\Phi,$$

$$\int_0^{2\pi} \mathcal{F}_1\Big|_{y_{1,2}=y_{10,20}} \cos\psi_1 \, d\psi_1 = \frac{3A_1}{4(\omega_{01}^2 - \omega_{02}^2)} \Bigg[\nu_1^2 (\omega_{02}^2 - \nu_1^2)\gamma_1 \left(\frac{A_1^2}{2} + A_2^2 + C_1^2 \right) - \nu_2^2 (\omega_{01}^2 - \nu_1^2)\gamma_2 \left(\frac{(\omega_{01}^2 - \nu_1^2)^2}{2c_1^2} A_1^2 + \frac{(\omega_{02}^2 - \nu_1^2)^2}{c_1^2} A_2^2 + C_2^2 \right) + (\omega_{02}^2 - \nu_1^2) \left(\nu_1^2 \gamma_1 C_1 - \frac{(\omega_{01}^2 - \nu_1^2)\gamma_2 \nu_2^2 C_2}{c_1} \right) A_2 \cos\Phi \Bigg],$$

$$\int_0^{2\pi} \mathcal{F}_2\Big|_{y_{1,2}=y_{10,20}} \sin\psi_2 \, d\psi_2 = \pm \frac{3(\omega_{01}^2 - \nu_1^2)}{4(\omega_{01}^2 - \omega_{02}^2)} \left(\nu_1^2 \gamma_1 C_1 - \frac{(\omega_{01}^2 - \nu_1^2)\gamma_2 \nu_2^2 C_2}{c_1} \right) A_1^2 \sin\Phi,$$

$$\int_0^{2\pi} \mathcal{F}_2\Big|_{y_{1,2}=y_{10,20}} \cos\psi_2 \, d\psi_2 = \frac{3A_2}{4(\omega_{01}^2 - \omega_{02}^2)} \Bigg[\nu_1^2 (\omega_{01}^2 - \nu_1^2)\gamma_1 \left(A_1^2 + \frac{A_2^2}{2} + C_1^2 \right) - \nu_2^2 (\omega_{02}^2 - \nu_1^2)\gamma_2 \left(\frac{(\omega_{01}^2 - \nu_1^2)^2}{c_1^2} A_1^2 + \frac{(\omega_{02}^2 - \nu_1^2)^2}{2c_1^2} A_2^2 + C_2^2 \right) + (\omega_{01}^2 - \nu_1^2) \left(\nu_1^2 \gamma_1 C_1 - \frac{(\omega_{01}^2 - \nu_1^2)\gamma_2 \nu_2^2 C_2}{c_1} \right) \frac{A_1^2}{A_2} \cos\Phi \Bigg]. \tag{12.70}$$

By substitution of (12.70) into (12.64) and (12.63) we obtain the following equations for steady-state values of the amplitudes and phases:

$$\delta_1(\omega_{02}^2 - \nu_1^2) - \delta_2(\omega_{01}^2 - \nu_1^2) = \frac{3}{4} \frac{\omega_{02}^2 - \nu_1^2}{\omega_1} \left(\nu_1^2 \gamma_1 C_1 - \frac{(\omega_{01}^2 - \nu_1^2)\gamma_2 \nu_2^2 C_2}{c_1} \right) A_2 \sin\Phi,$$

$$\delta_1(\omega_{01}^2 - \nu_1^2) - \delta_2(\omega_{02}^2 - \nu_1^2) = \pm \frac{3}{2} \frac{\omega_{01}^2 - \nu_1^2}{\omega_2 + \omega_{02}} \left(\nu_1^2 \gamma_1 C_1 - \frac{(\omega_{01}^2 - \nu_1^2)\gamma_2 \nu_2^2 C_2}{c_1} \right) \frac{A_1^2}{A_2} \sin\Phi, \tag{12.71}$$

$$\begin{aligned}
&\Delta_1(\omega_{01}^2-\omega_{02}^2)+\frac{3}{4}\left[\frac{\nu_1^2(\omega_{02}^2-\nu_1^2)}{\omega_{01}}\gamma_1\left(\frac{A_1^2}{2}+A_2^2+C_1^2\right)\right.\\
&\quad-\frac{\nu_2^2(\omega_{01}^2-\nu_1^2)}{c_1^2\omega_{01}}\gamma_2\left(\frac{(\omega_{01}^2-\nu_1^2)^2}{2}A_1^2+(\omega_{02}^2-\nu_1^2)^2A_2^2+m_1^2C_2^2\right)\\
&\quad\left.+\frac{\omega_{02}^2-\nu_1^2}{\omega_1}\left(\nu_1^2\gamma_1C_1-\frac{(\omega_{01}^2-\nu_1^2)\gamma_2\nu_2^2C_2}{c_1}\right)A_2\cos\Phi\right]=0,
\end{aligned} \tag{12.72}$$

$$\begin{aligned}
&\Delta_2(\omega_{01}^2-\omega_{02}^2)-\frac{3}{4}\left[\frac{\nu_1^2(\omega_{01}^2-\nu_1^2)}{\omega_{02}}\gamma_1\left(A_1^2+\frac{A_2^2}{2}+C_1^2\right)\right.\\
&\quad-\frac{\nu_2^2(\omega_{02}^2-\nu_1^2)}{c_1^2\omega_{02}}\gamma_2\left((\omega_{01}^2-\nu_1^2)^2A_1^2+\frac{(\omega_{02}^2-\nu_1^2)^2}{2}A_2^2+c_1^2C_2^2\right)\\
&\quad\left.-\frac{2(\omega_{01}^2-\nu_1^2)}{\omega_2+\omega_{02}}\left(\nu_1^2\gamma_1C_1-\frac{(\omega_{01}^2-\nu_1^2)\gamma_2\nu_2^2C_2}{c_1}\right)\frac{A_1^2}{A_2}\cos\Phi\right]=0,
\end{aligned}$$

where $\Delta_2=\pm(\Delta-2\Delta_1)$, $\Delta=\omega-2\omega_{01}\mp\omega_{02}$. As in the previous case, it follows from Eqs. (12.71), (12.72) that they possess nontrivial solutions only for $n_2=+1$, i.e., for $\omega=2\omega_1+\omega_2$.

Let us analyze the solution of Eqs. (12.71), (12.72) in the specific case discussed above, when $\nu_1=\nu_2=\nu$, $\delta_1=\delta_2=\delta$, $\gamma_1=\gamma_2=\gamma$, $c_1=c_2=c$. It follows from Eqs. (12.71) that in this case

$$\frac{A_1^2}{A_2^2}=\frac{\omega_2+\omega_{02}}{2\omega_1}. \tag{12.73}$$

Further, eliminating Φ from Eqs. (12.71), (12.72) and taking account of (12.73), we obtain the following equations for A_1 and ω_1:

$$\begin{aligned}
&\frac{\Delta-3\Delta_1}{\nu}-\frac{3\nu}{8}\left\{\gamma A_1^2\left[\frac{2}{\omega_{02}}-\frac{1}{\omega_{01}}+\frac{2\omega_1}{\omega_2+\omega_{02}}\left(\frac{1}{\omega_{02}}-\frac{2}{\omega_{01}}\right)\right]\right.\\
&\quad\left.+\left(\frac{1}{\omega_{02}}-\frac{1}{\omega_{01}}\right)\frac{(\nu^2-\omega^2)^2+c^2}{\left((\nu^2-\omega^2)^2-c^2\right)^2}\gamma B^2\right\}=0,
\end{aligned} \tag{12.74}$$

$$\begin{aligned}
&\left\{\frac{2\Delta_1}{\nu}-\frac{3\nu}{4\omega_{01}}\left[\left(1+\frac{4\omega_1}{\omega_2+\omega_{02}}\right)\gamma A_1^2+\frac{(\nu^2-\omega^2)^2+m^2}{\left((\nu^2-\omega^2)^2-m^2\right)^2}\gamma B^2\right]\right\}^2\\
&\quad+\frac{4\delta^2}{\nu^2}-\frac{9\gamma^2\nu^2}{8\omega_1}\frac{A_1^2B^2}{(\omega_2+\omega_{02})(\nu^2-\omega^2-c)^2}=0.
\end{aligned}$$

Examples of the dependencies of $\gamma A_{1,2}^2$ and of ω_1/ν on Δ/ν are illustrated in Fig. 12.10 a, b and c. It is seen from this figure that, in contrast to the previous case, in the case under consideration combination oscillations can be excited in hard manner only, i.e., at a finite initial perturbation. Excitation

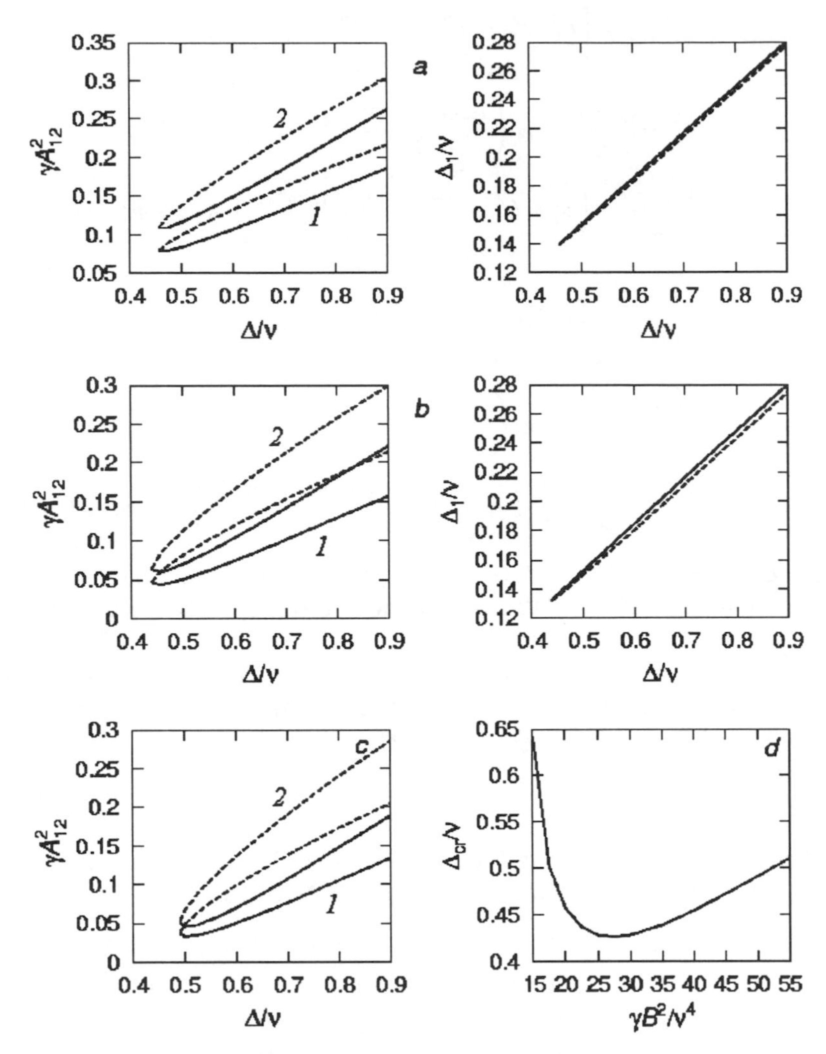

Fig. 12.10. Plots of γA_1^2 (curves 1), γA_2^2 (curves 2) and Δ_1/ν versus Δ/ν for $4\delta^2/\nu^2 = 0.005$, $c/\nu^2 = 0.3$ and (a) $\gamma B^2/\nu^4 = 20$, (b) $\gamma B^2/\nu^4 = 25$, (c) $\gamma B^2/\nu^4 = 50$; (d) plot of Δ_{cr}/ν versus $\gamma B^2/\nu^4$

of such oscillations is possible only for $B \geq B_{\text{cr}}$ and $|\Delta| \geq \Delta_{\text{cr}}$, where Δ_{cr} is a nonmonotonic function of the external force amplitude B (see Fig. 12.10 d).

12.5 Electro-mechanical vibrators and capacitative sensors of small displacements

A system in which combination resonance has a dominant role in the excitation of strong vibration is an electro-mechanical vibrator with a power supply circuit forming an oscillatory circuit (see Fig. 12.11). Such a vibrator was suggested and studied by Gabaraev et al. [96]. The equations of this

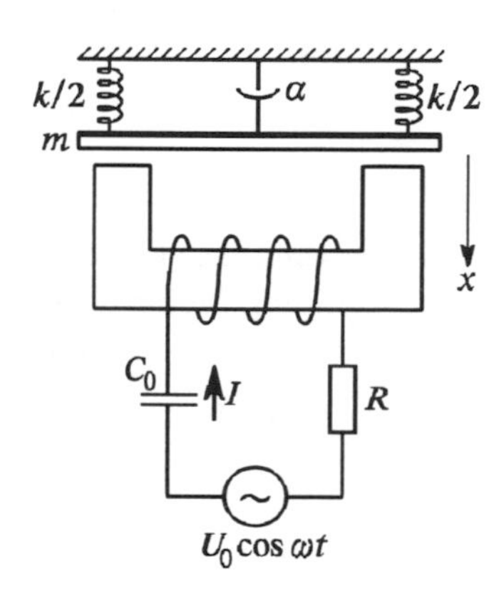

Fig. 12.11. Schematic image of the electro-mechanical vibrator

vibrator are Landa:1988(2),Landa:1991(2)

$$\frac{d^2}{dt^2}\Big(L(x)I\Big) + R\,\frac{dI}{dt} + \frac{I}{C_0} = U_0\omega \sin \omega t,$$
$$m\,\frac{d^2x}{dt^2} + \alpha\,\frac{dx}{dt} + kx = F(x,I), \tag{12.75}$$

where x is the plate displacement, m is the plate mass, $L(x)$ is the inductance of the coil with a core depending on the size of the clearance between the plate and the core, $F(x,I) = (I/2)\mathrm{d}\Phi/\mathrm{d}x$ is the ponderomotive force acting on the plate, and $\Phi = L(x)I$ is the magnetic flux. Following [96] we set $\mathrm{d}\Phi/\mathrm{d}x = I\mathrm{d}L/\mathrm{d}x$.

For small x we can represent the inductance $L(x)$ as

$$L(x) = L_0(1 + a_1 x), \tag{12.76}$$

where the sign and the value of the coefficient a_1 depend on the origin of reference of the variable x which is assumed to coincide with the plate equilibrium position. Taking account of (12.76) we obtain

$$F(x,I) = \frac{L_0 a_1}{2}\,I^2. \tag{12.77}$$

Similar to (12.59), the oscillations of the current in the oscillatory circuit are quasi-periodic with the fundamental frequencies ω and $\omega \pm j\nu$ ($j = 1, 2, \ldots$); ν is an unknown frequency of oscillations of the plate. Hence, we put

$$I = A \sin \Psi + y, \tag{12.78}$$

where $\Psi = \omega t + \psi$,

$$A = \frac{U_0}{L_0 \omega} \left[\left(1 - \frac{\Omega_0^2}{\omega^2} \right)^2 + \frac{4\delta_1^2}{\omega^2} \right]^{-1/2}, \quad \tan \psi = \frac{2\delta_1}{\omega(1 - \Omega_0^2/\omega^2)}, \tag{12.79}$$

$$\Omega_0 = \frac{1}{\sqrt{L_0 C_0}}, \quad \delta_1 = \frac{R}{2L_0}.$$

For sufficiently large values of the plate Q-factor determined by the ratio ν/δ_2, where $\delta_2 = \alpha/2m$ is the damping factor of the plate, the equations for x and y, in view of (12.76)–(12.78), can be written as

$$\ddot{x} + \nu^2 x = -\epsilon \left(2\delta_2 \dot{x} - 2\nu \Delta x - \frac{L_0 a_1}{2m} (A \sin \Psi + y)^2 \right), \tag{12.80}$$

$$\ddot{y} + 2\delta_1 \dot{y} + \Omega_0^2 y + a_1 \frac{\mathrm{d}^2}{\mathrm{d}t^2} \Big(x(A \sin \Psi + y) \Big) = 0, \tag{12.81}$$

where ϵ is a small parameter which in the final results should be put equal to unity, and $\Delta = \nu - \nu_0$ is the mistuning between the oscillation frequency ν and the natural frequency of the plate vibration $\nu_0 = \sqrt{k/m}$.

If $\delta_1 \gg \delta_2$ then the current in the oscillatory circuit becomes stationary much faster than the plate vibration. Therefore we can seek only a steady-state solution of Eq. (12.81). Moreover, if in solving Eq. (12.80) we restrict ourselves to terms of order ϵ only, then it is sufficient to solve Eq. (12.81) in the zero approximation with respect to ϵ. In this approximation one can set

$$x = B \cos \chi, \tag{12.82}$$

where $\chi = \nu t + \varphi$. A steady-state solution of Eq. (12.81), in view of (12.82), can be represented in the form of a trigonometric series involving an infinity of terms with combination frequencies $\omega + j\nu$, where $j = 0, \pm 1, \pm 2, \ldots$. We restrict ourselves to only those terms of this series which, being substituted into Eq. (12.80), give linear terms relative to B. It is not difficult to make sure that under this condition it will suffice to take into account only those terms with $j = \pm 1$. Hence a steady-state solution of Eq. (12.81) of interest to us can be sought in the form

$$y = B \Big(C_1 \cos(\Psi + \chi + \vartheta_1) + C_2 \cos(\Psi - \chi + \vartheta_2) \Big). \tag{12.83}$$

Substituting (12.83) and (12.82) into Eq. (12.81), we find

$$C_{1,2} = \frac{a_1(\omega \pm \nu)^2}{2\sqrt{\left((\omega \pm \nu)^2 - \Omega_0^2\right)^2 + 4\delta_1^2(\omega \pm \nu)^2}},$$

$$\tan\vartheta_{1,2} = -\frac{(\omega \pm \nu)^2 - \Omega_0^2}{2\delta_1(\omega \pm \nu)}. \tag{12.84}$$

If we further substitute (12.83), in view of (12.84), into Eq. (12.80) and assume the amplitude B and the phase φ to be slowly varying functions, we obtain the following truncated equations for B and φ:

$$\dot{B} = (\eta - \delta_2)B, \quad \dot{\varphi} = -\Delta + \frac{L_0 a_1}{4m\nu} A^2(C_1 \sin\vartheta_1 + C_2 \sin\vartheta_2), \tag{12.85}$$

where

$$\eta = \frac{L_0 a_1}{4m\nu} A^2(C_1 \cos\vartheta_1 - C_2 \cos\vartheta_2). \tag{12.86}$$

Substituting into (12.86) the expressions (12.84) calculated to first order with respect to ν/ω, we find

$$\eta \approx \frac{a_1^2\omega^4\left((\omega^2 - \Omega_0^2)(\omega^2 + 3\Omega_0^2) - 4\delta_1^2\omega^2\right)}{2L_0 m\left((\omega^2 - \Omega_0^2)^2 + 4\delta_1^2\omega^2\right)^3}\, \delta_1 U_0^2. \tag{12.87}$$

We see that Eqs. (12.85) closely resemble truncated van der Pol equations. This is why such systems we called *self-oscillatory systems with high-frequency energy sources* [179, 189]. In addition to the resemblance of the truncated equations to those for a self-oscillatory system, the vibrator under consideration possesses many properties which are typical of self-oscillatory systems, e.g., the property to be synchronized by an external periodic force or another similar vibrator (see Ch. 10).

By virtue of this analogy with self-oscillatory systems, we can speak about self-excitation of systems similar to the vibrator under consideration. It follows from (12.85) that the condition for the self-excitation is $\eta \geq \delta_2$; as seen from (12.87) this condition can be fulfilled only for $a_1 \neq 0$ and $\omega > \omega_{\mathrm{cr}} > \Omega_0$, where ω_{cr} is a certain critical value of the frequency ω. The latter inequality means that the self-excitation of oscillations is possible only on the right-hand slope of the resonance curve described by (12.79). This is confirmed experimentally. For a fixed value of ω the condition $\eta = \delta_2$ determines a critical value of the power source voltage U_0^*, beyond which the self-excitation of oscillations occurs. The dependence U_0^* on the frequency ω is of a nonmonotonic character and reaches its minimum for $\omega = \omega_m$, where ω_m is a certain value of ω. This dependence is demonstrated in Fig. 12.12 a for two specific cases: $\Omega_0 \gg \delta_1$ and $\Omega_0 \ll \delta_1$. In the first case $\omega_m \approx \Omega_0 + \delta_1/\sqrt{5}$, and in the second case $\omega_m \approx \sqrt{8}\,\delta_1$.

From the second equation of (12.85) and (12.84) we find the oscillation frequency in the neighborhood of the self-excitation boundary (when A^2 is determined by (12.87), (12.86)). It is

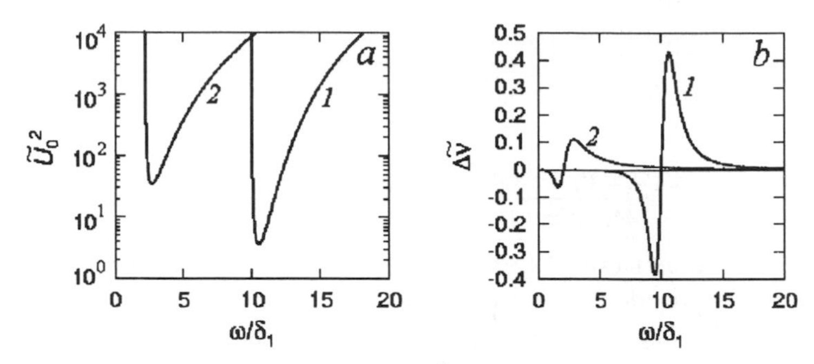

Fig. 12.12. Plots of (a) $\tilde{U}_0^2 = a_1^2 U_0^{*2}/2mL_0\delta_1^3\delta_2$ and (b) $\Delta\tilde{\nu} = 4(\nu - \nu_0)L_0 m\nu_0\delta_1^2/a_1^2 U_0^2$ versus ω/δ_1 for $\Omega_0/\delta_1 = 10$ (curves 1) and $\Omega_0 = 0$ (curves 2)

$$\nu = \nu_0 + \frac{a_1^2\omega^4(\omega^2 - \Omega_0^2)}{4L_0 m\nu_0\left((\omega^2 - \Omega_0^2)^2 + 4\delta_1^2\omega^2\right)^2} U_0^2. \tag{12.88}$$

The dependence of $\nu - \nu_0$ on ω is illustrated in Fig. 12.12 b for the same two specific cases.

When in the representation of $L(x)$ and in the calculation of the variable y nonlinear terms are taken into account, one can calculate the steady-state amplitude of the plate vibration and determine whether its excitation is soft or hard. This calculation is rather awkward, if not difficult in principle.

A system, which is similar to the electro-mechanical vibrator considered, is the so-called capacitative sensor of small displacements. Let a small ball be suspended by a spring from a wall and connected to a capacitative sensor of small displacements (see Fig. 12.13). The instability of such a sensor was

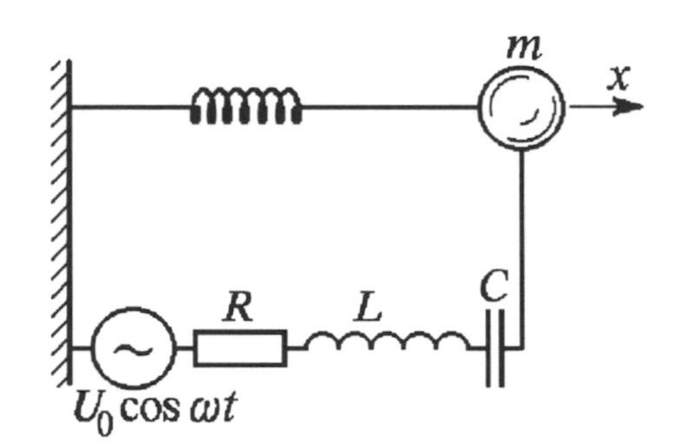

Fig. 12.13. Schematic image of a capacitative sensor of small displacements connected to a trial ball

considered by Braginsky et al. [54, 55, 57], who used it in so-called experiments with trial bodies. The capacitative sensor of small displacements is a capacitor, one of whose plates is connected to a body whose displacement should be measured. The capacitor is part of an oscillatory circuit which involves a source of alternating voltage. It has been found experimentally that vibration of the body, obstructing the measurements, appears in a certain range of the frequencies ω as the voltage U_0 exceeds a critical value.

The equations of the system depicted in Fig. 12.13 are

$$\ddot{q} + 2\delta_1 \dot{q} + \Omega_0^2 \frac{C_0}{C(x)} q = \frac{U_0}{L} \cos \omega t,$$
$$\ddot{x} + 2\delta_2 \dot{x} + \nu_0^2 x = -\frac{1}{m} F(x, q), \tag{12.89}$$

where q is the charge on the capacitor plates, $C(x) = C_0/(1 + x/d_0)$ is the capacitance when the displacement of the small ball is x, $C_0 = \epsilon_0 \epsilon S/d_0$ is the capacitance for $x = 0$, S is the capacitor plate area, ϵ is the permittivity of the dielectric between the capacitor plates, ϵ_0 is the permittivity of vacuum, d_0 is the distance between the capacitor plates for $x = 0$, $\Omega_0 = 1/\sqrt{LC_0}$ is the natural frequency of the oscillatory circuit for $x = 0$, m is the mass of the small ball, and

$$F(x, q) = \frac{\epsilon_0 \epsilon S q^2}{2(d_0 + x)^2 C^2(x)} = \frac{q^2}{2C_0 d_0}$$

is the force of attraction between the capacitor plates.[2]

Comparing Eqs. (12.89) and (12.75), we see that they differ only in nonlinear terms. Performing calculations similar to those above we obtain approximate equations for the amplitude B and the phase φ of the same form as Eqs. (12.85) with

$$\eta = \frac{\Omega_0^4 \left((\omega^2 - \Omega_0^2)(3\omega^2 + \Omega_0^2) + 4\delta_1^2 \omega^2 \right)}{2Lmd_0^2 \left((\omega^2 - \Omega_0^2)^2 + 4\delta_1^2 \omega^2 \right)^3} \delta_1 U_0^2. \tag{12.90}$$

From this it follows that the self-excitation of oscillations is possible in a certain range of frequencies ω which is located mainly on the right-hand slope of the resonance curve. This agrees with the results of [54, 55, 57] and is confirmed experimentally. It also follows from (12.90) that the lowest critical value of the voltage U_0 corresponding to the self-excitation threshold takes place for a certain value of ω ($\omega = \omega_m$). If the Q-factor for the oscillatory circuit is sufficiently large, i.e., $\Omega_0 \gg \delta_1$, then $\omega_m \approx \Omega_0 + \delta_1/\sqrt{5}$. This lowest voltage value is

$$U_{0\,\mathrm{min}} = \frac{24 d_0 \delta_1^2}{5} \sqrt{\frac{6Lm\delta_2}{\Omega_0 \sqrt{5}}}.$$

A similar self-excitation of mechanical oscillations of a trial body takes place for optical sensors of small displacements as well, in which the role

[2] The expressions for C_0, $C(x)$ and $F(x, q)$ are written in the plane capacitor approximation.

of a high-frequency source is played by light [57]. A similar effect was also observed experimentally for a torsional pendulum under the action of an ultra-high-frequency field by Braginsky et al. [56].

We note that the mechanism of self-excitation of oscillations considered above underlies the excitation of mechanical oscillations in many systems. For example, strong vibration of resonators filled with some kind of powerful radiation, in particular with electromagnetic radiation, can be explained by similar mechanism. Such vibration of resonators, used in powerful colliding beam accelerators, was observed by Karliner et al. [145].

13. Parametric excitation of oscillations

13.1 Parametrically excited nonlinear oscillators

The simplest equation for a nonlinear oscillator under an external action varying its natural frequency is

$$\ddot{x} + 2\delta\dot{x} + \omega_0^2\big(1 + f(t)\big)F(x) = 0, \tag{13.1}$$

where $f(t)$ is a function of time, and $F(x)$ is a nonlinear function of x.

An example of a similar system is a physical pendulum with a vertically vibrated axis of suspension. The equation of such a pendulum is

$$\ddot{\varphi} + \frac{H}{J}\dot{\varphi} + \frac{mb}{J}\big(g + f(t)\big)\sin\varphi = 0, \tag{13.2}$$

where J and m are the moment of inertia and mass of the pendulum, respectively, b is the distance between the center of mass and the axis of suspension, $H\dot{\varphi}$ is the moment of the friction force, g is the acceleration of gravity, and $f(t)$ is the acceleration of the axis of suspension.

In the case when $F(x) = x$, $\delta = 0$ and $f(t)$ is a periodic function, Eq. (13.1) is the Hill equation. The behavior of the solution of this equation is well studied (see, e.g., [302]). Some examples of the behavior of a damped nonlinear oscillator are given below. A number of examples of parametric excitation of nonlinear oscillators by noise are also considered.

13.1.1 Slightly nonlinear oscillator with small damping and small harmonic parametric action

In the case of slight nonlinearity, small damping and a small harmonic parametric action Eq. (13.1) can be written as

$$\ddot{x} + \omega_0^2 x = \epsilon(-2\delta\dot{x} - \omega_0^2 Bx\cos 2\omega t - \omega_0^2\gamma x^3). \tag{13.3}$$

According to the Krylov–Bogolyubov method, we can seek an approximate solution of Eq. (13.3) in the form

$$x = A\cos\psi + \epsilon u_1(A,\varphi,t) + \epsilon^2 u_2(A,\varphi,t) + \dots, \tag{13.4}$$

where $\psi = n\omega t + \varphi$,

$$\dot{A} = \epsilon f_1(A,\varphi) + \epsilon^2 f_2(A,\varphi) + \dots ,$$
$$\dot{\varphi} = -\Delta + \epsilon F_1(A,\varphi) + \epsilon^2 F_2(A,\varphi) + \dots , \tag{13.5}$$

and $\Delta = n\omega - \omega_0$ is the frequency mistuning. Substituting (13.4), (13.5) into Eq. (13.3) and equating the coefficients of ϵ we obtain the following equation for u_1:

$$\frac{\partial^2 u_1}{\partial t^2} + \omega_0^2 u_1 = \left(2f_1\omega_0 - \Delta A\frac{\partial F_1}{\partial \varphi}\right)\sin\psi + 2\delta\omega_0 A\sin\psi - \omega_0^2\gamma A^3\cos^3\psi$$
$$+ \left(2F_1\omega_0 A + \Delta\frac{\partial f_1}{\partial \varphi}\right)\cos\psi - \omega_0^2 AB\cos 2\omega t\cos\psi. \tag{13.6}$$

In the case of the main parametric resonance ($n = 1$), from the condition for the absence of resonance terms in the function $u_1(A,\varphi,t)$ we find

$$2f_1\omega_0 - \Delta A\,\frac{\partial F_1}{\partial \varphi} = -2\delta\omega_0 A + \frac{\omega_0^2 AB}{2}\sin 2\varphi,$$
$$2F_1\omega_0 A + \Delta\,\frac{\partial f_1}{\partial \varphi} = \frac{3}{4}\,\omega_0^2\gamma A^3 + \frac{\omega_0^2 AB}{2}\cos 2\varphi.$$

Solving these equations and substituting the solution found into (13.5), we obtain the following equations for the amplitude and the phase:

$$\dot{A} = -\,\delta A + \frac{\omega_0^2 AB}{4\omega}\sin 2\varphi, \quad \dot{\varphi} = -\,\Delta + \frac{3}{8}\,\omega_0\gamma A^2 + \frac{\omega_0^2 B}{4\omega}\cos 2\varphi. \tag{13.7}$$

Equations (13.7), even if they look like Eqs. (12.12), differ essentially from these equations. The steady-state solution of Eqs. (13.7) is

$$\gamma A^2 = \frac{8}{3}\left(\frac{\omega}{\omega_0} - 1 \pm \sqrt{\frac{\omega_0^2 B^2}{16\omega^2} - \frac{\delta^2}{\omega_0^2}}\right), \quad \sin 2\varphi = \frac{4\omega\delta}{\omega_0^2 B}. \tag{13.8}$$

From the expressions obtained it may be inferred that parametric excitation of oscillations is of a threshold character: it is possible for $B \geq B_{\text{th}} = 4\omega\delta/\omega_0^2 \approx 4\delta/\omega_0$ only. Plots of $|\gamma|A^2$ and φ versus ω/ω_0 constructed from (13.8) are shown in Fig. 13.1. It is seen from this figure that within the frequency range $\omega_1 < \omega < \omega_2$, where $\omega_{1,2} \approx \omega_0(1 \mp \sqrt{B^2/16 - \delta^2/\omega_0^2}\,)$, oscillations are excited for as small initial conditions as wished. Outside of this frequency range hard excitation of oscillations only is possible; i.e., oscillations can be excited from a certain initial perturbation.

To find the excitation conditions for parametric resonance of the second kind (for $n = 2$), we have to bear in mind that in this case the frequency ω_0 is close to 2ω. Therefore the term responsible for the parametric excitation in Eq. (13.6) does not make a contribution to the equations for the amplitude and phase in the first approximation. Thus, we should consider the second approximation. Assuming δ, $\gamma \sim \epsilon^2$ we obtain the following equations for u_1 and u_2:

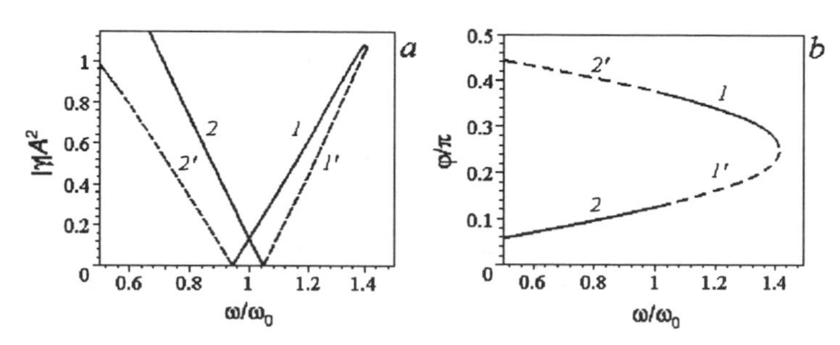

Fig. 13.1. Plots (a) of $|\gamma|A^2$ and (b) of φ versus ω/ω_0 for $B = 0.283$, $\omega_0/\delta = 20$ in the case of the main resonance. Dashed lines show the unstable parts of the dependencies

$$\frac{\partial^2 u_1}{\partial t^2} + \omega_0^2 u_1 = \left(2 f_1 \omega_0 - \Delta A \frac{\partial F_1}{\partial \varphi}\right) \sin\psi + \left(2 F_1 \omega_0 A + \Delta \frac{\partial f_1}{\partial \varphi}\right) \cos\psi - \omega_0^2 A B \cos 2\omega t \cos\psi, \tag{13.9}$$

$$\frac{\partial^2 u_2}{\partial t^2} + \omega_0^2 u_2 = \left(2 f_2 \omega_0 - \Delta A \frac{\partial F_2}{\partial \varphi}\right) \sin\psi + 2\delta\omega_0 A \sin\psi - \omega_0^2 \gamma A^3 \cos^3\psi + \left(2 F_2 \omega_0 A + \Delta \frac{\partial f_2}{\partial \varphi}\right) \cos\psi - \omega_0^2 B u_1 \cos 2\omega t. \tag{13.10}$$

It follows from Eq. (13.9) that $f_1 = F_1 = 0$ and

$$u_1(A, \varphi, t) = -\frac{AB}{2}\left(\cos\varphi - \frac{\omega_0^2}{16\omega^2 - \omega_0^2}\cos(2\psi - \varphi)\right). \tag{13.11}$$

Substituting further (13.11) into Eq. (13.10) and equating the coefficients of $\sin\psi$ and $\cos\psi$ we find the following equations for f_2 and F_2:

$$2 f_2 \omega_0 - \Delta A \frac{\partial F_2}{\partial \varphi} = -2\delta\omega_0 A - \frac{\omega_0^2 A B^2}{4}\sin 2\varphi,$$

$$2 F_2 \omega_0 A + \Delta \frac{\partial f_2}{\partial \varphi} = \frac{\omega_0^2(\omega_0^2 - 8\omega^2) A B^2}{2(16\omega^2 - \omega_0^2)} - \frac{\omega_0^2 A B^2}{4}\cos 2\varphi.$$

Solving these equations we find the equations for the amplitude and the phase in the second approximation:

$$\dot{A} = -\delta A - \frac{\omega_0^2 A B^2}{16\omega}\sin 2\varphi,$$

$$\dot{\varphi} = -\Delta + \frac{3}{8}\omega_0 \gamma A^2 + \frac{\omega_0(\omega_0^2 - 8\omega^2) B^2}{4(16\omega^2 - \omega_0^2)} - \frac{\omega_0^2 B^2}{16\omega}\cos 2\varphi. \tag{13.12}$$

From (13.12) it may be inferred that the threshold value of the amplitude B, for which parametric resonance of the second kind is possible, is equal to $4\sqrt{\delta\omega}/\omega_0 \approx 2\sqrt{2\delta/\omega_0}$. For the same values of parameters it is far beyond

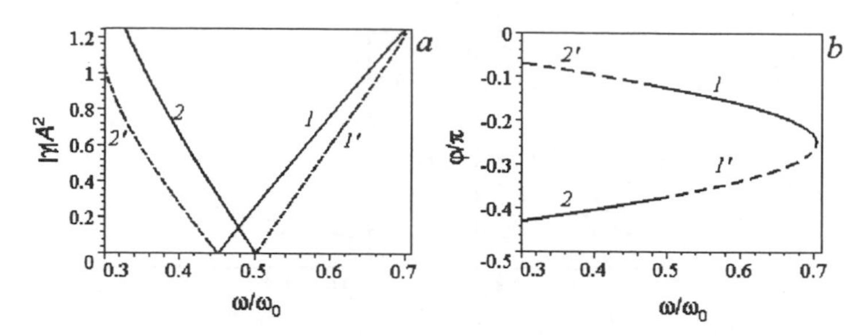

Fig. 13.2. Plots (a) of $|\gamma|A^2$ and (b) of φ versus ω/ω_0 for $B = 0.7$, $\omega_0/\delta = 20$ in the case of the second kind resonance ($n = 2$). Dashed lines correspond to unstable parts of the dependencies

that of the main parametric resonance. Plots of $|\gamma|A^2$ and φ versus ω/ω_0 constructed from Eqs. (13.12) for $\dot{A} = \dot{\varphi} = 0$ are shown in Fig. 13.2.

Calculation of the excitation conditions for parametric resonance of a higher kind (for $n > 2$) can be performed by using higher approximations only.

13.2 Chaotization of a parametrically excited nonlinear oscillator

Period-doubling bifurcations and transition to chaos were observed both in systems modeled by the equation of a parametically excited physical pendulum (see, e.g., [29, 234, 206, 207, 154, 155, 253]) and in systems described by the Duffing equation with a parametric action (see, e.g., [132, 133, 1, 335]). We dwell here only on some results of these works and focus our attention on the qualitative behavior of the system and on the spectral characteristics of oscillations.

Let us revert to Eq. (13.1) and set $\omega_0 = 1$, $F(x) = \sin x$ and $f(t) = B\cos 2\omega t$. Then we obtain the equation describing the motion of a pendulum with a harmonically vibrated axis of suspension:

$$\ddot{x} + 2\delta\dot{x} + \big(1 + B\cos 2\omega t\big)\sin x = 0. \tag{13.13}$$

As follows from the preceding section, for small δ, the boundary of stability of the equilibrium state for Eq. (13.13) is determined by the condition $B = B_{\text{th}} \approx 4\delta$, and for $B > B_{\text{th}}$ periodic oscillations of period 2π are excited. Numerical simulation of Eq. (13.13) shows that these oscillations remain stable up to a certain value $B = B_1$ depending on δ. So, $B_1 \approx 1.426$ for $\delta = 0.1$. For $B > B_1$ the rotation of the pendulum appears, with period equal to that for the vibration of the suspension axis, i.e., π. This rotation can be either clockwise or anti-clockwise, depending on the initial conditions. As B increases, each rotation undergoes period-doubling bifurcations. Beyond the cascade of the period-doubling bifurcations, when $B \approx 2.066$, two chaotic attractors, corresponding to irregular rotations of the pendulum, come into existence in the system phase space. These attractors are separated so that the direction of rotation depends on the initial conditions. Only for $B > 2.09$ the direction of rotation changes chaotically in the course of time. The mean frequency of changes of the rotation direction increases with increasing B. It should be noted that there are windows where the rotations become periodic.

An example of the irregular rotation of the pendulum is represented in Fig. 13.3 a. It is seen from this figure that the pendulum rotates irregular-

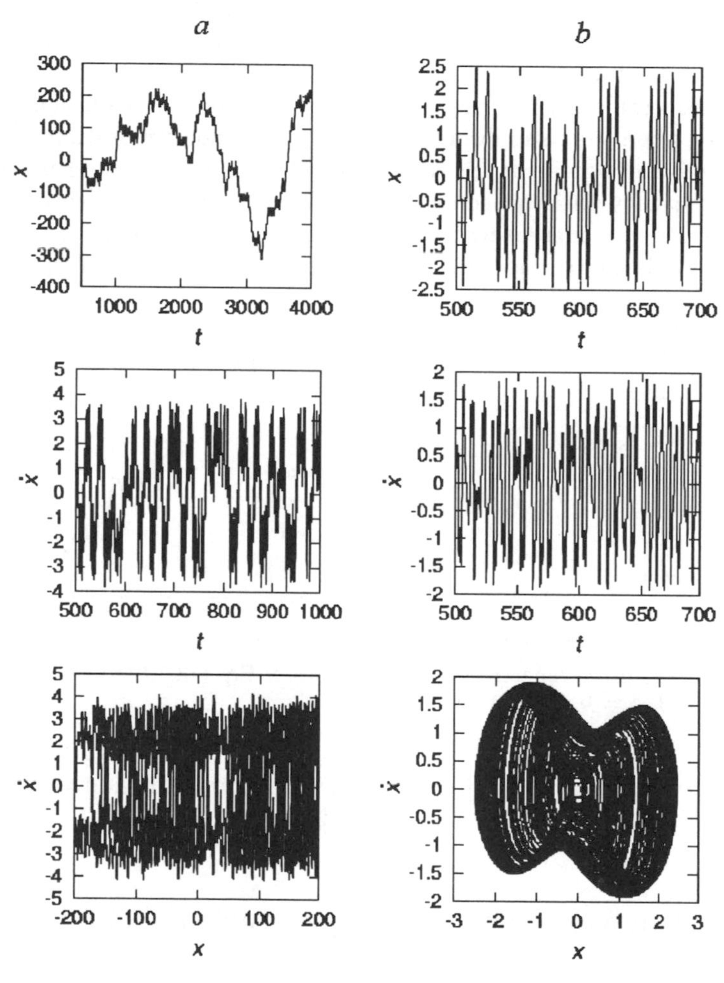

Fig. 13.3. The solution of Eq. (13.13) and the projections of phase portraits on the plane x, $\dot{x}$ for $\omega = 1$, $\delta = 0.1$: (a) $B = 3$, $\alpha = 0$ and (b) $B = 3.5$, $\alpha = 2$

ly and the direction of rotation changes chaotically. This rotation causes a considerable slow drift of the angle x.

To avoid rotations of the pendulum and obtain chaotic oscillations, we may include nonlinear friction in Eq. (13.13). Thus, we consider the equation

$$\ddot{x} + 2\delta(1 + \alpha\dot{x}^2)\dot{x} + (1 + B\cos 2\omega t)\sin x = 0, \tag{13.14}$$

where α is the nonlinear friction factor. The nonlinear friction, if it is of a sufficient value, results in cessation of rotation and in chaotic oscillations of the pendulum about its equilibrium position (see Fig. 13.3 b). The correlation dimension of the attractor associated with these chaotic pendulum oscillations is equal to 2.51 ± 0.05 for $B = 3$, $\alpha = 0$ and 2.09 ± 0.03 for $B = 3.5$, $\alpha = 2$. We see that the presence of nonlinear friction results in an essential decrease of the dimension.

Of considerable interest are the power spectra of the oscillations excited. In the case when the pendulum is excited by a harmonic vibration of the suspension axis and nonlinear friction is negligible, its power spectrum contains the low-frequency part caused by the slow drift of x; the power spectrum density decreases with increasing frequency but not monotonically (see Fig. 13.4 a). With nonlinear friction the low-frequency part of the power spectrum

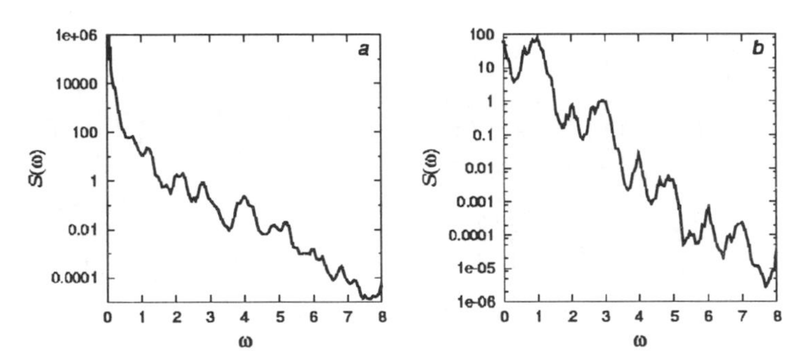

Fig. 13.4. The power spectra for the solutions of Eq. (13.13) for $\omega = 1$, $\delta = 0.1$, and (a) $B = 3$, $\alpha = 0$ and (b) $B = 3.5$, $\alpha = 2$

decreases substantially and distinct peaks at the frequencies divisible by the natural frequency come into existence (Fig. 13.4 b).

13.3 Parametric excitation of pendulum oscillations by noise

The problem of excitation of an oscillator under a parametric random action was first studied analytically by Stratonovich and Romanovskii as early as 1958 [317, 319] and later by Dimentberg in 1980 [70]. To obtain the limitation of the oscillation amplitude Stratonovich and Romanovskii took into account nonlinear friction. In fact, the inclusion of nonlinear friction is not necessary since the limitation of amplitude can occur owing to nonlinearity of the restoring force [190, 192]. However, the inclusion of nonlinear friction makes it possible to obviate random rotations of the pendulum through an angle divisible by 2π. These rotations make the analysis of the results obtained more difficult.

The equation for a pendulum with a randomly vibrated suspension axis with regard to nonlinear friction can be written in the form

$$\ddot{x} + 2\delta\big(1 + \alpha\dot{x}^2\big)\dot{x} + \omega_0^2\big(1 + \xi(t)\big)\sin x = 0, \tag{13.15}$$

where $\xi(t)$ is a comparatively wide-band random process with nonzero power spectrum density at the frequency $\omega = 2\omega_0$. We assume that the intensity of the vibration of the suspension axis is moderately small so that pendulum oscillations can be considered small to an extent that $\sin x$ can be presented as

$$\sin x \approx \big(1 - \gamma x^2\big)x, \tag{13.16}$$

where $\gamma = 1/6$.

An approximate analytical solution of the problem can be obtained on the assumptions that $\delta/\omega_0 \sim \epsilon$, $\gamma x^2 \sim \epsilon$, $\xi(t) \sim \sqrt{\epsilon}$, where ϵ is a certain small parameter which should be put equal to unity in the final results. With these assumptions Eq. (13.15), in view of (13.16), is conveniently written as

$$\ddot{x} + \omega_0^2 x = \epsilon\Big(-2\delta\big(1 + \alpha\dot{x}^2\big)\dot{x} + \omega_0^2\gamma x^3\Big) - \sqrt{\epsilon}\xi(t)\big(1 - \epsilon\gamma x^2\big)x. \tag{13.17}$$

Equation (13.17) can be solved by the Krylov–Bogolyubov method; to do this we set $x = A\cos\psi + \epsilon u_1 + \dots$, where $\psi = \omega_0 t + \phi$,

$$\dot{A} = \epsilon f_1 + \dots, \qquad \dot{\phi} = \epsilon F_1 + \dots, \tag{13.18}$$

$u_1, \dots, f_1, \dots, F_1, \dots$, are unknown functions. By using the Krylov–Bogolyubov method for stochastic equations (see [319, 166]) we find the expressions for these unknown functions f_1 and F_1. Substituting these expressions into Eqs. (13.18) we obtain

$$\begin{aligned} \dot{A} &= \left[-\delta\left(1 + \frac{3\omega_0^2}{4}\alpha A^2\right) + \frac{\omega_0}{2}\overline{g_1\left(\psi, \xi(t)\right)}\right] A, \\ \dot{\phi} &= \omega_0\overline{g_2\left(\psi(t), \xi(t)\right)}, \end{aligned} \tag{13.19}$$

where

$$g_1\big(\psi(t), \xi(t)\big) = \xi(t)\sin 2\psi(t), \quad g_2\big(\psi(t), \xi(t)\big) = \xi(t)\cos^2\psi(t);$$

the bar over an expression indicates its time averaging. As follows from [319], the Fokker–Planck equation associated with Eqs. (13.19) is

$$\frac{\partial w(A,\phi)}{\partial t} = -\frac{\partial}{\partial A}\Bigg\{\Bigg[-\delta\left(1+\frac{3\omega_0^2}{4}\alpha A^2\right)$$

$$+\frac{\omega_0^2}{2}\int_{-\infty}^{0}\left\langle\frac{\partial g_1(\psi,\xi(t))}{\partial\psi}\,g_2\big(\psi(t+\tau),\xi(t+\tau)\big)\right\rangle \mathrm{d}\tau\Bigg] A w(A,\phi)\Bigg\}$$

$$-\,\omega_0^2\int_{-\infty}^{0}\left\langle\frac{\partial g_2(\psi,\xi(t))}{\partial\psi}\,g_2\big(\psi(t+\tau),\xi(t+\tau)\big)\right\rangle \mathrm{d}\tau\,\frac{\partial w(A,\phi)}{\partial\varphi}$$

$$+\frac{K_1\omega_0^2}{8}\frac{\partial^2}{\partial A^2}\Big(A^2 w(A,\phi)\Big)+\frac{K_2\omega_0^2}{2}\frac{\partial^2 w(A,\phi)}{\partial\phi^2}, \tag{13.20}$$

where the angular brackets signify averaging over a statistical ensemble,

$$K_1=\frac{\kappa(2\omega_0)}{2},\quad K_2=\frac{1}{4}\left(\kappa(0)+\frac{\kappa(2\omega_0)}{2}\right), \tag{13.21}$$

and

$$\kappa(\omega)=\int_{-\infty}^{\infty}\langle\xi(t)\xi(t+\tau)\rangle\cos\omega\tau\,\mathrm{d}\tau$$

is the power spectrum density of the process $\xi(t)$ at the frequency ω.

Let us now calculate the integrals in Eq. (13.20) taking account of the expressions for g_1 and g_2. As a result we obtain

$$\int_{-\infty}^{0}\left\langle\frac{\partial g_1(\psi,\xi(t))}{\partial\psi}\,g_2\big(\psi(t+\tau),\xi(t+\tau)\big)\right\rangle \mathrm{d}\tau$$

$$=\frac{1}{2}\int_{-\infty}^{0}\langle\xi(t)\xi(t+\tau)\rangle\cos 2\omega_0\tau\,\mathrm{d}\tau=\frac{\kappa(2\omega_0)}{4}, \tag{13.22}$$

$$\int_{-\infty}^{0}\left\langle\frac{\partial g_2(\psi,\xi(t))}{\partial\psi}\,g_2\big(\psi(t+\tau),\xi(t+\tau)\big)\right\rangle \mathrm{d}\tau$$

$$=\frac{1}{4}\int_{-\infty}^{0}\langle\xi(t)\xi(t+\tau)\rangle\sin 2\omega_0\tau\,\mathrm{d}\tau\equiv M. \tag{13.23}$$

The value of M depends on the characteristics of the random process $\xi(t)$: if $\xi(t)$ is white noise then $M=0$, but if $\xi(t)$ has a finite correlation time, for example, in the case when its power spectrum density is

$$\kappa(\omega)=\frac{\beta^2\kappa(2\omega_0)}{(\omega-2\omega_0)^2+\beta^2},$$

then

$$M = -\frac{\beta\omega_0\kappa(2\omega_0)}{4\left(16\omega_0^2 + \beta^2\right)}.$$

It should be noted that the value of M is negative, resulting in a decrease of the mean oscillation frequency. This decrease is the more considerable the larger is the intensity of noise $\xi(t)$.

The following Langevin equations can be related to the Fokker–Planck equation (13.20), in view of (13.22), (13.23):

$$\dot{A} = \delta\left(\eta - \frac{3\omega_0^2}{4}\,\alpha A^2\right)A + \frac{\omega_0}{2}\,A\zeta_1(t), \quad \dot{\phi} = \omega_0^2 M + \omega_0\zeta_2(t), \tag{13.24}$$

where

$$\eta = \frac{\omega_0^2\kappa(2\omega_0)}{8\delta} - 1, \tag{13.25}$$

and $\zeta_1(t)$ and $\zeta_2(t)$ each are white noise of zero mean. The noise intensities are K_1 and K_2, respectively. As will be seen from the following, the parameter η characterizes the extent to which the noise intensity is in excess of its critical value.

The steady-state solution of Eq. (13.20), satisfying the condition for the probability flux to be equal to zero, is

$$w(A,\phi) = \frac{C}{2\pi A^2}\exp\left[\frac{2}{1+\eta}\left(\eta\ln A - \frac{aA^2}{2}\right)\right], \tag{13.26}$$

where $a = 3\alpha\omega_0^2/4$ is the nonlinear parameter. The constant C is determined from the normalization condition

$$\int_0^{2\pi}\int_0^{\infty} w(A,\phi)A\,\mathrm{d}A\,\mathrm{d}\phi = 1.$$

Upon integrating (13.26) with respect to ϕ, we find the expression for the probability density of the amplitude of oscillations:

$$w(A) = CA^{(\eta-1)/(1+\eta)}\exp\left(-\frac{aA^2}{1+\eta}\right). \tag{13.27}$$

From the normalization condition we find

$$C = 2\begin{cases}\left(\dfrac{a}{1+\eta}\right)^{\eta/(1+\eta)}\dfrac{1}{\Gamma(\eta/(1+\eta))} & \text{for } \eta \geq 0\\ 0 & \text{for } \eta \leq 0.\end{cases} \tag{13.28}$$

Hence,

$$w(A) = 2\begin{cases}\left(\dfrac{a}{1+\eta}\right)^{\eta/(1+\eta)}\dfrac{A^{(\eta-1)/(1+\eta)}}{\Gamma(\eta/(1+\eta))}\exp\left(-\dfrac{aA^2}{1+\eta}\right) & \text{for } \eta \geq 0\\ \delta(A) & \text{for } \eta \leq 0.\end{cases} \tag{13.29}$$

The fact that for $\eta \le 0$ the probability density of the amplitude turns out to be a δ-function is associated with neglecting the additive noise.[1]

Using (13.29) we can find $\langle A \rangle$ and $\langle A^2 \rangle$. They are

$$\langle A \rangle = \begin{cases} \sqrt{\dfrac{1}{a(1+\eta)}} \dfrac{\Gamma\big(\eta/(1+\eta)+1/2\big)}{\Gamma\big(\eta/(1+\eta)+1\big)} \eta & \text{for } \eta \ge 0 \\ 0 & \text{for } \eta \le 0, \end{cases}$$
$$\langle A^2 \rangle = \begin{cases} \dfrac{\eta}{a} & \text{for } \eta \ge 0 \\ 0 & \text{for } \eta \le 0. \end{cases} \tag{13.30}$$

It is seen from this that for $\eta > 0$ the parametric excitation of pendulum oscillations occurs under the effect of noise. This manifests itself in the fact that the mean values of the amplitude and the amplitude squared become different from zero. If an observer detects such oscillations and does not know the causes of their occurrence then the observer can draw the conclusion that he is viewing chaotic self-oscillations. The question naturally arises of whether or not the observer can distinguish between the process observed and chaotic self-oscillations. This problem will be discussed below.

13.3.1 The results of a numerical simulation of the oscillations of a pendulum with a randomly vibrated suspension axis

Since the theoretical results obtained are approximate and give no way of determining the shape of the pendulum oscillations, we studied solutions of Eq. (13.15) numerically for $\delta = 0.1$, $\omega_0 = 1$, $\alpha = 10$ [190]. The noise $\xi(t)$ was simulated by means of a Gaussian random source with the variance equal to unity. The sample frequency was set to 100 Hz. Thus, the spectral density of $\xi(t)$ was approximately uniform in the range from 0 to 100 Hz. This allows us to denote the noise intensity by κ, but not by $\kappa(2\omega_0)$.

In numerical experiments it is more convenient to calculate not the mean amplitude squared but the variance of the corresponding variable. It is evident that the dependencies of these values on noise intensity should be similar. Indeed, in the case when the amplitude A is a slowly changing function the variance is equal to $\langle A^2 \rangle/2$. An example of the dependence of the variance of x ($\sigma_x^2 = \langle x^2 \rangle$) on the noise intensity κ is given in Fig. 13.5. We see that in the vicinity of the excitation boundary, when κ is close to $\kappa_{cr} = 8\delta/\omega_0^2$, this dependence can be approximated by a straight line intersecting the abscissa at point κ_{cr}. The slope of this straight line is equal to 0.074. This differs little from that follows from the theoretical results presented above: according to

[1] Consideration of the effect of additive noise was undertaken by Dimentberg [70] and by Landa and Zaikin [194].

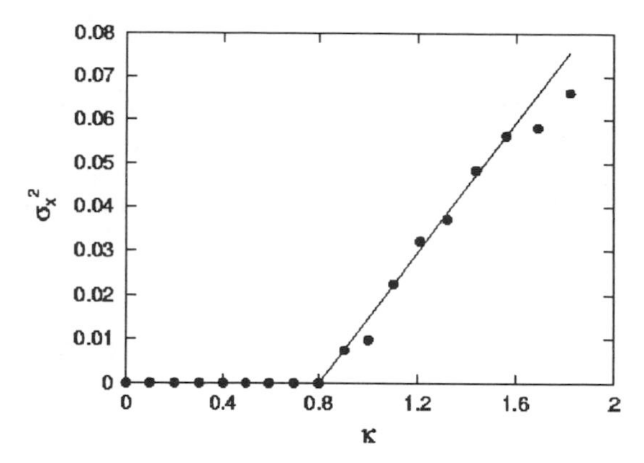

Fig. 13.5. An example of the dependence of σ_x^2 on the noise intensity κ found by numerical simulation of Eq. (13.15) for $\delta = 0.1$, $\omega_0 = 1$, $\alpha = 10$. The solid straight line is described by the equation $\sigma_x^2 = 0.074(\kappa - \kappa_{\text{cr}})$, where $\kappa_{\text{cr}} = 8\delta/\omega_0^2 = 0.8$

them $\sigma_x^2 \approx \langle A^2 \rangle/2 = \omega_0^2/(12\alpha\delta) \approx 0.083$. Away from the excitation boundary the growth rate of the variance somewhat decreases.

13.3.2 On-off intermittency

A distinguishing characteristic of the noise-induced oscillations considered is that they develop via so-called on-off intermittency [193, 195]. This can be seen, for example, from Fig. 13.6, where the plots $x(t)$ and $\dot{x}(t)$ found by numerical simulation of Eq. (13.15) are given. It is seen that close to the threshold of excitation the pendulum oscillates in the immediate vicinity of its equilibrium position over prolonged periods (so-called 'laminar' phases); these weak oscillations alternate with short strong bursts ('turbulent' phases). Away from the threshold the duration of laminar phases decreases and that of turbulent ones increases, and finally laminar phases disappear. The variance of the pendulum oscillations increases in this process.

The term 'on-off intermittency' was recently introduced by Platt et al. in [271], though a map associated with a similar type of intermittent behavior was first considered by Pikovsky [269] and then by Fujisaka and Yamada [94]. Of prime importance is the fact that on-off intermittency can take place not only in dynamical systems but in stochastic systems as well [116]. In [116] the statistical properties of on–off intermittency were obtained from the analysis of the map

$$x_{n+1} = a(1 + z_n)x_n + f(x_n),$$

where z_n is either a certain deterministic chaotic process or a random process, a is the bifurcation parameter, and $f(x_n)$ is a nonlinear function. For this map it was shown that the mean duration of laminar phases has to be proportional to a^{-1}.

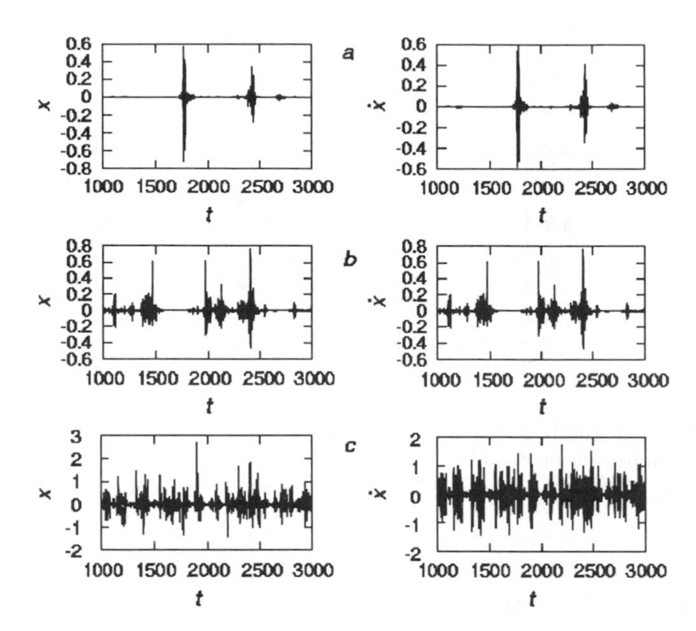

Fig. 13.6. Plots of x and $\dot{x}$ versus t found by numerical simulation of Eq. (13.15) for $\delta = 0.1$, $\omega_0 = 1$, $\alpha = 10$, (a) $\kappa = 0.81$, (b) $\kappa = 0.9025$ and (c) $\kappa = 1.69$

Let us calculate the mean duration of laminar phases for the pendulum, using Eqs. (13.24) and the Fokker–Planck equation (13.20) associated with them. We assume that the pendulum oscillates in a laminar phase if the oscillation amplitude A is not larger than a certain value ε. Then the mean duration of the laminar phase τ_ε is determined by the mean duration of a random walk-like motion of a representative point inside the circle of radius ε on the plane x, $\dot{x}$. This duration can be calculated (see, e.g., [163, 198]) using the steady-state solution of Eq. (13.20) with the boundary condition

$$w(A, \phi)|_{A=\varepsilon} = 0. \tag{13.31}$$

Because the value of ε is assumed to be small, we can neglect the term $(3/4)\delta\alpha\omega_0^2 A^2$ in Eq. (13.20). In so doing the solution of Eq. (13.20) with the boundary condition (13.31) is

$$w(A, \phi) = \frac{2G_0}{\delta(1-\eta)A}\left(\left(\frac{A}{\varepsilon}\right)^{(\eta-1)/(\eta+1)} - 1\right), \tag{13.32}$$

where G_0 is the value of the probability flow

$$G = \frac{\omega_0^2 \kappa(2\omega_0)}{8}\left(\frac{\eta A w}{\eta+1} - \frac{1}{2}\frac{\mathrm{d}}{\mathrm{d}A}(A^2 w)\right)$$

across any circumference inside the circle of radius ε.

The value of G_0 is determined from the normalization condition by integrating the expression (13.32) over the area of the circle of radius ε. As a result, we find

$$G_0^{-1} = \frac{2\pi\varepsilon}{\delta\eta} = \frac{16\pi\varepsilon}{\omega_0^2\big(\kappa(2\omega_0) - \kappa_{\rm cr}\big)}. \tag{13.33}$$

Since the representative point, touching the boundary of the circle, can either return with a certain probability p (if $\dot{A} < 0$) or intersect this boundary (if $\dot{A} < 0$), we obtain for the mean duration of the laminar phase τ_ϵ the following expression:

$$\tau_\varepsilon = G_0^{-1}(1-p)\sum_{j=1}^{\infty} jp^{j-1}.$$

Summarizing this series and taking account of the expression (13.33), we find

$$\tau_\varepsilon = G_0^{-1}(1-p)^{-1} = \frac{2\pi\varepsilon}{\delta\eta(1-p)} = \frac{16\pi\varepsilon}{\omega_0^2\big(\kappa(2\omega_0) - \kappa_{\rm cr}\big)(1-p)}. \tag{13.34}$$

For small η and ε, when $p \approx 1/2$ the mean duration of laminar phase is proportional to ε and inversely proportional to η. This result agrees quite well with [116].

Processing the results of the numerical simulation of Eq. (13.15) showed that the formula (13.34) with $p = 1/2$ is valid to a good approximation. This is demonstrated in Fig. 13.7, where the dependencies of τ_ε on $\kappa - \kappa_{\rm cr}$ for two values of ε are given. The corresponding theoretical dependencies are shown by solid lines.

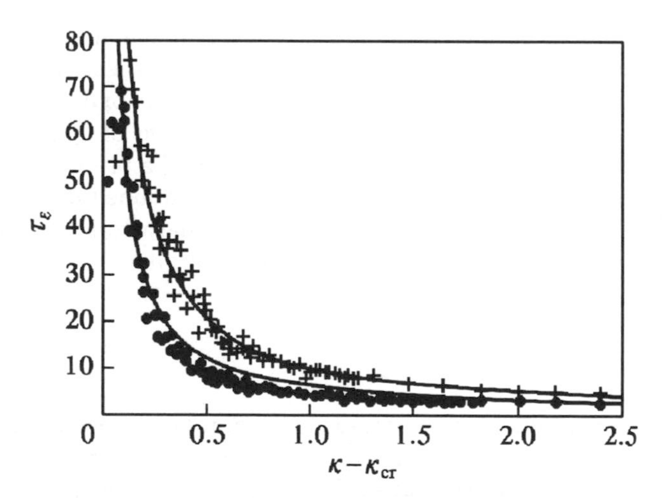

Fig. 13.7. The numerical and theoretical dependencies of τ_ε on $\kappa - \kappa_{\rm cr}$ for $\omega_0 = 1$, $\delta = 0.1$, $\alpha = 10$, $\varepsilon = 0.06$ (circles) and $\varepsilon = 0.1$ (crosses). Solid lines are theoretical dependencies

13.3.3 Correlation dimension

Inasmuch as the pendulum oscillations under consideration are conditioned by nothing but the noise, their dimension would be expected to be sufficiently large. However, the calculations of correlation dimension, performed by Landa and Zaikin [190] both in ordinary Takens' space and by using the well-adapted basis [181], have shown that the dimension is not large. The saturation of correlation dimension with increasing embedding space dimension points to this.[2] An example of the dependence of correlation dimension on the embedding space dimension is shown in Fig. 13.8 a. As the noise intensity

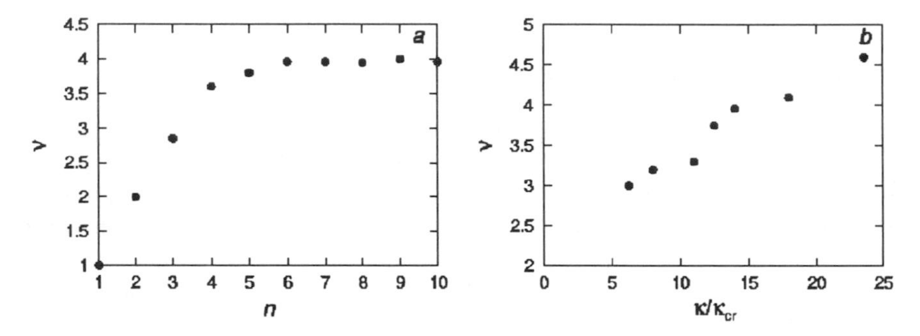

Fig. 13.8. The dependencies of the correlation dimension ν (a) on the embedding space dimension n and (b) on the relative spectral density κ/κ_{cr} for $\omega_0 = 1$, $\delta = 0.1$, $\alpha = 100$

increases the dimension increases only slowly, but it remains finite. The dependence of the correlation dimension ν on the relative spectral density κ/κ_{cr} is depicted in Fig. 13.8 b. So, the dimension gives no way of distinguishing between noise-induced oscillations and chaotic oscillations of dynamical origin. An example of such oscillations was considered in the preceding section. It should be particularly emphasized that the result obtained contradicts popular opinion that the dimension is precisely the characteristic which allows the chaotic oscillations in dynamical systems and random oscillations caused by noise to be distinguishable.

13.3.4 Power spectra

Similar to the pendulum with a harmonically vibrated suspension axis, in the case when the pendulum is excited by noise and nonlinear friction is negligible, its power spectrum contains a low-frequency part caused by the slow drift of x. The form of the spectrum depends almost not at all on the ratio κ/κ_{cr} (see Fig. 13.9 a and b). For any κ/κ_{cr} the spectrum practically de-

[2] It should be noted that the corresponding correlation integrals have no clearly defined linear part, making the exact evaluation of the dimension difficult.

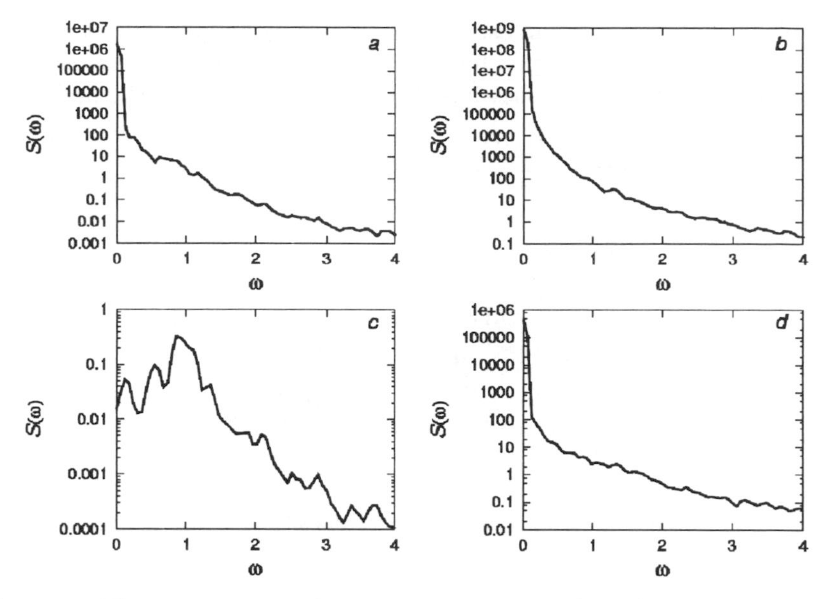

Fig. 13.9. The power spectra for the solutions of Eq. (13.15) for $\omega = 1$, $\delta = 0.1$: (a) $\kappa/\kappa_{\rm cr} = 1.08$, $\alpha = 0$, (b) $\kappa/\kappa_{\rm cr} = 16.2$, $\alpha = 0$, (c) $\kappa/\kappa_{\rm cr} = 1.08$, $\alpha = 10$, and (d) $\kappa/\kappa_{\rm cr} = 16.2$, $\alpha = 10$

creases monotonically and is reminiscent of the flicker noise spectrum. With nonlinear friction the low-frequency part of the power spectrum decreases significantly, and, if the noise intensity differs little from its critical value (Fig. 13.9 c), the spectrum has a peak located close to the natural frequency ω_0. As the noise intensity increases, this peak decreases and disappears eventually; the qualitative behavior of the spectrum becomes the same as for $\alpha = 0$ (see Fig. 13.9 d).

13.3.5 The Rytov–Dimentberg criterion

Let us revert to the question of whether or not one can distinguish between noise-induced oscillations and chaotic oscillations of dynamical origin. A similar question was first formulated by Rytov [294] and later by Dimentberg [69, 70], as applied to the problem of distinguishing between noise passed through a linear narrow-band filter and periodic but noisy self-oscillations. It was shown that in the case of noisy self-oscillations the probability density for instantaneous amplitude squared has to peak at a certain finite value of the amplitude, whereas for noise passed through a filter it has to be monotonically decreasing.

In the case of chaotic oscillations of dynamical origin the probability density for instantaneous amplitude squared would also be expected to peak at one or several values of the amplitude. We have verified this statement by an example of chaotic pendulum oscillations caused by periodic vibration of the suspension axis. The corresponding histogram for the probability density of instantaneous amplitude is shown in Fig. 13.10. We see that the probability

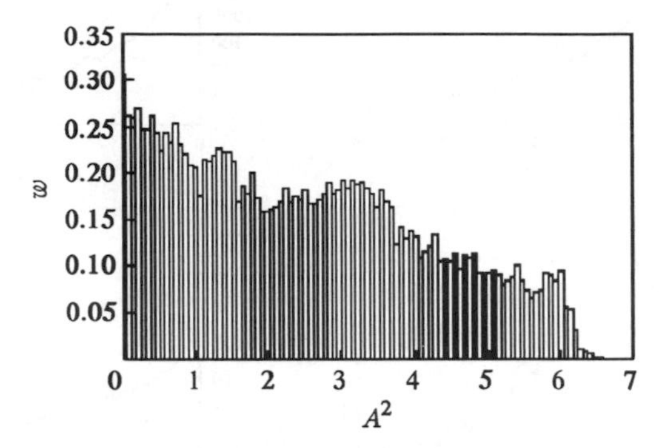

Fig. 13.10. The histogram for the probability density of instantaneous amplitude squared for $B = 3.5$, $\alpha = 2$

density is not monotonically decreasing with increasing amplitude but has several peaks only slightly defined.

It follows from the results presented in the first section that in the case of the parametric excitation of pendulum oscillations under the effect of random vibration of the suspension axis the probability density for the value aA^2 is $\tilde{w}(aA^2) = w(A)/2aA$, where $w(A)$ is determined by the expression (13.29). The dependence $\tilde{w}(aA^2)$ for $\kappa/\kappa_{\rm cr} = 1.25$ is shown in Fig. 13.11 a. We see that

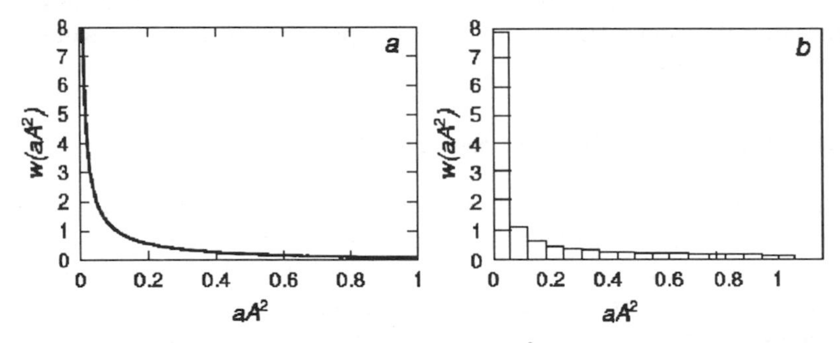

Fig. 13.11. (a) The theoretical dependence $\tilde{w}(aA^2)$ and (b) the histogram for the probability density of aA^2 calculated from the numerical solution of Eq. (13.15) for $\kappa/\kappa_{cr} = 1.25$

the probability density for the amplitude squared, calculated analytically, is monotonically decreasing with increasing amplitude. Similar results are also obtained from the data of a numerical simulation. The histograms of the probability density for aA^2 calculated from the numerical solution of Eq. (13.15) for $\kappa(2)/\kappa_{cr}(2) = 1.25$ is presented in Fig. 13.11 b.

Dimentberg also suggested another, more convenient, version of the criterion considered. In place of instantaneous amplitude, the probability density for the process $x(t)$ itself is analyzed. It is shown that if the probability density for $x > 0$ is not monotonically decreasing then the process $x(t)$ is self-oscillatory. But if the probability density for $x > 0$ is monotonically decreasing then the process $x(t)$ can be both self-oscillatory and noise passed through a filter. It should be noted that the author passes over in silence the fact that for using this criterion the probability density for x should be an even function.

We have verified the second version of the Dimentberg criterion for both noise-induced pendulum oscillations and chaotic oscillations caused by harmonic action. We have detected that this version is also usable. This is illustrated in Fig. 13.12. Thus, in spite of the essentially nonlinear transformation

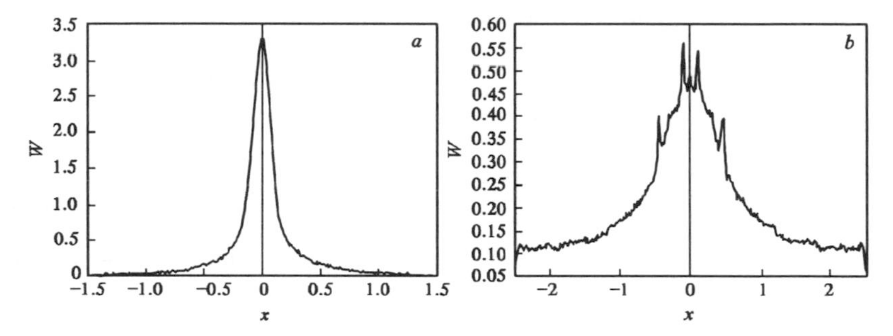

Fig. 13.12. The histograms of the probability density $W(x)$ (a) in the case of noise-induced oscillation for $\kappa/\kappa_{cr} = 14$ and (b) in the case of chaotic oscillations caused by harmonic vibration of the suspension axis for $B = 3.5$, $\alpha = 2$

of noise, in the case under consideration the Rytov–Dimentberg criterion is true. It is undeniable that the question of the veracity of this criterion in the general case is still an open question.

13.4 Parametric resonance in a system of two coupled oscillators

We consider excitation of periodic parametric oscillations in systems with two degrees of freedom using, as an example, two coupled oscillators described by the following equations:

$$\ddot{x}_1 + 2\delta_1\dot{x}_1 + \nu_1^2(1+\gamma_1 x_1^2)x_1 + c_1x_2 + (\mu_{11}x_1 - \mu_{12}x_2)\cos 2\omega t = 0,$$
$$\ddot{x}_2 + 2\delta_2\dot{x}_2 + \nu_2^2(1+\gamma_2 x_2^2)x_2 + c_2x_1 - (\mu_{21}x_1 - \mu_{22}x_2)\cos 2\omega t. \tag{13.35}$$

We assume that the coefficients $\delta_{1,2}$, $\gamma_{1,2}$, μ_{11}, μ_{12}, μ_{21} and μ_{22} are sufficiently small, so that Eqs. (13.35) can be rewritten as

$$\begin{aligned}\ddot{x}_1 + \nu_1^2 x_1 + c_1 x_2 &= \epsilon\Big(-2\delta_1\dot{x}_1 - \nu_1^2\gamma_1 x_1^3 - (\mu_{11}x_1 - \mu_{12}x_2)\cos 2\omega t\Big), \\ \ddot{x}_2 + \nu_2^2 x_2 + c_2 x_1 &= \epsilon\Big(-2\delta_2\dot{x}_2 - \nu_2^2\gamma_2 x_2^3 + (\mu_{21}x_1 - \mu_{22}x_2)\cos 2\omega t\Big),\end{aligned} \tag{13.36}$$

where ϵ is a small parameter.

We restrict ourselves to the cases of main parametric resonance, when the frequency of excited oscillations is equal to half the frequency of the parametric action, and of combination resonance, when parametric oscillations are excited with the frequencies ω_1 and ω_2 associated with the action frequency 2ω by the relation

$$2\omega = \omega_1 \pm \omega_2. \tag{13.37}$$

To analyze the main resonance, we use the Krylov–Bogolyubov method. A solution of Eq. (13.36) can be sought in the form

$$x_1 = A_1\cos\psi_1 + \epsilon u_1 + \dots, \quad x_2 = \frac{c_2}{\omega_{01}^2 - \nu_2^2}A_1\cos\psi_1 + \epsilon u_2 + \dots, \tag{13.38}$$

where $\psi_1 = \omega t + \varphi_1$,

$$\omega_{01} = \frac{\nu_1^2 + \nu_2^2}{2} + \sqrt{\frac{(\nu_1^2 - \nu_2^2)^2}{4} + c_1c_2}, \tag{13.39}$$

is one of the normal frequencies of the generative system, A_1 and φ_1 obey the equations

$$\dot{A}_1 = \epsilon f_1(A_1, \varphi_1) + \dots, \qquad \dot{\varphi}_1 = -\Delta_1 + \epsilon F_1(A_1, \varphi_1) + \dots, \tag{13.40}$$

$\Delta_1 = \omega - \omega_{01}$ is the frequency mistuning, and f_1, $F_1, \dots$, u_1, $u_2, \dots$ are unknown functions. Substituting (13.38) into (13.36) and equating the coefficients of ϵ, we obtain

$$\begin{aligned}\ddot{u}_1 + \nu_1^2 u_1 + c_1 u_2 &= \left(2\omega_{01}f_1 - A_1\Delta_1\frac{\partial F_1}{\partial\varphi_1}\right)\sin\psi_1 \\ &+ \left(2\omega_{01}A_1F_1 + \Delta_1\frac{\partial f_1}{\partial\varphi_1}\right)\cos\psi_1 + 2\delta_1\omega_{01}A_1\sin\psi_1 \\ &- \nu_1^2\gamma_1 A_1^3\cos^3\psi_1 - \left(\mu_{11} - \mu_{12}\frac{c_2}{\omega_{01}^2 - \nu_2^2}\right)A_1\cos 2\omega t\cos\psi_1, \\ \ddot{u}_2 + \nu_2^2 u_2 + c_2 u_1 &= \left[\left(2\omega_{01}f_1 - A_1\Delta_1\frac{\partial F_1}{\partial\varphi_1}\right)\sin\psi_1\right. \\ &\left.+ \left(2\omega_{01}A_1F_1 + \Delta_1\frac{\partial f_1}{\partial\varphi_1}\right)\cos\psi_1 + 2\delta_2\omega_{01}A_1\sin\psi_1\right]\frac{c_2}{\omega_{01}^2 - \nu_2^2} \\ &- \nu_2^2\gamma_2\frac{c_2^3}{(\omega_{01}^2 - \nu_2^2)^3}A_1^3\cos^3\psi_1 - \left(\mu_{21} - \mu_{22}\frac{c_2}{\omega_{01}^2 - \nu_2^2}\right)A_1\cos 2\omega t\cos\psi_1.\end{aligned} \tag{13.41}$$

We seek a solution of Eqs. (13.41) in the form of a Fourier series as

$$u_{1,2} = B_{1,2} \cos \psi_1 + C_{1,2} \sin \psi_1 + \dots . \tag{13.42}$$

By substituting (13.41) into (13.40) and equating the coefficients of $\cos \psi_1$ and $\sin \psi_1$ we obtain the equations for $B_{1,2}$ and $C_{1,2}$:

$$\begin{aligned}
(\nu_1^2 - \omega_{01}^2)B_1 + c_1 B_2 &= 2\omega_{01} A_1 F_1 + \Delta_1 \frac{\partial f_1}{\partial \varphi_1} - \frac{3}{4}\nu_1^2 \gamma_1 A_1^3 \\
&\quad - \frac{1}{2}\left(\mu_{11} - \mu_{12}\frac{c_2}{\omega_{01}^2 - \nu_2^2}\right) A_1 \cos 2\varphi_1 , \\
c_2 B_1 + (\nu_2^2 - \omega_{01}^2) B_2 &= \left(2\omega_{01} A_1 F_1 + \Delta_1 \frac{\partial f_1}{\partial \varphi_1}\right)\frac{c_2}{\omega_{01}^2 - \nu_2^2} \\
&\quad - \frac{3}{4}\nu_2^2 \gamma_2 \frac{c_2^3}{(\omega_{01}^2 - \nu_2^2)^3} A_1^3 + \frac{1}{2}\left(\mu_{21} - \mu_{22}\frac{c_2}{\omega_{01}^2 - \nu_2^2}\right) A_1 \cos 2\varphi_1 ,
\end{aligned} \tag{13.43}$$

$$\begin{aligned}
(\nu_1^2 - \omega_{01}^2)C_1 + c_1 C_2 &= 2\omega_{01} f_1 - A_1 \Delta_1 \frac{\partial F_1}{\partial \varphi_1} + 2\delta_1 \omega_{01} A_1 \\
&\quad - \frac{1}{2}\left(\mu_{11} - \mu_{12}\frac{c_2}{\omega_{01}^2 - \nu_2^2}\right) A_1 \sin 2\varphi_1 , \\
c_2 C_1 + (\nu_2^2 - \omega_{01}^2) C_2 &= \left(2\omega_{01} f_1 - A_1 \Delta_1 \frac{\partial F_1}{\partial \varphi_1}\right. \\
&\quad \left. + 2\delta_2 \omega_{01} A_1\right)\frac{c_2}{\omega_{01}^2 - \nu_2^2} + \frac{1}{2}\left(\mu_{21} - \mu_{22}\frac{c_2}{\omega_{01}^2 - \nu_2^2}\right) A_1 \sin 2\varphi_1 .
\end{aligned} \tag{13.44}$$

These equations allow us both to find f_1 and F_1 and to determine the corrections to the eigenvector. Since the determinants of the systems of equations (13.43) and (13.44) are equal to zero, the equations for f_1 and F_1 are found from the compatibility conditions. These equations are

$$\begin{aligned}
2\omega_{01} A_1 F_1 + \Delta_1 \frac{\partial f_1}{\partial \varphi_1} &= \frac{3}{4}\omega_{01}^2 \Gamma_1 A_1^3 + M_1 A_1 \cos 2\varphi_1 , \\
2\omega_{01} f_1 - A_1 \Delta_1 \frac{\partial F_1}{\partial \varphi_1} &= -2\omega_{01} \tilde{\delta}_1 A_1 + M_1 A_1 \sin 2\varphi_1 ,
\end{aligned} \tag{13.45}$$

where

$$\tilde{\delta}_1 = \frac{\delta_1(\omega_{01}^2 - \nu_2^2) + \delta_2(\omega_{01}^2 - \nu_1^2)}{2\omega_{01}^2 - \nu_1^2 - \nu_2^2} ,$$

$$M_1 = \frac{1}{2(2\omega_{01}^2 - \nu_1^2 - \nu_2^2)}\big(\mu_{11}(\omega_{01}^2 - \nu_2^2) + \mu_{22}(\omega_{01}^2 - \nu_1^2) - \mu_{12}c_2 - \mu_{21}c_1\big),$$

$$\Gamma_1 = \frac{\nu_1^2\gamma_1(\omega_{01}^2 - \nu_2^2) + \nu_2^2\gamma_2 c_2^2(\omega_{01}^2 - \nu_1^2)/(\omega_{01}^2 - \nu_2^2)^2}{\omega_{01}^2(2\omega_{01}^2 - \nu_1^2 - \nu_2^2)} .$$

A solution of Eqs. (13.45) is

$$f_1 = -\tilde{\delta}_1 A_1 + \frac{M_1}{\omega + \omega_{01}} A_1 \sin 2\varphi_1,$$

$$F_1 = \frac{3}{8}\omega_{01}\Gamma_1 A_1^2 + \frac{M_1}{\omega + \omega_{01}} A_1 \cos 2\varphi_1. \tag{13.46}$$

From (13.40) and (13.46) we obtain the equations for steady-state values of the amplitude A_1 and the phase φ_1:

$$\tilde{\delta}_1 = \frac{M_1}{\omega + \omega_{01}} \sin 2\varphi_1, \quad \frac{3}{8}\omega_{01}\Gamma_1 A_1^2 = \Delta_1 - \frac{M_1}{\omega + \omega_{01}} \cos 2\varphi_1. \tag{13.47}$$

It follows from Eqs. (13.47) that

$$\sin 2\varphi_1 = \frac{\tilde{\delta}_1(\omega + \omega_{01})}{M_1},$$

$$\frac{3}{8}\Gamma_1 A_1^2 = \frac{\Delta_1}{\omega_{01}} \pm \frac{1}{\omega_{01}}\sqrt{\frac{M_1^2}{(\omega + \omega_{01})^2} - \tilde{\delta}_1^2}. \tag{13.48}$$

It can easily be shown that the solution of Eqs. (13.48) is real only if $M_1 \geq 2\omega_{01}\tilde{\delta}_1$.

Another solution of Eqs. (13.36) can be obtained if we take, as a generative solution, the following:

$$x_1 = A_2 \cos \psi_2, \quad x_2 = \frac{c_2}{\omega_{02}^2 - \nu_2^2} A_2 \cos \psi_2,$$

where $\psi_2 = \omega t + \varphi_2$, and

$$\omega_{02} = \frac{\nu_1^2 + \nu_2^2}{2} - \sqrt{\frac{(\nu_1^2 - \nu_2^2)^2}{4} + c_1 c_2}, \tag{13.49}$$

is the second normal frequency of the generative system. Performing calculations similar to those presented above, we find the following equations for φ_2 and A_2:

$$\sin 2\varphi_2 = \frac{\tilde{\delta}_2(\omega + \omega_{02})}{M_2},$$

$$\frac{3}{8}\Gamma_2 A_2^2 = \frac{\Delta_2}{\omega_{02}} \pm \frac{1}{\omega_{02}}\sqrt{\frac{M_2^2}{(\omega + \omega_{02})^2} - \tilde{\delta}_2^2}, \tag{13.50}$$

where

$$\Delta_2 = \omega - \omega_{02}, \quad \tilde{\delta}_2 = \frac{\delta_1(\omega_{02}^2 - \nu_2^2) + \delta_2(\omega_{02}^2 - \nu_1^2)}{2\omega_{02}^2 - \nu_1^2 - \nu_2^2},$$

$$M_2 = \frac{1}{2(2\omega_{02}^2 - \nu_1^2 - \nu_2^2)} \big(\mu_{11}(\omega_{02}^2 - \nu_2^2) + \mu_{22}(\omega_{02}^2 - \nu_1^2) - \mu_{12}c_2 - \mu_{21}c_1\big),$$

$$\Gamma_2 = \frac{\nu_1^2\gamma_1(\omega_{02}^2 - \nu_2^2) + \nu_2^2\gamma_2 c_2^2(\omega_{02}^2 - \nu_1^2)/(\omega_{02}^2 - \nu_2^2)^2}{\omega_{02}^2(2\omega_{02}^2 - \nu_1^2 - \nu_2^2)}.$$

The condition for the existence of a real solution of Eqs. (13.50) is $M_2 \geq 2\omega_{02}\tilde{\delta}_2$.

It should be noted that in the case of identical oscillators, when $\nu_1 = \nu_2 = \nu$, $c_1 = c_2 = c$, the first of the solutions found corresponds to in-phase oscillations, and the second one corresponds to anti-phase oscillations.

Examples of the dependencies of γA_1^2 and γA_2^2 on $\Delta/\nu = \omega/\nu - 1$, calculated from the formulas (13.48), (13.50) for $\nu_1 = \nu_2 = \nu$, $\gamma_1 = \gamma_2 = \gamma$, $c_1 = c_2 = c$, $\delta_1 = \delta_2 = \delta$, are shown in Fig. 13.13. It is seen that for

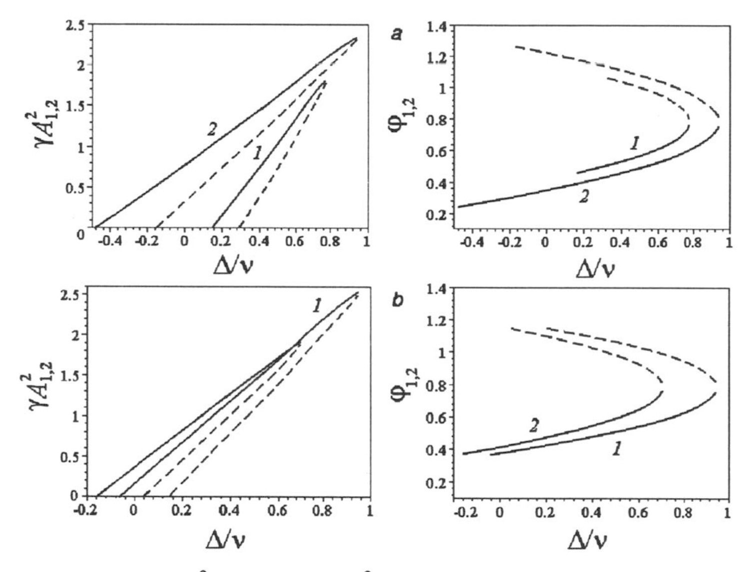

Fig. 13.13. Plots of γA_1^2 (curve 1), γA_2^2 (curve 2), φ_1 (curve 1) and φ_2 (curve 2) versus Δ/ν for $\nu_1 = \nu_2 = \nu$, $\gamma_1 = \gamma_2 = \gamma$, $c_1 = c_2 = c$, $\delta_1 = \delta_2 = \delta$, $\delta/\nu = 0.1$, $M_1/\nu^2 = 0.3$, $M_2/\nu^2 = 0.265$, (a) $c/\nu^2 = 0.5$ and (b) $c/\nu^2 = 0.1$. The dashed lines show unstable parts of the dependencies

$c/\nu^2 = 0.5$ two resonance curves are separated, whereas for $c/\nu^2 = 0.1$ they partially overlap.

We now turn our attention to combination parametric resonance, when oscillations are excited with frequencies ω_1 and ω_2 related to the parametric action frequency by the formula (13.37). To consider this case, it is convenient in Eqs. (13.36) to go to the normal coordinates y_1 and y_2 by

$$x_1 = y_1 + \frac{\omega_{02}^2 - \nu_2^2}{c_2} y_2, \quad x_2 = \frac{\omega_{01}^2 - \nu_1^2}{c_1} y_1 + y_2. \tag{13.51}$$

By substituting (13.51) into Eqs. (13.36) we obtain the following equations for y_1 and y_2:

$$\begin{aligned}\ddot{y}_1 + \omega_{01}^2 y_1 = \frac{\omega_{01}^2 - \nu_2^2}{\omega_{01}^2 - \omega_{02}^2} \epsilon \Bigg[&-2\left(\delta_1 + \frac{\omega_{01}^2 - \nu_1^2}{\omega_{01}^2 - \nu_2^2}\delta_2\right)\dot{y}_1 \\ &- 2\frac{\omega_{02}^2 - \nu_2^2}{c_2}(\delta_1 - \delta_2)\dot{y}_2 - \nu_1^2\gamma_1\left(\dot{y}_1 + \frac{\omega_{02}^2 - \nu_2^2}{c_2} y_2\right)^3 \\ &- \frac{\omega_{01}^2 - \nu_1^2}{c_2}\nu_2^2\gamma_2\left(y_1\frac{\omega_{01}^2 - \nu_1^2}{c_1} + y_2\right)^3 - (M_{11}y_1 - M_{12}y_2)\cos 2\omega t\Bigg],\end{aligned} \tag{13.52}$$

$$\begin{aligned}\ddot{y}_2 + \omega_{02}^2 y_2 = \frac{\omega_{01}^2 - \nu_2^2}{\omega_{01}^2 - \omega_{02}^2} \epsilon \Bigg[&2\frac{\omega_{01}^2 - \nu_1^2}{c_1}(\delta_1 - \delta_2)\dot{y}_1 \\ &- 2\left(\frac{\omega_{02}^2 - \nu_2^2}{\omega_{02}^2 - \nu_1^2}\delta_1 + \delta_2\right)\dot{y}_2 - \frac{\omega_{02}^2 - \nu_2^2}{c_1}\nu_1^2\gamma_1\left(y_1 + \frac{\omega_{02}^2 - \nu_2^2}{c_2} y_2\right)^3 \\ &- \nu_2^2\gamma_2\left(\frac{\omega_{01}^2 - \nu_1^2}{c_1} y_1 + y_2\right)^3 + (M_{21}y_1 - M_{22}y_2)\cos 2\omega t\Bigg],\end{aligned}$$

where

$$\begin{aligned}M_{11} &= \mu_{11} - (\omega_{01}^2 - \nu_1^2)\left(\frac{\mu_{12}}{c_1} + \frac{\mu_{21}}{c_2}\right) + \frac{\omega_{01}^2 - \nu_1^2}{\omega_{01}^2 - \nu_2^2}\mu_{22}, \\ M_{12} &= -\frac{\omega_{02}^2 - \nu_2^2}{c_2}(\mu_{11} - \mu_{22}) + \mu_{12} - \frac{(\omega_{02}^2 - \nu_2^2)^2}{c_2^2}\mu_{21}, \\ M_{21} &= -\frac{\omega_{02}^2 - \nu_2^2}{c_1}(\mu_{11} - \mu_{22}) - \frac{(\omega_{01}^2 - \nu_1^2)^2}{c_1^2}\mu_{12} + \mu_{21}, \\ M_{22} &= \frac{\omega_{02}^2 - \nu_2^2}{\omega_{02}^2 - \nu_1^2}\mu_{11} + (\omega_{01}^2 - \nu_1^2)\left(\frac{\mu_{12}}{c_1} + \frac{\mu_{21}}{c_2}\right) + \mu_{22}.\end{aligned}$$

A solution of Eqs. (13.52) can be sought in the form

$$y_1 = A_1\cos(\omega_1 t + \varphi_1) + \ldots, \quad y_2 = A_2\cos(\omega_2 t + \varphi_2) + \ldots.$$

By using the regular Krylov–Bogolyubov technique free from conditions for smallness of the mistunings $\Delta_1 = \omega_1 - \omega_{01}$ and $\Delta_2 = \omega_2 - \omega_{02}$ we obtain, as a first approximation, the following equations for A_1, A_2, φ_1 and φ_2:

$$\begin{aligned}\dot{A}_1 &= \frac{\omega_{01}^2 - \nu_2^2}{\omega_{01}^2 - \omega_{02}^2}\left(-(\delta_1 + \alpha\delta_2)A_1 - \frac{M_{12}}{2(\omega_1 + \omega_{01})}A_2\sin\Phi\right), \\ \dot{A}_2 &= \frac{\omega_{01}^2 - \nu_2^2}{\omega_{01}^2 - \omega_{02}^2}\left(-(\alpha\delta_1 + \delta_2)A_2 \mp \frac{M_{21}}{2(\omega_2 + \omega_{02})}A_1\sin\Phi\right),\end{aligned} \tag{13.53}$$

$$\dot{\varphi}_1 = -\Delta_1 + \frac{\omega_{01}^2 - \nu_2^2}{\omega_{01}^2 - \omega_{02}^2}\left(\frac{3\nu_1^2\gamma_1}{8\omega_{01}}\left(A_1^2 + 2\alpha\frac{c_1}{c_2}A_2^2\right)\right.$$
$$\left. + \frac{3\nu_2^2\gamma_2}{8\omega_{01}}\alpha\left(\alpha\frac{c_2}{c_1}A_1^2 + 2A_2^2\right) - \frac{M_{12}}{2(\omega_1+\omega_{01})}\frac{A_2}{A_1}\cos\Phi\right),$$
$$\dot{\varphi}_2 = -\Delta_2 + \frac{\omega_{01}^2 - \nu_2^2}{\omega_{01}^2 - \omega_{02}^2}\left(\frac{3\nu_1^2\gamma_1}{8\omega_{02}}\alpha\left(2A_1^2 + \alpha\frac{c_1}{c_2}A_2^2\right)\right.$$
$$\left. + \frac{3\nu_2^2\gamma_2}{8\omega_{02}}\left(2\alpha\frac{c_2}{c_1}A_1^2 + A_2^2\right) - \frac{M_{21}}{2(\omega_2+\omega_{02})}\frac{A_1}{A_2}\cos\Phi\right), \tag{13.54}$$

where $\Phi = \varphi_1 \pm \varphi_2$, $\alpha = (\omega_{01}^2 - \nu_1^2)/(\omega_{01}^2 - \nu_2^2) = (\omega_{02}^2 - \nu_2^2)/(\omega_{02}^2 - \nu_1^2)$.

It follows from Eqs. (13.53) that in the stationary case

$$\frac{A_2^2}{A_1^2} \equiv \beta = \pm\frac{M_{21}}{M_{12}}\frac{\omega_1+\omega_{01}}{\omega_2+\omega_{02}}\frac{\delta_1+\alpha\delta_2}{\alpha\delta_1+\delta_2}, \tag{13.55}$$

$$\sin^2\Phi = \frac{4(\delta_1+\alpha\delta_2)(\alpha\delta_1+\delta_2)(\omega_1+\omega_{01})(\omega_2+\omega_{02})}{|M_{12}M_{21}|}. \tag{13.56}$$

We see from the expression (13.55) that in the case when M_{12} and M_{21} have the same signs, combination resonance is possible only for the relation between the frequencies of the form $2\omega = \omega_1 + \omega_2$ (when in the expression (13.55) there is the '+' sign), but if the signs of M_{12} and M_{21} are opposite then only the relation $2\omega = \omega_1 - \omega_2$ is possible. From (13.56) we find the condition for the existence of the combination resonance:

$$\frac{|M_{12}M_{21}|}{16\omega_{01}\omega_{02}(\delta_1+\alpha\delta_2)(\alpha\delta_1+\delta_2)} \geq 1.$$

Setting $\dot{A}_1 = \dot{A}_2 = \dot{\varphi}_1 = \dot{\varphi}_2 = 0$ in Eqs. (13.53), (13.54), we obtain four equations for steady-state values of A_1, A_2, Φ and ω_1 (the frequency ω_2 is related to ω_1 by (13.37)) as functions of the action frequency ω. It is convenient to take the frequency mistuning $\Delta = \omega_1 - \omega_{01} \pm (\omega_2 - \omega_{02}) = \Delta_1 + \Delta_2 \text{sign} M_{12}/M_{21}$ in place of ω. Examples of the dependencies of $A_{1,2}^2$ and of Δ_1/ν_1 on the relative mistuning Δ/ν_1 are depicted in Fig. 13.14 in two cases: when M_{12} and M_{21} are of the same sign (curves 1, 2) and when M_{12} and M_{21} are of opposite sign (curves 1′, 2′). We see that in the first case the combination resonance exists mainly in a range of positive frequency mistunings, whereas in the second case it exists mainly in a range of negative frequency mistunings.

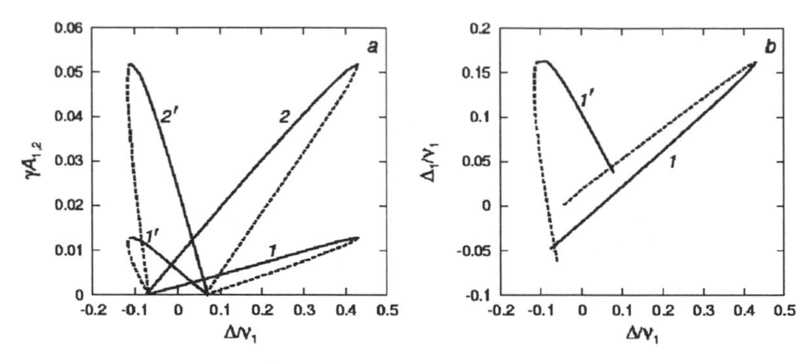

Fig. 13.14. Plots of (a) $A_{1,2}^2$ and (b) Δ_1/ν_1 versus Δ/ν_1 for $\nu_2/\nu_1 = \delta_2/\delta_1 = 2$, $\gamma_1 = \gamma_2 = 1$, $2\delta_1/\nu_1 = 0.1$, $c_1 = c_2 = 1.1\nu_1^2$, $M_{12} = M_{21} = 85\delta_1\nu_1$ (curves 1, 2) and $M_{12} = -M_{21} = 85\delta_1\nu_1$ (curves 1′, 2′). Dashed lines show unstable parts of the dependencies

13.5 Simultaneous forced and parametric excitation of an oscillator

13.5.1 Parametric amplifier

Let us first consider a linear oscillator under parametric and forced actions simultaneously. In particular, this problem is of interest because such an oscillator is a model of the so-called parametric amplifier used widely in electrical engineering.

Let the equation of the oscillator under consideration be

$$\ddot{x} + \omega_0^2 x = \epsilon\big(-2\delta\dot{x} - \omega_0^2 B_{\rm p} x \cos 2\omega_{\rm p} t + B\cos(\omega t + \chi)\big), \tag{13.57}$$

where ϵ is a small parameter, $B_{\rm p}$ and $2\omega_{\rm p}$ are the amplitude and frequency of the parametric action (of pumping), and B, ω and χ are the amplitude, frequency and phase of the external force (the signal). We assume that the signal frequency ω is close both to the natural frequency ω_0 and to the frequency $\omega_{\rm p}$, i.e., $|\omega - \omega_0|$, $|\omega - \omega_{\rm p}| \le \epsilon$. In this case a solution of Eq. (13.57) is conveniently sought in the form

$$x = A_1 \cos\omega t + A_2 \sin\omega t + \epsilon u_1(A_1, A_2, t) + \dots, \tag{13.58}$$

where A_1 and A_2 obey the equations

$$\dot{A}_1 = \epsilon f_1(A_1, A_2) + \dots, \quad \dot{A}_2 = \epsilon f_2(A_1, A_2) + \dots. \tag{13.59}$$

Substituting (13.58) into (13.57), taking account of (13.59) and requiring that the right-hand side of the equation for u_1 be free from resonance terms, we find f_1 and f_2:

$$
\begin{aligned}
f_1 &= -\delta A_1 - \Delta_1 A_2 + c(A_1 \sin 2\Delta_2 t - A_2 \cos 2\Delta_2 t) + \frac{B}{2\omega_0} \sin\chi, \\
f_2 &= -\delta A_2 + \Delta_1 A_1 - c(A_1 \cos 2\Delta_2 t + A_2 \sin 2\Delta_2 t) + \frac{B}{2\omega_0} \cos\chi,
\end{aligned}
\tag{13.60}
$$

where $\Delta_1 = \omega - \omega_0$, $\Delta_2 = \omega - \omega_{\mathrm{p}}$, $c = \omega_0 B_{\mathrm{p}}/4$. If $|\Delta_2| \ll \delta$ then a steady-state solution of Eqs. (13.59) can be found by putting $\dot{A}_1 = \dot{A}_2 = 0$, i.e., $f_1 = f_2 = 0$. Solving these equations, in view of (13.60), we find

$$
\begin{aligned}
A_1 &= \frac{B}{2\omega_0} \frac{(\delta + c\sin 2\Delta_2 t)\sin\chi - (\Delta_1 + c\cos 2\Delta_2 t)\cos\chi}{\delta^2 + \Delta_1^2 - c^2}, \\
A_2 &= \frac{B}{2\omega_0} \frac{(\delta - c\sin 2\Delta_2 t)\cos\chi + (\Delta_1 - c\cos 2\Delta_2 t)\sin\chi}{\delta^2 + \Delta_1^2 -^2 c}.
\end{aligned}
\tag{13.61}
$$

From this we can find the oscillation amplitude $A = \sqrt{A_1^2 + A_2^2}$ and the phase $\varphi = -\arctan(A_2/A_1)$:

$$
\begin{aligned}
A &= \frac{B}{2\omega_0} \frac{\sqrt{\delta^2 + \Delta_1^2 + c^2 - 2c\,(\delta\sin(2\Delta_2 t + \chi) - \Delta_1\cos(2\Delta_2 t + \chi))}}{\delta^2 + \Delta_1^2 - c^2}, \\
\varphi &= \arctan \frac{\Delta_1 \sin\chi + \delta\cos\chi - c\sin(2\Delta_2 t + \chi)}{\Delta_1\cos\chi - \delta\sin\chi + c\cos(2\Delta_2 t + \chi)}.
\end{aligned}
\tag{13.62}
$$

It can be shown that the solution (13.62) is stable if $c \le \sqrt{\delta^2 + \Delta_1^2}$; otherwise parametric oscillations with frequency ω_{p} are excited.

From (13.62) we can calculate the ratio of the amplitudes squared in the presence of pumping (A^2) and in the absence of pumping ($A_0^2 = A^2|_{c=0}$). This ratio is

$$
\begin{aligned}
\frac{A^2}{A_0^2} &= \frac{(\delta^2 + \Delta_1^2)(\delta^2 + \Delta_1^2 + c^2)}{(\delta^2 + \Delta_1^2 - c^2)^2} \\
&\quad - \frac{2c(\delta^2 + \Delta_1^2)\big(\delta\sin(2\Delta_2 t + \chi) - \Delta_1\cos(2\Delta_2 t + \chi)\big)}{(\delta^2 + \Delta_1^2 - c^2)^2}.
\end{aligned}
\tag{13.63}
$$

It is seen from this that for $\Delta_2 \neq 0$ the ratio (13.63) depends on time. Averaging this ratio over time we find

$$
\frac{\overline{A^2}}{A_0^2} = \frac{(\delta^2 + \Delta_1^2)(\delta^2 + \Delta_1^2 + c^2)}{(\delta^2 + \Delta_1^2 - c^2)^2} > 1.
$$

Thus, the average energy of the signal in the presence of pumping can be much greater than in its absence.

If $\Delta_2 = 0$ then the value of the ratio (13.63) depends on the signal phase χ. Extremal values of this ratio are attained for $\chi = \chi_0 = -\arctan(\delta/\Delta_1)$. These extremal values are

$$\frac{A^2}{A_0^2} = \frac{(\delta^2 + \Delta_1^2)\left(\delta^2 + \Delta_1^2 + c^2 \pm 2c\sqrt{\delta^2 + \Delta_1^2}\right)}{(\delta^2 + \Delta_1^2 - c^2)^2} .$$

It may be inferred from this expression that for $\Delta_2 = 0$ we can achieve a considerable gain in the signal value by an appropriate choice of the signal phase.

13.5.2 Regular and chaotic oscillations in a model of childhood infections

We consider here the effect of simultaneous forced and parametric actions on a nonlinear oscillator by the example of a standard epidemiological model for the description of seasonal variations of the incidence of childhood diseases such as chickenpox, measles, mumps and rubella [68, 300, 258, 199]. The model includes four components: (1) susceptibles (S); (2) exposed but not yet infective (E); (3) infective (I); 4) recovered and immune (R). For this reason the model under consideration is often called SEIR. Relationships between these components are illustrated schematically in Fig. 13.15. The relative

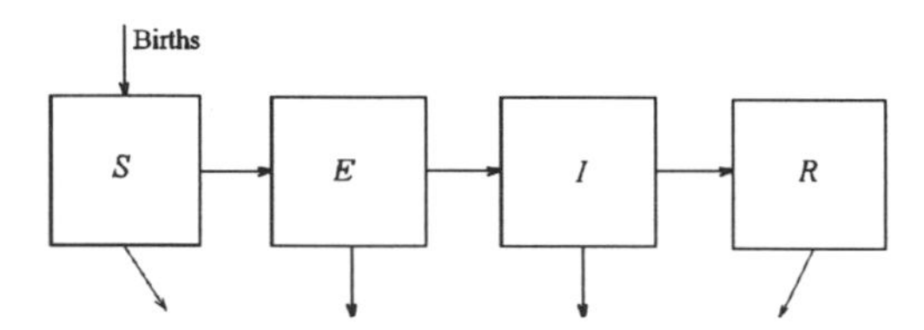

Fig. 13.15. Diagram illustrating mutual relations between different components in the SEIR model

number of children S susceptible to infection increases with total number of children and decreases both owing to the fact that a section of the group remains unexposed and because a section of them falls into the category of the exposed but not yet infective (E). Some of the children exposed remains noninfective, whereas others fall into the category of the infective (I). In its turn, a group of the infective children do not fall sick and another part, having had the disease, recover and fall into the fourth category (R). Taking account of the fact that the total number of children under consideration is constant, the model equations can be written as

$$\begin{aligned} \dot{S} &= m(1 - S) - bSI, \\ \dot{E} &= bSI - (m + a)E, \qquad (13.64) \\ \dot{I} &= aE - (m + g)I, \end{aligned}$$

$$\dot{R} = gI - mR, \qquad (13.65)$$

where $1/m$ is the average expectancy time, $1/a$ is the average latency period, $1/g$ is the average infection period, b is the contact rate (the average number of susceptibles contacted infection annually). Let us note that Eqs. (13.64) do not include the variable R; hence these equations can be considered independently of Eq. (13.65).

Equations (13.64) were first considered by Dietz [68], who assumed that the contact rate b varies periodically with the period equal to one year and found analytically a periodic solution of the equations considered. Later these equations were studied in detail by Olsen and Schaffer [258] and Engbert and Drepper [80]. It was shown that periodic variations of the contact rate can result not only in periodic oscillations of childhood infections but in chaotic ones as well.

It is easily shown that for a time-independent contact rate $b = \text{const} = b_0$ Eqs. (13.64) have, depending on the parameters, either one (for $ab_0 \leq (m + a)(m + g)$) or two (for $ab_0 > (m + a)(m + g)$) singular points: one of them has the coordinates $S = 1$, $E = I = 0$ and another (if it exists) has the coordinates

$$S_0 = \frac{(m+a)(m+g)}{ab_0}, \quad E_0 = \frac{m}{m+a} - \frac{m(m+g)}{ab_0},$$
$$I_0 = \frac{am}{(m+a)(m+g)} - \frac{m}{b_0}. \tag{13.66}$$

In the case that there is only one singular point it is stable, whereas in the case that both of the singular points exist the first of them is aperiodically unstable and the second is stable. These cases are said to correspond to extinction of epidemics and endemic equilibrium, respectively.

It is shown in [258] that the values of the model parameters most closely corresponding to the estimates made for childhood diseases in first world countries are $m = 0.02\,\text{year}^{-1}$, $a = 35.84\,\text{year}^{-1}$, $g = 100\,\text{year}^{-1}$, $b_0 = 1800\,\text{year}^{-1}$. For these parameters Eqs. (13.64) have two singular points. In our studies we have used these same parameter values.

If the parameter b oscillates with time then the variables S, E and I oscillate too, and these oscillations are executed about the stable singular point with coordinates (13.66). Therefore, it is convenient to substitute into Eqs. (13.64) the new variables $x = S/S_0 - 1$, $y = E/E_0 - 1$, and $z = I/I_0 - 1$. Putting $b = b_0\big(1 + b_1 f(t)\big)$, where $f(t)$ is a function describing the shape of the contact rate variation, let us rewrite Eqs. (13.64) in the variables x, y, z:

$$\begin{aligned}
\dot{x} + mx &= -b_0 I_0 \Big[\big(1 + b_1 f(t)\big)\left(x + z + xz\right) + b_1 f(t)\Big], \\
\dot{y} + (m+a)y &= (m+a)\Big[\big(1 + b_1 f(t)\big)\left(x + z + xz\right) + b_1 f(t)\Big], \\
\dot{z} + (m+g)z &= (m+g)y.
\end{aligned} \tag{13.67}$$

In Eqs. (13.67) the term $b_1 f(t)$ can be considered as an external action upon the system. We see from (13.67) that this action is not only multiplicative, i.e., parametric, but also additive, i.e., forcing as well. It should be noted that, owing to quadratic nonlinearity, the forcing action can cause a strong response of the system even in the absence of resonance.

For $b_1 = 0$ and for small initial deviations from the equilibrium state $x = 0$, $y = 0$, $z = 0$ the system executes damped oscillations which are close to harmonic ones in shape (Fig. 13.16 a). The frequency of these oscillations

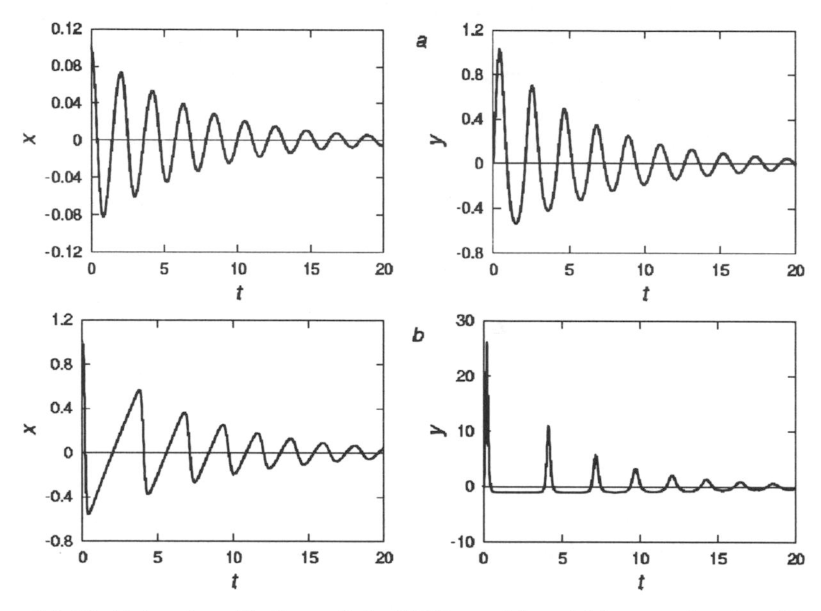

Fig. 13.16. Natural oscillations of the SEIR model variables x and y for $y(0) = 0$, $z(0) = 0$, (a) $x(0) = 0.1$ and (b) $x(0) = 1$. The values of the parameters are determined by (13.67). The shape of the variable z is similar to that of y

$\omega_0 \approx \pi$. As the initial deviations increase, the natural oscillations of the system become close to disconnected, as exemplified by Fig. 13.16 b. The frequency of the natural oscillations decreases with increasing amplitude.

Periodic variation of the contact rate. As mentioned above, in [68, 258] it was assumed that, owing to seasonal variations of environmental conditions, the contact rate b depends periodically on time with a period equal to one year, viz., $f(t) = \cos \omega t$, where $\omega = 2\pi$. We emphasize that the frequency of the contact rate variation is about twice the natural frequency of small free oscillations of the model variables, ω_0.

It was shown that the periodic variation of the parameter b causes the appearance of either periodic or chaotic oscillations of the variables S, E and I. For very small b_1 the oscillations excited are close to harmonic at the frequency of the action ω (Fig. 13.17 a). For a certain value of b_1 a period-doubling bifurcation occurs that is associated with a parametric mechanism of the oscillation excitation. As b_1 increases the main frequency of the oscillations

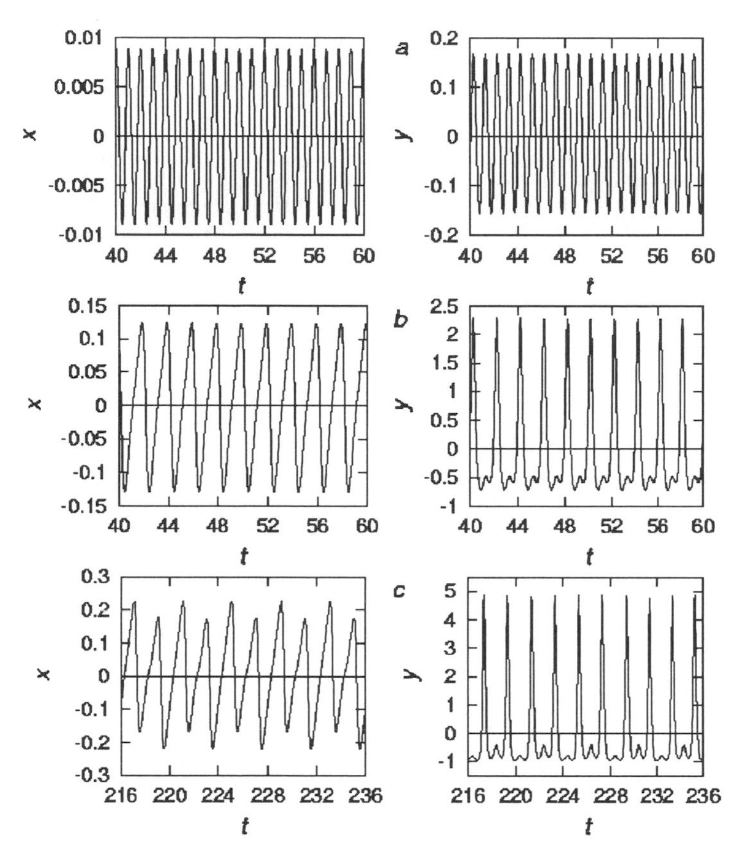

Fig. 13.17. The time dependencies of the SEIR model variables x and y in the case of the periodic variation of the contact rate for (a) $b_1 = 0.03$, (b) $b_1 = 0.1$ and (c) $b_1 = 0.26$

remains equal to $\omega/2$ and the shape of the oscillations of the variable x approaches a saw-tooth (Fig. 13.17 b). On further increasing b_1 another period-doubling bifurcation takes place, (Fig. 13.17 c) and then a drastic transition to chaos, accompanied by a drastic increase in the oscillation variance, occurs. We note that chaotic oscillations for $b_1 = 0.28$ were first found numerically by Olsen and Schaffer [258]. For this value of b_1 the time dependencies of x and y, and the projection of the phase trajectory on the x, y–plane found by numerical simulation of Eqs. (13.67) are shown in Fig. 13.18 a. It should be noted that these oscillations closely resemble the experimental data in their form. This fact seemingly justifies the model with a periodic variation of the contact rate. However, as seen in Fig. 13.8 b, similar results can also be obtained for a random variation of the contact rate. The similarity between the shapes of the oscillations is associated with the fact that the variation of the contact rate only induces a transition to an oscillatory state, whereas

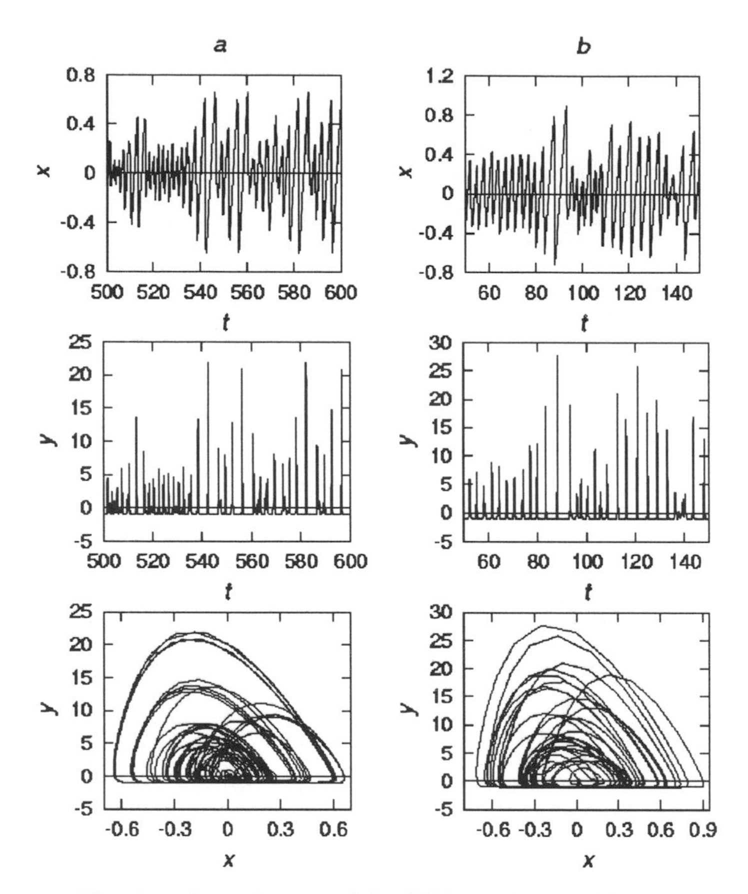

Fig. 13.18. The time dependencies of the SEIR model variables x and y, and the projection of the phase trajectory on the x, y–plane for (a) $f(t) = \cos 2\pi t$, $b_1 = 0.28$ and (b) $f(t) = \chi(t)$, $b_1 = 0.235$

the shape of the induced oscillations is mainly determined by the intrinsic properties of the system, which also manifest themselves in the shape of its natural oscillations.

The evolution of the power spectra of the oscillations in the case of a periodic variation of the contact rate is illustrated in Fig. 13.19. We see that for $b_1 = 0.03$ the spectral density does peak at the frequency ω whereas, for $b_1 = 0.1$, it peaks at the frequency $\omega/2$. For $b_1 = 0.26$ the power spectrum contains the fourth subharmonic, and for $b_1 = 0.28$, when the oscillations are chaotic, the spectrum becomes continuous with a maximum at the frequency $\omega/4$.

To clarify the physical mechanisms of the excitation of oscillations let us change Eqs. (13.67) somewhat so that the amplitudes of parametric and forcing actions can be varied independently. Namely, let us rewrite Eqs. (13.67)

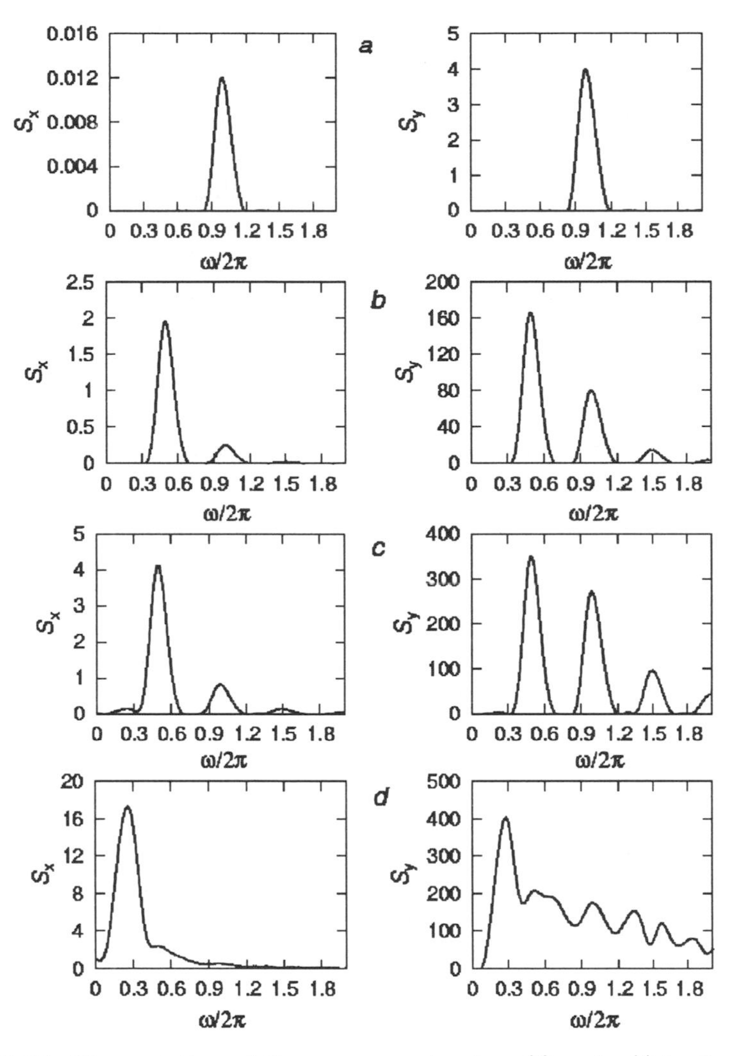

Fig. 13.19. The evolution of the power spectra for $x(t)$ and $y(t)$ in the case of harmonic variation of the contact rate: $b_1 = 0.03$ (a), $b_1 = 0.1$ (b), $b_1 = 0.26$ (c) and $b_1 = 0.28$ (d). We see that the spectra for $y(t)$ are wider than those for $x(t)$

as

$$\begin{aligned}
\dot{x} + mx &= -b_0 I_0 \Big[\big(1 + b_1 f(t)\big)\big(x + z + xz\big) + b_2 f(t)\Big], \\
\dot{y} + (m+a)y &= (m+a)\Big[\big(1 + b_1 f(t)\big)\big(x + z + xz\big) + b_2 f(t)\Big], \\
\dot{z} + (m+g)z &= (m+g)y.
\end{aligned} \tag{13.68}$$

First we consider the case when the additive action is absent, i.e., $b_2 = 0$, $b_1 \neq 0$. In this case oscillations are excited only beyond a certain critical value of the parameter b_1. This is the characteristic property of parametric excitation of oscillations. For $b_1 > 0.75$ the solution becomes unstable and goes to infinity.

In the case when the action is only additive, i.e., $b_1 = 0$, $b_2 \neq 0$, oscillations are excited even for values of b_2 as small as is wished. However, for $b_2 < b_2^{(cr)}$, where $b_2^{(cr)} \approx 0.0885$, the amplitude of these oscillations is very small and their frequency is equal to ω. For $b_2 > b_2^{(cr)} \approx 0.09$ the rate of change of the amplitude increases rapidly. The drastic increase in the rate of change of the amplitude is associated with the appearance of subharmonic resonance. Indeed, the main frequency of the oscillations excited, for $b_2 > b_2^{(cr)}$, becomes equal to $\omega/2$. For $b_2 > 0.15$ the solution, as in the case of parametric excitation, becomes unstable and goes to infinity.

The computation of Eqs. (13.68) shows that the combined effect of parametric and forcing actions, which was described above, results in the stabilization of the solution for moderately large b_1.

Random variation of the contact rate. From a physical standpoint, an assumption of random variation of the contact rate seems more justified than a periodic variation. It is evident that $b(t)$ has to be a sufficiently wide-band random process for which the spectral density peaks at the frequency corresponding to one reciprocal year. Starting from this assumption we have simulated numerically Eqs. (13.67) with $f(t) = \chi(t)$, where $\chi(t)$ is a random process which is a solution of the equation

$$\ddot{\chi} + 2\pi\dot{\chi} + 6\pi^2\chi = k\xi(t), \tag{13.69}$$

$\xi(t)$ is white noise, and k is a factor which we choose such that the variance of $\chi(t)$ is equal to 1/2. The plots of $\chi(t)$ and its spectral density are shown in Fig. 13.20.

The results of numerical simulation of Eqs. (13.67) with $f(t) = \chi(t)$ are shown in Fig. 13.18 b for the same values of the parameters as in Fig. 13.18 a, but for $b_1 = 0.235$. The latter was chosen so that the variance of $x(t)$ would be approximately the same as for $f(t) = \cos 2\pi t$, $b_1 = 0.28$. It is seen from

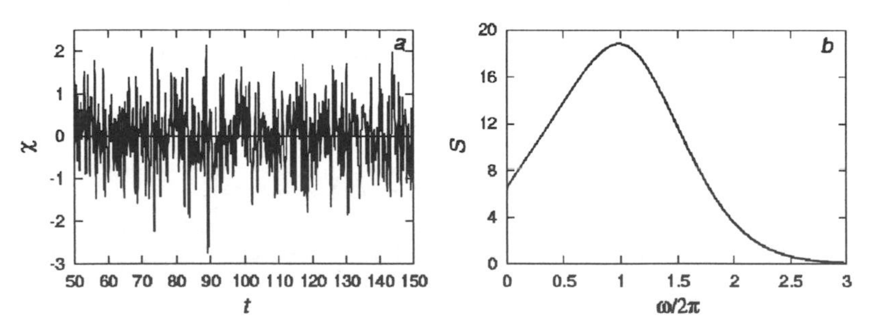

Fig. 13.20. Plots of (a) $\chi(t)$ and (b) of its spectral density $S(\omega)$. We see that the spectral density peaks at the frequency 2π

this figure that the noise-induced oscillations differ very slightly in their form from those for the case of a harmonic variation of the contact rate. The power spectra of the oscillations induced by the random and harmonic variations of the contact rate are also similar: both of them peak at the frequency $\omega/4$ (compare Fig. 13.19 d and 13.21 d). It should be noted that in the case of

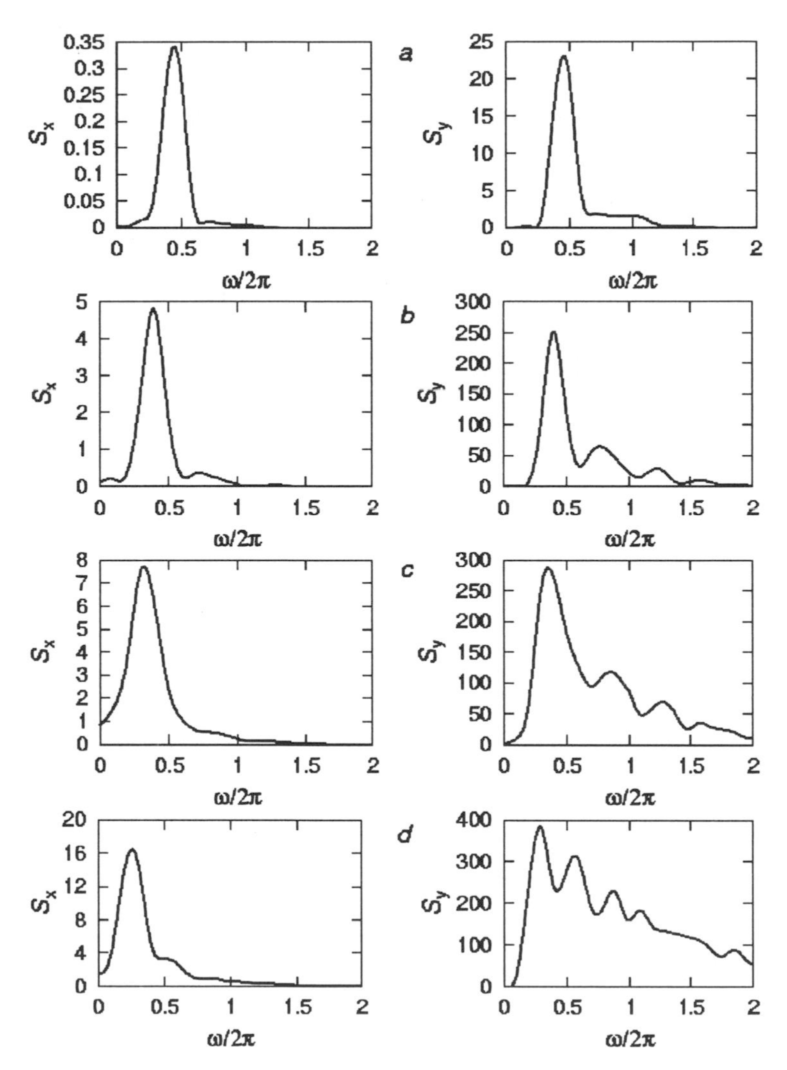

Fig. 13.21. The evolution of the power spectra for the SEIR model variables $x(t)$ (left column) and $y(t)$ (right column) in the case of random variation of the contact rate: $b_1 = 0.03$ (a), $b_1 = 0.1$ (b), $b_1 = 0.2$ (c) and $b_1 = 0.235$ (d). We see that, as for the case of periodic variation of the contact rate, the spectra for $y(t)$ are wider than those for $x(t)$

random variation of the contact rate the power spectra are always continuous, even if have a number of maxima (see Fig. 13.21). As b_1 increases the main maximum of the spectrum shifts to lower frequencies.

To study the mechanisms of the excitation of oscillations in more detail, we have attempted to simulate Eqs. (13.68), (13.69) with $f(t) = \chi(t)$ and put $b_2 = 0$, $b_1 \neq 0$ and vice versa. It is found that in the case of only multiplicative action the excitation of oscillations occurs via on-off intermittency as for a pendulum with a randomly vibrated suspension axis (see above). The critical value of the parameter b_1 is approximately equal to 0.095, i.e., it is essentially larger than in the case of harmonic variation of the contact rate. An example of oscillations of the variables x and y for $b_1 = 0.099$, illustrating the on-off intermittency, is given in Fig. 13.22. It should be noted that in the case

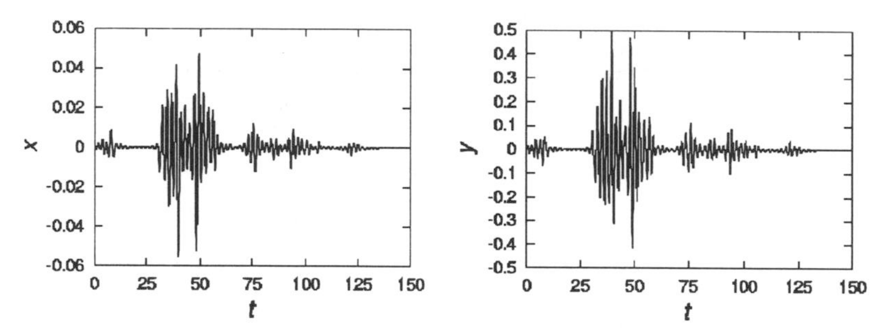

Fig. 13.22. An example of oscillations of the SEIR model variables x and y for $b_1 = 0.099$ in the case of purely multiplicative random action

of purely parametric random action the solution goes to infinity even for $b_1 = 0.1$.

In the case when $b_1 = 0$, $b_2 \neq 0$, the system behaves in a similar manner: the solution goes to infinity for $b_2 \geq 0.08$.

14. Changes in the dynamical behavior of nonlinear systems induced by high-frequency vibration or by noise

A rich variety of problems concerning changes in the dynamical behavior and in the properties of nonlinear systems under high-frequency vibration are set out in the book by Blekhman [50]. The author suggested pooling these problems to form a new division of mechanics under the name 'Vibrational Mechanics'. In this chapter we consider some other similar problems arosed in recent years and amplify our consideration by the influence of noise. The chapter is written in collaboration with I.I. Blekhman.

14.1 The appearance and disappearance of attractors and repellers induced by high-frequency vibration or noise

In this section the appearance and disappearance of attractors and repellers induced by either high-frequency vibration or noise is considered by the example of a pendulum with a vibrated axis of suspension. Stabilization of the upper equilibrium position of such a pendulum caused by vertical high-frequency vibration of its suspension axis has long been known [81, 203, 52, 135, 141, 142, 46, 32]. In [192, 198] it is shown that a similar phenomenon also takes place if the pendulum suspension axis vibrates randomly.

Let us consider a physical pendulum with a suspension axis harmonically vibrated in a certain direction making an angle γ with the vertical (Fig. 14.1). The equation of motion of such a pendulum is

$$\ddot{\varphi} + \frac{H}{J}\dot{\varphi} + \frac{mb}{J}\Big(g\sin\varphi + \nu_0^2 a\cos\nu_0 t\sin(\varphi-\gamma)\Big) = 0, \tag{14.1}$$

where J and m are the moment of inertia and the mass of the pendulum respectively, b is the distance between the center of mass and the suspension axis, $H\dot{\varphi}$ is the moment of the friction force, and a and ν_0 are the amplitude and frequency of the suspension axis vibration, respectively. By scaling time this equation can be put in the form:

$$\ddot{\varphi} + 2\beta\dot{\varphi} + \sin\varphi + B\cos\omega t\sin(\varphi-\gamma) = 0. \tag{14.2}$$

We assume that $\omega \gg 1$, β.

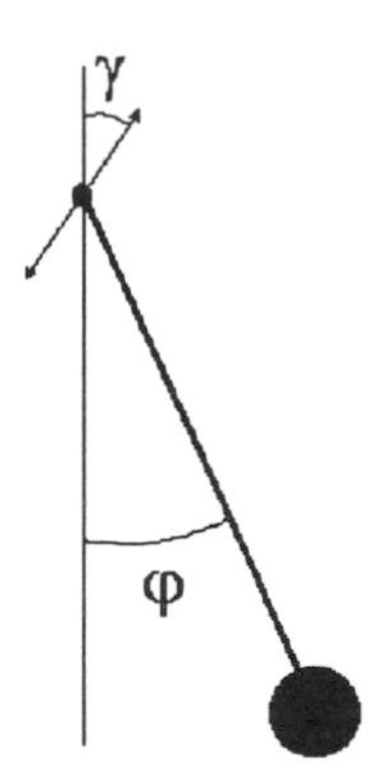

Fig. 14.1. Schematic image of a pendulum with a suspension axis vibrated in a certain direction making an angle γ with the vertical

A solution of Eq. (14.2) can be represented as the superposition of 'slow' oscillations $\chi(t)$ and 'fast' small vibration $\psi(t)$:

$$\varphi(t) = \psi(t) + \chi(t), \tag{14.3}$$

where $|\psi(t)| \ll 1$, $\dot{\psi}(t) \sim \omega$ and $\dot{\chi}(t) \sim 1$. Substituting (14.3) into Eq. (14.2), separating 'fast' and 'slow' terms and retaining only linear terms with respect to ψ, we find the following equations:

$$\ddot{\psi} + 2\beta\dot{\psi} + \psi\cos\chi + B\cos\omega t\sin(\chi - \gamma) = 0, \tag{14.4a}$$

$$\ddot{\chi} + 2\beta\dot{\chi} + \sin\chi + B\cos(\chi - \gamma)\overline{\psi(t)\cos\omega t} = 0. \tag{14.4b}$$

Equation (14.4a) can be solved on the assumption that χ is unchanged with time. In so doing an approximate stationary solution of this equation is

$$\psi(t) \approx \frac{B\sin(\chi - \gamma)}{\omega^2}\cos\omega t. \tag{14.5}$$

It follows from (14.5) that

$$\overline{\psi(t)\cos\omega t} = \frac{B\sin(\chi - \gamma)}{2\omega^2}. \tag{14.6}$$

By substituting (14.6) into Eq. (14.4b) we obtain the following equation for $\chi(t)$:

$$\ddot{\chi} + 2\beta\dot{\chi} + \sin\chi + \frac{B^2}{4\omega^2}\sin 2(\chi - \gamma) = 0. \tag{14.7}$$

Equation (14.7) has four equilibrium states which are described by the equation

$$\sin\chi + \frac{B^2}{4\omega^2}\sin 2(\chi - \gamma) = 0. \tag{14.8}$$

This equation can be solved analytically only in the case when $\gamma \ll 1$. In this case we find

$$\chi_1 \approx \frac{B^2}{B^2 + 2\omega^2}\gamma, \quad \chi_2 \approx \pi + \frac{B^2}{B^2 - 2\omega^2}\gamma,$$
$$\chi_{3,4} = \pm \arccos\left(-\frac{2\omega^2}{B^2}\right) + \frac{8\omega^4 - B^4}{4\omega^4 - B^4}\gamma. \tag{14.9}$$

If vibration is vertical, χ_1 goes into the lower equilibrium position $\chi = 0$, and χ_2 goes into the upper equilibrium position $\chi = \pi$; $\chi_{3,4}$ are unstable equilibrium positions which arise under the action of vibration: they exist only if $B^2 > 2\omega^2$.

Let us further consider the stability of the equilibrium states found, starting from Eq. (14.7). For this purpose we linearize Eq. (14.7) with respect to a small deviation $\alpha = \chi - \chi_j$, where χ_j is one of the equilibrium positions. The linearized equation is

$$\ddot{\alpha} + 2\beta\dot{\alpha} + \left(\cos\chi_j + \frac{B^2}{2\omega^2}\cos 2(\chi_j - \gamma)\right)\alpha = 0. \tag{14.10}$$

It is easily seen from Eq. (14.10) that the equilibrium position χ_1 remains stable for any B (in the framework of Eq. (14.10)). The stability condition for the equilibrium position χ_2 is

$$B^2 \geq 2\omega^2. \tag{14.11}$$

At the condition (14.11), between the stable states χ_1 and χ_2 there are the unstable states χ_3 and χ_4.

For comparison, let us consider the stability of the pendulum equilibrium positions by starting from the initial equation (14.2). For this purpose we linearize Eq. (14.2) relative to a small deviation $y = \varphi - \varphi_0$ from the equilibrium position under consideration and substitute τ/ω in place of t. As a result we obtain the following equation:

$$\frac{d^2y}{d\tau^2} = \epsilon\left(-2\beta\frac{dy}{d\tau} - \left(\epsilon\cos\varphi_0 + \frac{B}{\omega}\cos\tau\cos(\varphi_0 - \gamma)\right)y\right), \tag{14.12}$$

where $\epsilon = 1/\omega$ is a small parameter. As will be seen from the following, in the vicinity of the boundary of stability the parameter B is of order ω.

A solution of Eq. (14.12) corrected for τ less than or of order 1 may be sought as a power series in ϵ, viz.,

$$y = y_0 + \epsilon y_1 + \epsilon^2 y_2 + \dots . \tag{14.13}$$

Substituting (14.13) into (14.2), assuming $\beta \sim \epsilon$, and equating the coefficients of ϵ^0, ϵ^1, ϵ^2, ... we find y_0, y_1, y_2, ...:

$$
\begin{aligned}
y_0 &= C_1\tau + C_2, \\
y_1 &= -\frac{B}{\omega}\Big(C_1(2\sin\tau - \tau\cos\tau) - C_2\cos\tau\Big)\cos(\varphi_0 - \gamma), \\
y_2 &= -\left(C_1\frac{\tau^3}{6} + C_2\frac{\tau^2}{2}\right)\cos\varphi_0 - C_1\frac{\beta\tau^2}{\epsilon} + \frac{B^2}{4\omega^2}\Bigg[\left(-\frac{5}{4}\sin 2\tau \right. \\
&\quad \left. + \frac{\tau}{2}\cos 2\tau - \frac{\tau^3}{3}\right)C_1 + \left(\frac{\cos 2\tau}{2} - \tau^2\right)C_2\Bigg]\cos^2(\varphi_0 - \gamma), \\
&\dots,
\end{aligned}
\tag{14.14}
$$

where C_1, C_2 are arbitrary constants determined by the initial conditions.

According to the Floquet theory [91], for the determination of the stability of the equilibrium state we have to know two fundamental particular solutions of Eq. (14.12) for $\tau = 2\pi$. We denote them by $y^{(1)}(\tau)$ and $y^{(2)}(\tau)$. If the solution $y^{(1)}(\tau)$ satisfies the initial conditions $y^{(1)}(0) = 1$, $\dot{y}^{(1)}(0) = 0$, and the solution $y^{(2)}(\tau)$ satisfies the initial conditions $y^{(2)}(0) = 0$, $\dot{y}^{(2)}(0) = 1$, then, as can be shown, the stability condition sought for is $\max|\mu_{1,2}| \le 1$, where

$$
\mu_{1,2} = \frac{y^{(1)}(2\pi) + \dot{y}^{(2)}(2\pi)}{2} \pm \sqrt{\dot{y}^{(1)}(2\pi)y^{(2)}(2\pi)}
\tag{14.15}
$$

are the so-called *multipliers*. It follows from (14.13), (14.14) that

$$
\begin{aligned}
y^{(1)}(2\pi) &= 1 + 2\pi^2\epsilon^2\left(1 - \frac{B^2}{2\omega^2}\cos^2(\varphi_0 - \gamma)\right) + o(\epsilon^3), \\
\dot{y}^{(1)}(2\pi) &= 2\pi\epsilon^2\left(1 - \frac{B^2}{2\omega^2}\cos^2(\varphi_0 - \gamma)\right) + o(\epsilon^3), \\
y^{(2)}(2\pi) &= 2\pi + o(\epsilon), \\
\dot{y}^{(2)}(2\pi) &= 1 + 2\pi^2\epsilon^2\left(1 - \frac{B^2}{2\omega^2}\cos^2(\varphi_0 - \gamma) - \frac{2\beta}{\pi\epsilon}\right) + o(\epsilon^3).
\end{aligned}
\tag{14.16}
$$

By substituting (14.16) into (14.15) we obtain the stability condition (14.11), which can be written as

$$
B \ge B_{\text{cr}} = \sqrt{2}\,\omega. \tag{14.17}
$$

We note that condition (14.11) is independent of the pendulum damping factor. If the damping factor is nonzero and $B^2 > 2\omega^2$ then the pendulum upper equilibrium position is asymptotically stable. After a small deviation from the equilibrium position the pendulum will execute damped quasi-periodic oscillations with two fundamental frequencies ω and

$$
\Omega = \frac{|\mathrm{Im}\,\mu_{1,2}|}{2\pi\epsilon} = \sqrt{\frac{B^2}{2\omega^2} - 1}
\tag{14.18}
$$

(the expression for the 'natural' frequency Ω can be found from (14.15), (14.16)). As B increases, the frequency Ω increases too and, for a certain value B which we denote by B^*, Ω attains the value $\omega/2$. For this value of Ω the upper equilibrium position becomes unstable again. This loss of stability is similar to that for the lower equilibrium position and takes place for about the same value of B. It can be shown that the mentioned loss of stability is associated with the passage of one of the multipliers through the value -1; the latter causes excitation of oscillations with the frequency $\omega/2$. The upper boundary of stability (B^*) and the dependence of the amplitude of oscillations A on $B - B^*$ close to the excitation threshold can be estimated if in Eq. (14.2) we neglect the unit compared to $B\cos\omega t$ (this corresponds to neglecting the gravitational force) and expand $\sin\varphi$ as a power series in the vicinity of $\varphi = \gamma$ or $\varphi = \pi + \gamma$, restricting ourselves to cubic terms. As a result we obtain the equation

$$\ddot{y} + 2\beta\dot{y} \pm B\left(y - \frac{y^3}{6}\right)\cos\omega t = 0. \tag{14.19}$$

A solution of Eq. (14.19) can be sought in the form $y = A\sin(\omega t/2 + \phi)$, where A and ϕ are slowly varying functions. Substituting this solution into Eq. (14.19), we obtain the following approximate equations for A and ϕ:

$$\begin{aligned} \dot{A} &= \mp\frac{BA}{2\omega}\left(1 - \frac{A^2}{12}\right)\sin 2\phi - \beta A, \\ \dot{\phi} &= -\frac{\omega}{4} \mp \frac{B}{2\omega}\left(1 - \frac{A^2}{6}\right)\cos 2\phi. \end{aligned} \tag{14.20}$$

It follows from Eqs. (14.20) that, for $\beta \ll \omega$, the steady-state value of $\sin 2\phi$ is small and therefore the steady-state value of A^2 can be approximately found from the second equation of (14.20) for $\cos 2\phi = 0$. The expression for A^2 is conveniently written as

$$A^2 \approx 6(1 - B^*/B), \tag{14.21}$$

where $B^* \approx \omega^2/2$. It follows from (14.17) that $B^* \gg B_{\text{cr}}$.

Numerical computations of Eq. (14.12) [253] have shown that for $B > B^*$, depending on the initial conditions, the pendulum can both oscillate with the period equal to $4\pi/\omega$ about either the lower or the upper equilibrium position and rotate in one or the other direction with a frequency equal to the frequency of vibration of the suspension axis.

Stabilization of the upper equilibrium position of the pendulum caused by high-frequency vibration of its suspension axis signifies as if another attractor appears in the pendulum phase space. This attractor is induced by the high-frequency vibration. In addition, two repellers, which are unstable equilibrium states, are also induced by the high-frequency vibration.

Recently it was shown [190, 192, 198] that similar birth of the induced attractors and repellers can occur as a result not only of regular but also of random, sufficiently high-frequency, vibration of the pendulum suspension axis. In the case of vertical random vibration of the suspension axis the equation describing the oscillations of the pendulum can be written as

$$\ddot{\varphi} + 2\beta\dot{\varphi} + \big(1 + \zeta(t)\big)\sin\varphi = 0, \tag{14.22}$$

where $\zeta(t)$ is the acceleration of the pendulum suspension axis in terms of the acceleration of gravity. Let $\zeta(t)$ be colored narrow-band noise described by the equation

$$\ddot{\zeta} + 2\alpha\dot{\zeta} + \omega^2\zeta = \xi(t), \tag{14.23}$$

where $\xi(t)$ is white noise of intensity κ, $1 \ll \alpha \ll \omega$. In this case the correlation function of the process $\zeta(t)$ is

$$\langle\zeta(t)\zeta(t+\tau)\rangle \approx \sigma^2 \mathrm{e}^{-\alpha\tau}\cos\omega\tau, \tag{14.24}$$

where $\sigma^2 = \kappa/(2\omega^2\alpha)$ is the variance of $\zeta(t)$.

Let us first show directly that the pendulum upper equilibrium state becomes stable if σ^2 is sufficiently large. If the power spectrum of the random process $\zeta(t)$ does not contain components in zones of parametric resonance, fluctuations of the variable φ caused by the random vibration of the suspension axis are small. Putting $\varphi = \chi + \psi$, where $\chi = \langle\varphi\rangle$ and $\psi \ll \chi$, we obtain from (14.22)

$$\begin{aligned} &\ddot{\chi} + 2\beta\dot{\chi} + \sin\chi + \cos\chi\langle\zeta(t)\psi\rangle = 0, \\ &\ddot{\psi} + 2\beta\dot{\psi} + \cos\chi\,\psi + \zeta(t)\sin\chi = 0. \end{aligned} \tag{14.25}$$

A steady-state solution of Eqs. (14.25), having the form

$$\chi = \pi, \quad \psi = 0, \tag{14.26}$$

corresponds to the upper equilibrium position of the pendulum, in whose stability we are interested. To investigate this stability, we can linearize Eqs. (14.25) with respect to small deviations from the solution (14.26). The linearized equations are

$$\ddot{y} + 2\beta\dot{y} - y - \langle\zeta(t)\psi\rangle = 0, \tag{14.27a}$$

$$\ddot{\psi} + 2\beta\dot{\psi} - \psi - y\zeta(t) = 0, \tag{14.27b}$$

where $y = \chi - \pi$. A steady-state solution of Eq. (14.27b) is

$$\psi(t) = \frac{1}{2\sqrt{1+\beta^2}} \int_{-\infty}^{t} \Big(\mathrm{e}^{p_1(t-t')} - \mathrm{e}^{p_2(t-t')}\Big) y(t')\zeta(t')\,\mathrm{d}t',$$

where $p_{1,2} = -\beta \pm \sqrt{1+\beta^2}$ are the roots of the characteristic equation $p^2 + 2\beta p - 1 = 0$. From here we find

$$\langle\zeta(t)\psi(t)\rangle = \frac{1}{2\sqrt{1+\beta^2}} \int_{-\infty}^{t} \Big(\mathrm{e}^{p_1(t-t')} - \mathrm{e}^{p_2(t-t')}\Big) y(t')\langle\zeta(t)\zeta(t')\rangle\,\mathrm{d}t'. \tag{14.28}$$

Putting $t' - t = \tau$ in this expression and taking into account that the value y does not vary significantly during the correlation time of the random process $\zeta(t)$, we rewrite (14.28) in the following form:

$$\langle \zeta(t)\psi(t)\rangle = -\frac{1}{2\sqrt{1+\beta^2}}\, y(t) \int_0^\infty \left(e^{-p_1\tau} - e^{-p_2\tau}\right)\langle \zeta(t)\zeta(t+\tau)\rangle \, d\tau. \tag{14.29}$$

Substituting (14.24) into this expression and calculating the integral we obtain

$$\langle \zeta(t)\psi(t)\rangle = -\frac{\sigma^2\left(\omega^2 - (p_1+\alpha)(p_2+\alpha)\right)}{\left(\omega^2 + (p_1+\alpha)^2\right)\left(\omega^2 + (p_2+\alpha)^2\right)}\, y(t). \tag{14.30}$$

Because $\omega \gg 1, \beta, \alpha$, we have

$$\langle \zeta(t)\psi(t)\rangle \approx -\sigma^2\omega_2\, y(t). \tag{14.31}$$

Substituting (14.31) into (14.27a) we obtain the following approximate equation for $y(t)$:

$$\ddot{y} + 2\beta\dot{y} + \Omega_0^2 y = 0, \tag{14.32}$$

where $\Omega_0 = \sqrt{\sigma^2/\omega^2 - 1}$ is the natural frequency of small oscillations of the pendulum about the upper equilibrium position. It follows from (14.32) that the mean deviation of the pendulum from its upper equilibrium position decays, i.e., the equilibrium position is stable, if the frequency Ω_0 is real. This condition is fulfilled if

$$\frac{\sigma^2}{\omega^2} \geq 1. \tag{14.33}$$

Let us show further that at the condition (14.33) an additional maximum and two minima appear in the probability distribution for φ. For this we assume that weak additive white noise $\xi(t)$ acts upon the pendulum. In this case Eq. (14.22) becomes

$$\ddot{\varphi} + 2\beta\dot{\varphi} + \left(1 + \zeta(t)\right)\sin\varphi = \xi_a(t). \tag{14.34}$$

It follows from Eq. (14.34) that there is no stationary probability distribution for the variables φ and $\dot{\varphi}$. Therefore we use the 'slow' variable χ and the 'fast' variable ψ, which, for $\xi_a(t) = 0$, are described by Eqs. (14.25). Taking account of the fact that $\zeta(t)$ is a narrow-band random process, which can be represented as

$$\zeta(t) = \zeta_1(t)\cos\omega t + \zeta_2(t)\sin\omega t, \tag{14.35}$$

where $\zeta_1(t)$ and $\zeta_2(t)$ are 'slow' variables, we can put

$$\psi = A(t)\cos\omega t + B(t)\sin\omega t, \tag{14.36}$$

where $A(t)$ and $B(t)$ are also 'slow' variables. Substituting $\varphi(t) = \chi(t) + \delta\varphi(t)$, in view of (14.36), into Eq. (14.34), taking account of (14.35), and equating the slowly varying components and the coefficients of $\cos\omega t$ and $\sin\omega t$, we obtain the following equations for χ, A and B:

$$\ddot{\chi} + 2\beta\dot{\chi} + \sin\chi + \frac{\cos\chi}{2}\langle A\zeta_1 + B\zeta_2\rangle = \xi_a(t), \tag{14.37}$$

$$\begin{aligned} (\omega^2 - \cos\chi)A - 2\beta\omega B &= \zeta_1 \sin\chi, \\ 2\beta\omega A + (\omega^2 - \cos\chi)B &= \zeta_2 \sin\chi. \end{aligned} \tag{14.38}$$

Because $\omega \gg 1$, β, we find from Eqs. (14.38)

$$A \approx \frac{\zeta_1}{\omega^2}\sin\chi, \quad B \approx \frac{\zeta_2}{\omega^2}\sin\chi. \tag{14.39}$$

Substituting further (14.39) into Eq. (14.37) we obtain for $\chi(t)$ the following equation:

$$\ddot{\chi} + 2\beta\dot{\chi} + \sin\chi + \frac{\sigma^2}{2\omega^2}\sin 2\chi = \xi_a(t). \tag{14.40}$$

Here it is taken into account that $\langle\zeta_1^2 + \zeta_2^2\rangle = 2\sigma^2$.

The Fokker–Planck equation for the probability density $w(\chi, \dot{\chi}, t)$ associated with Eq. (14.40) is conveniently written as [319]

$$\frac{\partial w}{\partial t} = -\left[\dot{\chi}\frac{\partial w}{\partial \chi} - \left(\sin\chi + \frac{\sigma^2}{2\omega^2}\sin 2\chi\right)\frac{\partial w}{\partial \dot{\chi}}\right] + \left[2\beta\frac{\partial(\dot{\chi}w)}{\partial\dot{\chi}} + \frac{\kappa_a}{2}\frac{\partial^2 w}{\partial\dot{\chi}^2}\right], \tag{14.41}$$

where κ_a is the intensity of the noise $\xi_a(t)$. As shown in [319], a steady-state solution of Eq. (14.41) can be found by equating each from the braces on the right-hand side to zero. As a result, we obtain

$$w(\chi, \dot{\chi}) = C\exp\left(-\frac{2\beta\dot{\chi}^2}{\kappa_a}\right)w(\chi), \tag{14.42}$$

where

$$w(\chi) = \exp\left[\frac{4\beta}{\kappa_a}\left(\cos\chi + \frac{\sigma^2}{4\omega^2}\cos 2\chi\right)\right], \tag{14.43}$$

and C is the normalization constant. It is easily seen from (14.42) that the function $w(\chi, \dot{\chi})$ has four extrema for $\dot{\chi} = 0$ and $\chi = 0$, $\chi = \pi$, $\chi = \pm\arccos(-\omega^2/\sigma^2)$. The last of them exist only if $\sigma^2 > \omega^2$, i.e., they are induced by the random vibration of the pendulum suspension axis. Under this condition the probability density $w(\chi)$ has two maxima (for $\chi = 0$ and $\chi = \pi$) and two minima (for $\chi = \pm\arccos(-1/\omega^2\sigma^2)$). If $\sigma^2 < \omega^2$ then the probability density $w(\chi)$ has only one maximum (for $\chi = 0$) and one minimum (for $\chi = \pi$). Thus, the random vibration considered causes multistability. An example of the transformation of the probability density $w(\chi)$ with increasing σ^2 is shown in Fig. 14.2 for $\kappa_a = 4$.

We note that many other examples where additional peaks appear in the probability density under the influence of multiplicative noise are considered by Horsthemke and Lefever in their book [124].

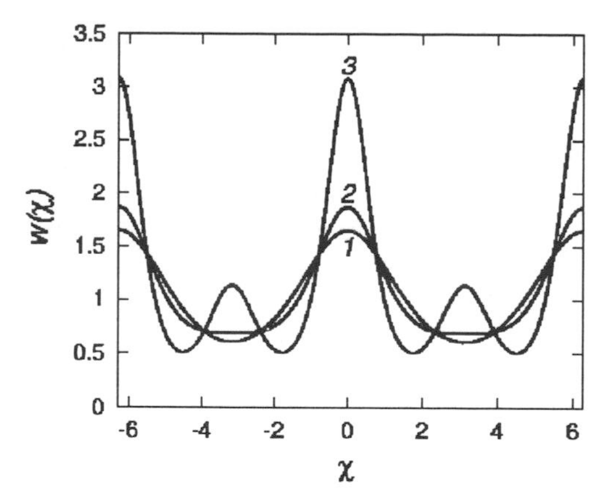

Fig. 14.2. An example of the transformation of the function $w(\chi)$, determined by the expression (14.43), with increasing σ^2 for $\kappa_a = 4$ and $\sigma^2/\omega^2 = 0$, 1 and 5 for curves 1–3, respectively

14.2 Vibrational transport and electrical rectification

Vibrational transport has been known for a long time (see, e.g., [45, 112, 101, 245, 314, 50]). In the following section we show that there is a certain analogy between classical vibrational transport and noise-induced transport of Brownian particles which in recent years has attracted considerable interest from many scientists, for the most part in the context of different biological and chemical problems (see, e.g., [268, 324, 219, 39, 74, 113]). A physical experiment demonstrating the possibility of such transport in a ratchet-like potential field created by a laser beam is described in [83]. In [2] it was experimentally shown that directed motion of a particle can be induced merely by turning on and off a periodic asymmetric potential (more recently, this phenomenon has become known as *flashing ratchet*). Similar experiments are also presented in [102]. Some experiments concerning the horizontal transport of granular materials placed on a saw-tooth-shaped base by means of vertical vibration are described in [67].

Systems in which noise-induced transport of Brownian particles occurs are often called *stochastic ratchets* by analogy with mechanical device 'ratchet-and-pawl' described and considered by Feynman [88]. Feynman showed that in the case of thermodynamical equilibrium the ratchet on average is at rest — it advances and retreats by an equal number of teeth on the wheel — as it must be because of the Second Law of Thermodynamics.

It should be noted that similar phenomena were also discussed by Smoluchowski [312] well before Feynman. The 'ratchet-and-pawl' device constitutes a mechanical rectifier. It is similar in essence to an electrical rectifier. However, as is often the case, the problems associated with electrical rectification of fluctuations were discussed independently of the ratchet problems

[58, 218, 3, 138, 233, 318]. In [218, 3] it was found that in the simplest electrical rectifier, consisting of capacitor and diode, the capacitor can be charged without an external source, at the expense of only thermal fluctuations. This paradoxical result cast some doubt on the applicability of the Second Law of Thermodynamics to the phenomenon considered [233]. As far back as 1950, however, considering diode as a nonlinear resistor, Brillouin [58] showed that, for the Second Law of Thermodynamics to apply, a shift of the voltage-current characteristic of the nonlinear resistor must be taken into account. Stratonovich [318, 320] established, on a certain model of diode, that such a shift does indeed occur due to fluctuations of the current, and he calculated it. With this shift, the mean value of the voltage drop across the capacitor and the mean current in the circuit were found to vanish for the case of thermodynamical equilibrium.

In the last few years much attention has been attracted to the problem associated with the separation of particles of different mass or size. This problem has roots originating in mechanical engineering [103, 370, 50]. In connection with this problem, studies of different models of ratchet devices giving flux reversals as the mass or size of the particles change, are very important [26, 236, 40, 27, 197, 198, 201].

14.2.1 Vibrational transport

Let us now consider the simplest model of vibrational transport [50] and show that such transport can occur owing to both periodic and random vibration. Let a body of mass m lie on a horizontal plane vibrating in the direction of axis x. We assume that a force of dry friction between the body and plane has different values for $\dot{x} > 0$ and $\dot{x} < 0$. This is possible if the surface of the plane is rough. Then we can write the following equation for the motion of the body:

$$m\dot{y} = -f(y) + F(t), \tag{14.44}$$

where $y = \dot{x}$, $F(t)$ is the inertial force due to vibration of the plane,

$$f(y) = \begin{cases} mga_1 & \text{for } y > 0, \\ -mga_2 & \text{for } y < 0, \end{cases} \tag{14.45}$$

and $a_{1,2}$ are the friction factors.

In mechanical engineering it is usual to consider a harmonically vibrated plane, i.e., $F(t) = mB \sin \omega t$. In this case the availability or lack of transport are determined by the value of B and the difference between a_1 and a_2. If $B < \min ga_{1,2}$ then the body, being at rest for $t = 0$, remains immobile for all t. In the case of $ga_1 < B < ga_2$ the body moves towards the right during the time lapse between $t_1 + nT$ and $t_2 + nT$, where $t_1 = (1/\omega) \arcsin(ga_1/B)$, t_2 is determined by the equation $(B/\omega)(\cos \omega t_1 - \cos \omega t_2) = ga_1(t_2 - t_1)$, n is an integer and $T = 2\pi/\omega$; during the remainder of the time the body is at rest. It is evident that

$$\overline{y} = \frac{1}{T}\int_0^T y(t)\,\mathrm{d}t > 0.$$

Thus, the body moves on average, though no constant forces act upon it; in the process the motion occurs in the direction of less resistance offered by the friction force. In the case of most interest, when $B > \max ga_{1,2}$, the body moves towards both the right and the left, but on average it moves in the direction of less resistance as before. Let us consider this case in more detail in the time interval $0 \le t \le T$. A solution of Eq. (14.44), in view of (14.45), is

$$y(t) = \begin{cases} y(0) + ga_2 t + \dfrac{B}{\omega}(1 - \cos\omega t) & \text{for } 0 \le t \le t_1, \\ -ga_1(t - t_1) + \dfrac{B}{\omega}(\cos\omega t_1 - \cos\omega t) & \text{for } t_1 \le t \le t_2, \\ ga_2(t - t_2) + \dfrac{B}{\omega}(\cos\omega t_2 - \cos\omega t) & \text{for } t_2 \le t \le T, \end{cases} \quad (14.46)$$

where $y(0)$, t_1 and t_2 are determined by the following equations:

$$\begin{aligned} &y(0) + ga_2 t_1 + \frac{B}{\omega}(1 - \cos\omega t_1) = 0, \\ &-ga_1(t_2 - t_1) + \frac{B}{\omega}(\cos\omega t_1 - \cos\omega t_2) = 0, \\ &ga_2(T - t_2) + \frac{B}{\omega}(\cos\omega t_2 - 1) - y(0) = 0. \end{aligned} \quad (14.47)$$

An example of the plot of $y(t)$ is given in Fig. 14.3.

It follows from (14.46), (14.47) that

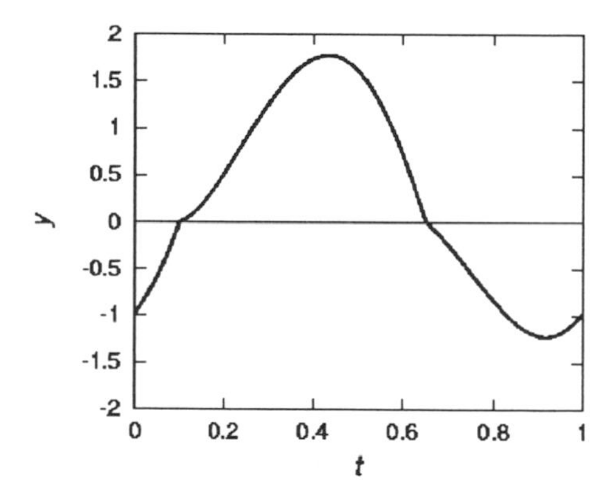

Fig. 14.3. The plot of $y(t)$ for $ga_1 = 5.024$, $ga_2 = 2\pi$, $B/\omega = 2$ and $\omega = 2\pi$; for these values of the parameters $t_1 = 0.0971137$, $t_2 = 0.65267$, $y(0) = -0.970434$, $\overline{y} \approx 0.2461$, $B_0/\omega \approx 1.4175$

$$\overline{y} = \frac{1}{\omega}\sqrt{B^2 - \frac{\pi^2 g^2 a_1^2 a_2^2}{(a_1+a_2)^2}\left(\sin\frac{\pi a_2}{a_1+a_2}\right)^{-2}}\,\cos\frac{\pi a_1}{a_1+a_2}. \tag{14.48}$$

It is evident that this expression is valid for $B \geq B_0$, where

$$B_0 = \frac{\pi g a_1 a_2}{a_1+a_2}\left(\sin\frac{\pi a_2}{a_1+a_2}\right)^{-1}.$$

It can be seen from (14.48) that, as one would expect, $\overline{y}$ is equal to zero for $a_1 = a_2$, positive for $a_1 < a_2$ and negative for $a_1 > a_2$. It is important that $\overline{y}$ is independent of the mass m of the body.

Let us consider further the case of random vibration of the plane. We put $F(t) = m\xi(t)$, where $\xi(t)$ is sufficiently wide-band noise of intensity K with zero mean value. We will assume that K is much more than the intensity of thermal noise; otherwise the force of dry friction should be shifted (see the next subsection) and, as a consequence, vibrational transport should be absent. In this case we can use the Fokker–Planck equation associated with the Langevin equation (14.44). The stationary solution of this equation satisfying the condition for the probability flux to be zero is

$$w(y) = C\exp\left(\frac{2}{K}\int_0^y f(y')\,\mathrm{d}y'\right), \tag{14.49}$$

where the constant C is determined from the normalization condition. It is

$$C = \left(\int_{-\infty}^{\infty}\exp\left(\frac{2}{K}\int_0^y f(y')\,\mathrm{d}y'\right)\mathrm{d}y\right)^{-1}. \tag{14.50}$$

Using the expressions (14.49), (14.50) we can find the mean value of y:

$$\begin{aligned}\langle\varphi(y)\rangle = &\int_{-\infty}^{\infty} y\exp\left(\frac{2}{K}\int_0^y f(y')\,\mathrm{d}y'\right)\mathrm{d}y\\ &\times\left(\int_{-\infty}^{\infty}\exp\left(\frac{2}{K}\int_0^y f(y')\,\mathrm{d}y'\right)\mathrm{d}y\right)^{-1}.\end{aligned} \tag{14.51}$$

If $f(y)$ is described by the expression (14.45) then

$$\langle y\rangle = \frac{K(a_2 - a_1)}{2ga_1a_2}. \tag{14.52}$$

So, in this case we obtain the same result as for harmonic vibration: for $a_1 = a_2$ the body is on average at rest, whereas for $a_1 \neq a_2$ the body moves on average in the direction of less resistance. This result is similar to the rectification of fluctuations (see below). However, there is a dissimilarity from the case of harmonic vibration: in the latter the effect is of threshold character, whereas in the case of random vibration the transport can occur for the noise intensity as small as is wished.

14.2.2 Rectification of fluctuations

Let us consider the simplest electrical rectifier circuit shown in Fig. 14.4. Taking into account the shift of the diode current-voltage characteristic we

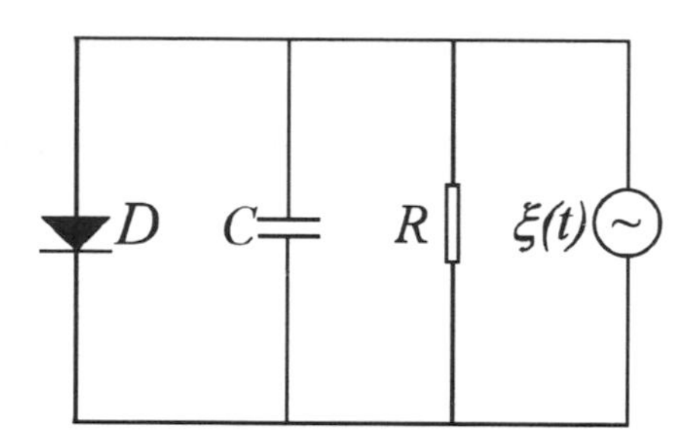

Fig. 14.4. Schematic image of an electrical rectifier

can write the following equation for the voltage drop across the diode V:

$$\dot{V} = -F(V - V_0) - V/\tau + \xi(t), \tag{14.53}$$

where $CF(V - V_0)$ is the current flowing through the diode, $\tau = RC$ is the relaxation time, and $\xi(t)$ is white noise of intensity K.

For simplicity we set $F(v)$ in the form

$$F(v) = \begin{cases} a_1 v & \text{for } v > 0, \\ a_2 v & \text{for } v < 0. \end{cases}$$

The value of V_0 can be calculated by using the technique suggested by Stratonovich [320]. In so doing we find

$$V_0 = \sqrt{\frac{K_0 \tau (1 + a_1 \tau)(1 + a_2 \tau)}{\pi}} \frac{(a_1 - a_2)\tau}{\sqrt{1 + a_1 \tau} + \sqrt{1 + a_2 \tau}}, \tag{14.54}$$

where K_0 is the intensity of thermal fluctuations.

By solving the corresponding Fokker–Planck equation we obtain the following expression for the stationary probability density:

$$w(V) = C \begin{cases} \exp\left(-\dfrac{(1 + a_1 \tau)(V - V_0)^2}{K\tau}\right) & \text{for } V > V_0, \\ \exp\left(-\dfrac{(1 + a_2 \tau)(V - V_0)^2}{K}\tau\right) & \text{for } V < V_0, \end{cases} \tag{14.55}$$

where

$$C = \frac{2\sqrt{(1 + a_1 \tau)(1 + a_2 \tau)}}{\sqrt{\pi K \tau}\,(\sqrt{1 + a_1 \tau} + \sqrt{1 + a_2 \tau})}$$

is the normalization constant.

If the intensity of noise $\xi(t)$ is greater than, or equal to, K_0 then

$$\langle V\rangle = \sqrt{\frac{\tau}{\pi}}\left(\sqrt{K}-\sqrt{K_0}\right)\left(a_2-a_1\right)\tau\,\frac{\sqrt{(1+a_1\tau)(1+a_2\tau)}}{\sqrt{1+a_1\tau}+\sqrt{1+a_2\tau}}. \qquad (14.56)$$

From here it follows that $\langle V\rangle$ does be equal to zero for $K = K_0$ and not equal to zero for $K > K_0$. The sign of $\langle V\rangle$ is determined by the sign of the difference $a_2 - a_1$. Thus, as would be expected, for $K > K_0$ and $a_1 \neq a_2$ we obtain the rectification of fluctuations, i.e., directed motion of electrons caused by noise.

14.3 Noise-induced transport of Brownian particles (stochastic ratchets)

More often than not, consideration of noise-induced transport is restricted to the so-called overdamped case, when the mass of the Brownian particle can be neglected and its motion is described by a first order differential equation of the form

$$\gamma\dot{x} + f(x) = \varphi(t) + \zeta(x,t) + \sqrt{\gamma}\xi(t), \qquad (14.57)$$

where γ is the viscous friction factor, $f(x)$ is a periodic function of x possessing an asymmetry, $\varphi(t)$ is a regular periodic force, $\zeta(x,t)$ is a random process with zero mean value, and $\sqrt{\gamma}\xi(t)$ is white noise of intensity γK imitating thermal fluctuations. The process $\zeta(x,t)$ can be either given or described by additional equations. The function $f(x)$ is most commonly set in a form corresponding to a saw-tooth potential $U(x) = \int_0^x f(x)\,\mathrm{d}x$ shown in Fig. 14.5. In this case

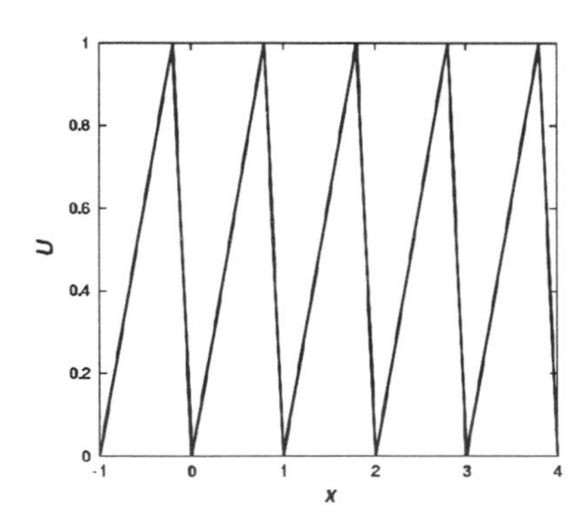

Fig. 14.5. An example of a saw-tooth potential $U(x)$

$$f(x) = \begin{cases} a_1 & \text{for } nL < x < nL + x_1, \\ -a_2 & \text{for } nL - x_2 < x < nL, \end{cases} \tag{14.58}$$

where $n = 0, \pm 1, \pm 2, \ldots$, and $L = x_1 + x_2$ is the period of the function $f(x)$. It is easily shown that, in the absence of the disturbances $\varphi(t)$, $\zeta(x,t)$ and $\xi(t)$, the points $x = nL$ and $x = nL + x_1 = (n+1)L - x_2$ correspond to stable and unstable equilibrium states, respectively. If there is noise then transitions from one stable state to another can occur. Directional motion of the particle will take place if the probabilities of transitions in opposite directions are different.

It is usual to distinguish two types of ratchet devices [39, 73, 74, 113, 301]: (a) $\zeta(x,t)$ is a random force independent of x; and (b) $\zeta(x,t)$ depends on x. In its turn, the latter can be also divided into two classes: (i) those where $\zeta(x,t) = f(x)\chi(t)$, which is to say that the height of the potential barrier fluctuates [39]; and (ii) where $\zeta(x,t)$ is a random function of t and x [301]. We note that flashing ratchets belong to the first class.

We consider below the one-dimensional motion of a Brownian particle in a viscous medium described by the following equation:

$$m\ddot{x} + \gamma\dot{x} + f(x) = \varphi(t) + \zeta(x,t) + \sqrt{\gamma}\,\xi(t), \tag{14.59}$$

where m is the particle mass, $f(x)$ is a potential force, and $\varphi(t)$, $\zeta(x,t)$ and $\xi(t)$ are the same that in Eq. (14.57). For $m \to 0$ this equation reduces to Eq. (14.57).

14.3.1 Noise-induced transport of light Brownian particles in a viscous medium with a saw-tooth potential

Here we consider the case when viscous friction in the medium is sufficiently large and mass of the particle is sufficiently small, so that the motion of the particle can be described approximately by Eq. (14.57), where $f(x)$ is determined by (14.58). In addition, we assume that $\zeta(x,t) = f(x)\xi_1(t)$, where $\xi_1(t)$ is white noise of zero mean value and of intensity K_1 which is uncorrelated with $\xi(t)$. The Fokker–Planck equation associated with Eq. (14.57) at the specified conditions is [319]

$$\frac{\partial w}{\partial t} = -\frac{1}{\gamma}\left\{\frac{\partial}{\partial x}\left[\left(\varphi(t) - f(x) + \frac{K_1}{2\gamma} f(x) f'(x)\right) w(x,t)\right] - \frac{1}{2}\frac{\partial^2}{\partial x^2}\Big(K(x)w(x,t)\Big)\right\}, \tag{14.60}$$

where $K(x) = K + f^2(x)K_1/\gamma$. Because $f(x)$ is a periodic function of x, $w(x,t)$ is also a periodic function of x. Thus Eq. (14.60) only needs to be solved within the interval from $-x_2$ to x_1.

Let us show that the statistical average of the particle velocity $\dot{x}$ is determined by the relationship

$$\langle \dot{x} \rangle = \int_{-x_2}^{x_1} G(x,t)\,\mathrm{d}x, \tag{14.61}$$

where

$$G(x,t) = -\frac{1}{2\gamma}\frac{\partial K(x)w(x,t)}{\partial x} + F(x,t)w(x,t) \tag{14.62}$$

can be treated as the instantaneous probability flux,

$$F(x,t) = \frac{1}{\gamma}\left(\varphi(t) - f(x) + \frac{K_1}{2\gamma}f(x)f'(x)\right).$$

Averaging Eq. (14.57) over statistical ensemble and taking into account that the random process $\xi(t)$ has zero mean value and $\langle\zeta(x,t)\rangle = (K_1/2\gamma)f(x)f'(x)$, [1] we obtain

$$\langle \dot{x} \rangle = \langle F(x,t) \rangle = \int_{-x_2}^{x_1} F(x,t)w(x,t)\,\mathrm{d}x.$$

According to (14.62), this expression can be rewritten as

$$\langle \dot{x} \rangle = \int_{-x_2}^{x_1} \left(G(x,t) + \frac{1}{2\gamma}\frac{\partial K(x)w(x,t)}{\partial x}\right)\mathrm{d}x.$$

By virtue of the spatial periodicity of the functions $w(x,t)$ and $K(x)$ we obtain the formula (14.61). Averaging (14.61) over time we have

$$\overline{\langle \dot{x} \rangle} = \int_{-x_2}^{x_1} \overline{G(x,t)}\,\mathrm{d}x, \tag{14.63}$$

where

$$\overline{G(x,t)} = \lim_{T\to\infty}\frac{1}{T}\int_0^T G(x,t)\,\mathrm{d}t.$$

So, we find that the mean particle velocity is proportional to the probability flux averaged over both space and time.

It should be emphasized that the mean particle velocity depends on the viscous friction factor γ, which in turn depends on the size and shape of the particle.[2] This dependence allows us to separate particles of different size and shape.

Let us calculate $\overline{\langle \dot{x} \rangle}$ in three specific cases:

---

[1] The mean value of $\zeta(x,t)$ is distinct from zero because of the correlation between $\xi_1(t)$ and x [319].

[2] For sphere-shaped particles this dependence is described by Stokes' law, according to which the factor γ is proportional to the particle radius.

(1) The case of a regular force.

We now consider the case when $K_1 = 0$ and $\varphi(t) \neq 0$. If the function $\varphi(t)$ is sufficiently slow, we can use the so-called quasi-stationary approximation for solving Eq. (14.60), i.e., neglect the term $\partial w/\partial t$. In this approximation we obtain from (14.60) the following equation for $w(x,t)$:

$$\Big(\varphi - f(x)\Big)w(x,t) - \frac{K}{2}\frac{\partial w(x,t)}{\partial x} = \gamma G(t), \tag{14.64}$$

where $G(t)$ is the probability flux in the direction of the axis x in the instant t. Solving Eq. (14.64) with account taken of (14.58) we find

$$w(x,t) = \begin{cases} \left(C(t) - \dfrac{\gamma G(t)}{q_1}\right)\exp\left(\dfrac{2q_1 x}{K}\right) + \dfrac{\gamma G(t)}{q_1} & \text{for } 0 \leq x \leq x_1, \\ \left(C(t) - \dfrac{\gamma G(t)}{q_2}\right)\exp\left(\dfrac{2q_2 x}{K}\right) + \dfrac{\gamma G(t)}{q_2} & \text{for } -x_2 \leq x \leq 0, \end{cases} \tag{14.65}$$

where

$$q_{1,2} = \varphi(t) \mp a_{1,2}, \tag{14.66}$$

and $C(t)$ is an arbitrary function of t. From the periodicity condition of the function $w(x,t)$ we find the relation between $G(t)$ and $C(t)$:

$$C(t) = \left(\frac{\gamma}{q_1 q_2}\,\frac{q_2\exp\big(2U_0\varphi/Ka_1\big) - q_1\exp\big(-2U_0\varphi/Ka_2\big)}{\exp\big(2U_0\varphi/Ka_1\big) - \exp\big(-2U_0\varphi/Ka_2\big)} - \frac{(q_2 - q_1)\exp\big(2U_0/K\big)}{\exp\big(2U_0\varphi/Ka_1\big) - \exp\big(-2U_0\varphi/Ka_2\big)}\right)G(t),$$

where $U_0 = a_1 x_1 = a_2 x_2$ is the height of the potential barrier. The probability flux $G(t)$ can be found from the normalization condition for the probability density $w(x,t)$. In so doing we obtain

$$G^{-1}(t) = \gamma U_0\left(\frac{1}{a_1 q_1} + \frac{1}{a_2 q_2}\right) + \frac{\gamma K(a_1 + a_2)^2}{2q_1^2 q_2^2}\exp\left(-\frac{2U_0}{K}\right) \times \frac{\Big(\exp\big(2U_0\varphi/Ka_1\big) - \exp\big(2U_0/K\big)\Big)\Big(\exp\big(-2U_0\varphi/Ka_2\big) - \exp\big(2U_0/K\big)\Big)}{\exp\big(2U_0\varphi/Ka_1\big) - \exp\big(-2U_0\varphi/Ka_2\big)}. \tag{14.67}$$

In the most interesting case when φ is sufficiently small, namely when

$$\max\varphi \ll K/L, \tag{14.68}$$

we find

$$G(t) = G_0\varphi(t) + G_1\varphi^2(t) + \dots, \tag{14.69}$$

where

$$G_0 = \frac{U_0 a_1 a_2}{\gamma K^2 (a_1 + a_2) \sinh^2(U_0/K)},$$
$$G_1 = G_0 \frac{a_2 - a_1}{a_1 a_2} \left(\frac{U_0^2}{K^2 \sinh^2(U_0/K)} + \frac{U_0}{K \tanh(U_0/K)} - 2 \right). \tag{14.70}$$

Let us analyze the result obtained. If the condition (14.68) is fulfilled and $\varphi(t) \equiv B_0 = \text{const}$ (in addition to $f(x)$ a constant force acts upon the particle), then

$$\overline{\langle \dot{x} \rangle} \approx \frac{U_0^2}{\gamma K^2 \sinh^2(U_0/K)} B_0, \tag{14.71}$$

i.e., the particle moves in the direction of this constant force no matter what the relation between a_1 and a_2. We see from (14.71) that in the absence of thermal fluctuations, when $K \to 0$, $\overline{\langle \dot{x} \rangle} \to 0$, i.e., transport of the particle is impossible despite the presence of the constant force. This is because the force is small and cannot by itself push the particle over the potential barrier. In another specific case, when $\varphi(t) = B \cos \omega t$, where B is sufficiently small, we obtain

$$\overline{\langle \dot{x} \rangle} \approx \frac{U_0^2 (a_2 - a_1) B^2}{2\gamma K^2 a_1 a_2 \sinh^2(U_0/K)} \left(\frac{U_0^2}{K^2 \sinh^2(U_0/K)} + \frac{U_0}{K \tanh(U_0/K)} - 2 \right), \tag{14.72}$$

i.e., the particle moves in the direction of the slower rate of potential change. It is easy to verify that, in the absence of thermal fluctuations, transport of the particle cannot occur, just as in the case of a small constant force.

It follows from (14.63) and (14.67) that the mean particle velocity is inversely proportional to γ, i.e., particles of greater size move slower.

Examples of the dependencies of $v \equiv \gamma \overline{\langle \dot{x} \rangle} / B^2$ on K/U_0 described by the formulas (14.63), (14.67) are shown in Fig. 14.6 for a number of values of B. We see that these dependencies are of radically different kinds for $B < \min(a_1, a_2)$ and $B > \min(a_1, a_2)$. In the first case, for $K/U_0 \to 0$, i.e., in the absence of thermal fluctuations, $v \to 0$. In the second case v tends to a certain finite value as $K/U_0 \to 0$, which can be calculated from the theory of vibrational transport (see the preceding section). For $B < 0.5$ the dependencies found are almost coincident with those described by the approximate formula (14.72). In this case the averaged particle velocity is maximal for $K/U_0 \approx 0.43$. If the ratio K/U_0 is either very small or very large, noise-induced transport is not feasible.

The results obtained can be explained in the following manner. Noise-induced transport can occur if fluctuational transitions through each potential barrier are more frequent in one direction than in another. Because the probability of the transition through a certain potential barrier depends only on its height and the noise intensity, transport is impossible in the absence

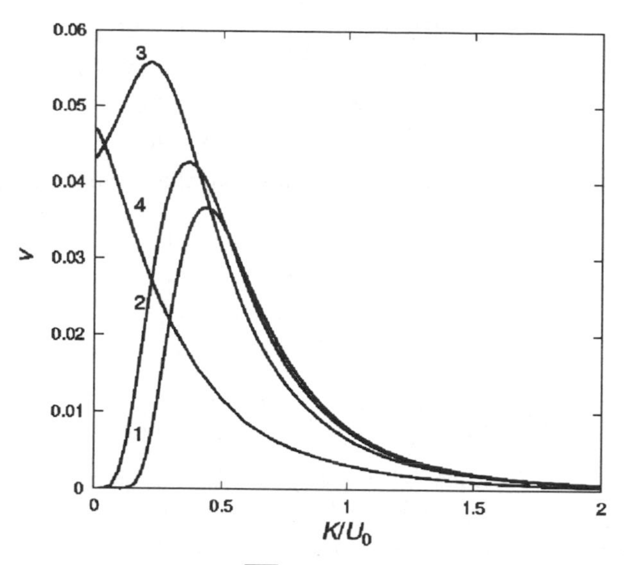

Fig. 14.6. Dependencies of $v \equiv \gamma\overline{\langle\dot{x}\rangle}/B^2$ on K/U_0 as described by Eqs. (14.63), (14.67) for $a_1 = 1.25$, $a_2 = 5$, $x_1 = 0.8$, $x_2 = 0.2$: $B = 0.1$ (curve 1); $B = 1$ (curve 2); $B = 2$ (curve 3); $B = 5$ (curve 4)

of the additional force $\varphi(t)$. In the case of a constant force the result is self-evident because the heights of the potential barriers for the particle moving to the right and to the left are different. The case of an alternating force is more complicated. During one half-period the right potential barrier is lowered to $U_0 - Bx_1$, whereas the left one rises to $U_0 + Bx_2$. During the next half-period the right potential barrier rises to $U_0 + Bx_1$, whereas the left one is lowered to $U_0 - Bx_2$. Because the lowering of the potential barrier plays a dominant role, the particle moves, on average, in the direction of larger lowering of the potential barrier.

(2) The case of random modulation of the potential barrier height.

For simplicity assume that in Eq. (14.57) the regular force $\varphi(t)$ is absent. In this case a stationary solution of the Fokker–Planck equation (14.60) satisfying the continuity condition for $x = 0$ is

$$w(x) = \begin{cases} -\dfrac{\gamma G}{a_1}\left[1 - \exp\left(-\dfrac{2a_1 x}{K^{(1)}}\right)\right] + C\exp\left(-\dfrac{2a_1 x}{K^{(1)}}\right) \\ \qquad\qquad\qquad\qquad \text{for} \quad 0 < x < x_1, \\ \dfrac{\gamma G}{a_2}\left[1 - \exp\left(\dfrac{2a_2 x}{K^{(2)}}\right)\right] + C\exp\left(\dfrac{2a_2 x}{K^{(2)}}\right) \\ \qquad\qquad\qquad\qquad \text{for} \quad -x_2 < x < 0, \end{cases} \tag{14.73}$$

where $K^{(1,2)} = K + K_1 a_{1,2}^2/\gamma$.

From the periodicity condition for the function $w(x)$ we find the relation between G and C:

$$\gamma G \left\{ a_1 \left[1 - \exp\left(-\frac{2U_0}{K^{(2)}} \right) \right] + a_2 \left[1 - \exp\left(-\frac{2U_0}{K^{(1)}} \right) \right] \right\}$$
$$= C a_1 a_2 \left[\exp\left(-\frac{2U_0}{K^{(1)}} \right) - \exp\left(-\frac{2U_0}{K^{(2)}} \right) \right]. \tag{14.74}$$

Taking account of (14.73) and (14.74), and from the normalization condition, we find G:

$$G = \frac{2a_1^2 a_2^2}{\gamma(a_1 + a_2)} \left[\exp\left(-\frac{2U_0}{K^{(1)}} \right) - \exp\left(-\frac{2U_0}{K^{(2)}} \right) \right]$$
$$\times \left\{ (a_1 + a_2)(K + K_1 a_1 a_2) \left[1 - \exp\left(-\frac{2U_0}{K^{(1)}} \right) \right] \left[1 - \exp\left(-\frac{2U_0}{K^{(2)}} \right) \right] \right.$$
$$\left. - 2U_0(a_2 - a_1) \left[\exp\left(-\frac{2U_0}{K^{(1)}} \right) - \exp\left(-\frac{2U_0}{K^{(2)}} \right) \right] \right\}^{-1}. \tag{14.75}$$

It follows from (14.75) that $\langle \dot{x} \rangle = GL \neq 0$ only if $a_1 \neq a_2$ and $K_1 \neq 0$. The dependencies of $v \equiv \gamma \langle \dot{x} \rangle$ on $K_1/(\gamma U_0)$ for different values of K/U_0, and on K/U_0 for a number of fixed values of $K_1/(\gamma U_0)$, are shown in Fig. 14.7. It is

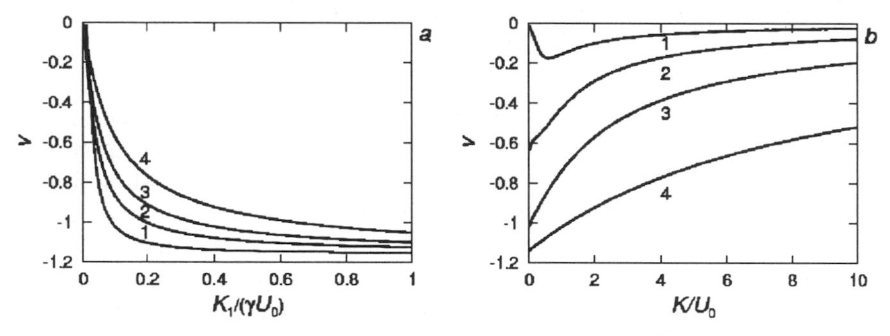

Fig. 14.7. Dependencies of $v \equiv \gamma\langle\dot{x}\rangle$: (a) on $K_1/(\gamma U_0)$ for $K/U_0 = 0$, 0.5, 1 and 2 for curves 1–4 respectively; and (b) on K/U_0 for $K_1/(\gamma U_0) = 0.01$, 0.035, 0.1 and 0.4 for curves 1–4 respectively. In both cases, $a_1 = 1.25$ and $a_2 = 5$

interesting that the particle moves on average in the direction of the greater rate of the potential change. We draw attention to the fact that, for random modulation of the potential barrier, the directional motion of particles is possible even in the absence of thermal fluctuations ($K = 0$).

(3) The case of an additional random force with large correlation time.

Fluctuational transport of a Brownian particle in a viscous medium induced by thermal noise and a correlated random force that is a Markov process was studied in [73]. However, concrete results were obtained only for dichotomous and 'kangaroo'-like processes. Here we consider this problem for the case where the correlated random force is the so-called Ornstein–Uhlenbeck process [345]. We can then write the following equations of motion:

$$\gamma \dot{x} + f(x) = y + \sqrt{\gamma}\xi(t), \tag{14.76}$$

$$\dot{y} = -\beta y + \xi_1(t), \tag{14.77}$$

where $f(x)$ is determined by the expression (14.58), and $\xi(t)$ and $\xi_1(t)$ are uncorrelated white noises with zero mean values and intensities equal to K and K_1, respectively. The stationary probability density of the variable y is independent of x and can be easily calculated from the Fokker–Planck equation associated with Eq. (14.77). It is equal to

$$p(y) = \sqrt{\frac{\beta}{\pi K_1}} \exp\left(-\frac{\beta y^2}{K_1}\right). \tag{14.78}$$

Let us calculate now the conditional probability density of the variable x for a fixed value of y. In the quasi-stationary approximation, which is valid for sufficiently large correlation time of the process $y(t)$, this probability density $w(x|y)$ satisfies the following Fokker–Planck equation:

$$G(y) = -\frac{f(x) - y}{\gamma} w(x|y) - \frac{K}{2\gamma} \frac{\partial w(x|y)}{\partial x}, \tag{14.79}$$

where $G(y)$ is the probability flux for a fixed value of y. A solution of Eq. (14.79) is

$$w(x|y) = \begin{cases} \dfrac{\gamma G(y)}{y - a_1} + \left(C(y) - \dfrac{\gamma G(y)}{y - a_1}\right) \exp\left(\dfrac{2(y - a_1)}{K} x\right) \\ \qquad\qquad \text{for} \quad 0 < x < x_1, \\ \dfrac{\gamma G(y)}{y + a_2} + \left(C(y) - \dfrac{\gamma G(y)}{y + a_2}\right) \exp\left(\dfrac{2(y + a_2)}{K} x\right) \\ \qquad\qquad \text{for} \quad -x_2 < x < 0, \end{cases} \tag{14.80}$$

where $C(y)$ is an arbitrary function of y. From the periodicity condition of the function $w(x|y)$ we find a relation between $G(y)$ and $C(y)$:

$$\begin{aligned} C(y) = \frac{\gamma G(y)}{q_1 q_2} &\left[q_2 \exp\left(\frac{2U_0 y}{K a_1}\right) - q_1 \exp\left(-\frac{2U_0 y}{K a_2}\right) \right. \\ &\left. - (a_1 + a_2) \exp\left(\frac{2U_0}{K}\right) \right] \left[\exp\left(\frac{2U_0 y}{K a_1}\right) - \exp\left(-\frac{2U_0 y}{K a_2}\right) \right]^{-1}, \end{aligned} \tag{14.81}$$

where $q_{1,2} = y \mp a_{1,2}$. Taken together, (14.81) and the normalization condition yield the probability flux $G(y)$:

$$\begin{aligned} G^{-1}(y) = \gamma U_0 &\left(\frac{1}{a_1 q_1} + \frac{1}{a_2 q_2} \right) - \frac{K(a_1 + a_2)^2}{2 q_1^2 q_2^2} \exp\left(-\frac{2U_0}{K}\right) \\ &\times \left[\exp\left(\frac{2U_0}{K}\right) - \exp\left(-\frac{2U_0 y}{K a_2}\right) \right] \left[\exp\left(\frac{2U_0 y}{K a_1}\right) - \exp\left(\frac{2U_0}{K}\right) \right] \\ &\times \left[\exp\left(\frac{2U_0 y}{K a_1}\right) - \exp\left(-\frac{2U_0 y}{K a_2}\right) \right]^{-1}. \end{aligned} \tag{14.82}$$

Taking into account that $L = x_1 + x_2 = U_0/a_1 + U_0/a_2$, in the quasi-stationary approximation we find

$$\langle \dot{x} \rangle = \langle G(y) \rangle L = \frac{a_1 + a_2}{a_1 a_2} \langle G(y) \rangle U_0, \tag{14.83}$$

where

$$\langle G(y) \rangle = \int_{-\infty}^{\infty} G(y) p(y) \, \mathrm{d}y, \tag{14.84}$$

where $p(y)$ is determined by (14.78). The dependencies of $v \equiv \gamma \langle \dot{x} \rangle$ on K_1/β calculated numerically for $a_1 = 1.25$, $a_2 = 5$ and different values of K/U_0 are shown in Fig. 14.8 a. It is seen from this figure that for a fixed value

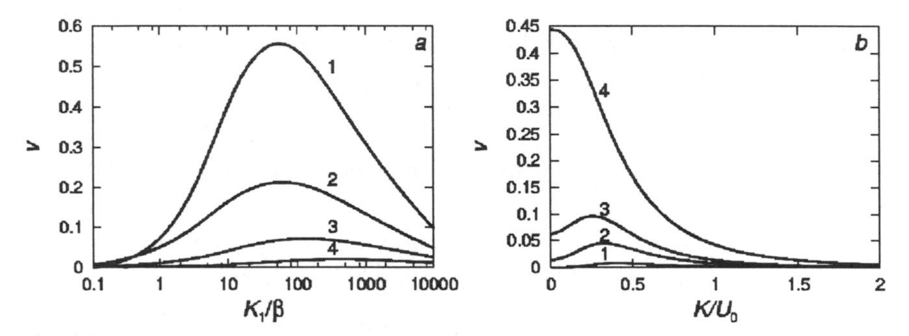

Fig. 14.8. Dependencies of $v \equiv \gamma \langle \dot{x} \rangle$: (a) on K_1/β for $K/U_0 = 0.1$, 0.5, 1 and 2 for curves 1–4 respectively; (b) on K/U_0 for $K_1/\beta = 0.2$, 1, 2 and 10 for curves 1–4 respectively. In both cases, $U_0 = 1$, $L = 1$, $a_1 = 1.25$ and $a_2 = 5$

of K/U_0 the value of v first increases as K_1/β increases and then slowly decreases again, approaching zero as $K_1/\beta \to \infty$. The peak of v is located at the greater values of K_1/β, the greater is K/U_0. The dependencies of v on K/U_0 for a fixed value of K_1/β, shown in Fig. 14.8 b, are of a somewhat different form. They display maxima at a certain value of $K/U_0 \neq 0$ only for K_1/β less than a critical value, whereas for greater K_1/β the dependencies become monotonically decreasing.

14.3.2 The effect of the particle mass

To solve the problem on the effect of the particle mass on the noise-induced transport it is necessary to find a solution of two-dimensional Fokker–Planck equation. In [26, 27] such a solution was found numerically by using the so-called matrix continued fraction technique. Tackling this problem analytically is possible in two limiting cases only: when the particle mass is very small and when it is sufficiently large. We consider both these cases.

The case of small particle masses. We revert now to Eq. (14.59) assuming that m is sufficiently small, viz.,

$$m \max f'(x) \ll \gamma^2. \tag{14.85}$$

Putting, for simplicity, $\zeta(x,t)$ to be equal to zero, we rewrite Eq. (14.59) in the form

$$\mu\ddot{x} + \dot{x} + \frac{f(x)}{\gamma} = \frac{\varphi(t)}{\gamma} + \frac{1}{\sqrt{\gamma}}\xi(t), \tag{14.86}$$

where $\mu = m/\gamma$. In the case when the function $f(x)$ is differentiable and μ is sufficiently small, we can obtain an approximate one-dimensional Fokker–Planck equation for the probability density of the variable x, much as this was done by Stratonovich [319]. The derivation of such an equation for $\varphi(t) =$ const is given in Appendix A (see Eq. (A.20)). For $\mu = 0$ the equation found is the exact Fokker–Planck equation corresponding to the Langevin equation (14.57).

We can use Eq. (A.20) for solving our problem in the quasi-stationary approximation. A stationary solution of Eq. (A.20) is described by Eq. (A.21). Setting in Eq. (A.21) $\epsilon^2 = \mu$, and retaining the terms up to the order 4, inclusive, with respect to μ, we find the following quasi-stationary solution of the equation for $w(x,t)$:

$$\begin{aligned} w(x,t) = \exp\left(-\frac{2\big(U(x)-\varphi(t)x\big)}{K}\right)\Bigg\{C(t) - \frac{2\gamma G(t)}{K}\int\limits_0^x \Bigg(1 - \mu f'(x') \\ - \mu^2\big(F_1(x) - 2\varphi(t)f''(x)\big) - \mu^3\left[F_2(x) - \frac{2}{K}F_3(x)\varphi(t) + \frac{7}{2}\varphi^2(t)f'''(x)\right] \\ - \mu^4\Bigg[F_4(x) - \frac{2}{K}F_5(x)\varphi(t) + \left(\frac{K}{16}f^{V}(x) + \frac{71}{12}\big(3f(x)-\varphi(t)\big)f^{IV}(x)\right. \\ \left. + \frac{83}{2}f'''(x)f'(x) + \frac{105}{4}\big(f''(x)\big)^2\right)\varphi^2(t)\Bigg]\Bigg) \\ \times \exp\left(\frac{2\big(U(x')-\varphi x'\big)}{K}\right)\mathrm{d}x'\Bigg\}, \end{aligned} \tag{14.87}$$

where $G(t)$ is the probability flux at a fixed instant of t,

$$\begin{aligned} F_1(x) &= \frac{3K}{4}f'''(x) + 2f(x)f''(x) + \big(f'(x)\big)^2, \\ F_2(x) &= \frac{29K^2}{48}f^{V}(x) + \frac{K}{8}\Big(23f(x)f^{IV}(x) + 52f'(x)f'''(x) \\ &\quad + 31\big(f''(x)\big)^2\Big) + \frac{7}{2}f^2(x)f'''(x) + 12f(x)f'(x)f''(x) + 2\big(f'(x)\big)^3, \\ F_3(x) &= \frac{23K^2}{16}f^{IV}(x) + \frac{7K}{2}f(x)f'''(x) + 6Kf'(x)f''(x), \end{aligned}$$

$$
\begin{aligned}
F_4(x) &= \frac{143K^3}{288} f^{VII}(x) + \frac{K^2}{144}\Big(485 f(x) f^{VI}(x) + 1677 f^V(x) f'(x) \\
&+ 3081 f^{IV}(x) f''(x) + 1820\big(f'''(x)\big)^2\Big) + \frac{K}{16}\Big(123 f^2(x) f^V(x) \\
&+ 713 f(x) f^{IV}(x) f'(x) + 1147 f(x) f'''(x) f''(x) + 760 f'''(x)\big(f'(x)\big)^2 \\
&+ 919\big(f''(x)\big)^2 f'(x)\Big) + \frac{71}{12} f^3(x) f^{IV}(x) + \frac{83}{2} f^2(x) f'''(x) f'(x) \\
&+ \frac{129}{2} f(x) f''(x)\big(f'(x)\big)^2 + \frac{105}{4} f^2(x)\big(f''(x)\big)^2 + 5\big(f'(x)\big)^4, \\
F_5(x) &= \frac{485K^3}{288} f^{VI}(x) + \frac{123K^2}{16} f(x) f^V(x) + \frac{K^2}{32}\Big(713 f^{IV}(x) f'(x) \\
&+ 1147 f'''(x) f''(x)\Big) + \frac{71K}{8} f^2(x) f^{IV}(x) + \frac{83K}{2} f(x) f'''(x) f'(x) \\
&+ \frac{129K}{4} f''(x)\big(f'(x)\big)^2 + \frac{105K}{4} f(x)\big(f''(x)\big)^2.
\end{aligned}
$$

From here on we will assume that $\varphi(t)$ is small, namely, $|\varphi(t)| \ll K/L$. From the periodicity condition of the function $w(x,t)$ we can find a relationship between $C(t)$ and $G(t)$. Taking into account that $G(t)$ is of order $\varphi(t)$ we obtain

$$
\begin{aligned}
\varphi(t)C(t) &= \frac{\gamma G(t)}{L}\left(1 + \frac{L\varphi(t)}{K}\right)\Big(I_1(\varphi) - \mu I_2(\varphi) - \mu^2 I_3(\varphi) \\
&\times \Big(I_1(\varphi) - \mu I_2(\varphi) - \mu^2 I_3(\varphi) - \mu^3 I_4(\varphi) - \mu^4 I_5(\varphi)\Big), \qquad (14.88)
\end{aligned}
$$

where $I_j(\varphi) = I_{j0} - I_{j1}\varphi$,

$$
\begin{aligned}
&I_{10} = I[1], \quad I_{11} = \frac{2}{K} I[x], \quad I_{20} = I[f'(x)], \quad I_{21}(t) = \frac{2}{K} I[x f'(x)], \\
&I_{30} = I[F_1(x)], \quad I_{31} = \frac{2}{K} I[x F_1(x) + K f''(x)], \\
&I_{40} = I[F_2(x)], \quad I_{41} = \frac{2}{K} I[x F_2(x) + F_3(x)], \quad I_{50} = I[F_4(x)], \\
&I_{51} = \frac{2}{K} I[x F_4(x) + F_5(x)], \quad I[F(x)] = \int_0^L F(x)\exp\left(\frac{2U(x)}{K}\right) dx.
\end{aligned}
$$

Taking account of (14.87), we obtain from the normalization condition another relationship between $C(t)$ and $G(t)$. It can be written as

$$
C(t) I_6(\varphi)\varphi(t) - \frac{2\gamma G(t)\varphi(t)}{K}\Big(I_7 - \mu I_8 - \mu^2 I_9 - \mu^3 I_{100} - \mu^4 I_{110}\Big) = \varphi(t), \qquad (14.89)
$$

where $I_6(\varphi) = I_{60} + I_{61}\varphi$,

$$I_{60} = \int_0^L \exp\left(-\frac{2U(x)}{K}\right) \mathrm{d}x, \quad I_{61} = \frac{2}{K}\int_0^L x \exp\left(-\frac{2U(x)}{K}\right) \mathrm{d}x,$$
$$I_7 = II[1], \quad I_8 = II[f(x)], \quad I_9 = II[F_1(x)], \quad I_{100} = II[F_2(x)],$$
$$I_{110} = II[F_4(x)],$$
$$II[F(x)] = \int_0^L \int_0^x F(x') \exp\left(\frac{2\big(U(x') - U(x)\big)}{K}\right) \mathrm{d}x' \, \mathrm{d}x.$$

Solving Eqs. (14.88), (14.89) we find $G(t)$. It can be written as

$$\begin{aligned} G(t) &= G_{01}\Big(1 + \mu\, M_{10} + \mu^2 M_{20} + \mu^3 M_{30} + \mu^4 M_{40}\Big)\varphi(t) \\ &+ \Big[G_{02}\Big(1 + \mu M_{10} + \mu^2 M_{20} + \mu^3 M_{30} + \mu^4 M_{40}\Big) \\ &+ \mu G_{01}\Big(M_{11} + \mu M_{21} + \mu^2 M_{31} + \mu^3 M_{41}\Big)\Big]\varphi^2(t), \end{aligned} \tag{14.90}$$

where

$$G_{01} = \frac{L}{\gamma I_{10} I_{60}}, \quad G_{02} = G_{01}\left[\frac{I_{11}}{I_{10}} - \frac{I_{61}}{I_{60}} - \frac{L}{K}\left(1 - \frac{2I_7}{I_{10}I_{60}}\right)\right], \tag{14.91}$$

$$\begin{aligned} M_{10} &= \frac{I_{20}}{I_{10}}, \quad M_{11} = -M_{10}\left[\frac{I_{21}}{I_{20}} - \frac{I_{11}}{I_{10}} + \frac{2L}{KI_{60}}\left(\frac{I_8}{I_{20}} - \frac{I_7}{I_{10}}\right)\right], \\ M_{20} &= M_{10}^2 + \frac{I_{30}}{I_{10}}, \quad M_{30} = 2M_{10}M_{20} - M_{10}^3 + \frac{I_{40}}{I_{10}}, \\ M_{21} &= 2M_{10}M_{11} - \frac{I_{30}}{I_{10}}\left[\frac{I_{31}}{I_{30}} - \frac{I_{11}}{I_{10}} + \frac{2L}{KI_{60}}\left(\frac{I_9}{I_{30}} - \frac{I_7}{I_{10}}\right)\right], \\ M_{31} &= 2\big(M_{10}M_{21} + M_{20}M_{11}\big) - 3M_{10}^2 M_{11} \\ &\quad - \frac{I_{40}}{I_{10}}\left[\frac{I_{41}}{I_{40}} - \frac{I_{11}}{I_{10}} + \frac{2L}{KI_{60}}\left(\frac{I_{100}}{I_{40}} - \frac{I_7}{I_{10}}\right)\right], \\ M_{40} &= M_{10}\big(2M_{30} - M_{10}M_{20}\big) + \big(M_{20} - M_{10}^2\big)^2 + \frac{I_{50}}{I_{10}}, \\ M_{41} &= M_{10}\big(2M_{31} - 2M_{20}M_{11} - M_{10}M_{21}\big) + 2M_{11}M_{30} + 2\big(M_{20} - M_{10}^2\big) \\ &\quad \times\big(M_{21} - 2M_{10}M_{11}\big) - \frac{I_{50}}{I_{10}}\left[\frac{I_{51}}{I_{50}} - \frac{I_{11}}{I_{10}} + \frac{2L}{KI_{60}}\left(\frac{I_{110}}{I_{50}} - \frac{I_7}{I_{10}}\right)\right]. \end{aligned} \tag{14.92}$$

If $\varphi(t) = B\cos\omega t$, then

$$\begin{aligned} \overline{\langle \dot{x} \rangle} &\approx \frac{B^2 L}{2}\Big[G_{02} + \mu(G_{01}M_{11} + G_{02}M_{10}) + \mu^2(G_{01}M_{21} + G_{02}M_{20}) \\ &+ \mu^3(G_{01}M_{31} + G_{02}M_{30}) + \mu^4(G_{01}M_{41} + G_{02}M_{40})\Big]. \end{aligned} \tag{14.93}$$

As an example, we set the function $f(x)$ proportional to the first two terms of the Fourier series for $f(x)$ determined by (14.58), viz.

$$f(x) = \sum_{n=1}^{2} \frac{1.1}{n\pi} \left((a_1 + a_2) \sin \frac{2\pi n(x + x_0)}{L} - a_1 \sin \frac{2\pi n(x + x_0 - x_1)}{L} \right.$$
$$\left. - a_2 \sin \frac{2\pi n(x + x_0 + x_2)}{L} \right), \qquad (14.94)$$

where x_0 is chosen so that $f(0) = 0$. Plots of the functions

$$U(x) = -\sum_{n=1}^{2} \frac{1.1L}{2n^2\pi^2} \left[(a_1 + a_2) \cos \frac{2\pi n(x + x_0)}{L} \right.$$
$$- a_1 \cos \frac{2\pi n(x + x_0 - x_1)}{L}$$
$$\left. - (a_1 + a_2) \left(\cos \frac{2\pi n x_0}{L} - \cos \frac{2\pi n(x_0 - x_1)}{L} \right) \right], \qquad (14.95)$$

$f(x)$ and $f'(x)$ for $x_1 = 0.8$, $x_2 = 0.2$ and $x_0 \approx 0.073$ are shown in Fig. 14.9.

Let us consider further two specific cases: (i) the function $f(x)$ is independent of the particle mass and (ii) $f(x)$ is proportional to the particle mass. The latter takes place, for example, when the potential $U(x)$ is a gravitational one.

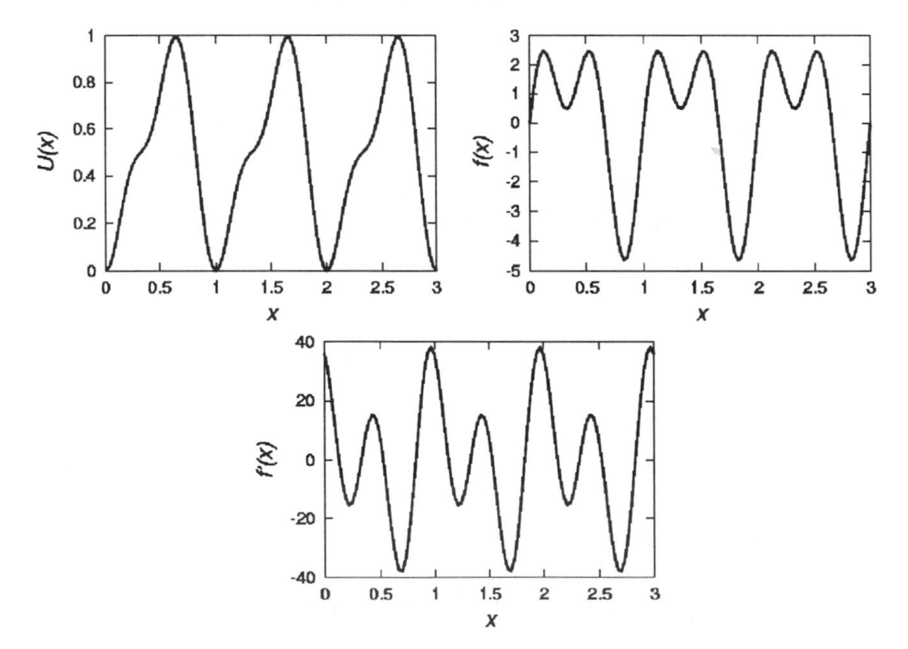

Fig. 14.9. Plots of the functions $U(x)$, $f(x)$ and $f'(x)$ as determined by Eqs. (14.94) and (14.95) for $x_1 = 0.8$, $x_2 = 0.2$ and $x_0 \approx 0.073$

(i) For $f(x)$ described by the formula (14.94) the dependencies of $v \equiv \gamma\overline{\langle\dot{x}\rangle}/B^2$ on K/U_0 for $\mu = 0$ and $\mu = 0.005$ are shown in Fig. 14.10. We see that up to the fourth approximation the flux reversal does not occur for any values of K/U_0. Only in the fourth approximation, for moderately large values of K/U_0 ($K/U_0 > 5$), it takes place.

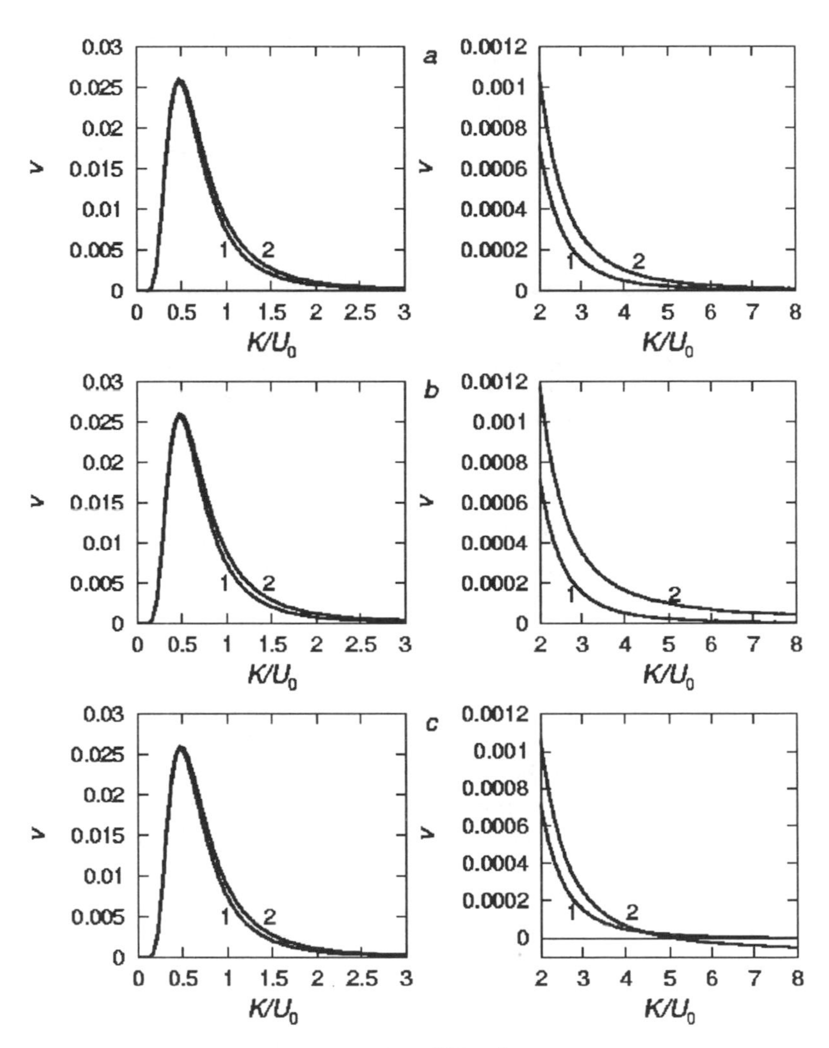

Fig. 14.10. The dependencies of $v \equiv \gamma\overline{\langle\dot{x}\rangle}/B^2$ on K/U_0 for $f(x)$ described by the formula (3.19) for $\mu = 0$ and $\mu = 0.005$ (curves 1 and 2): (a) in the second approximation, (b) in the third approximation, and (c) in the fourth approximation; the results in the range $0 \leq K/U_0 \leq 3$ are represented on the left, and the results in the range $2 \leq K/U_0 \leq 8$ are represented on the right

It is evident that the dependence of the mean velocity of a particle on its mass can be used for the separation of particles of different masses. Examples of the dependencies of $v \equiv \gamma\overline{\langle \dot{x} \rangle}/B^2$ on μ for two values of K/U_0 calculated in the second, third and fourth approximations are illustrated in Fig. 14.11. We see that the difference between the results is essential even for small μ. The

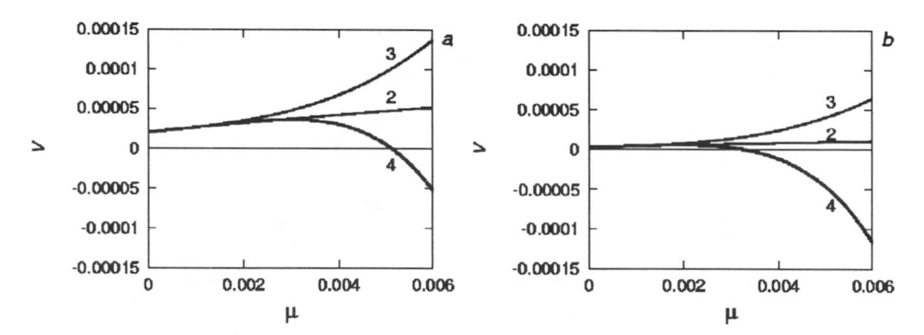

Fig. 14.11. The dependencies of v on μ for (a) $K/U_0 = 5$ and (b) $K/U_0 = 8$. The results obtained in the second, third and fourth approximations with respect to μ are labeled 2, 3 and 4, respectively

significant difference between the results obtained in different approximations for $\mu > 0.002$ points to the fact that the series on the left-hand side of Eq. (A.19) converges only asymptotically. This is why the conclusion about the flux reversal calls for further investigation.

(ii) In the case of a gravitational potential, when a particle moves along a hilly vibrated surface with periodic structure, we can put $U(x) = \mu\tilde{U}(x)$, $f(x) = \mu\tilde{f}(x)$, ... It should be emphasized that in this case the particle flux is not zero only for $\mu > 0$. For definiteness, we, as before, set the function $f(x)$ to be proportional to the first two terms of the Fourier series for $f(x)$ corresponding to a saw-tooth potential, namely, we assume that $f(x)$ is described by (14.94). Performing manipulations much as in the foregoing item, we find the following series for the mean particle velocity:

$$\overline{\langle \dot{x} \rangle} \approx \frac{B^2 L}{2}\,\mu\Big[\tilde{G}_{02} + \mu G_{01}\tilde{M}_{11} + \mu^2(\tilde{G}_{02}\tilde{M}_{10} + G_{01}\tilde{M}_{21})$$
$$+\mu^3(\tilde{G}_{02}\tilde{M}_{20} + G_{01}\tilde{M}_{31}) + \mu^4(\tilde{G}_{02}\tilde{M}_{30} + G_{01}\tilde{M}_{41}) + \ldots\Big], \quad (14.96)$$

where $\tilde{G}_{02} = G_{02}/\mu$, G_{01} and G_{02} are determined by (14.91),

$$\tilde{M}_{10} = \frac{M_{10}}{\mu}, \quad \tilde{M}_{11} = \frac{M_{11}}{\mu}, \quad \tilde{M}_{20} = \frac{I_{300}}{I_{10}}, \quad \tilde{M}_{21} = -\frac{I_{310}}{I_{10}} + \frac{I_{11}I_{300}}{I_{10}^2}$$
$$-\frac{2L}{KI_{10}I_{60}}\left(I_{90} - \frac{I_{70}I_{300}}{I_{10}}\right), \quad \tilde{M}_{30} = \tilde{M}_{10}^2 + \frac{I_{301}}{I_{10}} + \frac{I_{400}}{I_{10}},$$

$$\tilde{M}_{31} = 2\tilde{M}_{10}\tilde{M}_{11} - \frac{I_{311}}{I_{10}} + \frac{I_{11}I_{301}}{I_{10}^2} - \frac{2L}{KI_{10}I_{60}}\left(I_{91} - \frac{I_{70}I_{301}}{I_{10}}\right)$$
$$-\frac{I_{410}}{I_{10}} + \frac{I_{11}I_{400}}{I_{10}^2} - \frac{2L}{KI_{10}I_{60}}\left(I_{1000} - \frac{I_{70}I_{400}}{I_{10}}\right),$$
$$\tilde{M}_{41} = 2\left(\tilde{M}_{10}\tilde{M}_{21} + \frac{I_{300}}{I_{10}}\tilde{M}_{11}\right) - \frac{I_{411}}{I_{10}} + \frac{I_{11}I_{401}}{I_{10}^2} - \frac{2L}{KI_{10}I_{60}}\left(I_{1001}\right.$$
$$\left.-\frac{I_{70}I_{401}}{I_{10}}\right) - \frac{I_{510}}{I_{10}} + \frac{I_{11}I_{500}}{I_{10}^2} - \frac{I_{21}I_{500}}{I_{10}I_{60}} - \frac{2L}{KI_{10}I_{60}}\left(I_{1100} - \frac{I_{70}I_{500}}{I_{10}}\right),$$

where

$$I_{300} = \frac{3K}{4}\tilde{I}[\tilde{f}'''(x)], \quad I_{310} = 2\tilde{I}[\tilde{f}''(x)] + \frac{3}{2}\tilde{I}[x\tilde{f}'''(x)], \quad I_{301} = \tilde{I}\big[F_6(x)\big],$$
$$I_{311} = \frac{2}{K}\tilde{I}\big[xF_6(x)\big], \quad I_{90} = \frac{3K}{4}\tilde{I}I[\tilde{f}'''(x)], \quad I_{91} = \tilde{I}I\big[F_6(x)\big],$$
$$I_{400} = \frac{29K^2}{48}\tilde{I}[\tilde{f}^V(x)], \quad I_{401} = \frac{K}{8}\tilde{I}\big[F_7(x)\big], \quad I_{410} = \frac{K}{24}\tilde{I}[29xf^V(x)$$
$$+69\tilde{f}^{IV}(x)], \quad I_{411} = \frac{1}{4}\tilde{I}\big[xF_7(x) + 4\big(7\tilde{f}(x)\tilde{f}'''(x) + 12\tilde{f}'(x)\tilde{f}''(x)\big)\big],$$
$$I_{1000} = \frac{29K^2}{48}\tilde{I}I[\tilde{f}^V(x)], \quad I_{1001} = \frac{K}{8}\tilde{I}I\big[F_7(x)\big],$$
$$I_{500} = \frac{143K^3}{288}\tilde{I}[\tilde{f}^{VII}(x)], \quad I_{501} = \frac{K^2}{144}\tilde{I}\big[485\tilde{f}(x)\tilde{f}^{VI}(x)$$
$$+1677\tilde{f}^V(x)\tilde{f}'(x) + 3081\tilde{f}^{IV}(x)\tilde{f}''(x) + 1820\big(\tilde{f}'''(x)\big)^2\big],$$
$$I_{510} = \frac{K^2}{144}\tilde{I}[143x\tilde{f}^{VII}(x) + 485\tilde{f}^{VI}(x)],$$
$$I_{1100} = \frac{143K^3}{288}\tilde{I}I[f^{VII}(x)], \quad \tilde{I}[F(x)] = \int_0^L F(x)\exp\left(\frac{2\mu\tilde{U}(x)}{K}\right)\,dx,$$
$$\tilde{I}I[F(x)] = \int_0^L\int_0^x F(x')\exp\left(\frac{2\mu(\tilde{U}(x') - \tilde{U}(x))}{K}\right)\,dx'\,dx,$$
$$F_6(x) = 2\tilde{f}(x)\tilde{f}''(x) + \big(\tilde{f}'(x)\big)^2,$$
$$F_7(x) = 23\tilde{f}(x)\tilde{f}^{IV}(x) + 52\tilde{f}'(x)\tilde{f}'''(x) + 31\big(\tilde{f}''(x)\big)^2.$$

The dependencies of $v \equiv \gamma\overline{\langle\dot{x}\rangle}/B^2$ on $K/\tilde{U}_0$ for $\mu = 0.05$, 0.1, 0.2 and 0.25 are shown in Fig. 14.12 in different approximations with respect to μ. We see that for $\mu = 0.05$ (a) the influence of different approximations on the behavior of the dependencies under consideration is very small; only for $K/U_0 \approx 0.23$ flux reversal is found in the fifth approximation, but the values of reversed velocity are very small. For $\mu = 0.1$ (b) the flux reversal is also found only in the fifth approximation, but the values of reversed velocity are

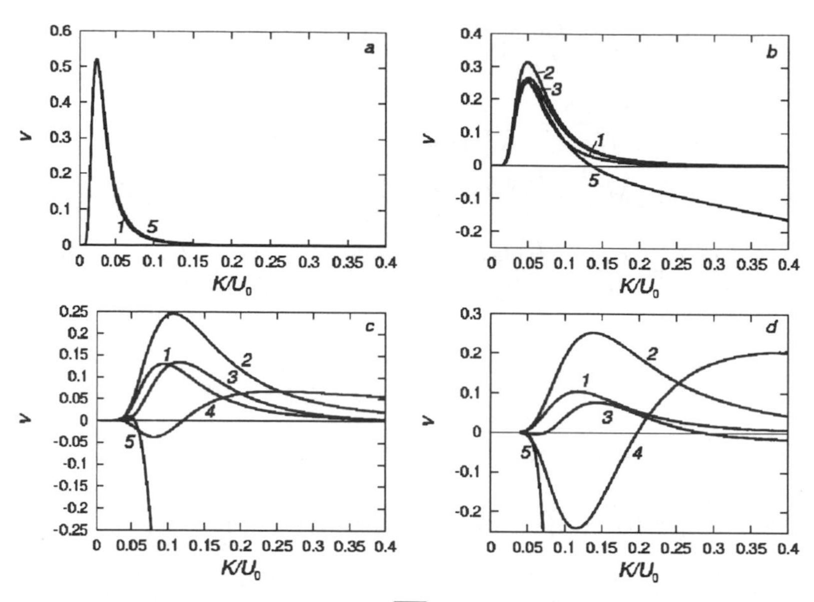

Fig. 14.12. The dependencies of $v \equiv \gamma\overline{\langle \dot{x} \rangle}/B^2$ on $K/\tilde{U}_0$ for $\tilde{f}(x)$ described by the formula (3.19) for (a) $\mu = 0.05$, (b) $\mu = 0.1$, (c) $\mu = 0.2$ and (d) $\mu = 0.25$ in the first (curves 1), second (curves 2), third (curves 3), fourth (curves 4) and fifth (curves 5) approximations

moderately large. For larger values of μ (c and d) flux reversal is found for lower approximations as well.

The dependencies of v on μ for a number of values of $K/\tilde{U}_0$ are given in Fig. 14.13, also in different approximations with respect to μ. We see that the behavior of these dependencies is rather complicated and, except for very small $K/\tilde{U}_0$ (e.g., for $K/\tilde{U}_0 < 0.01$ all of these dependencies nearly coincide), depends strongly on the number of the approximation. Nevertheless, we can conclude that the flux reversal occurs as the particle mass increases. Thus, we have found that, for not very small intensities of noise, a light particle moves, on average, in the direction of smaller steepness of slope, whereas a heavy particle moves, on average, in the opposite direction.

The case of large particle masses. We consider here another limiting case when the particle mass is moderately large, so that the quality factor Q of the corresponding oscillator is also sufficiently large.[3] For the sake of simplicity we assume that in Eq. (14.59) the function $f(x)$ is determined by (14.58) and $\zeta(x, t) \equiv 0$.

The problem is to calculate the particle velocity $\dot{x}$ averaged over both statistical ensemble and time. To solve this problem Eq. (14.59) is conveniently rewritten in terms of variables x and E [319, 198], where E is the energy which is described by

[3] The expression for Q will be derived below.

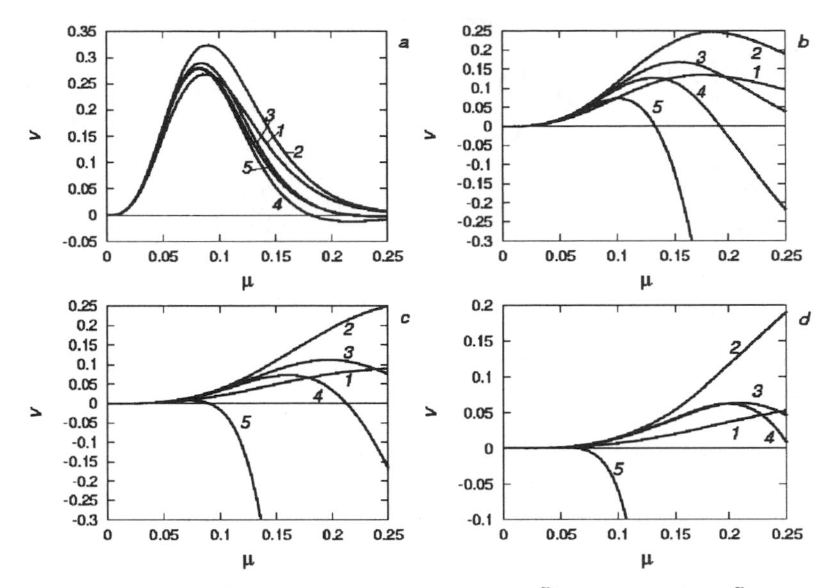

Fig. 14.13. The dependencies of v on μ for (a) $K/\tilde{U}_0 = 0.05$, (b) $K/\tilde{U}_0 = 0.1$, (c) $K/\tilde{U}_0 = 0.15$ and (d) $K/\tilde{U}_0 = 0.2$. The curves 1, 2, 3, 4 and 5 correspond to the first, second, third, fourth and fifth approximations, respectively

$$E = \frac{m\dot{x}^2}{2} + U(x) - x\varphi(t). \tag{14.97}$$

If $\dot{\varphi}(t)$ is sufficiently small, and the damping factor $\delta \equiv \gamma/(2m)$ much less than the the least frequency of natural oscillations $\omega(U_0)$, then E is a slowly varying function of time.

Multiplying both sides of Eq. (14.59) by $\dot{x}$ we obtain the following exact equation for E:

$$\dot{E} = -\gamma\dot{x}^2 - x\dot{\varphi}(t) + \sqrt{\gamma}\,\dot{x}\xi(t). \tag{14.98}$$

This equation in combination with

$$\dot{x} = \pm\sqrt{\frac{2\Big(E - U(x) + x\varphi(t)\Big)}{m}} \tag{14.99}$$

are completely equivalent to Eq. (14.59).

The following two Fokker–Planck equations for the probability densities $w_{1,2}(x, E, t)$ correspond to two pairs of the Langevin equations (14.98), (14.99):

$$\frac{\partial w_{1,2}(x,E,t)}{\partial t} = \mp \frac{\partial}{\partial x}\left(\sqrt{\frac{2\bigl(E-U(x)+x\varphi(t)\bigr)}{m}}\, w_{1,2}(x,E,t)\right)$$
$$+\frac{\partial}{\partial E}\left[\left(\frac{2\gamma\bigl(E-U(x)+x\varphi(t)\bigr)}{m} + x\dot{\varphi}(t) - \frac{\gamma K}{2m}\right) w_{1,2}(x,E,t)\right]$$
$$+\gamma K \frac{\partial^2}{\partial E^2}\left(\frac{E-U(x)+x\varphi(t)}{m}\, w_{1,2}(x,E,t)\right). \qquad (14.100)$$

It follows from Eq. (14.99) that the statistical average of the particle velocity is

$$\langle \dot{x}\rangle = \int\limits_{0}^{U_0(1-\varphi(t)/a_1)} \int\limits_{x_{\min}(E,t)}^{x_{\max}(E,t)} \sqrt{\frac{2\bigl(E-U(x)+x\varphi(t)\bigr)}{m}}\, w_1(x,E,t)\,\mathrm{d}x\,\mathrm{d}E$$
$$- \int\limits_{0}^{U_0(1+\varphi(t)/a_2)} \int\limits_{x_{\min}(E,t)}^{x_{\max}(E,t)} \sqrt{\frac{2\bigl(E-U(x)+x\varphi(t)\bigr)}{m}}\, w_2(x,E,t)\,\mathrm{d}x\,\mathrm{d}E, \qquad (14.101)$$

where U_0 is the value of the potential $U(x)$ at points corresponding to unstable equilibrium states (the value of the potential at points corresponding to stable equilibrium states is taken as zero); $x_{\min}(E,t) = -E/\bigl(a_2+\varphi(t)\bigr)$ and $x_{\max}(E,t) = E/\bigl(a_1-\varphi(t)\bigr)$ are the slowly varying extreme values taken by x during the oscillations with energy E, and they are equal to the roots of the equation $U(x) - x\varphi(t) = E$.

Let us represent $w_{1,2}(x,E,t)$ in the form of the product of $w_{1,2}(E,t)$ and the conditional probability density $w_{1,2}(x|E,t)$:

$$w_{1,2}(x,E,t) = w_{1,2}(E,t)w_{1,2}(x|E,t). \qquad (14.102)$$

The conditional probability density is described by the following equation [319]

$$\frac{\partial w_{1,2}(x|E,t)}{\partial t} = \mp \frac{\partial}{\partial x}\left(\sqrt{\frac{2\bigl(E-U(x)+x\varphi(t)\bigr)}{m}}\, w_{1,2}(x|E,t)\right). \qquad (14.103)$$

Since, according to the assumption, E is a slowly varying function of time, then $w_{1,2}(x|E,t)$ has time to follow variations of E. This is why we can take as $w_{1,2}(x|E,t)$ steady-state solutions of Eqs. (14.103) [319].

We restrict ourselves by the case when $\dot{\varphi}$ is negligibly small and the quasi-stationary approximation is valid. In this case we can put in Eqs. (14.103) $\partial w_{1,2}(x|E,t)/\partial t = 0$. Then solutions of Eqs. (14.103) are

$$w_1(x|E,t) = w_2(x|E,t) = N(E,t) \begin{cases} \sqrt{\dfrac{m}{2\Big(E - U(x) + x\varphi(t)\Big)}} \\ \text{for } U(x) - x\varphi(t) \le E, \\ 0 \\ \text{for } U(x) - x\varphi(t) > E. \end{cases} \tag{14.104}$$

The slowly varying function $N(E,t)$ is determined from the normalization condition:

$$N(E,t) = \left(\int\limits_{x_{\min}(E,t)}^{x_{\max}(E,t)} \sqrt{\frac{m}{2\Big(E - U(x) + x\varphi(t)\Big)}}\, \mathrm{d}x \right)^{-1} \equiv \frac{\omega(E,t)}{\pi}, \tag{14.105}$$

where

$$\begin{aligned} \omega(E,t) &= \pi \left(\int\limits_{x_{\min}(E,t)}^{x_{\max}(E)} \sqrt{\frac{m}{2\Big(E - U(x) + x\varphi(t)\Big)}}\, \mathrm{d}x \right)^{-1} \\ &= \frac{\pi\Big(a_1 - \varphi(t)\Big)\Big(a_2 + \varphi(t)\Big)}{a_1 + a_2} \sqrt{\frac{1}{2mE}} \end{aligned} \tag{14.106}$$

is the slowly varying frequency of oscillations with energy E. It follows from (14.106) that the condition $\delta \ll \omega(U_0)$, whereby the suggested technique is valid, is equivalent to

$$Q = \frac{\pi a}{\gamma} \sqrt{\frac{2m}{U_0}} \gg 1, \tag{14.107}$$

where Q is the quality factor, $a = a_1 a_2/(a_1 + a_2)$.

Substituting (14.102) into (14.101), in view of (14.104) and (14.105), and taking into account the expressions for $x_{\min}(E,t)$, $x_{\max}(E,t)$ and $\omega(E,t)$, we find

$$\begin{aligned} \langle \dot{x} \rangle = \frac{\pi}{\sqrt{2m}} \Bigg(&\int\limits_0^{U_0(1-\varphi(t)/a_1)} \sqrt{E}\, w_1(E,t)\, \mathrm{d}E \\ &- \int\limits_0^{U_0(1+\varphi(t)/a_2)} \sqrt{E}\, w_2(E,t)\, \mathrm{d}E \Bigg) . \end{aligned} \tag{14.108}$$

Thus, for calculation of the mean particle velocity it is sufficient to know the one-dimensional probability densities $w_{1,2}(E,t)$. Below we show that $w_{1,2}(E,t)$ each is described by a one-dimensional Fokker–Planck equation.

Substituting (14.102) into Eq. (14.100), in view of (14.104) and (14.105), and integrating this equation over x from $x_{\min}(E,t)$ to $x_{\max}(E,t)$, we find the equations for $w_{1,2}(E,t)$:

$$\frac{\partial w_{1,2}(E,t)}{\partial t} = \frac{\partial}{\partial E}\left\{\left[\omega(E,t)J(E,t)\left(\gamma + \frac{\pi m\Big(a_2 - a_1 + 2\varphi(t)\Big)}{\Big(a_1 - \varphi(t)\Big)\Big(a_2 + \varphi(t)\Big)}\dot{\varphi}(t)\right)\right.\right.$$
$$\left.\left. - \frac{\gamma K}{2m}\right] w_{1,2}(E,t)\right\} + \frac{\gamma K}{2}\frac{\partial^2}{\partial E^2}\Big(\omega(E,t)J(E,t)w_{1,2}(E,t)\Big), \qquad (14.109)$$

where

$$J(E,t) = \frac{1}{\pi}\int\limits_{x_{\min}(E,t)}^{x_{\max}(E,t)} \sqrt{\frac{2\Big(E - U(x) + x\varphi(t)\Big)}{m}}\,\mathrm{d}x$$
$$= \frac{2(a_1 + a_2)}{3\pi\big(a_1 - \varphi(t)\big)\big(a_2 + \varphi(t)\big)}\sqrt{\frac{2E}{m}}\,E \qquad (14.110)$$

is the action [203].

Like Eq. (14.103), in the quasi-stationary approximation we can put in Eqs. (14.109) $\partial w_{1,2}(E,t)/\partial t = 0$ and ignore $\dot{\varphi}(t)$. In this approximation solutions of Eqs. (14.109) are conveniently written as

$$w_{1,2}(E,t) = \frac{C_{1,2}(t)}{U_0}\sqrt{\frac{E}{U_0}}\exp\left(-\frac{2E}{K}\right), \qquad (14.111)$$

where $C_{1,2}(t)$ are found from the normalization conditions

$$C_{1,2}(t) = \left(\int\limits_0^{1\mp\varphi(t)/a_{1,2}} \sqrt{y}\,\mathrm{e}^{-2U_0 y/K}\,\mathrm{d}y\right)^{-1}. \qquad (14.112)$$

The expressions for $C_{1,2}(t)$ are

$$C_{1,2}(t) = \frac{8U_0}{K}\left[\sqrt{\frac{2\pi K}{U_0}}\,\mathrm{erf}\left(\sqrt{\frac{2\big(a_{1,2} \mp \varphi(t)\big)U_0}{a_{1,2}K}}\right)\right.$$
$$\left. - 4\sqrt{\frac{a_{1,2} \mp \varphi(t)}{a_{1,2}}}\exp\left(-2\,\frac{\big(a_{1,2} \mp \varphi(t)\big)U_0}{a_{1,2}K}\right)\right]^{-1}. \qquad (14.113)$$

Substituting (14.111) into (14.108) and taking account of (14.107) we find

$$\langle \dot{x} \rangle = \frac{\pi^2 a K^2}{4\gamma Q U_0^2} \left\{ C_1(t) \left[1 - \left(\frac{2(a_1 - \varphi(t))U_0}{a_1 K} + 1 \right) \exp\left(- \frac{2(a_1 - \varphi(t))U_0}{a_1 K} \right) \right] \right.$$
$$\left. - C_2(t) \left[1 - \left(\frac{2(a_2 + \varphi(t))U_0}{a_2 K} + 1 \right) \exp\left(- \frac{2(a_2 + \varphi(t))U_0}{a_2 K} \right) \right] \right\}, \tag{14.114}$$

where $C_{1,2}(t)$ are determined by (14.113).

For small $\varphi(t)$ we can expand (14.114), in view of (14.113), as a power series in $\varphi(t)$. Retaining the terms up to $\varphi^2(t)$, inclusive, we obtain

$$\langle \dot{x} \rangle = \frac{\pi^2 \varphi(t)}{\gamma Q} \left\{ 8 \left[2 \left(\exp\left(\frac{2U_0}{K} \right) - 1 \right) - \sqrt{\frac{2\pi U_0}{K}} \operatorname{erf}\left(\sqrt{\frac{2U_0}{K}} \right) \exp\left(\frac{2U_0}{K} \right) \right] \right.$$
$$\times \left[\sqrt{\frac{2\pi K}{U_0}} \operatorname{erf}\left(\sqrt{\frac{2U_0}{K}} \right) \exp\left(\frac{2U_0}{K} \right) - 4 \right]^{-2}$$
$$+ \frac{4(a_2 - a_1)U_0 \varphi(t)}{K a a_1 a_2} \left[8 \left(2 \exp\left(\frac{2U_0}{K} \right) - 1 \right) + \frac{4K}{U_0} \left(\exp\left(\frac{2U_0}{K} \right) - 1 \right) \right.$$
$$+ \sqrt{\frac{2\pi K}{U_0}} \operatorname{erf}\left(\sqrt{\frac{2U_0}{K}} \right) \exp\left(\frac{2U_0}{K} \right) \left(\frac{K}{U_0} - \frac{8U_0}{K} - 10 \right.$$
$$\left. + \left(4 - \frac{K}{U_0} \right) \exp\left(\frac{2U_0}{K} \right) \right) - 2\pi \left(2 - \frac{K}{U_0} \right) \operatorname{erf}^2\left(\sqrt{\frac{2U_0}{K}} \right) \exp\left(\frac{4U_0}{K} \right) \Bigg]$$
$$\left. \times \left[\sqrt{\frac{2\pi K}{U_0}} \operatorname{erf}\left(\sqrt{\frac{2U_0}{K}} \right) \exp\left(\frac{2U_0}{K} \right) - 4 \right]^{-3} \right\}. \tag{14.115}$$

In the case when $\varphi(t) = B \cos \omega t$ the expression (14.115) should be averaged over time. The averaged velocity is

$$\overline{\langle \dot{x} \rangle} = \frac{2\pi^2 U_0 (a_2 - a_1) B^2}{a a_1 a_2 \gamma K Q} \left[4 \left(\left(4 + \frac{K}{U_0} \right) \exp\left(\frac{2U_0}{K} \right) - 2 - \frac{K}{U_0} \right) \right.$$
$$+ \sqrt{\frac{2\pi K}{U_0}} \operatorname{erf}\left(\sqrt{\frac{2U_0}{K}} \right) \exp\left(\frac{2U_0}{K} \right) \left(\frac{K}{U_0} - \frac{8U_0}{K} - 10 \right.$$
$$\left. + \left(4 - \frac{K}{U_0} \right) \exp\left(\frac{2U_0}{K} \right) \right) - 2\pi \left(2 - \frac{K}{U_0} \right) \operatorname{erf}^2\left(\sqrt{\frac{2U_0}{K}} \right) \exp\left(\frac{4U_0}{K} \right) \Bigg]$$
$$\times \left[\sqrt{\frac{2\pi K}{U_0}} \operatorname{erf}\left(\sqrt{\frac{2U_0}{K}} \right) \exp\left(\frac{2U_0}{K} \right) - 4 \right]^{-3} . \tag{14.116}$$

It is seen that v depends on m not directly but only through the quality factor Q. The dependencies of $v = \dfrac{a a_1 a_2 \gamma}{2\pi^2 (a_2 - a_1) B^2} \overline{\langle \dot{x} \rangle}$ on K/U_0 for two values of Q

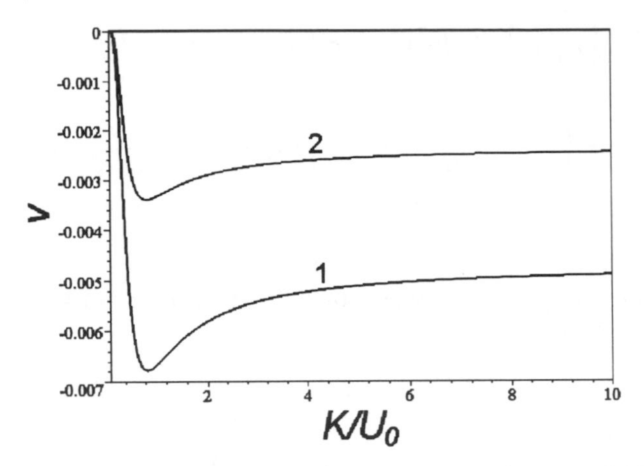

Fig. 14.14. The dependencies of v on K/U_0 in the case of a potential independent of the particle mass for $Q = 5$ and 10 (curves 1 and 2, respectively)

are shown in Fig. 14.14). It should be noted that the flux velocity decreases in inverse proportion to the square root of the particle mass.

Next we consider another specific case when the saw-tooth potential field is formed by a gravitational force. At the conditions specified above we can use the expression (14.116) by substituting into it $m\tilde{U}_0$, $m\tilde{a}$ and $m\tilde{a}_{1,2}$ in place of U_0, a and $a_{1,2}$, respectively. Taking into account that

$$Q = \frac{\pi \tilde{a} m}{\gamma}\sqrt{\frac{2}{\tilde{U}_0}} \equiv m\tilde{Q},$$

we can express Q in terms of m. As a result we obtain

$$\begin{aligned}\overline{\langle \dot{x} \rangle} = \frac{2\pi^2 \tilde{U}_0 (\tilde{a}_2 - \tilde{a}_1) B^2}{\tilde{a}\tilde{a}_1\tilde{a}_2\gamma m^2 K \tilde{Q}} \Bigg[4\Bigg(\left(4 + \frac{K}{mU_0}\right) \exp\left(\frac{2mU_0}{K}\right) - 2 - \frac{K}{mU_0} \Bigg) \\ + \sqrt{\frac{2\pi K}{m\tilde{U}_0}} \operatorname{erf}\left(\sqrt{\frac{2m\tilde{U}_0}{K}}\right) \exp\left(\frac{2m\tilde{U}_0}{K}\right) \Bigg(\frac{K}{m\tilde{U}_0} - \frac{8m\tilde{U}_0}{K} - 10 \\ + \left(4 - \frac{K}{m\tilde{U}_0}\right) \exp\left(\frac{2m\tilde{U}_0}{K}\right)\Bigg) - 2\pi\left(2 - \frac{K}{m\tilde{U}_0}\right) \operatorname{erf}^2\left(\sqrt{\frac{2m\tilde{U}_0}{K}}\right) \\ \times \exp\left(\frac{4m\tilde{U}_0}{K}\right)\Bigg] \left[\sqrt{\frac{2\pi K}{m\tilde{U}_0}} \operatorname{erf}\left(\sqrt{\frac{2m\tilde{U}_0}{K}}\right) \exp\left(\frac{2m\tilde{U}_0}{K}\right) - 4\right]^{-3}.\end{aligned} \tag{14.117}$$

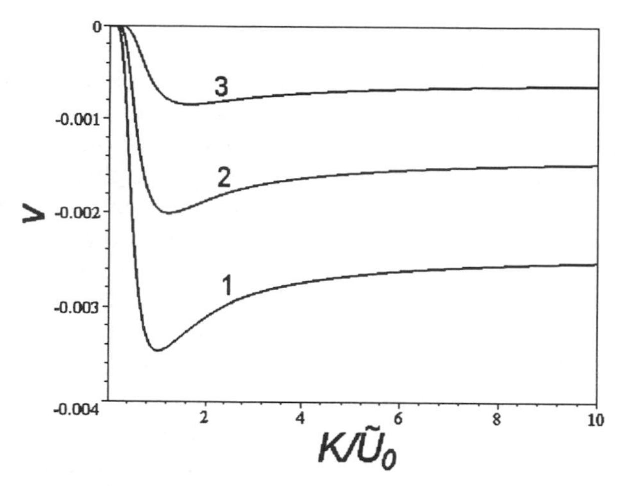

Fig. 14.15. The dependencies of v on $K/\tilde{U}_0$ in the case of a gravitational potential for $\tilde{Q} = 5$, $m = 1.25$, 1.5 and 2 (curves 1, 2 and 3, respectively)

The dependencies of $v = \dfrac{\tilde{a}\tilde{a}_1\tilde{a}_2\gamma}{2\pi^2(\tilde{a}_2 - \tilde{a}_1)B^2}\overline{\langle\dot{x}\rangle}$ on $K/\tilde{U}_0$ for a number of fixed values of m are presented in Fig. 14.15. It is seen that in this case the dependencies found are similar those for the potential independent of particle mass, but the decrease of the flux velocity obeys a more complicated law.

Thus, we have shown that in the case of large particle masses the particle flux is directed oppositely as compared with the case of small masses. This takes place both in the case of a potential independent of mass and in the case of a gravitational potential. The results obtained, on the one hand, reinforce those obtained above and showing that the flux reversal occurs for a certain particle mass, and, on the other, show that the number of the flux reversals can be only odd.

14.4 Stochastic and vibrational resonances: similarities and distinctions

In the last few years, a substantial body of journal papers and reviews has evolved which are devoted to the so-called stochastic resonance phenomenon (see, e.g., [243, 244, 365, 98, 16]). Usually this phenomenon is considered by the example of an overdamped bistable oscillator described by the equation

$$\dot{x} + f(x) = A\cos\omega t + \xi(t), \tag{14.118}$$

where $A\cos\omega t$ is a weak input signal, $f(x) = \mathrm{d}U(x)/\mathrm{d}x$, $U(x)$ is a symmetric double-well potential, $\xi(t)$ is white noise of intensity K, i.e., $\langle\xi(t)\xi(t+\tau)\rangle = K\delta(\tau)$. The simplest example of a symmetric double-well potential is

$$U(x) = -x^2/2 + x^4/4. \tag{14.119}$$

It is known that the response Q of this system to the input signal is a non-monotone function of the noise intensity K and peaks at a certain value of $K = K_{\mathrm{m}}$. The gain factor Q is defined as the ratio of the amplitude of the output signal component at the frequency ω to the amplitude A of the input signal. The dependence of Q on K at a fixed value of ω closely resembles a resonance dependence for a driven oscillator. Therefore many researchers suppose that the resonance occurs when the noise intensity is such that the signal frequency is equal to half the mean frequency of fluctuational transitions from one steady state to another determined by the so-called Kramers rate (regarding fluctuational transitions see, e.g., [163, 198]). However, this is not the case because the corresponding dependence of Q on ω at a fixed value of K is not of resonance character but monotonically descending.

We believe that the reason for such non-monotone character of the dependence $Q(K)$ lies in the non-monotone dependence of the system effective stiffness on the noise intensity K. By effective stiffness is meant the following. With respect to the output signal component $s_\omega(t)$ at the frequency ω, Eq. (14.118) can be linearized. In this case the complex amplitude B of the output signal is determined by the equation

$$\mathrm{i}\omega B + c(A, \omega, K)B = A, \tag{14.120}$$

where $c(A, \omega, K)$ is a complex quantity. The real part of this quantity $c_{\mathrm{r}}(A, \omega, K)$ may be treated as the proper effective stiffness, whereas its imaginary part $c_{\mathrm{i}}(A, \omega, K)$ may be treated as the additional damping factor. The effective stiffness determines the bandwidth of a low-frequency filter described by Eq. (14.118). The inference about the non-monotone change of the effective stiffness in response to noise is consistent with Klimontovich's idea [153] that stochastic resonance in an overdamped bistable oscillator is explained by the change of the bandwidth of the corresponding low-frequency filter.

It is evident that the indicated change of the effective stiffness can be induced not only by noise but also by a high-frequency force [202]. Below we consider both of these cases and compare the results.

To find the effective stiffness, we should stepwisely linearize the nonlinear term x^3. First of all we should perform statistical linearization [274, 267]. For this we substitute $\lambda(t, A, \omega, K)x$ in place of x^3, where $\lambda(t, A, \omega, K)$ is determined from the condition for minimum of the mean-square error

$$\epsilon = \langle (x^3 - \lambda x)^2 \rangle. \tag{14.121}$$

In (14.121) $\langle\,\rangle$ denote statistical average. We find from (14.121)

$$\lambda(t, A, \omega, K) = \frac{\langle x^4(t) \rangle}{\langle x^2(t) \rangle} = \frac{\int_{-\infty}^{\infty} x^4 w(x,t)\,\mathrm{d}x}{\int_{-\infty}^{\infty} x^2 w(x,t)\,\mathrm{d}x}, \tag{14.122}$$

where $w(x,t)$ is the steady probability density of x which is a periodic function of t. Further we should perform harmonic linearization of the term $\big(\lambda(t,A,\omega,K)-1\big)s_\omega(t)$, i.e., represent it in the form $c(A,\omega,K)s_\omega(t)$. For this we represent $s_\omega(t)$ and $\lambda(t)$ as $s_\omega(t) = B\cos\omega t$ and $\lambda(t) = \lambda_0 + \lambda_c\cos 2\omega t + \lambda_s \sin 2\omega t$. Then

$$c_r(A,\omega,K) = \lambda_0 - 1 + \frac{\lambda_c}{2}, \quad c_i(A,\omega,K) = -\frac{\lambda_s}{2}. \tag{14.123}$$

Unfortunately, to calculate $w(x,t)$ we must solve the nonstationary Fokker–Planck equation corresponding to the Langevin equation (14.118). However, finding even an approximate solution of this equation is a very difficult problem. That is why for calculations of $c(A,\omega,K)$ we will use other methods.

A solution of Eq. (14.120) can be written as

$$B = Q(A,\omega,K)A\exp\big(\mathrm{i}\psi(A,\omega,K)\big),$$

where

$$Q(A,\omega,K) = \frac{1}{\sqrt{c_r^2(A,\omega,K) + \big(\omega + c_i(A,\omega,K)\big)^2}},$$
$$\psi(A,\omega,K) = -\arctan\frac{\omega + c_i(A,\omega,K)}{c_r(A,\omega,K)}. \tag{14.124}$$

It follows from (14.124) that c_r and c_i are related to $Q(A,\omega,K)$ and $\psi(A,\omega,K)$ by

$$c_r(A,\omega,K) = \frac{\cos\psi(A,\omega,K)}{Q(A,\omega,K)}, \quad c_i(A,\omega,K) = -\frac{\sin\psi(A,\omega,K)}{Q(A,\omega,K)} - \omega. \tag{14.125}$$

Thus, we can calculate c_r and c_i directly from the results of the numerical simulation of Eq. (14.118). Another technique for approximate calculations of these quantities will be considered below.

It should be noted that all known results of the numerical calculations of $Q(A,\omega,K)$ starting from the direct numerical simulation of Eq. (14.118) are based on the analysis of power spectra of $x(t)$ in which, as known [319], discrete components at the signal frequency ω must be present. But in real spectra, for very small signal amplitudes, these components do not stand out sharply against the continuous spectrum. Furthermore, it is impossible by means of power spectra to calculate the phase shift $\psi(A,\omega,K)$. Therefore we suggest another technique for numerical calculations of $Q(A,\omega,K)$ and $\psi(A,\omega,K)$. This technique is based on the principle of a so-called synchronous detector. It consists in the following. We calculate the sine and cosine components of the output signal by

$$B_s = \frac{2}{nT}\left\langle\int_0^{nT} x(t)\sin\omega t\,\mathrm{d}t\right\rangle, \quad B_c = \frac{2}{nT}\left\langle\int_0^{nT} x(t)\cos\omega t\,\mathrm{d}t\right\rangle, \tag{14.126}$$

where $T = 2\pi/\omega$, n is an integer, and the angular brackets denote averaging over the ensemble of different realizations of $\xi(t)$.

14.4.1 Stochastic resonance in an overdamped oscillator

With the technique described above we have calculated $Q(K)$ and $\psi(K)$ for different values of A and ω by numerical simulation of Eq. (14.118). The results are illustrated in Fig. 14.16 a and b. We see from these figures that the value of K_m is the greater the smaller is A and the greater is ω. The dependencies of K_m on A and on ω are shown in Fig. 14.16 c on the left and on the right, respectively. From the data found we can calculate the real and imaginary parts of the effective stiffness by formulas (14.125). The results are given in Fig. 14.17.

The complex effective stiffness can be approximately found analytically in the following manner. Let us seek a solution of Eq. (14.118) in the form of the sum of signal $s(t) = \langle x(t) \rangle$ and of noise $n(t)$. Here $\langle\,\rangle$ denote the average over the statistical ensemble. It is evident that $\langle n(t) \rangle = 0$. From Eq. (14.118) we find the following equations for $s(t)$ and $n(t)$:

$$\dot{s} - s + s^3 + 3m_2 s + m_3 = A\cos\omega t, \tag{14.127}$$

$$\dot{n} - n + 3s^2 n + 3s(n^2 - m_2) + n^3 - m_3 = \xi(t), \tag{14.128}$$

where $m_k = \langle n^k \rangle$ is the kth moment of the noise n.

The Fokker–Planck equation associated with Eq. (14.128) is

$$\frac{\partial w}{\partial t} = -\frac{\partial}{\partial n}\Big[(n - n^3 - 3s^2 n - 3s(n^2 - m_2) + m_3)w\Big] + \frac{K}{2}\frac{\partial^2 w}{\partial n^2}. \tag{14.129}$$

We will seek a solution of Eq. (14.129) in the form of a finite number of terms of the so-called *Edgeworth expansion* [319]:

$$w(n,t) = \frac{1}{\sqrt{2\pi m_2}}\left[1 + \sum_{j=3}^{N} \frac{b_j}{j!(2m_2)^{j/2}} H_j\left(\frac{n}{\sqrt{2m_2}}\right)\right] \exp\left(-\frac{n^2}{2m_2}\right), \tag{14.130}$$

where $H_j(z)$ is a Hermite polynomial, $m_2(t)$ and $b_j(t)$ are unknown functions; $b_j(t)$ are the so-called quasi-moments [161], which are related to the moments $m_k(t)$ by $b_3(t) = m_3(t)$, $b_4(t) = m_4(t) - 3m_2^2(t)$, ... It is known that the moments m_3 and m_4 determine, respectively, the skewness and excess for the corresponding probability distribution.[4] It should be noted that the representation of the solution in the form (14.130) means that we use the Galerkin method [97] for solving Eq. (14.129).

[4] The quantities $\gamma_a = m_3/m_2^{3/2}$ and $\gamma_e = m_4/m_2^2$ are called, respectively, *the coefficients of skewness and of excess*.

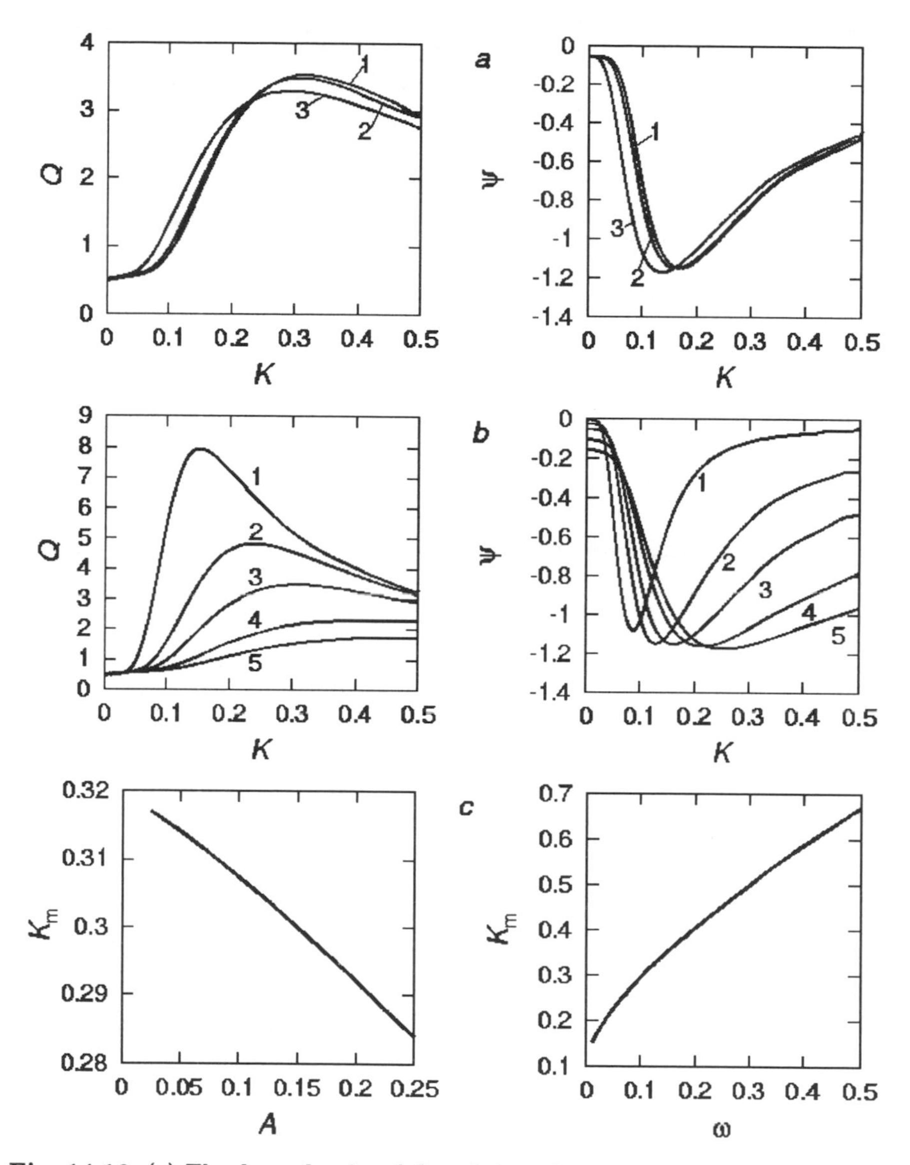

Fig. 14.16. (a) The dependencies of Q and ψ on the noise intensity K for $\omega = 0.1$ and $A = 0.025$, $A = 0.1$ and $A = 0.2$ (curves 1, 2 and 3, respectively); (b) the same for $A = 0.1$, $\omega = 0.01$, $\omega = 0.05$, $\omega = 0.1$, $\omega = 0.2$ and $\omega = 0.3$ (curves 1, 2, 3, 4 and 5, respectively); (c) the plots of K_m versus A (on the left) and versus ω (on the right)

The equations for the unknown functions $m_2(t)$, $m_3(t)$, $b_4(t)$, ..., $b_N(t)$ can be derived by using well known properties of the Hermite polynomials. Substituting $w(n,t)$ into Eq. (14.129), multiplying the latter in turn by

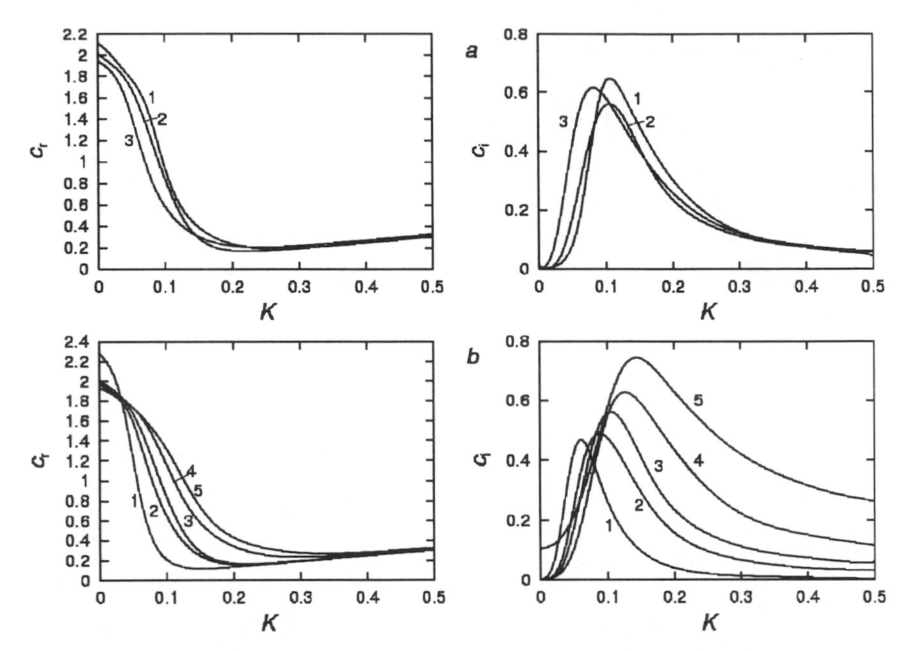

Fig. 14.17. (a) The dependencies of c_r and c_i on the noise intensity K for $\omega = 0.1$ and $A = 0.025$, $A = 0.1$ and $A = 0.2$ (curves 1, 2 and 3, respectively); (b) the same for $A = 0.1$, $\omega = 0.01$, $\omega = 0.05$, $\omega = 0.1$, $\omega = 0.2$ and $\omega = 0.3$ (curves 1, 2, 3, 4 and 5, respectively)

$H_2\big(n/\sqrt{2m_2}\big)$, $H_3\big(n/\sqrt{2m_2}\big)$, $H_4\big(n/\sqrt{2m_2}\big)$, ..., and integrating over $y = n/\sqrt{2m_2}$ from $-\infty$ to ∞, we obtain the following equations:

$$
\begin{aligned}
\dot{m}_2 &= 2\Big((1 - 3s^2 - 3m_2)m_2 - 3sm_3 - b_4\Big) + K \\
\dot{m}_3 &= 3\Big((1 - 3s^2 - 9m_2)m_3 - 6sm_2^2 - 3sb_4 - b_5\Big), \\
\dot{b}_4 &= 4\Big((1 - 3s^2 - 12m_2)b_4 - 18sm_2m_3 - 6m_2^3 + m_3^2 - 3sb_5 - b_6\Big), \\
\dot{b}_5 &= 5\Big((1 - 3s^2 - 15m_2)b_5 - 24sm_2b_4 + 5m_3b_4 - 36m_2^2m_3 \\
&\quad + 12sm_3^2 - 3sb_6 - b_7\Big), \\
\dot{b}_6 &= 6\Big((1 - 3s^2 - 18m_2)b_6 - 30sm_2b_5 + m_3b_5 - 60m_2^2b_4 \\
&\quad + 5b_4^2 + 18sm_3b_4 - 60sm_2^2m_3 - 3sb_7 - b_8\Big), \\
&\ldots .
\end{aligned}
\tag{14.131}
$$

Equations (14.131) should be solved in combination with Eq. (14.127).

The results obtained for $A = 0.1$, $\omega = 0.1$, $N = 4$ are demonstrated in Fig. 14.18, where the corresponding dependencies found by numerical simulation of the initial equation (14.118) are also given. It is seen that the results

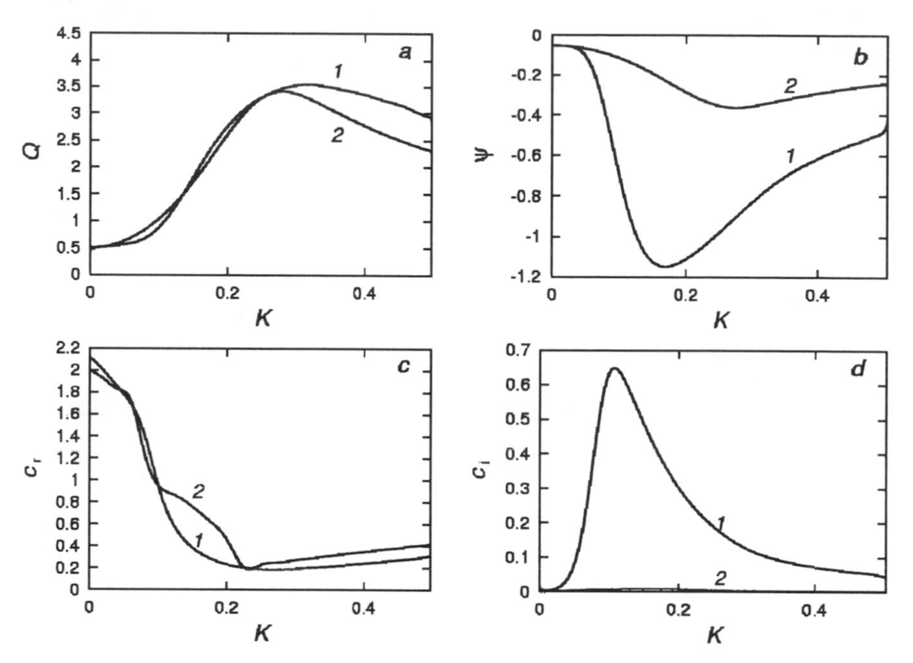

Fig. 14.18. The dependencies of (a) Q, (b) ψ, (c) c_r and (d) c_i on the noise intensity K for $\omega = 0.1$ and $A = 0.1$ found by numerical simulation of Eq. (14.118) (curves 1) and of Eq. (14.127) in combination with the first three equations (14.131) (curves 2)

obtained by the two methods for $Q(K)$ and $c_r(K)$ are in close agreement, whereas for $\psi(K)$ and $c_i(K)$ they differ markedly in a quantitative sense.

To support our conclusion about the mechanism of stochastic resonance, we simulated the equation for a nonlinear monostable oscillator

$$\dot{x} - x + x^3 - 0.4 = A\cos\omega t + \xi(t). \qquad (14.132)$$

This equation has only one singular point $x \approx 1.1597$. Numerical simulation of Eq. (14.132) shows that, as differentiated from the bistable oscillator, the response of this system to the input signal is essentially independent of the signal amplitude. An example of the dependencies of Q, ψ, c_r and c_i on the noise intensity K is given in Fig. 14.19 for $A = 0.1$, $\omega = 0.1$. We see that these dependencies possess extrema much as for the bistable oscillator. The corresponding theoretical dependencies, which are obtained from the first five equations (14.131) and the equation for $s(t)$

$$\dot{s} - s + s^3 + 3m_2 s + m_3 - 0.4 = A\cos\omega t, \qquad (14.133)$$

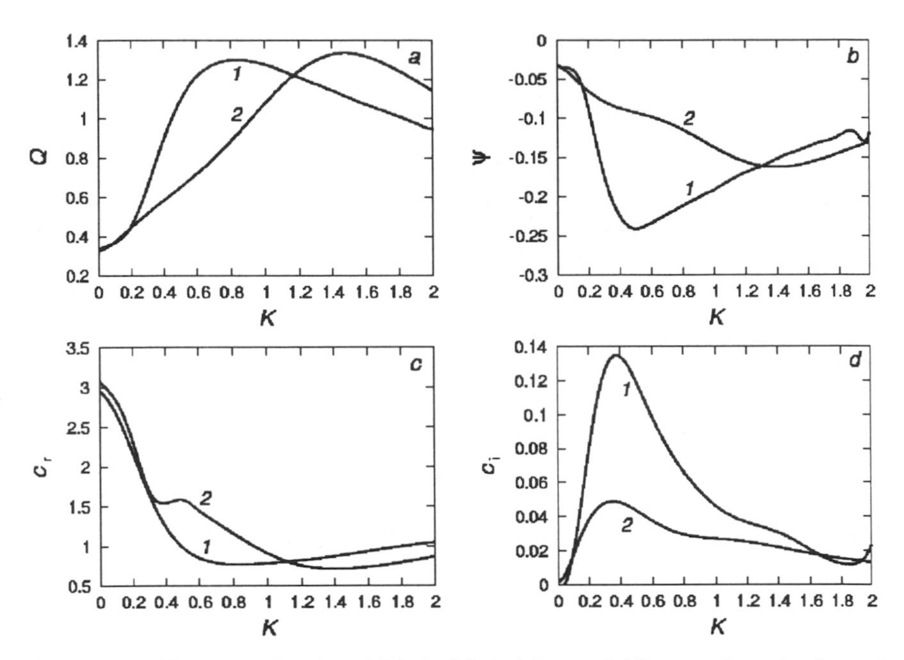

Fig. 14.19. The dependencies of (a) Q, (b) ψ, (c) c_r and (d) c_i on the noise intensity K for $\omega = 0.1$ and $A = 0.1$ found by numerical simulation of Eq. (14.132) (curves 1) and of the first five equations (14.131) in combination with Eq. (14.133) (curves 2)

are presented in the same figure. It can be seen that numerical and theoretical dependencies coincide qualitatively but differ quantitatively.

14.4.2 Vibrational resonance in an overdamped oscillator

As mentioned above, the change of the complex effective stiffness can be induced not only by noise but also by a high-frequency force. We consider this case first by the example of an overdamped bistable oscillator described by the equation

$$\dot{x} - x + x^3 = A \cos \omega t + C \cos \Omega t, \tag{14.134}$$

where $\Omega \gg \omega$. Computing Eq. (14.134) and extracting the cosine and sine constituents of the output signal at the frequency ω by using the technique described in the preceding section, we find the dependencies of Q and ψ on C for a fixed value of Ω and different values of A and ω. Examples of such dependencies are given in Fig. 14.20. In addition, in Fig. 14.20 c are shown the dependencies of the value $C_{\rm m}$ corresponding to the maximum of gain factor Q on A at a fixed value of ω and on ω at a fixed value of A. We see that the dependencies found coincide qualitatively with those for noise.

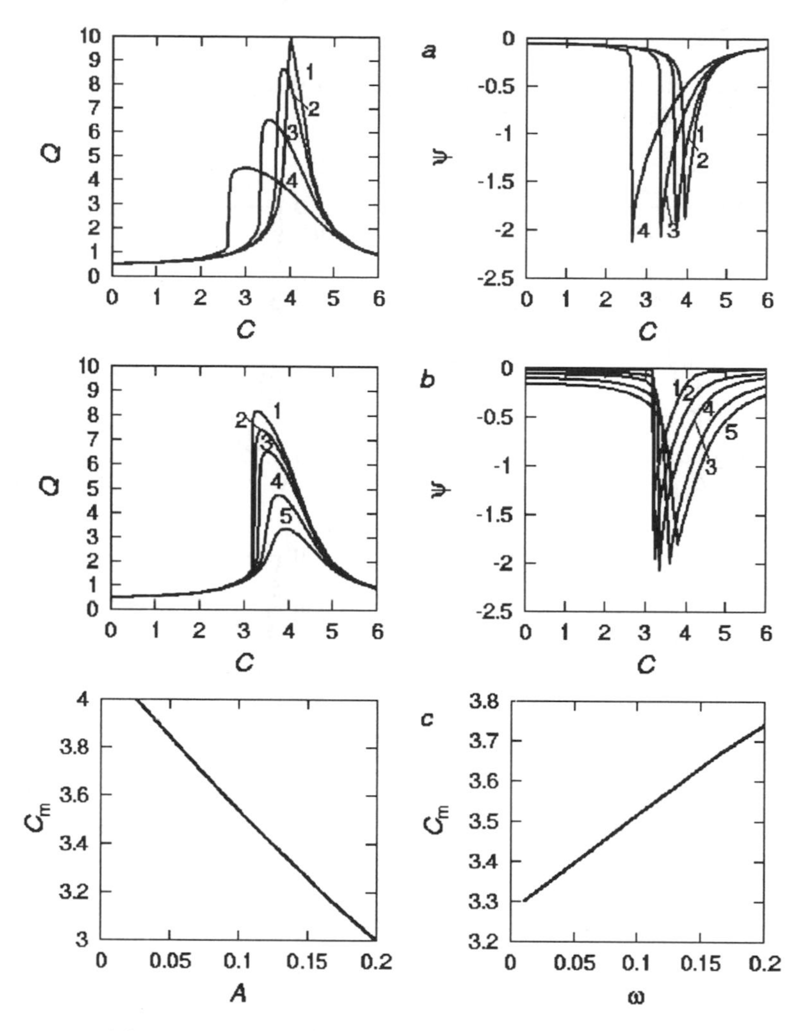

Fig. 14.20. (a) The dependencies of Q and ψ on the high-frequency vibration amplitude C for $\Omega = 5$, $\omega = 0.1$ and $A = 0.025$, $A = 0.05$, $A = 0.1$ and $A = 0.2$ (curves 1, 2, 3 and 4, respectively); (b) the same for $A = 0.1$, $\omega = 0.01$, $\omega = 0.05$, $\omega = 0.1$, $\omega = 0.2$ and $\omega = 0.3$ (curves 1, 2, 3, 4 and 5, respectively); (c) the plots of C_{m} versus A at $\omega = 0.1$ (on the left) and versus ω at $A = 0.1$ (on the right)

This strengthens our assumption that the reason for the phenomenon under consideration is the change of the effective stiffness.

14.4.3 Stochastic and vibrational resonances in a weakly damped bistable oscillator. Control of resonance

Let us consider a weakly damped bistable oscillator described by the equation

$$\ddot{x} + 2\delta\dot{x} - x + x^3 = A\cos\omega t + \xi(t). \tag{14.135}$$

We note that an example of such an oscillator is a pendulum placed between the opposite poles of a magnet (see Ch. 4).

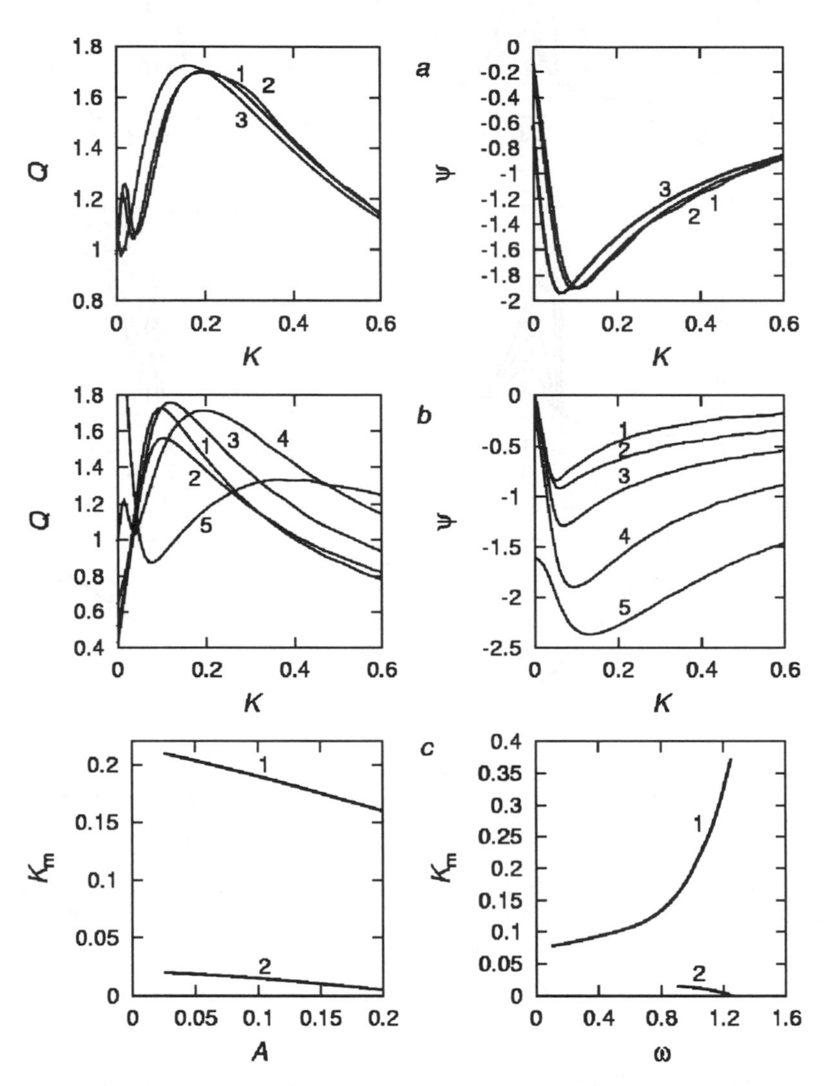

Fig. 14.21. (a) The dependencies of Q and ψ on the noise intensity K for $\delta = 0.1$, $\omega = 1$ and $A = 0.025$, $A = 0.1$ and $A = 0.2$ (curves 1, 2 and 3, respectively); (b) the same for $A = 0.1$, $\omega = 0.25$, $\omega = 0.5$, $\omega = 0.75$, $\omega = 1$ and $\omega = 1.25$ (curves 1, 2, 3, 4 and 5, respectively); (c) the plots of K_{m} versus A at $\omega = 1$ (on the left) and versus ω at $A = 0.1$ (on the right) for both the maxima (curves 1 and 2)

The change of the effective stiffness, with increasing noise intensity (or the amplitude of high-frequency vibration), means that the effective natural frequency of the oscillator described by Eq. (14.135) has to change too. It follows from this that the response of such an oscillator to the input signal $A \cos \omega t$ has to depend on the noise intensity. Numerical simulation of Eq. (14.135) shows that this is indeed so. The results of the simulation are represented in Fig. 14.21 for different values of A and ω. We see that the dependencies of Q and ψ on the noise intensity for different values of A and a fixed value of ω and for different values of ω and a fixed value of A are distinct from those for an overdamped oscillator: over a certain range of ω (in the vicinity of the frequency of small free oscillations) these dependencies have two maxima. For one of them the plots of K_{m} versus A for a fixed value of ω and versus ω for a fixed value of A are of the same character as for an overdamped oscillator (curves 1), whereas for the other maximum, existing for small K, the plot of K_{m} versus ω is decreasing but not increasing as for an overdamped oscillator (curves 2). The maximum existing for small K is caused by the not great decrease of natural frequency due to oscillations inside one of the potential wells, whereas the other maximum is caused by the more significant decrease of natural frequency due to jumps from one well to another.

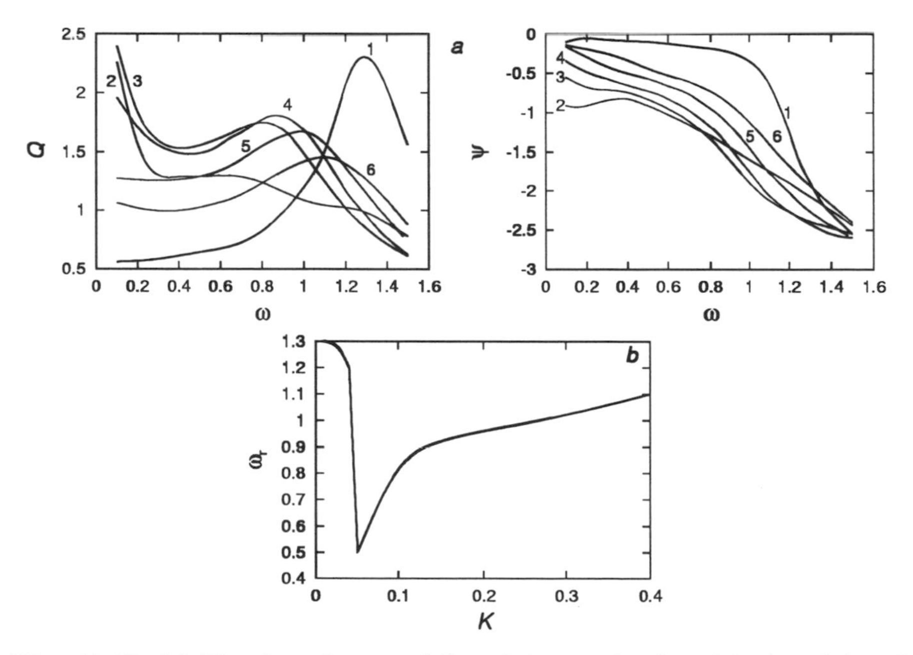

Fig. 14.22. (a) The dependencies of Q and ψ on ω for $\delta = 0.1$, $A = 0.1$ and $K = 0.01$, $K = 0.05$, $K = 0.1$, $K = 0.15$, $K = 0.25$ and $K = 0.4$ (curves 1, 2, 3, 4, 5 and 6, respectively); (b) the dependence of ω_{r} on K

Figure 14.22 a illustrates the dependencies $Q(\omega)$ and $\psi(\omega)$ for a fixed value of A and different values of K. For low frequencies these dependencies are similar to those for an overdamped oscillator, but for the frequencies close to the frequency of small free oscillations there is essential distinction: the dependencies are of a resonance character. The resonance frequency ω_r depends on K; as K increases ω_r first abruptly decreases but then it slowly increases (Fig. 14.22 b).

Similar results are obtained if we use high-frequency vibration in place of noise (see Figs. 14.23 and 14.24). We see that the behavior of the dependencies of Q and ψ on the high-frequency vibration amplitude C essentially depends on the frequency ω (compare Fig. 14.23 a and b): as for the case of noise, there are two resonances for ω close to the frequency of small free oscillations (Fig. 14.23 b). The mechanism of these two resonances is the same as for noise.

As C or K increases, the resonance frequency first decreases and then increases (see Figs. 14.24 c and 14.22 b). However, in the case of noise the decrease of the resonance frequency occurs significantly more sharply than in the case of high-frequency vibration. Thus, we can control the resonance frequency ω_r by changing either the noise intensity or the amplitude of high-frequency vibration. It follows from Figs. 14.22 b and 14.24 c that the effectiveness and smoothness of the control are more in the case of the high-frequency vibration than of noise.

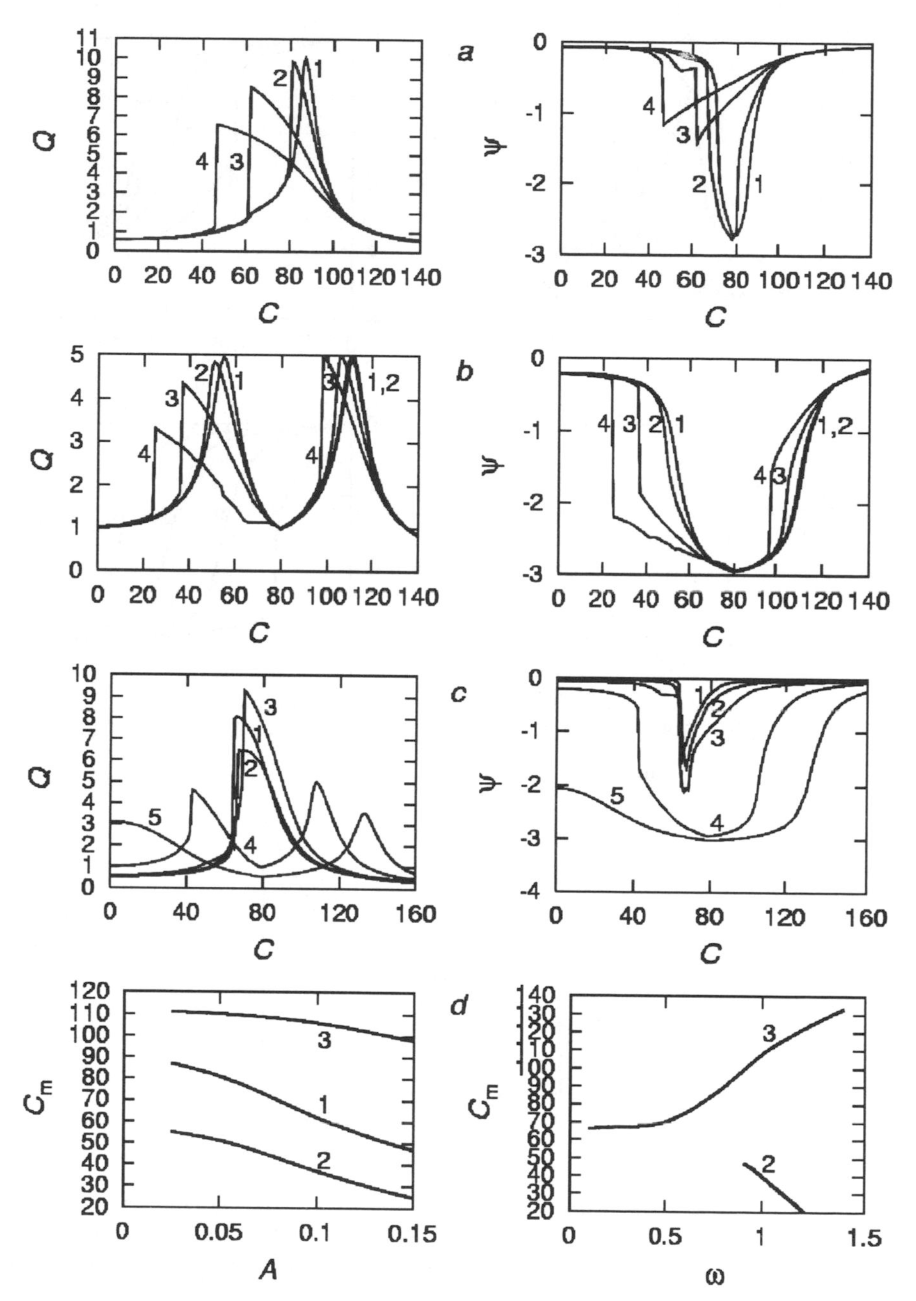

Fig. 14.23. (a, b) The dependencies of Q and ψ on the high-frequency vibration amplitude C for $\delta = 0.1$, $\Omega = 9.842$, (a) $\omega = 0.5$, (b) $\omega = 1$, and $A = 0.025$, $A = 0.05$, $A = 0.1$ and $A = 0.15$ (curves 1, 2, 3 and 4, respectively); (c) the same for $A = 0.08$, $\omega = 0.1$, $\omega = 0.25$, $\omega = 0.5$, $\omega = 1$ and $\omega = 1.4$ (curves 1, 2, 3, 4 and 5, respectively); (d) the plots of $C_{\rm m}$ versus A (on the left) for $\omega = 0.5$ (curve 1) and $\omega = 1$ (curves 2 and 3), and versus ω at $A = 0.08$ (on the right)

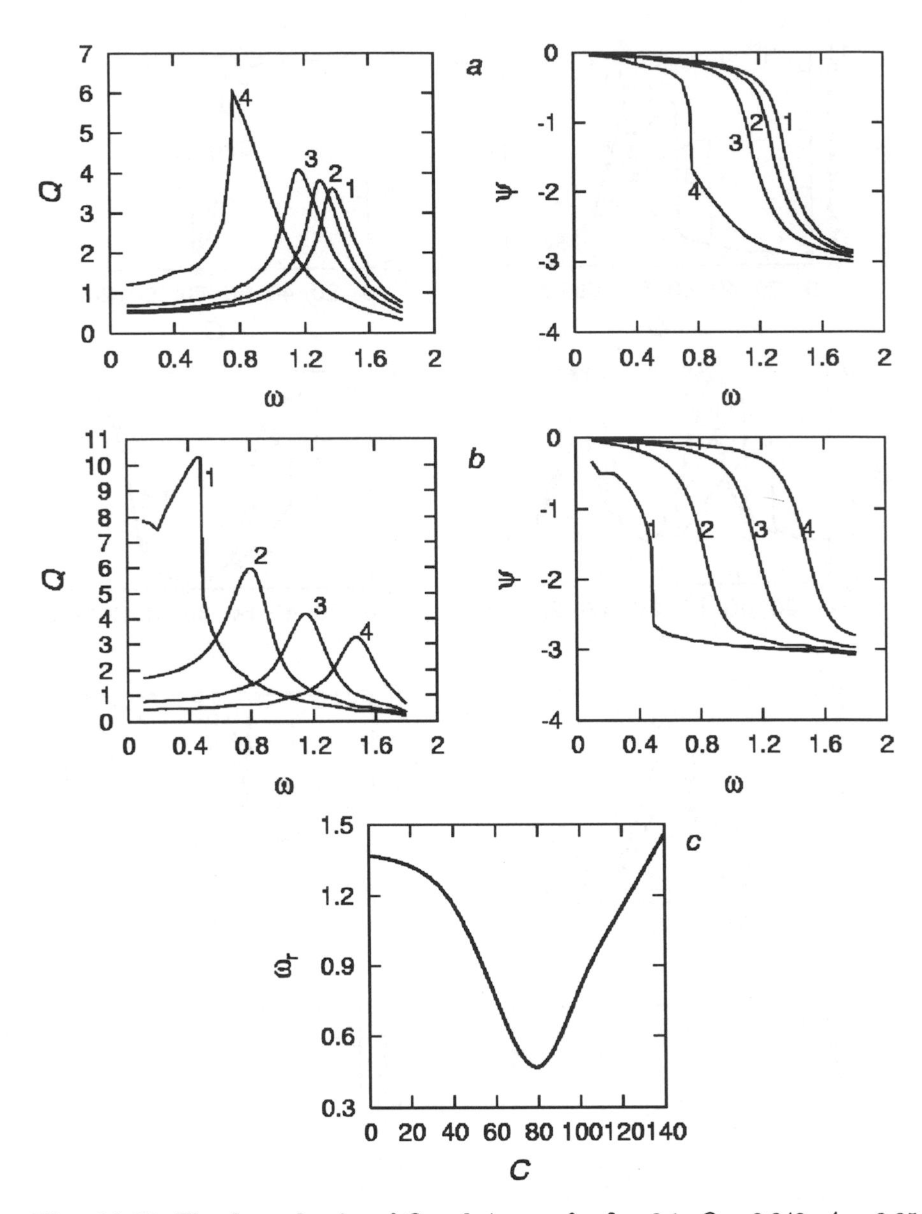

Fig. 14.24. The dependencies of Q and ψ on ω for $\delta = 0.1$, $\Omega = 9.842$, $A = 0.05$ and (a) $C = 0$, $C = 25$, $C = 40$ and $C = 60$ (curves 1, 2, 3 and 4, respectively), (b) $C = 80$, $C = 100$, $C = 120$ and $C = 140$ (curves 1, 2, 3 and 4, respectively); (c) plot of the resonance frequency ω_r versus C

A. Derivation of the approximate equation for the one-dimensional probability density

Let us consider an equation

$$\epsilon^2 \ddot{x} + \dot{x} + F(x) = \xi(t), \tag{A.1}$$

where $\xi(t)$ is white noise of zero mean and intensity K. Equation (A.1) can be rewritten in the form of two equations of the first order:

$$\epsilon \dot{x} = y, \quad \epsilon \dot{y} = -\frac{y}{\epsilon} - F(x) + \xi(x, t). \tag{A.2}$$

The two-dimensional Fokker–Planck equation associated with Eqs. (A.2) is

$$\epsilon^2 \frac{\partial w(x, y, t)}{\partial t} = -\epsilon \left(y \frac{\partial w}{\partial x} - F(x) \frac{\partial w}{\partial y} \right) + \frac{\partial (yw)}{\partial y} + \frac{K}{2} \frac{\partial^2 w}{\partial y^2}. \tag{A.3}$$

Let us seek a solution of Eq. (A.3) in the form of the following expansion:

$$w(x, y, t) = \sum_{n=0}^{\infty} \epsilon^n w_n(x, t) Y_n(y), \tag{A.4}$$

where $Y_n(y)$ are the eigenfunctions of the boundary-value problem described by the equation

$$\frac{K}{2} \frac{\mathrm{d}^2 Y}{\mathrm{d}y^2} + \frac{\mathrm{d}(yY)}{\mathrm{d}y} + \lambda Y = 0 \tag{A.5}$$

with the boundary conditions $Y(\pm\infty) = 0$. As can easily be shown, the eigenvalues of this problem $\lambda_n = n$, where $n = 0, 1, 2, \ldots$, and the eigenfunctions can be expressed in terms of the Hermite polynomials $H_n(z)$ as

$$Y_n(y) = \frac{(-1)^n}{\sqrt{\pi K 2^n n!}} \mathrm{e}^{-y^2/K} H_n \left(\frac{y}{\sqrt{K}} \right). \tag{A.6}$$

Substituting into (A.6) the expression for the Hermite polynomial we obtain

$$Y_n(y) = \sqrt{\frac{K^{n-1}}{\pi 2^n n!}} \frac{\mathrm{d}^n}{\mathrm{d}y^n} \left(\mathrm{e}^{-y^2/K} \right). \tag{A.7}$$

It can be shown that the functions $Y_n(y)$ satisfy the following orthogonality and normalization conditions:

$$\int_{-\infty}^{\infty} \frac{Y_n(y)Y_m(y)}{Y_0(y)}\, \mathrm{d}y = \delta_{nm}. \tag{A.8}$$

We substitute (A.4) into Eq. (A.3) taking into account the following relationships:[1]

$$\begin{aligned} \frac{\mathrm{d}Y_n(y)}{\mathrm{d}y} &= \sqrt{\frac{2(n+1)}{K}}\, Y_{n+1}(y), \\ yY_n(y) &= -\sqrt{\frac{K}{2}}\Big(\sqrt{n+1}Y_{n+1}(y) + \sqrt{n}Y_{n-1}(y)\Big), \\ \frac{\mathrm{d}\Big(yY_n(y)\Big)}{\mathrm{d}y} &= -\Big(\sqrt{(n+1)(n+2)}Y_{n+2}(y) + nY_n(y)\Big). \end{aligned} \tag{A.9}$$

As a result, we find

$$\begin{aligned} \epsilon^2 \sum_{n=0}^{\infty} \epsilon^n Y_n \frac{\partial w_n}{\partial t} = \sum_{n=0}^{\infty} \epsilon^n \Bigg[&\epsilon\sqrt{\frac{K}{2}}\Big(\sqrt{n+1}Y_{n+1} + \sqrt{n}Y_{n-1}\Big)\frac{\partial w_n}{\partial x} \\ &+ \epsilon F(x)\sqrt{\frac{2(n+1)}{K}}\, Y_{n+1}w_n - nY_n w_n \Bigg]. \end{aligned} \tag{A.10}$$

Equating the terms of $Y_n(y)$ with the same subscripts, we obtain the following equations:

$$\begin{aligned} \epsilon^2 \frac{\partial w_n}{\partial t} &= \sqrt{\frac{K}{2}}\left(\sqrt{n}\,\frac{\partial w_{n-1}}{\partial x} + \epsilon^2\sqrt{n+1}\,\frac{\partial w_{n+1}}{\partial x}\right) \\ &\quad + F(x)\sqrt{\frac{2n}{K}}\, w_{n-1} - n w_n. \end{aligned} \tag{A.11}$$

For $n \leq 4$ these equations are

$$\frac{\partial w_0}{\partial t} = \sqrt{\frac{K}{2}}\frac{\partial w_1}{\partial x}, \tag{A.12}$$

$$\epsilon^2 \frac{\partial w_1}{\partial t} = \sqrt{\frac{K}{2}}\frac{\partial w_0}{\partial x} + \epsilon^2\sqrt{K}\,\frac{\partial w_2}{\partial x} + \sqrt{\frac{2}{K}}\,F(x)w_0 - w_1, \tag{A.13}$$

$$\epsilon^2 \frac{\partial w_2}{\partial t} = \sqrt{K}\,\frac{\partial w_1}{\partial x} + \epsilon^2\sqrt{\frac{3K}{2}}\frac{\partial w_3}{\partial x} + \frac{2}{\sqrt{K}}\,F(x)w_1 - 2w_2, \tag{A.14}$$

$$\epsilon^2 \frac{\partial w_3}{\partial t} = \sqrt{\frac{3K}{2}}\frac{\partial w_2}{\partial x} + \epsilon^2\sqrt{2K}\,\frac{\partial w_4}{\partial x} + \sqrt{\frac{6}{K}}\,F(x)w_2 - 3w_3, \tag{A.15}$$

$$\epsilon^2 \frac{\partial w_4}{\partial t} = \sqrt{2K}\,\frac{\partial w_3}{\partial x} + \epsilon^2\sqrt{\frac{5K}{2}}\frac{\partial w_5}{\partial x} + 2\sqrt{\frac{2}{K}}\,F(x)w_3 - 4w_4. \tag{A.16}$$

[1] These relationships follow from the properties of Hermite polynomials.

Putting in Eqs. (A.11) $w_i = w_{i0} + \epsilon^2 w_{i1} + \epsilon^4 w_{i2} + \epsilon^6 w_{i3} + \ldots$ $(i = 1, 2, 3, \ldots)$, we can find sequentially the functions w_{10}, w_{11}, w_{12}, ..., w_{1n}, ... The calculations show that for $n \geq 1$ these functions can be expessed as

$$w_{1n} = \Phi_1^{(n)} \frac{\partial^{2n-2} w_{10}}{\partial x^{2n-2}} + \Phi_2^{(n)} \frac{\partial^{2n-3} w_{10}}{\partial x^{2n-3}} + \ldots + \Phi_{2n-1}^{(n)} w_{10}, \tag{A.17}$$

where $\Phi_k^{(n)}$ are functions of $F(x)$ and its derivatives, and

$$w_{10} = \sqrt{\frac{2}{K}} \left(\frac{K}{2} \frac{\partial w_0}{\partial x} + F(x) w_0 \right). \tag{A.18}$$

Substituting

$$w_1 = \sum_{n=0}^{\infty} \epsilon^{2n} w_{1n} \tag{A.19}$$

into Eq. (A.12) and using the fact that

$$w(x,t) = \int_{-\infty}^{\infty} w(x,y,t) \, \mathrm{d}y = w_0(x,t),$$

we obtain the following equation for $w(x,t)$:

$$\frac{\partial w}{\partial t} = \sqrt{\frac{K}{2}} \sum_{n=0}^{\infty} \epsilon^{2n} \frac{\partial w_{1n}}{\partial x}. \tag{A.20}$$

If ϵ^2 is sufficiently small then the series (A.19) converges and Eq. (A.20) is the exact one-dimensional equation for the probability density $w(x,t)$.

In a stationary case Eq. (A.20) becomes

$$\sqrt{\frac{K}{2}} \sum_{n=0}^{\infty} \epsilon^{2n} w_{1n} = -G, \tag{A.21}$$

where G is the probability flux. In this case the derivatives of w_{10}, which are contained in the expressions for w_{1n}, should be expanded, in their turn, as power series in ϵ^2:

$$\frac{\partial w_{10}}{\partial x} = -\epsilon^2 \Big[F''(x) + \epsilon^2 \Big(\frac{3K}{4} F^{IV}(x) + 2F(x)F'''(x) + 6F'(x)F''(x) \Big) + \ldots \Big] w_{10},$$

$$\frac{\partial^2 w_{10}}{\partial x^2} = -\epsilon^2 \Big[F'''(x) + \epsilon^2 \Big(\frac{3K}{4} F^{V}(x) + 2F(x)F^{IV}(x) + 8F'(x)F'''(x) + 6\big(F''(x)\big)^2 \Big) + \ldots \Big] w_{10},$$

....

References

1. Afraimovich V.S., Rabinovich M.I., Ugodnikov A.D. (1983) Critical points and 'phase transitions' in the stochastic behaviour of a nonautonomous anharmonic oscillator. JETP Lett. 38:72–75
2. Ajdari A., Prost J. (1992) Movement induit par un potential de basse symétrie: dielectrophorèse pulsée. C. R. Acad. Sci. Paris 315:1635–1639
3. Alkemade C.T.J. (1958) On the problem of Brownian motion of nonlinear systems. Physica 24:1029–1034
4. Andronov A.A., Vitt A.A. (1934) On the mathematical theory of self-oscillatory systems with two degrees of freedom (in Russian). ZhTF 4:122–136
5. Andronov A.A., Khaykin S.E. (1937) Theory of Oscillations (in Russian). Gostekhizdat, Moscow
6. Andronov A.A., Leontovich E.A. (1939) Certain causes of the dependence of limit cycles on parameters (in Russian). Uch. zap. GGU, vyp. 6, No 3, pp. 3–33
7. Andronov A.A. (1956) Collected Works (in Russian). Izd-vo AN SSSR, Moscow
8. Andronov A.A., Vitt A.A., Khaykin S.E. (1966) Theory of Oscillations. Pergamon Press, Oxford. [Russian original: Fizmatgiz, Moscow, 1959]
9. Anischenko V.S., Astakhov V.V. (1983) Bifurcations in an auto-stochastic generator with regular external excitation. Sov. Phys.–Tech. Phys. 28:1326–1329
10. Anischenko V.S., Letchford T.E., Safonova M.A. (1984) Stochasticity and the disruption of quasiperiodic motion due to doubling in a system of coupled generators. Radiophys. and Quantum Electron. 27:381–390
11. Anischenko V.S., Letchford T.E., Safonova M.A. (1985) Synchronization effects and bifurcations of synchronous and quasiperiodic oscillations in a nonautonomous generator. Radiophys. and Quantum Electron. 28:766–776
12. Anischenko V.S. (1987) Dynamical Chaos—Basic concepts. Teubner Texte, Leipzig, B. 14
13. Anischenko V.S. (1988) Dynamical Chaos in Physical Systems: Experimental Investigation of Self-Oscillating Circuits. Teubner Texte, Leipzig, B. 22
14. Anischenko V.S. (1990) Complex Oscillations in Simple Systems (in Russian). Nauka, Moscow
15. Anischenko V.S. (1995) Dynamical Chaos—Models and Experiments. World Scientific, Singapore
16. Anishchenko V.S. et al. (1999) Stochastic resonance: noise enhanced order. Phys. – Usp. 39:1128–1168
17. Arnold V.I. (1978) Mathematical Methods in Classical Mechanics. Springer, Berlin, Heidelberg. [Russian original: Nauka, Moscow, 1974]
18. Appleton E.V. (1922) The automatic synchronization of triode oscillator. Proc. Cambridge Philos. Soc. (Math. Phys. Sci.) 21
19. Aslamazov A.G., Larkin A.I. (1969) Josephson's effect in point superconductive contacts (in Russian). Pis. ZhETF 9:150–154

20. Babitzky V.I., Landa P.S. (1982) Self-excited vibrations in systems with inertial excitation. Sov. Phys.–Dokl. 27:826–827
21. Babitzky V.I., Landa P.S. (1984) Auto-oscillation systems with inertial self-excitation. ZAMM 64:329–339
22. Babitzky V.I., Landa P.S., Olkhovoy A.F., Perminov S.M. (1986) Stochastical behaviour of auto-oscillation systems with inertial self-excitation. ZAMM 66:73–81
23. Bakman M.E., Teodorchik K.F. (1936) Rectification of alternating current by oscillating thin string (in Russian). ZhTF 6:298–301
24. Bakscis B.P., Rimatis V.K. (1986) Automatic assembly by means of pneumatic hammers (in Russian). In: Avtomatizatsiya Sborochnykh Protsessov, Izd. Rizhsk. Polytechn. Inst., Riga, pp. 39–43
25. Bardakhanov S.P., Lygdenov V.S. (1990) Coherent structures in wake beyond a slightly streamlined body and sound generation under conditions of resonance(in Russian). Izv. SO AN SSSR, ser. tekhn. nauk, No 2:36–40
26. Bartussek R., Hänggi P., Kissner J.G. (1994) Periodically rocked thermal ratchets. Europhys. Lett. 28:459–464
27. Bartussek R. et al. (1997) Ratchets driven by harmonic and white noise. Physica D 109:17-23
28. Basov N.G., Letokhov V.S. (1968) Optical standards of frequency (in Russian). Usp. Fiz. Nauk 96:585–598
29. Batalova Z.S. (1967) On the motion of a rotor under the action of an external harmonic force (in Russian). Mekh. Tvyordogo Tela No 2:66–73
30. Bautin N.N. (1939) On the theory of synchronization (in Russian). ZhTF 9:510–513
31. Bautin N.N. (1955) Dynamical models of free clock movements (in Russian). In: To Memory of A.A. Andronov. Izd-vo AN SSSR, Moscow, pp. 109–172
32. Bellman R.E., Bentsman J., Meerkov S.M. (1986) Vibrational control of nonlinear systems: Vibrational stabilizability (1), Vibrational controllability and transient behavior (2). IEEE Trans. Autom. Control AC-31, No 8
33. Benettin G., Galgani L. (1979) Lyapunov characteristic exponents and stochasticity. In: Intrinsic Stochasticity in Plasma, ed. Laval G., Gressilon D. Orsay, Les éditions de physique courtaboeuf, pp. 93–114
34. Benettin G., Galgani L., Giorgilli A., Strelcyn J.M. (1980) Lyapunov characteristic exponents for smooth dynamical systems and for Hamiltonian systems; a method for computing all of them, P.1,2. Meccanica 15:9–20, 21–30
35. Berg van Den J., Moll J. (1955) Zur Anatomie des menschlichen musculus vocalis. Z. Anat. Entwickl. 118:465–470
36. Berg van Den J., Moolenar-Bije A.J., Damste P.H. (1958) Oesophageal speech. Folia Phoniatr. 10:66–83
37. Bezaeva L.G., Kaptsov L.N., Landa P.S. (1986) Synchronization threshold as a criterion of stochasticity in a generator with inertial nonlinearity. Sov. Phys. – Tech. Phys. 31:1105–1107
38. Bezaeva L.G., Kaptsov L.N., Landa P.S. (1987) Investigation of random modulation of the oscillations of a generator with an inertial nonlinearity under parametric external action. Sov. J. Commun. Technol. & Electron. 32:169–171
39. Dean Astumian R., Bier M. (1994) Fluctuation driven ratchets: molecular motors. Phys. Rev. Lett. 72:1766–1769
40. Bier M. (1996) Reversals of noise induced flow. Phys. Lett. A 211:12–18
41. Birkhoff G.D. (1908) On the asymptotic character of the solutions of certain linear differential equations containing a large parameter. Trans. Am. Math. Soc. 9:219–231

42. Birkhoff G.D. (1927) Dynamical Systems. Am. Math. Soc., Colloquium publications, vol. 9, New York
43. Bishop R.E.D., Hassan A.Y. (1964) The lift and drag forces on a circular cylinder oscillating in a flowing fluid. Proc. R. Soc. (London) A277:51–75
44. Bisplinghoff R.L., Ashley H., Halfman R.L. (1955) Aeroelasticity. Addison-Wesley, Cambridge MA
45. Blekhman I.I., Dzhanelidze G.Yu. (1964) Vibrational Movement (in Russian). Nauka, Moscow
46. Blekhman I.I. (1971) Synchronization of Dynamical Systems (in Russian). Nauka, Moscow
47. Blekhman I.I. (1988) Synchronization in science and technology. ASME Press, New York. [Russian original: Nauka, Moscow, 1981]
48. Blekhman I.I., Landa P.S., Rosenblum M.G. (1995) Synchronization and chaotization phenomena in oscillatory and rotatory dynamical systems. In: Nonlinear Dynamics: New Theoretical and Appllied Results, ed. Awrejcewicz J. Akademie Verlag, Berlin, pp. 17-54
49. Blekhman I.I., Landa P.S., Rosenblum M.G. (1995) Synchronization and chaotization in interacting dynamical systems. Appl. Mech. Rev. 48:733–752
50. Blekhman I.I. (2000) Vibrational Mechanics. World Scientific, Singapore. [Russian original: Nauka, Moscow, 1994]
51. Blumina L.Kh., Fedyaevsky K.K. (1969) Studies of the influence of forced oscillations of a cylinder in an air flow on the mechanism of vortex separation (in Russian). Izv. AN SSSR, Mekh. Zhidk. Gaza, No 1:118–119
52. Bogolyubov N.N. (1950) Perturbation Theory in Nonlinear Mechanics (in Russian). Sborn. Inst. Stroit. Mekh. AN USSR, No 14:9–34
53. Bogolyubov N.N., Mitropol'sky Yu.A. (1961) Asymptotic Methods in the Theory of Nonlinear Oscillations. Gordon and Breach, New York
54. Braginsky V.B. and Minakova I.I. (1964) Influence of the system of measuring small displacement on dynamical properties of mechanical oscillatory systems (in Russian). Vestn. MGU, Ser. 3, No 1:83–85
55. Braginsky V.B. (1970) Physical Experiments with Trial Bodies (in Russian). Nauka, Moscow, pp. 29–34
56. Braginsky V.B., Manukin A.B., Tikhonov M.Yu. (1970) Study of dissipative ponderomotive effects of electromagnetic radiation. Sov. Phys. - JETP 31:829–832
57. Braginsky V.B., Manukin A.B. (1974) Measuring Small Forces in Physical Experiments (in Russian). Nauka, Moscow, pp. 29–36
58. Brillouin L. (1950) Can the rectifier become a thermodynamical demon? Phys. Rev. 78:627–628
59. Broad D. (1979) The new theories of vocal fold vibration. In: Speech and Language: Advances in Basic Research and Practice, ed. Lass N. Academic Press, New York
60. Broomhead D.S., King G.P. (1986) Extracting qualitative dynamics from experimental data. Physica D 20:217–236
61. Budinsky N., Bounties T. (1983) Stability of nonlinear modes and chaotic properties of $1D$ Fermi–Pasta–Ulam lattices. Physica D 8:445–452
62. Bulgakov B.V. (1954) Oscillations (in Russian). Gostekhizdat, Moscow
63. Bumjalene S., Piragas K., Cenis A. (1990) Study of strange attractor's dimension and amplitude synchronization threshold of chaotic self-oscillations of photocurrent in n-type $Ge(Ni)$. Sov. Phys. – Semicond. 24:945–948
64. Butenin N.V., Neimark Yu.I., Fufaev N.A. (1987) Introduction to the Theory of Nonlinear Oscillations (in Russian). Nauka, Moscow

65. Cartwright M.L., Littlewood J.E. (1945) On nonlinear differential equations of the second order. I. The equation $\ddot{y} - k(1 - y^2)\dot{y} + y = b\lambda k \cos(\lambda t + \alpha)$, k large. J. London Math. Soc. 20:180–189
66. Chirikov B.V., Shepelyansky D.L. (1981) Stochastic oscillations of classical Yang–Mills fields (in Russian). Pis. ZhETF, 34:171–175
67. Derényi I., Tegzes P., Vicsek T. (1998) Collective transport in locally asymmetric periodic structures. Chaos 8:657–664
68. Dietz K. (1976) The incidence of infectious diseases under the influence of seasonal fluctuations. Lect. Notes Biomath. 11:1–15
69. Dimentberg M.F. (1969) On the problem of distinguishing between random forced oscillations and self-oscillations. Mekh. Tvyordogo Tela, No 6
70. Dimentberg M.F. (1980) Nonlinear Stochastic Problems of Mechanical Oscillations (in Russian). Nauka, Moscow
71. Dmitriev A.S., Kislov V.Ya. (1982) Strange attractor in a nonautonomous van der Pol equation. Radio Eng. Elektron. Phys. 27:154–156
72. Dmitriev A.S., Kislov V.Ya., Spiro A.G. (1983) Chaotic oscillations in a nonautonomous generator with a reactive nonlinearity. Radio Eng. Elektron. Phys. 28:115–124
73. Doering C.R., Horsthemke W., Riordan J. (1994) Nonequilibrium fluctuation-induced transport. Phys. Rev. Lett. 72:2984–2987
74. Doering C.R. (1995) Randomly rattled ratchets. Nuovo Cimento Soc. Ital. Fiz. 17D:685–697
75. Druzhilovskaya T.Ya., Neimark Yu.I. (1982) Stochastic self-oscillations of a nonlinear oscillator with impact energy absorber. J. Appl. Math. Mech. 46:740–745
76. Duffing G. (1918) Erzwungene Schwingungen bei veränderlicher Eigenfrequenz und ihre technische Bedeutung. Braunschweig, Vieweg
77. Dykman G.I., Landa P.S., Neimark Yu.I. (1991) Synchronizing the chaotic oscillations by external force. Chaos, Solitons & Fractals 1:339–353
78. Ebeling W., Erdmann U., Dunkel J., Jenssen M. (2000) Nonlinear dynamics and fluctuations of dissipative Toda chains. J. Stat. Phys. 101, No 12
79. Eckhaus W. (1979) Asymptotic Analysis of Singular Perturbations. North-Holland, Amsterdam
80. Engbert R., Drepper F.R. (1994) Chance and chaos in population biology – models of recurrent epidemics and food chain dynamics. Chaos, Solitons & Fractals 4:1147–1169
81. Erdelyi A. (1934) Über die kleinen Schwingungen eines Pendels mit oszillierenden Aufhangepunkt. ZAMM 14:
82. Esche R. (1952) Untersuchung der Schwingungskavitation in Flüssigkeiten. Akust. Beih. 4:208–234
83. Faucheux L.P. et al. (1995) Optical thermal ratchet. Phys. Rev. Lett. 74:1504–1507
84. Fedyaevsky K.K., Blyumina L.Kh. (1977) Hydrodynamics of separated flows around bodies (in Russian). Mashinostroyenie, Moscow
85. Feigenbaum M.J. (1978) Quantitative universality for a class of nonlinear transformations. J. Stat. Phys. 19:25–52
86. Feigenbaum M.J. (1979) The onset spectrum of turbulence. Phys. Lett. 74a:375–378
87. Fermi E., Pasta I., Ulam S. (1955) Studies of non-linear problems. Los Alamos Sci. Lab. Rep. LA–1940
88. Feynman R.P., Leighton R.B., Sands M. (1963) The Feunman Lectures on Physics. Addison-Wesley Publ., Inc., Reading, Massachusetts, Palo Alto, London, vol. 1, Chap. 46.

89. Flanagan J.L. (1972) Speech Analysis, Synthesis and Perception. Springer, New York
90. Flashka H. (1974) On the Toda lattice II. Prog. Theor. Phys. 51:703–716
91. Floquet G. (1883) Sur les équations diférentielles linéarites à coefficients périodiques. Ann. sci. Éc. Norm. Supér., ser. 2 12:47–48
92. Fröman N., Fröman P.O. (1965) JWKB Approximation. North-Holland, Amsterdam
93. Fujisaka H. (1984) Theory of diffusion and intermittency in chaotic system. Prog. Theor. Phys. 71:513–523
94. Fujisaka H., Yamada T. (1985) A new intermittency in coupled dynamical systems. Prog. Theor. Phys. 74:918–921
95. Fukunaga K. (1972) Introducton to Statistical Pattern Recognition. Academic Press, New York
96. Gabaraev F.A., Emelyanov M.A., Kryukov B.I. (1987) Dynamics of electromagnetic vibrators with asynchronous excitation. Asynchronous self-parametric excitation (in Russian). Vibrotechnika 2(59):14–21
97. Galerkin B.G. (1952) Collected Works, vol. 1 (in Russian). Izd-vo AN SSSR, Moscow
98. Gammaitoni L. et al. (1998) Stochastic resonance. Rev. Mod. Phys. 70:223–287
99. Gaponov-Grekhov A.V., Rabinovich M.I., Starobinets I.M. (1984) Dynamical model of the spatial development of turbulence. JETP Lett. 39:688–691
100. Göber M., Herzel H., Graf H.-F. (1992) Dimension analysis of El Ninõ/-Southern oscillation time series. Ann. Geophys. 10:729–734
101. Goncharevich I.F. (1972) Dynamics of Vibrational Transport (in Russian). Nauka, Moscow
102. Gorre-Talini L., Spatz J.P., Silberzan P. (1998) Dielectrophoretic ratchets. Chaos 8:650–656
103. Gortinsky V.V., Demsky A.B., Boriskin M.A. (1973) Processes of Separation at Plants for Working up Corn (in Russian). Kolos, Moscow
104. Grasman J., Nijmeijer H., Veling E.J.M. (1984) Singular perturbations and a mapping on an interval for the forced van der Pol relaxation oscillator. Physica D 13:195–210
105. Grassberger P. (1981) On the Hausdorff dimension of fractal attractors. J. Stat. Phys. 26:173–179.
106. Grassberger P., Procaccia I. (1983) Measuring the strangeness of strange attractors. Physica D 9:189–208.
107. Grassberger P. (1983) Generalized dimension of strange attractors. Phys. Lett. 97A:227–231
108. Grassberger P. (1988) Generalized dimension of strange attractors. Phys. Rev. A 38:1649–1652
109. Gribkov D.A. et al. (1994) Reconstruction of differential equations for autostochastic systems from a time series for a single dynamical variable of the process (in Russian). ZhTF 64:1–12
110. Grossman E.P. (1937) Flutter (in Russian). Trudy TSAGI, No 284
111. Guckenheimer J. (1980) Symbolic dynamics and relaxation oscillations. Physica D 1:227–235
112. Gutman I. (1968) Industrial Uses of Mechanical Vibrations. London
113. Hänggi P., Bartussek R. (1996) In: Nonlinear Physics and Complex Systems – Current Status and Future Trends, Lect. Notes in Phys., vol. 476, ed. Parisi J., Müller S.C., Zimmermann W. Springer-Verlag, Berlin, pp. 294–308
114. Hausdorff F. (1918) Dimension und Äußeres Maß. Math. Ann. 79:157–179
115. Hayashi C. (1964) Nonlinear Oscillations in Physical Systems. McGraw-Hill, New York

116. Heagy J.F., Platt N., Hammel S.M. (1994) Characterization of on–off intermittency. Phys. Rev. E 49:1140–1150
117. Henon M., Heiles C. (1964) The applicability of the third integral of motion; some numerical experiments. Astron. J. 69:73–79
118. Henon M. (1974) Integrals of the Toda lattice. Phys. Rev. B 9:1921–1923
119. Hentschel H.G.E., Procaccia I. (1983) The infinite number of generalized dimensions of fractals and strange attractors. Physica D 8:435–444
120. Herzel H., Kurths J., Landa P.S., Rosenblum M.G. (1989) New aspects of detecting chaos in a time series. In: Irreversible Processes and Selforganization, ed. Ebeling W., Ulbricht H., B. 23. Teubner, Leipzig, pp. 65–75
121. Herzel H. (1993) Bifurcations and chaos in voice signals. Appl. Mech. Rev. 46:399–413
122. Hirsch J.E., Huberman B.A., Scalapino D.J. (1982) Theory of intermittency. Phys. Rev. A 25:519–532
123. Hopf E. (1942) Abzweigung einer periodischen Lösung von einer stationaren Lösung eines differential Systems. In: Berl. Math.-Phys. Sachsische Akad. Wiss., B. 94. Leipzig, pp. 1–22
124. Horsthemke W., Lefever R. (1984) Noise-Induced Transitions. Springer-Verlag, Berlin
125. Hu B., Rudnick J. (1982) Exact solutions to the Feigenbaum renormalization-group equations for intermittency. Phys. Rev. Lett. 48:1645–1648
126. Hugenii C. (1673) Horoloqium Oscilatorium. Parisiis, France; The Pendulum Clock or Geometrical Demonstrations Concerning the Motion of Pendula as Applied to Clocks. Iowa State University Press, 1986.
127. Hurwitz A. (1895) Über die Bedingungen, unter welchen eine Gleichung nur Wurzeln mit negativen reelen Theilen besitzt. Math. Ann. 46:273–291
128. Idel'chik I.E. (1986) Handbook of Hydraulic Resistances. Hemisphere, Washington. [Russian original: Mashinostroenie, Moscow, 1975]
129. Il'chenko M.A., Rudenko A.N., Epshtein V.L. (1980) Generation of vortex noise in flow past a profile in a channel. Sov. Phys. – Acoust. 26:400–405
130. Il'chenko M.A., Rudenko A.N., Selin N.I. (1982) Distinctive features of the excitation of oscillations in flow past a profile in a channel. Sov. Phys. – Acoust. 28:135–137
131. Ishizaka K., Flanagan J.L. (1972) Synthesis of voiced sounds from a two-mass model of the vocal cords. Bell Syst. Tech. J. 51:1233–1268
132. Ito H. (1979) Successive subharmonic bifurcations and chaos in a nonlinear Mathieu equation. Prog. Theor. Phys. Jpn 61:815–824
133. Izrailev F.M., Rabinovich M.I., Ugodnikov A.D. (1981) Approximate description of three-dimensional dissipative systems. Phys. Lett. 86A:321–325
134. Janke E., Emde F., Lösch F. (1960) Tafeln Höherer Funktionen. Teubner, Stuttgart
135. Jeffereys H.A.B.S. (1950) Methods of Mathematical Physics. Cambridge
136. Kaidanovsky N.L., Khaikin S.E. (1933) Mechanical relaxation oscillations (in Russian). ZTF 3:91–109
137. Kalyanov E.V., Lebedev M.N. (1985) Stochastic oscillations in a system of coupled generators with inertial nonlinearity. Sov. J. Commun. Technol. Electron. 30:105–108
138. Van Kampen N.G. (1958) Thermal fluctuations in a nonlinear system. Phys. Rev. 110:319–323
139. Kaneko K. (1983) Doubling of torus. Prog. Theor. Phys. Jpn 69:1806–1810
140. Kaneko K. (1984) Oscillations and doubling of torus. Prog. Theor. Phys. Jpn 72:202–215

141. Kapitsa P.L. (1951) Dynamic stability of a pendulum with vibrated suspension axis (in Russian). ZhETF 21:588–597
142. Kapitsa P.L. (1951) Pendulum with vibrated suspension (in Russian). Usp. Fiz. Nauk 44:7–20
143. Kaplan J.L. and Yorke J.A. (1979) Chaotic behavior of multi-dimensional difference equations. Lect. Notes in Math., Berlin, Springer, vol. 730, pp. 204–227.
144. Kaptsov L.N. (1975) Appearance of spike mode operation in a nonautonomous generator with inertial nonlinearity. Radio Eng. Elektron. Phys. 20:45–49
145. Karliner M.M., Shapiro V.E., Shekhtman I.A. (1966) Instability of resonator walls under action of ponderomotive forces from electromagnetic field. Sov. Phys. – Tech. Phys. 11:1501–1511
146. Karman Th. von (1911) Über den Mechanismus des Widerstandes, den ein bewegter in einer Flüssigkeit erfährt. Göttingen Nachr. pp. 178–179
147. Katsuo N. (1961) Experimental study on sonoluminescence and ultrasonic cavitation. J. Phys. Soc. Jpn 16:1450–1459
148. Khokhlov R.V. (1954) On the capture theory for a small amplitude of the external force (in Russian). DAN SSSR 97:411–415
149. Klibanova I.M. (1969) Mutual synchronization of generators with an integer ratio of frequencies (in Russian). Radiofizika 11:1676–1681
150. Klibanova I.M., Malakhov A.N., Mal'tsev A.A. (1971) Fluctuations in multi-frequency generators (in Russian). Radiofizika 14:173–188
151. Klimontovich Yu.L., Kuryatov V.N., Landa P.S. (1966) On wave synchronization in a gas laser with a ring cavity. Sov. Phys. – JETP 24:1–7
152. Klimontovich Yu.L., Landa P.S., Lariontsev E.G. (1967) Stability of opposing-wave regime in a ring gas laser. Sov. Phys. – JETP 25:1076–1084
153. Klimontovich Yu.L. (1999) What are stochastic filtration and stochastic resonance? Phys. – Usp. 39:1169–1178
154. Koch B.P. et al. Experimental evidence for chaotic behavior of a parametrically forced pendulum. Phys. Lett. 96A:219–224
155. Koch B.P., Leven R.W. Subharmonic and homoclinic bifurcations in a parametrically forced pendulum. Physica D 16:1–13
156. Koenigs (1883) Recherches sur les substitutions uniformes. Bull. math.
157. Koenigs (1884) Recherches sur les equations functionelles. Ann. Ec. Norm.
158. Kostin I.K., Romanovsky Yu.M. (1972) Fluctuations in systems of many coupled generators (in Russian). Vestn. MGU, Ser. III, 13:698–674
159. Krylov N.M., Bogolyubov N.N. (1947) Introduction to Nonlinear Mechanics. Princeton University Press, New Jersey. [Russian original: Izd-vo AN USSR, Kiev, 1937]
160. Kuryatov V.N., Landa P.S., Lariontsev E.G. (1968) Frequency responses of a ring laser on vibrating ground. Radiophys. Quantum Electron. 11:1040–1046
161. Kuznetsov P.I., Stratonovich R.L., Tikhonov V.I. (1965) Quasi-moment functions in the theory of random processes. In: Nonlinear Transformations of Stochastic Processes. Pergamon Press, Oxford, pp. 59–63
162. Kuznetsov Yu.I., Landa P.S., Olkhovoy A.F., Perminov S.M. (1985) Relationship between the amplitude threshold of synchronization and entropy in stochastic self-excited systems. Sov. Phys. – Dokl. 30:221–223
163. Landa P.S., Stratonovich R.L. (1962) On the theory of fluctuational transitions of different systems from one stable state to another (in Russian). Vestn. MGU (Physics and Astronomy) No 1:33–45
164. Landa P.S., Tarankova N.D. (1975) Voltage-current responses of Josephson junctions. Radio Eng. Elektron. Phys. 20:82–87

165. Landa P.S., Tarankova N.D. (1976) Synchronization of a generator with modulated natural frequency. Radio Eng. Elektron. Phys. 21:34–38
166. Landa P.S. (1980) Self-Oscillations in Systems with Finite Number of Degrees of Freedom (in Russian). Nauka, Moscow
167. Landa P.S., Stratonovich R.L. (1982) Stationary probability distribution for one of the simplest strange attractors. Sov. Phys. – Dokl. 27:1032–1034
168. Landa P.S. (1983) Self-Oscillations in Continuous Systems (in Russian). Nauka, Moscow
169. Landa P.S., Olkhovoy A.F., Perminov S.M. (1983) Stochastic self-oscillations in physical systems with inertial self-excitation. Radiophys. & Quantum Electron. 26:422–427
170. Landa P.S., Stratonovich R.L. (1984) Probability characteristics of stochastic oscillations of an adjustable pendulum. Mech. Solids No 4:22–27
171. Landa P.S., Perminov S.M. (1985) Interaction of periodic and stochastic self-oscillations. Radiophys. & Quantum Electron. 28:284–287
172. Landa P.S., Stratonovich R.L. (1987) Theory of intermittency. Radiophys. & Quantum Electron. 30:53–57
173. Landa P.S. (1987) Influence of noise on the transition to chaos through intermittency. Moscow Univ. Phys. Bull. 42:19–25
174. Landa P.S. and Perminov S.M. (1987) Synchronization of chaotic oscillations in the Mackey–Glass system (in Russian). Radiofizika 30:437–439
175. Landa P.S., Chetverikov V.I. (1988) On the evaluation of the maximum Lyapunov exponent from a single experimental time series. Sov. Phys. – Tech. Phys. 33:263–268
176. Landa P.S. (1988) Transformers of high-frequency electrical oscillations in low-frequency mechanical vibrations (in Russian). Mashinovedeniye, No 6:90–95
177. Landa P.S., Rosenblum M.G. (1989) Method for evaluating the embedding dimension of an attractor from experimental results. Sov. Phys. – Tech. Phys. 34:6–10.
178. Landa P.S., Rosenblum M.G. (1989) A comparison of methods of constructing a phase space and determining the dimension of an attractor from experimental data. Sov. Phys. – Tech. Phys. 34:1229–1232
179. Landa P.S., Duboshinsky Ya.B. (1989) Self-oscillatory systems with high-frequency energy sources. Sov. Phys. – Usp. 32:723–731
180. Landa P.S., Rudenko O.V., Bakscis B.P. (1990–1991) Excitation of body self-oscillations and sound generation in gas or fluid flows Opt. & Acoust. Rev. 1:277–290
181. Landa P.S., Rosenblum M.G. (1991) Time series analysis for system identification and diagnostics. Physica D 48:232–254
182. Landa P.S. (1991) Electro-mechanical transformers of self-oscillatory kind (in Russian). Vibrotechnika No 66:129–153
183. Landa P.S., Rosenblum M.G. (1991) Study of chaotic oscillations of a bubble in liquid on action of a high-frequency sound field (in Russian). In: Trudy XI Vsesoyuznoi Akusticheskoi konf., sec. B. Moscow, pp. 133–136
184. Landa P.S., Bakscis B.P. (1992) The calculation of self-oscillations of an air cushioned body (in Russian). Vibrotekhnika No 68(4):11–18
185. Landa P.S., Rosenblum M.G. (1992) Inaccuracy in quantification of chaotic motion. Chaos, Solitons & Fractals 2:251–258
186. Landa P.S., Rosenblum M.G. (1992) Synchronization of chaotic self-oscillatory systems. Sov. Phys. – Dokl. 37:237–239
187. Landa P.S., Rosenblum M.G. (1993) Synchronization and chaotization of oscillations in coupled self-oscillating systems. Appl. Mech. Rev. 46:414–426

188. Landa P.S. (1996) A possible mechanism of the synchronization of oscillations in quasi-conservative systems (in Russian). Izv. RAN, Mekh. Tvyordogo Tela No 5:25–28
189. Landa P.S. (1996) Nonlinear Oscillations and Waves in Dynamical Systems. Kluwer Academic, Dordrecht
190. Landa P.S., Zaikin A.A. (1996) Noise-induced phase transitions in a pendulum with a randomly vibrating suspension axis. Phys. Rev. E 54:3535–3544
191. Landa P.S. (1997) Universality of oscillation theory laws. Types and role of mathematical models. Discrete Dyn. in Nat. and Soc. 1:99–110
192. Landa P.S., Zaikin A.A. (1997) Nonequilibrium noise-induced phase transitions in simple systems. JETP 84:197–208
193. Landa P.S. et al. (1997) Control of noise-induced oscillations of a pendulum with a randomly vibrating suspension axis. Phys. Rev. E 56: 1465–1470
194. Landa P.S., Zaikin A.A. (1998) Noise-induced phase transitions in nonlinear oscillators. In: AIP Conference Proceedings 465 (Computing Anticipatory Systems, CASYS'98, Liege, Belgium, 1998), pp. 419–433
195. Landa P.S. et al. (1998) On-off intermittency phenomena in a pendulum with a randomly vibrating suspension axis. Chaos, Solitons & Fractals 9:157-169
196. Landa P.S. (1998) Vocal folds as a vibro-impact system. In: Dynamics of Vibro-Impact Systems (Proc. of the Euromech Coll. 15–18 September 1998), ed. Babitsky V.I. Springer Berlin, Heidelberg, pp. 1–10
197. Landa P.S. (1998) Noise-induced transport of Brownian particles with consideration for their mass. Phys. Rev. E 58:1325–1333
198. Landa P.S., McClintock P.V.E. (2000) Changes in the dynamical behavior of nonlinear systems induced by noise. Phys. Rep. 323:1–80
199. Landa P.S., Rabinovitch A. (2000) Exhibition of intrinsic properties of certain systems in response to external disturbances. Phys. Rev. E 61:1829–1838
200. Landa P.S., Gribkov D.A., Kaplan A.Ya. (2000) Oscillatory processes in biological systems. In: Nonlinear Phenomena in Physical & Biological Sciences, ed. Malik S.K. Bangalohr, India
201. Landa P.S., Zaikin A.A., Shimansky-Geier L. (2000) Effect of the potential shape and of a Brownian particle mass on noise-induced transport. Chaos, Solitons & Fractals (in press)
202. Landa P.S., McClintock P.V.E. (2000) Vibrational resonance. J. Phys. A 33:L433–L438
203. Landau L.D., Lifshitz E.M. (1969) Mechanics. Pergamon Press, Oxford
204. Landau L.D. and Lifshitz E.M. (1965) The Theory of Elasticity (in Russian). Nauka, Moscow
205. Lauterborn W. (1990) Acoustic chaos. In: Frontiers of Nonlinear Acoustics. Proc. of the 12th ISNA, London, pp. 64–79.
206. Leven R.W., Koch B.P. (1981) Chaotic behavior of a parametrically excited damped pendulum. Phys. Lett. 86A:71–74
207. Leven R.W. et al. (1985) Experiments on periodic and chaotic motions of a parametrically forced pendulum. Physica D 16:371–384
208. Levi M. (1981) Qualitative analysis of the periodically forced relaxation oscillations. Mem. Am. Math. Soc. vol. 32, No 244
209. Levinson N. (1949) A second order differential equation with singular solutions. Ann. Math. 50:127–153
210. Lewis F.M. (1932) Vibration during acceleration through a critical speed. Trans. ASME No 3.
211. Lighthill M.J. (1952) On sound generated aerodynamically. I. General theory. Proc. R. Soc. A211:564–587

212. Lighthill M.J. (1954) On sound generated aerodynamically. II. Turbulence as a source of sound. Proc. R. Soc. A222:1–32
213. Liljencrants J. (1991) Numerical simulations of glottal flow. In: Proc. EUROSPEECH, Genova
214. Lorenz E.N. (1963) Deterministic nonperiodic flow. J. Atmos. Sci. 20:130–141
215. Lotka A.J. (1920) Undamped oscillations derived from the law of mass action. J. Am. Chem. Soc. 42:1595–1599
216. Lyapunov A.M. (1950) A General Problem of the Stability of Motion (in Russian). Gostekhizdat, Moscow
217. Lyapunov A.M. (1954–1956) Collected Works (in Russian), vols. 1,2. Izd-vo AN SSSR, Moscow
218. MacDonald D.K.C. (1957) Brownian movement. Phys. Rev. 108:541–545
219. Magnasco M.O. (1993) Forced thermal ratchets. Phys. Rev. Lett. 71:1477–1481
220. Makarov V., Ebeling W., Velarde M. (2000) Soliton-like waves on dissipative Toda lattices. Int. J. Bifurc. & Chaos 10, No 5
221. Malafeev V.M., Polyakova M.S., Romanovsky Yu.M. (1970) On the synchronization process in a chain of oscillators coupled via conductivity (in Russian). Radiofizika 13:936–941
222. Malakhov A.N., Mal'tsev A.A. (1971) The spectral line width in a system of N mutually synchronized oscillators (in Russian). DAN SSSR 196:1065–1067
223. Malkin I.G. (1956) Some Problems of the Theory of Nonlinear Oscillations (in Russian). Gostekhizdat, Moscow
224. Manakov S.V. (1974) On the problem of full integrability and stochastization in discrete dynamical systems (in Russian). ZhETF 67:543–555
225. Mandelbrot B.B. (1977) Fractals: Form, Chance and Dimension. Freeman, San Francisco
226. Mandelshtam L.I., Papaleksi N.D. (1934) On justification of a method of approximate solving differential equations (in Russian). ZhETF 4:117–121
227. Mandelshtam L.I. and Papaleksi N.D. (1947) On the phenomena of resonances of the nth kind (in Russian). Collected Works, vol. 2. Izd-vo AN SSSR, Moscow, pp. 13–62
228. Mandelshtam L.I. and Papaleksi N.D. (1947) On the theory of asynchronous excitation (in Russian). Collected Works, vol. 2. Izd-vo AN SSSR, Moscow, pp. 70–84
229. Mandelshtam L.I. (1955) Lectures on Oscillations (1930–1932), Collected Works, vol. 4 (in Russian). Izd-vo AN SSSR, Moscow
230. Mañé R. (1981) On the dimension of the compact invariant sets of certain nonlinear maps. Lect. Notes Math. Springer, Berlin, Heidelberg 898:230–242.
231. Manevich L.I., Mikhlin Yu.V., Pilipchuk V.N. (1989) Method of Normal Oscillations for Essentially Nonlinear Systems (in Russian). Nauka, Moscow
232. Marchenko Yu.I. (1967) Mutual synchronization of self-oscillatory systems having regard to delay of coupling forces (in Russian). Radiofizika 10:1533–1538
233. Marek A. (1959) A note to recent theories of Brownian motion in nonlinear systems. Physica 25:1358–1367
234. McLaughlin J.B. (1981) Period-doubling bifurcations and chaotic motion for a parametrically forced pendulum. J. Stat. Phys. 24:375–388
235. Meacham L.A. (1938) The bridge-stabilized oscillator. Proc. IRE 26:1278–1294
236. Millonas M., Dykman M.I. (1994) Transport and current reversal in stochastically driven ratchets. Phys. Lett. A 185:65–69
237. Mischenko E.F., Rozov N.Kh. (1975) Differential Equations with a Small Parameter and Relaxation Oscillations (in Russian). Nauka, Moscow

238. Mitropol'sky Yu.A. (1955) Nonstationary Processes in Nonlinear Oscillatory Systems (in Russian). Izd-vo AN USSR, Kiev
239. Mitropol'sky Yu.A. (1971) The Averaging Method in Nonlinear Mechanics (in Russian). Naukova Dumka, Kiev
240. Mitropol'sky Yu.A., Lopatin A.K. (1988) Group-Theoretic Approach to Asymptotic Methods of Nonlinear Mechanics (in Russian). Naukova Dumka, Kiev
241. Moiseev N.N. Asymptotic Methods for Nonlinear Mechanics (in Russian). Nauka, Moscow
242. Mosekilde E., Mouritsen O.G. (eds) (1995) Modeling the Dynamics of Biological Systems. Springer, Berlin
243. Moss F. (1994) Stochastic resonance: From the ice ages to the monkey's ear. In: Contemporary Problems in Statistical Physics, ed. Weiss G.H. SIAM, Philadelphia, p. 205
244. Moss F., Pierson D., O'Gorman D. (1994) Stochastic resonance: Tutorial and update. Bifurcation & Chaos 6:1383–1397.
245. Nagaev R.F. (1978) Periodic Regimes of Vibrational Transport (in Russian). Nauka, Moscow
246. Nagaev R.F. (1985) Mechanical Processes with Repeating Damped Impacts (in Russian). Nauka, Moscow
247. Nauenberg M., Rudnick J. (1981) Universality and the power spectrum at the onset of chaos. Phys. Rev. B 24:493–495
248. Nayfeh A.H. (1981) Introduction to Perturbation Techniques. John Wiley, New York
249. Neimark Yu.I., Aronovich G.V. (1955) On the conditions of self-excitation of singing flame (in Russian). ZhETF 28:567–578
250. Neimark Yu.I. (1972) The Method of Point Mappings in the Theory of Nonlinear Oscillations (in Russian). Nauka, Moscow
251. Neimark Yu.I. (ed.) (1972) Pattern Recognition and Medical Diagnostics (in Russian). Nauka, Moscow
252. Neimark Yu.I. (1978) Dynamical Systems and Controlled Processes (in Russian). Nauka, Moscow
253. Neimark Yu.I., Landa P.S. (1992) Stochastic and Chaotic Oscillations. Kluwer Academic, Dordrecht. [Russian original: Nauka, Moscow, 1987]
254. Neimark Yu.I. (1990) A mathematical model of interaction between producers, product and managers (in Russian). In: Dynamics of Systems (dynamics, stochasticity, bifurcations). Izd-vo GGU, Gor'kiy, pp. 84–89
255. Neimark Yu.I. (1991) Some problems of the qualitative theory of vibrations. Adv. Mech. 14:87–102
256. Nicolaevsky E.S., Shchur L.N. (1983) Intersection of separatrices of periodical trajectories and non-integrability of the classical Yang–Mills equations. Preprint ITEP–3, Moscow.
257. Novikov S., Manakov S.V., Pitaevsky L.P., Zakharov V.E. (1984) Theory of Solitons—The Inverse Scattering Method. Plenum, New York
258. Olsen L.F. and Schaffer W.M. (1990) Chaos versus noisy periodicity: Alternative hypothesis for childhood epidemics. Science 249:499–504
259. O'Malley J. (1974) Introduction to Singular Perturbations. Academic Press, New York
260. Panovko Ya.G., Gubanova I.I. (1964) Stability and Oscillations of Elastic Systems (in Russian). Nauka, Moscow
261. Parker S., Chua L.O. (1983) A computer-assisted study of forced relaxation oscillations. IEEE Trans. Circuits Syst. CAS-30:518–533

262. Parlitz U., English V., Cheffczyk C., Lauterborn W. (1990) Bifurcation structure of bubble oscillators. J. Acoust. Soc. Am. 88:1061–1077
263. Parygin V.N. (1956) Mutual synchronization of three coupled oscillators in the case of weak coupling (in Russian). Radiotekh. & Elektron. 1:197–204
264. Pavlikhina M.A., Smirnov L.P. (1958) Vortex wake in the case of the flow around an oscillating cylinder (in Russian). Izv. AN SSSR, OTN, No 8:124–127
265. Pecora L., Carroll T. (1990) Synchronization in chaotic systems. Phys. Rev. Lett. 64:821–824
266. Penner D.I. et al. (1974) Parametric thermo-mechanical oscillations (in Russian). In: Some Problems of Exciting Undamped Oscillations. Izd. Gos. Pedag. Inst., Vladimir, pp. 168–183
267. Pervozvansky A.A. (1962) Random processes in nonlinear control systems (in Russian). Fizmatgiz, Moscow
268. Peskin C., Odell G., Oster G. (1993) Cellular motions and thermal fluctuations—the Brownian ratchet. Biophys. J. 65:316–324
269. Pikovsky A.S. (1984) On the interaction of strange attractors. Z. Phys. B 55:149–154
270. Pinegin S.V., Tabachnikov Yu.B., Sipenkov I.E. (1982) Static and Dynamic Characteristics of Gasostatic Supports (in Russian). Nauka, Moscow
271. Platt N., Spiegel E.A., Tresser C. (1993) On-off intermittency: a mechanism for bursting. Phys. Rev. Lett. 70:279–282
272. Poincaré H. Sur les Courbes Définies par les Equations Différentielles. Gauthier-Villars, Paris
273. Poincaré H. Les méthodes nouvelles de la méchanique celeste, I (1892), II (1893), III (1899). Gauthier-Villars, Paris
274. Popov E.P., Pal'tov I.P. (1960) Approximate methods of the investigation of nonlinear control systems (in Russian). Fizmatgiz, Moscow
275. Potapov A.I., Stupin V.V. (1985) Thermo-parametric excitation of nonlinear oscillations of a string (in Russian). Probl. Matem. & Teor. Fiz. No 5:142–146
276. Poznyak E.L. (1970) On the faults of the small parameter method in the problems of self-oscillations in systems with two degrees of freedom (in Russian). Proc. of V Int. Conf. on Nonlinear Oscillations, vol. 3, p. 618. Izd-vo Instituta Matematiki AN USSR, Kiev
277. Rayleigh (Strutt J.W.) (1899) The explanation of certain acoustical phenomena. Scientific papers, Cambridge University Press 1:348–354; Acoustical observations. II. Ibid. 1:402–414
278. Rayleigh (Strutt J.W.) (1917) On the pressure developed in a liquid during the collapse of a spherical cavity. Philos. Mag. 34:93–98
279. Rayleigh (Strutt J.W.) (1945) The Theory of Sound. Dover, New York
280. Renyi A. (1970) Probability Theory. North-Holland, Amsterdam
281. Reynolds O. (1886) On the flow of gases. Phil. Mag. & J. Sci. 21:185–199
282. Richardson L.F. (1961) The problem of contiguity: an appendix of statistics of deadly quarrels. Gen. Syst. Yearbook 6:139–187
283. Romanovsky Yu.M. (1972) Mutual synchronization of many self-oscillatory systems coupled by a common medium (in Russian). Radiofizika 15:718–723
284. Romanovsky Yu.M., Stepanova N.V., Chernavsky D.S. (1975) Mathematical Models in Biophysics (in Russian). Nauka, Moscow
285. Romanovsky Yu.M., Stepanova N.V., Chernavsky D.S. (1984) Mathematical Biophysics (in Russian). Nauka, Moscow
286. Rosenberg R.M. (1962) The normal modes of nonlinear n-degree-of-freedom systems. J. Appl. Mech. Trans. ASME 29:7–14

287. Rosenberg R.M. (1966) On nonlinear vibrations of systems with many degrees of freedom. In: Adv. Appl. Mech., Academic Press, New York, vol. 9, pp. 156–243
288. Rosenblum M.G. (1993) A characteristic frequency of chaotic dynamical system. Chaos, Solitons & Fractals 3:617–626
289. Rössler O.E. (1976) An equation for continuous chaos. Phys. Lett. 57A:397–398
290. Routh E.J. (1892) Dynamics of a system of rigid bodies. Adv. Part. London
291. Rubanik V.P. (1963) On the mutual synchronization of self-oscillatory systems with multiple frequencies (in Russian). Radiotekhnika 6:278–285
292. Ruelle D., Takens F. (1971) On the nature of turbulence. Commun. Math. Phys. 20:167–192
293. Rytov S.M. (1935) Resonance of the nth kind in a system with two degrees of freedom in the case of strong couling (in Russian). ZhTF 5:3–14
294. Rytov S.M. (1966) Introduction to Statistical Radiophysics (in Russian). Nauka, Moscow
295. Sanders J.A., Verhulst F. (1985) Averaging Methods in Nonlinear Dynamical Systems. Springer, New York
296. Savinov G.V. (1953) Self-oscillatory systems with strongly expressed nonlinearity (in Russian). Vestn. MGU, ser. fiz.-mat., No 6:77–79
297. Savinov G.V. (1953) Self-oscillation in essentially nonlinear systems (in Russian). DAN SSSR 89:995–999
298. Sbitnev V.I. (1980) Stochasticity in a system of two coupled vibrators (in Russian). In: Nonlinear Waves. Stochasticity and Turbulence. IPF AN SSSR, Gorkii, pp. 46–56
299. Sbitnev V.I. (1984) Stochasticity in a system of coupled oscillators (in Russian). In: Proc. of IX Int. Conf. on Nonlinear Oscillations, vol. 3. Naukova Dumka, Kiev, pp. 477–479
300. Schaffer W.M. (1985) Order and chaos in ecological systems. Ecology 66:93–106
301. Schimansky-Geier L., Kschicho M., Fricke T. (1997) Flux of particles in sawtooth media. Phys. Rev. Lett. 79:3335–3338
302. Schmidt G. (1975) Parametererregte Schwngungen. Deutscher Verlag der Wissenschaften
303. Schmidt G., Seisl M. (1993) Subharmonic vibrations and chaos in forced nonlinear oscillators. ZAMM 2:93–107
304. Sekerskaya E.N. (1935) Regenerative receiver with hard regime (in Russian). ZhTF 5:253–261
305. Sharkovsky A.N. (1964) Co-existence of cycles of a continuous mapping of the line into itself (in Russian). Ukr. Mat. Zh. 26:61–71
306. Shimizu T. (1979) Analytic form of the simplest limit cycle in the Lorenz model. Physica A 97:383–398
307. Sidorova G.A. (1971) Mutual synchronization of two relaxation generators (in Russian). Vestn. MGU, Ser. 3, 12:655–661
308. Sidorova G.A. (1972) Synchronization of a relaxation generator by external triangular voltage (in Russian). Vestn. MGU, Ser. 3, 13:474–479
309. Skibarko A.P., Strelkov S.P. (1934) Qualitative study of processes in a complicated oscillator (On the van der Pol hysteresis) (in Russian). ZhTF 4:158–165
310. Skupoy V.F., Kopylov V.P. (1979) Synchronization of FM-generator (in Russian). Radiotekh. & Elektron. 24:1374–1379

311. Smirnova O.A., Stepanova N.V. (1971) A mathematical model of oscillations in the process of infectious immunity (in Russian). In: Oscillatory Processes in Biological and Chemical Systems. Izd. NCBI AN SSSR, Puschino-na-Oke, pp. 247–251
312. Smoluchowski M. (1912) Experimentell Nachweisbare der ublichen Thermodynamik wiedersprechende Molekularphanomene. Phys. Z. 13:1069–1080
313. Sorokin V.N. (1985) The Theory of Speech Production (in Russian). Radio i Svyaz', Moscow
314. Spivakovsky A.O., Goncharevich I.F. (1983) Vibrational and Wave Transporting Machines (in Russian). Nauka, Moscow
315. Stevens K.N. (1977) Physics of laryngeal behavior and larynx modes. Phonetica 34:264–279
316. Story B.H. and Titze I.R. (1995) Voice simulation with a body-cover model of the vocal folds. J. Acoust. Soc. Am. 97:1249–1260
317. Stratonovich R.L., Romanovskii Yu.M. (1958) Parametric effect of a random force on linear and nonlinear oscillatory systems. In: Nonlinear Transformations of Stochastic Processes. Pergamon Press, Oxford (1965), pp. 322–326
318. Stratonovich R.L. (1960) On the paradox in the theory of thermal fluctuations of nonlinear resistors (in Russian). Vestn. MGU No 4:99–102
319. Stratonovich R.L. (1961) Topics in the Theory of Random Noise, vol. 1., Gordon and Breach, New York (1963), vol. 2, Gordon and Breach, New York (1967). [Russian original: Selected Problems of Fluctuation Theory in Radioengineering, Sov. Radio, Moscow, 1961]
320. Stratonovich R.L. (1992) Nonlinear Nonequilibrium Thermodynamics. Springer, Berlin, Heidelberg, pp. 80–81. [Russian original: Nauka, Moscow, 1985]
321. Strelkov S.P. (1933) The Froude pendulum (in Russian). ZhTF 3: 563–570
322. Strelkov S.P. (1964) Introduction to the Oscillation Theory (in Russian). Nauka, Moscow
323. Strouhal V. von (1878) Über eine besondere Art der Tonerregung. Ann. Phys. 5:216–251
324. Svoboda K. et al. (1993) Direct observation of kinesin stepping by optical trapping interferometry. Nature 365:721–727
325. Szemplinska-Stupnicka W. (1969) On the phenomenon of the combination type resonance in nonlinear two-degree-of-freedom systems. Int. J. Non-Linear Mech. 4:335–359
326. Takens F. (1981) Detecting strange attractors in turbulence. Lect. Notes in Math. Springer, Berlin, Heidelberg, No 898, pp. 366–381
327. Tamarkin Ya.D. (1917) On Some General Problems of Theory of Differential Equations and on Expansion of Functions into Series (in Russian). Petrograd
328. Teodorchik K.F. (1937) Thermo-mechanical self-oscillatory systems (in Russian). Radiotekhnika No 6:5–15
329. Teodorchik K.F. (1943) On the theory of synchronization of relaxation generators (in Russian). DAN SSSR 40:63–66
330. Teodorchik K.F. (1945) Theory of synchronization of relaxation self-oscillatory systems (in Russian). J. of Physics, Moscow, 9:139–146
331. Teodorchik K.F. (1945) Generators with inertial nonlinearity (in Russian). DAN SSSR 50:191–192
332. Teodorchik K.F. (1946) Self-oscillatory systems with inertial nonlinearity (in Russian). ZhTF 16:845–854
333. Teodorchik K.F. (1946) On the theory of sinusoidal oscillations in systems with a great number of degrees of freedom (in Russian). DAN SSSR 52:33–35
334. Teodorchik K.F. (1952) Self-Oscillatory Systems (in Russian). Gostekhizdat, Moscow

335. Thompson J.M.T., McRobie F.A. (1993) Indeterminate bifurcations and the global dynamics of driven oscillators. In: Proc. 1st European Nonlinear Oscillations Conf. Hamburg, August 16–20, ed. Kreuzer E., Schmidt. G. Akademie Verlag, Berlin, pp. 107–128
336. Tikhonov A.N. (1948) Dependence of solutions of differential equations on a small parameter (in Russian). Mat. Sborn. 22(64):193–203
337. Titze I.R., Talkin D.T. (1979) A theoretical study of the effects of various laryngeal configurations on the acoustics of phonation. J. Acoust. Soc. Am. 66:60–74
338. Titze I.R. (ed.) (1992) Vocal Fold Physiology: New Frontiers in Basic Science. Singular, San Diego
339. Toda M. (1970) Waves in nonlinear lattice. Prog. Theor. Phys. Suppl. No 45:174–200
340. Toda M. (1975) Studies of a nonlinear lattice. Phys. Rep. 18:1–123
341. Toda M. (1981) Theory of Nonlinear Lattices. Springer, Berlin
342. Toda M. (1983) Nonlinear Waves and Solitons. Kluwer Acad., Dordrecht
343. Ueda Y., Akamatsu N. (1981) Chaotically transitional phenomena in the forced negative-resistance oscillator. IEEE Trans. Circuits Syst. CAS–28:217–223
344. Ueda Y., Thomsen J.S., Rasmussen J., Mosekilde E. (1993) Behavior of the solution to Duffing's equation for large forcing amplitudes. In: Proc. 1st European Nonlinear Oscillations Conf. Hamburg, August 16–20, ed. Kreuzer E., Schmidt. G. Akademie Verlag, Berlin, pp. 149–166
345. Uhlenbeck G.E., Ornstein L.S. (1930) On the theory of Brownian motion. Phys. Rev. 36:823–841
346. Utkin G.M. (1956) Self-oscillatory systems with two degrees of freedom in the case of multiple frequencies (in Russian). Radiotekhnika 11, No 10:66–72
347. Utkin G.M. (1957) Mutual synchronization of generators with multiple frequencies (in Russian). Radiotekh. & Elektron. 2:44–49
348. Vallis G.K. (1986) El Ninõ : A chaotic dynamical system? Science 232:243–245
349. Vallis G.K. (1988) Conceptual models of El Ninõ and the Southern Oscillation. J. Geophys. Res. 93:13979–13991
350. Van der Pol B. (1920) A theory of the amplitude of free and forced triode vibration. Radio Rev. 1:701–725
351. Van der Pol B. (1922) On oscillation hysteresis in a triode generator with two degrees of freedom. Philos. Mag., Ser. 6, vol. 43, No 256
352. Van der Pol B. (1926) On relaxation oscillation. Philos. Mag. 2:978–992
353. Van der Pol B. (1960) Selected scientific papers. North-Holland, Amsterdam
354. Vasilyeva A.B., Butuzov V.F. (1973) Asymptotic Expansion of Solutions of Singular-Perturbated Equations (in Russian). Nauka, Moscow
355. Vasilyeva A.B., Butuzov V.F. (1990) Asymptotic Methods in the Theory of Singular Perturbations (in Russian). Vysshaya Shkola, Moscow
356. Vermel A.S. (1974) On the problem on thermo-resistive oscillations (in Russian). In: Some Problems of Exciting Undamped Oscillations. Izd. Gos. Pedag. Inst., Vladimir, pp. 159–167
357. Virnik Ya.Z., Kovalyov A.S., Lariontsev E.G. (1970) Natural frequency fluctuations in a laser with synchronized modes (in Russian). Radiofizika 12:1769–1776
358. Volosov V.M. (1962) Averaging in the systems of ordinary differential equations (in Russian). UMN 17, No 3:3–21
359. Volosov V.M., Morgunov B.I. (1971) The Averaging Method in the Theory of Nonlinear Oscillatory Systems (in Russian). Izd-vo MGU, Moscow

360. Volterra V. (1931) Leçons sur la Theorie Mathematique de la Lutte pour la Vie. Gauthier-Villars, Paris
361. Whitham G.B. (1965) A general approach to linear and nonlinear dispersive waves using a langrangian. J. Fluid Mech. 22:273–281
362. Whitham G.B. (1967) Variational methods and applications to water waves. Proc. R. Soc., Ser. A, Math. and Phys. Sci. 299:6–25
363. Whitham G.B. (1974) Linear and Nonlinear Waves. Wiley-Intersci., New York
364. Whittaker E.T. (1964) A Treatise on the Analytical Dynamics of Particles and Rigid Bodies. Cambridge University Press
365. Wiesenfeld K., Moss F. (1995) Stochastic resonance and the benefits of noise: From the ice ages to crayfish and SQUIDs. Nature 373:33–36
366. Wittenburg J. (1977) Dynamics of Systems of Rigid Bodies. Teubner, Stuttgart
367. Wolf A., Swift J.B., Swinney H.L., Vastano J.A. (1985) Determining Lyapunov exponents from a time series. Physica D 16:285–317.
368. Yamamoto T., Hayashi S. (1964) Combination tones of differential type in nonlinear vibratory systems. Bull. JSME 7:690–695
369. Yang C.N., Mills R.L. (1954). Conservation of isotopic spin and isotopic Gauge invariance. Phys. Rev. 96:191–195
370. Zaika P.M. (1967) Vibrational Machines for the Cleaning of Corn (in Russian). Mashinostroenie, Moscow
371. Zanadvorov P.N. (1972) Point map technique in synchronization problems (in Russian). Vestn. LGU No 10:71–74
372. Zeiger S.G. et al. (1974) Wave and Fluctuational Processes in Lasers (in Russian). Nauka, Moscow
373. Zhabotinsky M.E. (1950) Self-oscillatory systems with two degrees of freedom in the case of multiple frequencies. ZhETF 20:421–428
374. Zheleztsov N.A. (1958) On the theory of discontinuous oscillations in systems of the second order (in Russian). Radiofizika 1:67–74
375. Zheludev N.I., Makarov V.A., Matveeva A.V., Svirko Yu.P. (1984) The structure of chaos in excitation of a nonlinear oscillator by a harmonic external force (in Russian). Vestn. MGU Ser.3, vol. 25, No 5:106–109

Index

Action–angle variables 4, 46
Adiabatic invariants 62
Aeolian tones 196
Algorithm 40, 41
– Benettin 40
– Wolf 41
Asynchronous 171, 176, 179
– excitation 176, 179
– suppression 171
Attractors 27, 32, 35, 41
– reconstruction 41
– strange 32, 35
–– chaotic 35
–– stochastic 35
– simple 32
–– stable limit cycle 32
–– stable torus 32
Autosolitons 159

Beats 165
– frequency 165
Bifurcations 44, 94, 267
– Andronov 94
–– direct 94
–– reverse 94
– Hopf 44
–– direct 44
–– reverse 44
– period-doubling 267

Chains 71, 75, 78, 83, 84
– Fermi–Pasta–Ulam 83
– linear homogeneous 71
– nonlinear homogeneous 78
– periodically inhomogeneous 75
– Toda 78, 84
–– ring 84
Chaotization 106, 193, 220, 292
– energetic criterion 106
– of coupled generators 220
– of nonlinear oscillator 292
– of periodic self-oscillations 193, 220
Characteristic exponents 33
Clock movement mechanisms 89
Competitive quenching 207
Correlation integral 38
– generalized 38
Cycles 35
– nodal 35
– saddle 35

Dimension 36–38, 40, 41
– capacity 37
– correlation 37
– embedding 41
– fractal 37
– generalized 38
– Hausdorff 37
– Lyapunov 40
– of attractors and repellers 36
Dynamical system 4, 6
Dynamical variables 6

Edgeworth expansion 362
Electro-mechanical vibrators 221, 283
Electronic generators 89
El Niño phenomenon 120
Energy 4
Equations 3, 16–18, 28, 52, 56, 59, 87, 89, 91, 110, 111, 132, 289
– Bautin 91
– bubble in fluid 56
– characteristic 28
– Duffing 52
– Hamilton–Jacobi 3
– Hill 289
– in standard form 16, 17
– Lorenz 110, 111
–– gyrostat 111
– Lotka–Volterra 59
– Rayleigh 89
– Rössler 132
– truncated 18
– van der Pol 89

– Yang–Mills 87
Excitation 93, 94, 289
– hard 94
– parametric 289
– soft 93

Feigenbaum scenario 44
Feigenbaum constant 44
Fixed points 42, 43
– stability 43
Froude pendulum 89, 92

Generators 92, 98, 100, 191, 208
– driving and driven 208
– frictional 92
– Rayleigh relaxation 100
– van der Pol–Duffing 98
– with inertial nonlinearity 191
– – spike mode 191

Hausdorff measure 37
Heated wire 116
Helmholtz resonator 113

Inertial interaction 107
In-phase and anti-phase regimes 210,235
Instability 30
– aperiodic 30
– oscillatory 30
Integral manifolds 35
Interaction of generators 223, 225, 233
– of chaotic oscillations 225
– of periodic and chaotic oscillations 223
– of two coupled Rayleigh relaxation generators 233
intermittency 45

Karman's vortex wake 196
Koenigs theorem 43

Lamerey diagram 44
Lavrov's device 67
Limit sets 27
– attracting and repelling 27
Lyapunov exponents 33, 40
– generalized 40

Mathematical models 5
– aggregated 5
– 'black boxes' 5
– models of a phenomenon 5
– 'portraits' 5

Method 9, 10, 12, 16, 21, 257, 270, 362
– asymptotic 12
– averaging 16
– Galerkin 362
– Krylov–Bogolyubov 12, 257, 270
– of a small parameter 9
– of slowly time-varying amplitudes 9
– Poincaré 9
– van der Pol 10
– Witham 21
Models 120, 133, 146, 152, 268, 314
– epidemiological 314
– of immune reaction 133
– of singing flame 152
– of vocal folds 146
– Vallis 120, 268
multipliers 34, 326

Neimark pendulum 102, 103
Normal coordinates 10
Number of degrees of freedom 6

On-off intermittency 299
Oscillations 7, 49, 63, 65, 96, 255, 266, 267, 292, 295
– chaotic 266, 267, 292
– forced 255
– free 49
– isochronous 96
– natural 7, 49
– noise-induced 295
– normal 63, 65
– parametric 292
– self-excited 7
Oscillators 255, 267, 269
– coupled 269
– driven 255
– Duffing 267

Parametric amplifier 312
Passage through the resonance 264
Pendulum with a vibrated suspension axis 323
Phase space 6
Poincaré cutting surface 42
Poincaré rotation number 47
Point maps 42
– succession 42
– translation 42
Prandtl formula 125

Ratchet-and-pawl device 331
Ratchets 331, 336
– flashing 331

– stochastic 331, 336
Relaxation time 98
Renyi entropy 38
Repellers 27, 32
– complex 32
– simple 32
Resonances 162, 167, 257, 260, 262, 270, 275, 290, 305, 309
– combination 275
– main 162, 257, 270
– of the nth kind 167
– parametric 290, 305
– – combination 309
– – main 290
– – of the second kind 290
– subharmonic 260
– superharmonic 262
Ruelle–Takens scenario 45
Rytov–Dimentberg criterion 303

Self-oscillations 122
– air-cushioned body 122
Self-oscillatory systems 5, 89, 91, 107, 113, 116, 127, 137
– relaxation 91
– thermo-mechanical 113
– – heated wire 116
– Thomsonian 91
– with inertial excitation 107
– with inertial nonlinearity 127
– with one degree of freedom 89
– with one and a half degrees of freedom 107
– with two or more degrees of freedom 137
Sensors 286, 287
– capacitative 286
– optical 287
Separatrix 50
Singular points 27–30
– center 28
– focus 28, 29
– node 28, 29
– saddle 28
– saddle-focus 29
– saddle-node 30
– stable and unstable 27
Stability 27, 30, 33, 35
– asymptotic 27
– Lyapunov 27, 33
– orbital 33
– Poisson 35
– Routh–Hurwitz criterion 30
Succession function 43
Swing 108
Synchronization 161, 167, 173, 180, 188, 191, 195, 205, 216, 219, 225, 233–246
– at multiple frequencies 216
– by harmonics 173
– by quenching 167, 191
– mutual 205
– – of three generators 237, 239
– – of three identical generators 238
– – of two coupled Rayleigh relaxation generators 233
– of a generator with modulated natural frequency 180
– of chaotic self-oscillations 195, 225
– – definition 225
– of N coupled generators 244
– of pendulum clocks 246
– of periodic self-oscillations 161, 188, 205
– of three generators coupled in a chain 239
– parametric 219
– subharmonic 167
– threshold 195
– two mechanisms 167
Systems 2–6, 45, 61, 84, 263
– active 5
– conservative 4
– dissipative 2, 4
– dynamical 2
– – chaotic 6
– Hamiltonian 3
– Henon–Heiles 84
– integrable and nonintegrable 45
– passive 5
– with slowly time-varying natural frequency 61
– with slowly time-varying parameters 263

Techniques 41
– Broomhead and King 41
– Takens 41
– well-adapted basis 41
Tenso-resistive effect 117
Theory 33, 326
– Floquet 33, 326
Toda potential 54

Vibration of resonators 288
Vibrational transport 331

Windows of periodicity 45

Foundations of Engineering Mechanics

Series Editors: Vladimir I. Babitsky, Loughborough University
Jens Wittenburg, Karlsruhe University

Palmov	Vibrations of Elasto-Plastic Bodies (1998, ISBN 3-540-63724-9)
Babitsky	Theory of Vibro-Impact Systems and Applications (1998, ISBN 3-540-63723-0)
Skrzypek/ Ganczarski	Modeling of Material Damage and Failure of Structures Theory and Applications (1999, ISBN 3-540-63725-7)
Kovaleva	Optimal Control of Mechanical Oscillations (1999, ISBN 3-540-65442-9)
Kolovsky	Nonlinear Dynamics of Active and Passive Systems of Vibration Protection (1999, ISBN 3-540-65661-8)
Guz	Fundamentals of the Three-Dimensional Theory of Stability of Deformable Bodies (1999, ISBN 3-540-63721-4)
Alfutov	Stability of Elastic Structures (2000, ISBN 3-540-65700-2)
Morozov/ Petrov	Dynamics of Fracture (2000, ISBN 3-540-64274-9)
Astashev/ Babitsky/ Kolovsky	Dynamics and Control of Machines (2000, ISBN 3-540-63722-2)

Foundations of Engineering Mechanics

Svetlitsky — Statics of Rods
(2000, ISBN 3-540-67452-7)

Landa — Regular and Chaotic Oscillations
(2001, ISBN 3-540-41001-5)

Printing: Saladruck, Berlin
Binding: H. Stürtz AG, Würzburg